Chemical Kinetics and Dynamics

Second Edition

Jeffrey I. Steinfeld

Massachusetts Institute of Technology

Joseph S. Francisco

Purdue University

William L. Hase

Wayne State University

Prentice Hall
Upper Saddle River, New Jersey 07458

Library of Congress Cataloging-in-Publication Data

Steinfeld, Jeffrey I.
    Chemical kinetics and dynamics / Jeffrey I. Steinfeld, Joseph S.
  Francisco, William L. Hase. — 2nd ed.
        p.   cm.
    ISBN 0-13-737123-2
    1. Chemical kinetics. 2. Molecular dynamics. I. Francisco,
Joseph Salvadore. II. Hase, William L.   III. Title.
QD502.S74   1998   98-28315
541.3′94—DC21   CIP

Acquisitions Editor: Matthew Hart
Editorial Assistant: Betsy Williams
Executive Managing Editor: Kathleen Schiaparelli
Assistant Managing Editor: Lisa Kinne
Art Director: Jayne Conte
Cover Designer: Bruce Kenselaar
Manufacturing Manager: Trudy Pisciotti
Production Supervision/Composition: WestWords, Inc.
Cover Illustrations: WestWords, Inc.

 © 1999, 1989 by Prentice-Hall, Inc.
Simon & Schuster/A Viacom Company
Upper Saddle River, New Jersey 07458

Printed in the United States of America

10 9 8 7 6 5 4 3 2 1

ISBN 0-13-737123-3

Prentice-Hall International (UK) Limited, *London*
Prentice-Hall of Australia Pty. Limited, *Sydney*
Prentice-Hall Canada Inc., *Toronto*
Prentice-Hall Hispanoamericana, S.A., *Mexico*
Prentice-Hall of India Private Limited, *New Delhi*
Prentice-Hall of Japan, Inc., *Tokyo*
Simon & Schuster Asia Pte. Ltd., *Singapore*
Editora Prentice-Hall do Brasil, Ltda., *Rio de Janeiro*

Contents

Preface   ix

Chapter 1   Basic Concepts of Kinetics   1

1.1   Definition of the Rate of a Chemical Reaction   1
1.2   Order and Molecularity of a Reaction   3
1.3   Integrated Reaction Rate Laws   6
1.4   Determination of Reaction Order: Reaction Half-Lives   13
1.5   Temperature Dependence of Rate Constants: The Arrhenius Equation   14
1.6   Reaction Mechanisms, Molecular Dynamics, and the Road Ahead   17
        References   18
        Bibliography   18
        Problems   19

Chapter 2   Complex Reactions   22

2.1   Exact Analytic Solutions for Complex Reactions   22
2.2   Approximation Methods   37
2.3   Example of a Complex Reaction Mechanism: The Hydrogen + Halogen
        Reaction   41
2.4   Laplace Transform Method   47
2.5   Determinant (Matrix) Methods   52
2.6   Numerical Methods   55
2.7   Stochastic Methods   66
        References   72
        Bibliography   74
        Appendix 2.1   The Laplace Transform   74
        Appendix 2.2   Numerical Algorithms for Differential Equations   76
        Appendix 2.3   Stochastic Numerical Simulation of Chemical Reactions   77
        Problems   79

Chapter 3   Kinetic Measurements   87

3.1   Introduction   87
3.2   Techniques for Kinetic Measurements   89
3.3   Treatment of Kinetic Data   105
References   120
Problems   121

Chapter 4   Reactions in Solution   124

4.1   General Properties of Reactions in Solution   124
4.2   Phenomenological Theory of Reaction Rates   125
4.3   Diffusion-Limited Rate Constant   130
4.4   Slow Reactions   132
4.5   Effect of Ionic Strength on Reaction Between Ions   133
4.6   Linear Free-Energy Relationships   136
4.7   Relaxation Methods for Fast Reactions   140
References   143
Bibliography   143
Problems   144

Chapter 5   Catalysis   147

5.1   Catalysis and Equilibrium   147
5.2   Homogeneous Catalysis   148
5.3   Autocatalysis and Oscillating Reactions   151
5.4   Enzyme-Catalyzed Reactions   159
5.5   Heterogeneous Catalysis and Gas-Surface Reactions   163
References   167
Problems   168

Chapter 6   The Transition from the Macroscopic to the Microscopic Level   171

6.1   Relation between Cross Section and Rate Coefficient   171
6.2   Internal States of the Reactants and Products   174
6.3   Microscopic Reversibility and Detailed Balancing   174
6.4   The Microscopic–Macroscopic Connection   175
References   177
Bibliography   178
Problems   178

Chapter 7   Potential Energy Surfaces   179

7.1   Long-range Potentials   180
7.2   Empirical Intermolecular Potentials   183
7.3   Molecular Bonding Potentials   184

7.4   Internal Coordinates and Normal Modes of Vibration   187
7.5   Potential Energy Surfaces   190
7.6   Ab Initio Calculation of Potential Energy Surfaces   191
7.7   Analytic Potential Energy Functions   196
7.8   Experimental Determination of Potential Energy Surface   204
7.9   Details of the Reaction Path   206
7.10   Potential Energy Surfaces of Electronically Excited Molecules   207
       References   211
       Bibliography   213
       Problems   215

Chapter 8   Dynamics of Bimolecular Collisions   217

8.1   Simple Collision Models   217
8.2   Two-body Classical Scattering   222
8.3   Complex Scattering Processes   231
       References   249
       Problems   250

Chapter 9   Experimental Chemical Dynamics   255

9.1   Molecular Beam Scattering   255
9.2   State-Resolved Spectroscopic Techniques   263
9.3   Molecular Dynamics of the $H + H_2$ Reaction   266
9.4   State-to-state Kinetics of the $F + H_2$ Reaction   268
9.5   Warning: Information Overload!   276
       References   276
       Problems   278
       Appendix   The Master Equation   282
       References   286

Chapter 10   Statistical Approach to Reaction
                  Dynamics: Transition State Theory   287

10.1   Motion on the Potential Surface   287
10.2   Basic Postulates and Derivation of Transition State Theory   289
10.3   Dynamical Derivation of Transition State Theory   294
10.4   Quantum Mechanical Effects in Transition State Theory   297
10.5   Thermodynamic Formulation of Transition State Theory   300
10.6   Applications of Transition State Theory   302
10.7   Microcanonical Transition State Theory   310
10.8   Variational Transition State Theory   312
10.9   Experimental Observation of the Transition State Region   314
10.10 Critique of Transition State Theory   316
       References   319
       Bibliography   320
       Problems   321

Chapter 11   Unimolecular Reaction Dynamics   324

11.1   Formation of Energized Molecules   326
11.2   Sum and Density of States   329
11.3   Lindemann-Hinshelwood Theory of Thermal Unimolecular Reactions   334
11.4   Statistical Energy-dependent Rate Constant $k(E)$ 338
11.5   RRK Theory   340
11.6   RRKM Theory   343
11.7   Application of RRKM Theory to Thermal Activation   349
11.8   Measurement of $k(E)$ 351
11.9   Intermolecular Energy Transfer   356
11.10   Product Energy Partitioning   359
11.11   Apparent and Intrinsic non-RRKM Behavior   362
11.12   Classical Mechanical Description of Intramolecular Motion and Unimolecular Decomposition   365
11.13   Infrared Multiple-Photon Excitation   367
11.14   Mode Specificity   374
        References   377
        Bibliography   382
        Problems   383

Chapter 12   Dynamics Beyond the Gas Phase   390

12.1   Transition State Theory of Solution Reactions   390
12.2   Kramers' Theory and Friction   402
12.3   Gas-Surface Reaction Dynamics   407
        References   420
        Bibliography   421
        Problems   422

Chapter 13   Information-Theoretical Approach
                  to State-to-State Dynamics   424

13.1   Introduction   424
13.2   The Maximal-Entropy Postulate   424
13.3   Surprisal Analysis and Synthesis: Product State Distribution in Exothermic Reactions   432
13.4   Information-Theoretical Analysis of Energy Transfer Processes   437
13.5   Conclusion   449
        References   449
        Bibliography   451
        Problems   452

Chapter 14   Kinetics of Multicomponent
                  Systems: Combustion Chemistry   453

14.1   Introduction   453
14.2   The Hydrogen-Oxygen Reaction, an Explosive Combustion Process   453

14.3   The Methane Combustion Process   459
        References   469

**Chapter 15   Kinetics of Multicomponent
                Systems: Atmospheric Chemistry   420**

15.1   Physical Structure of the Atmosphere   470
15.2   Chemical Composition of the Atmosphere   472
15.3   Photochemistry in the Atmosphere   472
15.4   Catalytic Cycles Involving Stratospheric Ozone   476
15.5   Modeling Studies of the Atmosphere   488
15.6   Atmospheric Measurements   489
15.7   Current Understanding of Atmospheric Kinetics   491
15.8   Conclusion   493
        References   494
        Bibliography   494
        Problems   495

Appendix 1   Quantum Statistical Mechanics   499

Appendix 2   Classical Statistical Mechanics   500

Appendix 3   Data Bases in Chemical Kinetics   507

Index   509

Preface

The first edition of *Chemical Kinetics and Dynamics,* which appeared in 1989, was an attempt to combine the essential content of classical chemical kinetics with the new developments in molecular dynamics which had transformed both our understanding of and experimental approach to the study of chemical reaction dynamics. At that time, the principal focus of the study of chemical reactions had shifted from the macroscopic treatment of empirical kinetics to the microscopic, molecular viewpoint of chemical dynamics. The microscopic approach had stimulated new experimental and theoretical developments, making chemical dynamics one of the most active fields of physical chemistry. Most of the chemical kinetics textbooks then available emphasized either the macroscopic kinetics or microscopic dynamics aspect, often at the expense of the other. We set about to cover both older and newer aspects of chemical reaction processes as comprehensively as possible, with the aim of illustrating the interconnections between phenomenological chemical kinetics and molecular reaction dynamics. The treatment was intended to be accessible to advanced undergraduates who had completed an introductory course in physical chemistry and to beginning graduate students, while serving as an entry point into the large and ever-growing research literature on chemical kinetics and reaction dynamics.

This approach proved to be a successful one, as attested to by favorable reviews and several subsequent imitations. Since the first edition appeared, a number of significant developments have occurred in the study of chemical reaction dynamics, both at the fundamental molecular level and in applications to complex systems. One such development is vastly greater computational power, which makes possible accurate calculation of potential energy surfaces, solution of classical or quantum-mechanical equations of motion on such surfaces, and integration of large coupled sets of kinetic equations, such as are encountered in atmospheric chemistry. New experimental techniques include refinement of laser and molecular-beam techniques to probe state-to-state kinetics, reactant orientation, vector correlations of product molecule distributions, and the use of ultrashort laser pulses to obtain real-time information on chemical reactions, including attempts to characterize the elusive Transition State.

We have tried to incorporate a number of these new developments in the present edition, without, however, eliminating the core experimental and theoretical bases of chemical kinetics. The text begins with an exposition of the basic principles of chemical kinetics, followed by a description of current experimental and analytic techniques in kinetics. The treatment includes reactions in the gas phase, in liquids, and at catalytic surfaces. The transition to the microscopic level is made by introducing molecular scattering and potential energy surfaces. Following a treatment of statistical reaction rate theories (transition-state and RRKM treatments for bimolecular and unimolecular reactions, respectively), the analysis of multilevel and multicomponent kinetic systems is carried out using master equation and information-theoretical methods. The text concludes with a treatment of important, real-world complex kinetic systems, viz., atmospheric and combustion chemistry. In order to keep the revised edition at a manageable length, some of the topics considered in the first edition, such as Laplace Transforms and information theory, have been condensed (but not eliminated). New topics covered in this edition include experimental tests and theoretical models of the $H + H_2$ reaction; current understanding of stratospheric chemistry, including heterogeneous processes; and the aforementioned attempts at observation of the Transition State region by "Femtochemistry" and electron detachment spectroscopy.

The background assumed is a basic knowledge of physical chemistry and enough mathematics to be able to set up and solve systems of linear differential equations. When advanced techniques, such as Laplace Transforms, matrix methods, or information theory are utilized, a brief introduction is provided in the text. The book is intended to provide students with the necessary background to delve into current research topics using the journal and review literature. To this end, extensive chapter references and bibliographies are provided. There is also a large number of problems and exercises, some of which involve numerical procedures. Please note that an Instructor's Manual for this text is available, containing problem solutions, sample computer programs, and suggestions for optimal use of the text for several different course syllabi. The Instructor's Manual may be ordered directly from the publisher: request ISBN 0-13-080605-6.

This book is based on a series of lectures given over a number of years to students at the Massachusetts Institute of Technology, Wayne State University, and Purdue University. We would like to express our appreciation to the numerous reviewers, colleagues, and students who have provided material for incorporation into the text, as well as pointing out errors in the text, problems, and problem solutions of the first edition; all such corrections have been incorporated in the present version, we hope without introducing new errors. Some of these contributors are T. Baer, S. Chapman, B. J. Garrison, P. Gaspar, R. G. Gilbert, D. T. Gillespie, D. M. Hirst, J. T. Hynes, S. R. Leone, R. D. Levine, R. Lucchese, W. H. Miller, M. Molina, J. Parson, B. S. Rabinovitch, H. Rabitz, G. C. Schatz, D. G. Truhlar, and J. C. Tully. We would like to thank John Challice and Matthew Hart of Prentice Hall for their skillful management of the revision and publication process, and our colleagues, friends, and families for their forbearance while all this was going on.

J. I. STEINFELD
J. S. FRANCISCO
W. L. HASE

C H A P T E R   1

Basic Concepts of Kinetics

"Nothing *is* in this world, because everything is in a state of becoming
something else"   —Heraclitus (540–480 B.C.E.), *Theaetetus*

Among the most familiar characteristics of a material system is its capacity for *chemical change.* In a chemistry lecture, the demonstrator mixes two clear liquids and obtains a colored solid precipitate. Living organisms are born, grow, reproduce, and die. Even the formation of planetary rocks, oceans, and atmospheres consists of a set of chemical reactions. The time scale for these reactions may be anywhere from a few femtoseconds (10^{-15} sec) to geologic times (10^9 years, or 10^{+16} sec).

The science of *thermodynamics* deals with chemical systems at equilibrium, which by definition means that their properties do not change with time. Most real systems are not at equilibrium and undergo chemical changes as they seek to approach the equilibrium state. *Chemical kinetics* deals with changes in chemical properties in time. As with thermodynamics, chemical kinetics can be understood in terms of a continuum model, without reference to the atomic nature of matter. The interpretation of chemical reactions in terms of the interactions of atoms and molecules is frequently called *reaction dynamics*. A knowledge of the dynamic basis for chemical reaction has, in fact, permitted us to design and engineer reactions for the production of an enormous number of compounds which we now regard as essential in our technological society.

We begin our study of chemical kinetics with definitions of the basic observable quantities, which are the concentrations of the chemical components in a system, and how these concentrations change with time.

1.1 DEFINITION OF THE RATE OF A CHEMICAL REACTION

Chemical kinetics may be described as the study of chemical systems whose composition changes with time. These changes may take place in the gas, liquid, or solid phase of a substance. A reaction occurring in a single phase is usually referred to as a *homogeneous* reaction, while a reaction which takes place at an interface between two phases is known as a *heterogeneous* reaction. An example of the latter is the reaction of a gas adsorbed on the surface of a solid.

1

The chemical change that takes place in any reaction may be represented by a *stoichiometric equation* such as

$$aA + bB \rightarrow cC + dD \tag{1-1}$$

where a and b denote the number of moles of reactants A and B that react to yield c and d moles of products C and D. Various symbols are used in the expression which relates the reactants and products. For example, the formation of water from hydrogen and oxygen may be written as the balanced, irreversible chemical reaction

$$2H_2 + O_2 \rightarrow 2H_2O \tag{1-2}$$

In this simple example, the single arrow is used to indicate that the reaction proceeds from the left (reactant) side as written: water does not spontaneously decompose to form hydrogen and oxygen. A double arrow in the stoichiometric equation is often used to denote a reversible reaction, that is, one which can proceed in either the forward or the reverse direction; an example is

$$H_2 + I_2 \rightleftharpoons 2HI \tag{1-3}$$

While each of equations (1-2) and (1-3) describes an apparently simple chemical reaction, it so happens that neither of these reactions proceeds as written. Instead, the reactions involve the formation of one or more *intermediate* species, and include several steps. These steps are known as *elementary* reactions. An elementary reaction is one in which the indicated products are formed directly from the reactants, for example, in a direct collision between an A and B molecule; intuitively, they correspond to processes occurring at the molecular level. In the hydrogen-oxygen reaction, a key elementary reaction is the attack of oxygen atoms on hydrogen molecules given by

$$O + H_2 \rightarrow OH + H$$

while in the hydrogen-iodine reaction it is

$$2I + H_2 \rightarrow H_2I + I$$

The details of these reactions are discussed in sections 14.2 and 2.3.2, respectively. In the meantime, note here that they involve atoms (O, I), free radicals (OH), and/or unstable intermediates (H_2I); this is often the case with elementary reactions.

The change in composition of the reaction mixture with time is the *rate of reaction, R*. For reaction 1-1, the rate of consumption of reactants is

$$R = -\frac{1}{a}\frac{d[A]}{dt} = -\frac{1}{b}\frac{d[B]}{dt} \tag{1-4}$$

A standard convention in chemical kinetics is to use the chemical symbol enclosed in brackets for species concentration; thus, [X] denotes the concentration of X. The negative signs in equation (1-4) indicate that during the course of the reaction the concentration of reactants decreases as the reactants are consumed; conversely, a positive sign indicates that the concentration of products increases as those species are formed. Consequently, the rate of formation of products C and D can be written as

$$R = +\frac{1}{c}\frac{d[C]}{dt} = +\frac{1}{d}\frac{d[D]}{dt} \tag{1-5}$$

The factors, *a, b, c,* and *d* in equations (1-4) and (1-5) are referred to as the *stoichiometric coefficients* for the chemical entities taking part in the reaction. Since the concentrations of reactants and products are related by equation (1-1), measurement of the rate of change of any one of the reactants or products would suffice to determine the rate of reaction *R*. In reaction (1-2), the rate of reaction would be

$$R = -\frac{1}{2}\frac{d[H_2]}{dt} = -\frac{d[O_2]}{dt} = +\frac{1}{2}\frac{d[H_2O]}{dt} \tag{1-6}$$

A number of different units have been used for the reaction rate. The dimensionality of *R* is

$$[\text{amount of material}][\text{volume}]^{-1}[\text{time}]^{-1}$$

or

$$[\text{concentration}][\text{time}]^{-1}$$

The standard SI unit of concentration is mole per cubic decimeter, abbreviated mol dm^{-3}. In the older literature on kinetics, one frequently finds the equivalent unit mol liter^{-1} for reactions in solution, and mol cm^{-3} for gas phase reactions. The SI unit is preferred, and should be used consistently. Multiplying moles cm^{-3} by Avogadro's Number (6.022×10^{23}) gives the units molecules cm^{-3}, which is still extensively used and, indeed, is convenient for gas phase reactions.

A subcommittee of the International Union of Pure and Applied Chemistry chaired by Laidler has attempted to standardize units, terminology, and notation in chemical kinetics.[1] We have attempted to follow the subcommittee's recommendations in this text.

1.2   ORDER AND MOLECULARITY OF A REACTION

In virtually all chemical reactions that have been studied experimentally, the reaction rate depends on the concentration of one or more of the reactants. In general, the rate may be expressed as a function *f* of these concentrations,

$$R = f([A],[B]) \tag{1-7}$$

In some cases the reaction rate also depends on the concentration of one or more intermediate species, e.g., in enzymatic reactions (see chapter 5). In other cases the rate expression may involve the concentration of some species which do not appear in the stoichiometric equation (1-1); such species are known as *catalysts,* and will be discussed in chapter 5. In still other cases, the concentration of product molecules may appear in the rate expression.

The most frequently encountered functional dependence given by equation (1-7) is the rate's being proportional to a product of algebraic powers of the individual concentrations, i.e.,

$$R \propto [A]^m[B]^n \tag{1-8}$$

The exponents m and n may be integer, fractional, or negative. This proportionality can be converted to an equation by inserting a proportionality constant k, thus:

$$R = k[A]^m[B]^n \tag{1-9}$$

This equation is called a *rate equation* or *rate expression*. The exponent m is the *order* of the reaction with respect to reactant A, and n is the order with respect to reactant B. The proportionality constant k is called the *rate coefficient*. The overall order of the reaction is simply $p = m + n$. A generalized expression for the rate of a reaction involving K components is

$$R = k \prod_{i=1}^{K} c_i^{ni} \tag{1-10}$$

The product is taken over the concentrations of each of the K components of the reaction. The reaction order with respect to the ith component is n_i, $p = \Sigma_{i=1}^{K} n_i$ is the overall order of the reaction, and k is the rate constant.

In equation (1-10), k must have the units

$$[\text{concentration}]^{-(p-1)}[\text{time}]^{-1}$$

so for a second-order reaction, i.e., $m = n = 1$ in equation (1-8), the units would be [concentration]$^{-1}$[time]$^{-1}$, or dm^{3}mol^{-1}sec^{-1} in SI units. Note that the units of liter mol^{-1}sec^{-1} are frequently encountered in the older solution-kinetics literature, and cm^{3}mol^{-1}sec^{-1} or cm^{3}molecule^{-1}sec^{-1} are still encountered in the gas-kinetics literature.

Elementary reactions may be described by their *molecularity*, which specifies the number of reactants that are involved in the reaction step. If a reactant spontaneously decomposes to yield products in a single reaction step, given by the equation

$$A \rightarrow \text{products} \tag{1-11}$$

the reaction is termed *unimolecular*. An example of a unimolecular reaction is the dissociation of N_2O_4, represented by

$$N_2O_4 \rightarrow 2NO_2$$

If two reactants A and B react with each other to give products, i.e.,

$$A + B \rightarrow \text{products} \tag{1-12}$$

the reaction is termed *bimolecular*. An example of a bimolecular reaction would be a metathetical atom-transfer reaction such as

$$O + H_2 \rightarrow OH + H$$

or

$$F + H_2 \rightarrow HF + H$$

Both of these reactions are discussed in subsequent chapters.

Three reactants that come together to form products constitute a *termolecular* reaction. In principle, one could go on to specify the molecularity of four, five, etc., reactants involved in an elementary reaction, but such reactions have not been encountered in nature. The situation reflects the molecular bases of elementary reactions. A single, suitably energized molecule can decompose according to equation (1-11); such unimolecular processes are discussed in chapter 11. A collision between two molecules can lead to a bimolecular reaction according to equation (1-12); this is further discussed in chapters 8 and 10. At moderate to high gas pressures, termolecular processes can occur, such as three-body recombination, i.e.,

$$A + B + M \rightarrow AB + M \qquad (1\text{-}13)$$

However, physical processes involving simultaneous interaction of four or more independent particles are so rare in chemical kinetics as to be completely negligible.

For an elementary reaction, the molecularity and the overall order of the reaction are the same. Thus, a bimolecular elementary reaction is second order, a termolecular reaction third order, and so on. The reverse is not always true, however. For example, the hydrogen-iodine reaction (1-3) is second order in both directions, but bimolecular reactions between H_2 and I_2, and between two HI molecules, are thought not to occur. Instead, the reaction consists of several unimolecular, bimolecular, and possibly termolecular steps (see chapter 2).

A further distinction between molecularity and reaction order is that, while molecularity has only the integer values 0, 1, 2, and 3, order is an experimentally determined quantity which can take on noninteger values. In principle, these values could be any number between $-\infty$ and $+\infty$, but values between -2 and 3 are usually encountered in practice. Negative orders imply that the component associated with that order acts to slow down the reaction rate; such a component is termed an *inhibitor* for that reaction. Fractional values of the reaction order always imply a complex reaction mechanism (see section 1.6). An example of a fractional-order reaction is the thermal decomposition of acetaldehyde given by

$$CH_3CHO \xrightarrow{300-800°C} CH_4 + CO \qquad (1\text{-}14)$$

which has a 3/2 reaction order, i.e.,

$$\frac{d[CH_4]}{dt} = (\text{constant})[CH_3CHO]^{3/2} \qquad (1\text{-}15)$$

Similarly, under certain conditions the reaction of hydrogen with bromine

$$H_2 + Br_2 \rightarrow 2HBr \qquad (1\text{-}16)$$

has a 3/2 reaction order, first order in $[H_2]$ and 1/2 order in $[Br_2]$:

$$\frac{d[HBr]}{dt} = (\text{constant})[H_2]\,[Br_2]^{1/2} \qquad (1\text{-}17)$$

Under other conditions, reaction (1-16) can display an even more complicated behavior, *viz.*,

$$\frac{d[\text{HBr}]}{dt} = \frac{(\text{constant})[\text{H}_2]\,[\text{Br}_2]^{1/2}}{1 + (\text{constant}')[\text{HBr}]} \tag{1-18}$$

The constants in equation (1-15), (1-17), and (1-18) are clearly not identifiable with an elementary reaction, but instead are phenomenological coefficients obtained by fitting the rate expression to experimental data. Such coefficients are properly termed *rate coefficients*, rather than rate constants. The latter term should be reserved for the coefficients in rate expressions for elementary reactions, which follow a rate expression having the form of equation (1-10).

1.3 INTEGRATED REACTION RATE LAWS

Thus far, we have defined the rate of reaction in terms of concentrations, orders, and reaction rate constants. Next, we consider the time behavior of the concentration of reactants in reactions with simple orders. The time behavior is determined by integrating the rate law for a particular rate expression.

1.3.1 Zero-Order Reaction

The rate law for a reaction that is zero order is

$$R = -\frac{d[\text{A}]}{dt} = k[\text{A}]^0 = k \tag{1-19}$$

Zero-order reactions are most often encountered in heterogeneous reactions on surfaces (see chapter 5). The rate of reaction for this case is independent of the concentration of the reacting substance. To find the time behavior of the reaction, equation (1-19) is put into the differential form

$$d[\text{A}] = -k\,dt \tag{1-20}$$

and then integrated over the boundary limits t_1 and t_2. Assuming that the concentration for A at $t_1 = 0$ is $[\text{A}]_0$, and at $t_2 = t$ is $[\text{A}]_t$, equation (1-20) becomes

$$\int_{[\text{A}]_0}^{[\text{A}]_t} d[\text{A}] = -k \int_{t_1=0}^{t_2=t} dt \tag{1-21}$$

Hence,

$$[\text{A}]_t - [\text{A}]_0 = -k(t - 0) \tag{1-22}$$

Consequently, the integrated form of the rate expression for the zero-order reaction is

$$[\text{A}]_t = [\text{A}]_0 - kt \tag{1-23}$$

A plot of [A] versus time should yield a straight line with intercept $[\text{A}]_0$ and slope k.

1.3.2   First-Order Reactions

A first-order reaction is one in which the rate of reaction depends only on one reactant. For example, the isomerization of methyl isocyanide, CH_3NC, is a first-order unimolecular reaction:

$$CH_3NC \rightarrow CH_3CN \tag{1-24}$$

This type of equation can be represented symbolically as

$$A \rightarrow B$$

and the rate of disappearance of A can be written as

$$R = -\frac{1}{a}\frac{d[A]}{dt} = k[A]^1 \tag{1-25}$$

Note that the reaction is of order one in the reactant A. Thus, since only one A molecule disappears to produce one product B molecule, $a = 1$ and equation (1-25) becomes

$$-\frac{d[A]}{dt} = k[A] \tag{1-26}$$

Integration of equation 1-26 leads to

$$-\int \frac{d[A]}{dt} = k \int dt \tag{1-27}$$

$$-\ln[A]_t = kt + \text{constant}$$

If the boundary conditions are such that at $t = 0$ the initial value of $[A]$ is $[A]_0$, the constant of integration in equation (1-27) can be eliminated if we integrate over the boundary limits as follows:

$$-\int_{[A]_0}^{[A]_t} \frac{d[A]}{[A]} = k \int_0^t dt \tag{1-28}$$

This gives

$$-(\ln[A]_t - \ln[A]_0) = kt \tag{1-29}$$

and hence

$$-\ln[A]_t = kt - \ln[A]_0 \tag{1-30}$$

Thus, the constant in equation (1-27) is just

$$\text{constant} = -\ln[A]_0 \tag{1-31}$$

Equation (1-30) can be written in various forms. Some that are commonly used are

$$\ln\left(\frac{[A]_t}{[A]_0}\right) = -kt \tag{1-32a}$$

$$[A]_t = [A]_0 e^{-kt} \tag{1-32b}$$

and

$$\frac{[A]_t}{[A]_0} = e^{-kt} \tag{1-32c}$$

where e is the base of natural logarithms ($e = 2.718281828459045...$). These forms of the integrated rate expression for the first-order reaction are worth remembering. From the exponential form of equation (1-32b) and (1-32c), one can determine a *time constant* τ which is called the *decay time* of the reaction. This quantity is defined as the time required for the concentration to decrease to $1/e$ of its initial value $[A]_0$. The time τ is given by

$$\tau = \frac{1}{k} \tag{1-33}$$

In experimental determinations of the rate constant k, the integrated form of the rate law is often written in decimal logarithms as

$$\log_{10}[A]_t = \log_{10}[A]_0 - \frac{kt}{2.303} \tag{1-34}$$

and a semilog plot of $[A]_t$ versus t will yield a straight line with $-k/2.303$ as slope and $[A]_0$ as intercept.

1.3.3 Second-Order Reactions

There are two cases of second-order kinetics. The first is a reaction between two identical species, *viz.*,

$$A + A \rightarrow \text{products} \tag{1-35}$$

The rate expression for this case is

$$R = -\frac{1}{2}\frac{d[A]}{dt} = k[A]^2 \tag{1-36}$$

The second case is an overall second-order reaction between two unlike species, given by

$$A + B \rightarrow \text{products} \tag{1-37}$$

In this case, the reaction is first order in each of the reactants A and B and the rate expression is

$$R = -\frac{d[A]}{dt} = k[A][B] \tag{1-38}$$

Note the appearance of the stoichiometric coefficient $1/2$ in equation (1-36), but not in equation (1-38).

Let us consider the first case, given by equation (1-35) and (1-36). Although not an elementary reaction, the disproportionation of HI [equation (1-3)] is a reaction which is exactly second order in a single reactant. Another example is the recombination of two identical radicals, such as two methyl radicals:

$$2CH_3 \rightarrow C_2H_6 \tag{1-39}$$

We integrate the rate law, equation (1-36), to obtain

$$-\int_{[A]_0}^{[A]_t} \frac{d[A]}{[A]^2} = 2k \int_0^t dt \tag{1-40}$$

which gives

$$\frac{1}{[A]_t} = \frac{1}{[A]_0} + 2kt \tag{1-41}$$

A plot of the inverse concentration of A ($[A]^{-1}$) versus time should yield a straight line with slope equal to $2k$ and intercept $1/[A]_0$.

To integrate the rate law for the second case, equations (1-37) and (1-38), it is convenient to define a progress variable x which measures the progress of the reaction to products as

$$x = ([A]_0 - [A]_t) = ([B]_0 - [B]_t) \tag{1-42}$$

where $[A]_0$ and $[B]_0$ are the initial concentrations. The rate expression given by equation (1-38) can then be rewritten in terms of x as

$$\frac{dx}{dt} = k([A]_0 - x)([B]_0 - x) \tag{1-43}$$

To find the time behavior, we integrate equation (1-43) thus:

$$\int_{x(0)}^{x(t)} \frac{dx}{([A]_0 - x)\,([B]_0 - x)} = k \int_0^t dt \tag{1-44}$$

To solve the integral on the left-hand side of equation (1-44), we separate the variables and use the method of partial fractions:

$$\int \frac{dx}{([A]_0 - x)([B]_0 - x)}$$
$$= \int \frac{dx}{([A]_0 - [B]_0)\,([B]_0 - x)} - \int \frac{dx}{([A]_0 - [B]_0)([A]_0 - x)} \tag{1-45}$$

Solving the right-hand side of equation (1-45) and equating it to the left-hand side of equation (1-44), we obtain, as the solution to the rate expression for the second case,

$$\frac{1}{([A]_0 - [B]_0)} \ln \left(\frac{[B]_0 [A]_t}{[A]_0 [B]_t} \right) = kt \tag{1-46}$$

In this case the experimental data may be plotted in the form of the left-hand side of the equation against t.

1.3.4   Third-Order Reactions

From the definition of overall reaction order in equation (1-10), we see that there are three possible types of third-order reactions: (1) $3A \rightarrow$ products; (2) $2A + B \rightarrow$ products; and (3) $A + B + C \rightarrow$ products. In the first case, in which the rate law depends on the third power of one reactant, the rate expression is

$$R \equiv -\frac{1}{3}\frac{d[A]}{dt} = k[A]^3 \tag{1-47}$$

This rate law can be integrated readily to obtain the solution

$$-\frac{1}{2}\left(\frac{1}{[A]^2} - \frac{1}{[A]_0^2}\right) = 3kt \tag{1-48}$$

Rearranging gives

$$\frac{1}{[A]^2} = \frac{1}{[A]_0^2} + 6kt \tag{1-49}$$

A plot of the inverse squared concentration of A ($[A]^{-2}$) with time should yield slope $6k$ and intercept $1/[A]_0^2$.

The second case,

$$2A + B \rightarrow C \tag{1-50}$$

which is second order in reactant A and first order in reactant B, has the overall order 3. The rate law for this reaction is

$$R \equiv -\frac{1}{2}\frac{d[A]}{dt} = k[A]^2[B] \tag{1-51}$$

This rate expression can be integrated for two possible subcases. The first is when the concentration of B is so much greater than that of the reactant A ($[B] \gg [A]$), that the concentration of B does not change during the course of the reaction. Under this condition, the rate expression can be rewritten as

$$R \equiv -\frac{1}{2}\frac{d[A]}{dt} = k'[A]^2 \tag{1-52}$$

so that the third-order expression reduces to a "pseudo second-order" expression. The solution for this case is equation (1-41), i.e.,

$$\frac{1}{[A]_t} = \frac{1}{[A]_0} + 2k't \tag{1-53}$$

A plot of $[A]^{-1}$ vs. time, for a fixed $[B]$, should then yield slope $2k'$ and intercept $1/[A]_0$; but note that the resulting rate coefficient is a function of the concentration of B, that is,

$$k' = k[B] \tag{1-54}$$

Forgetting that this rate coefficient contains an added concentration term can lead to errors in interpretation of data. A simple example of this type of reaction is the three-body recombination process, such as $I + I + M \rightarrow I_2 + M$ and $O + O_2 + M \rightarrow O_3 + M$.

In these cases the third body acts to remove the excess energy from the recombining reactants, thereby stabilizing the molecular products.

When the initial concentrations of the dissimilar reactants A and B are comparable in magnitude, the rate law can be integrated by introducing a progress variable as we did in solving the second-order reaction case. Accordingly, we define a progress variable y by

$$[A]_t = [A]_0 - 2y$$

and

$$[B]_t = [B]_0 - y$$

and with this condition we can rewrite equation (1-51) in terms of y as

$$\frac{dy}{dt} = k([A]_0 - 2y)^2 ([B]_0 - y) \tag{1-55}$$

Upon rearranging, we obtain

$$\frac{dy}{([A]_0 - 2y)^2 ([B]_0 - y)} = k\,dt \tag{1-56}$$

This equation can be integrated by the method of partial fractions to yield

$$\frac{1}{([A]_0 - 2[B]_0)}\left(\frac{1}{[A_0]} - \frac{1}{[A]_t}\right) + \frac{1}{([A]_0 - 2[B]_0)^2}\ln\left(\frac{[A]_t[B]_0}{[A]_0[B]_t}\right) = kt \tag{1-57}$$

An example of such a reaction is the gas phase reaction between nitric oxide and oxygen[2]

$$2NO + O_2 \rightarrow 2NO_2 \tag{1-58}$$

The third type of third-order reaction is first order in three different components, i.e.,

$$A + B + C \rightarrow product \tag{1-59}$$

The rate law for this reaction is

$$R = \frac{d[A]}{dt} = k[A][B][C] \tag{1-60}$$

To solve for the integrated rate law expression in this case, we use the method of partial fractions as before. The solution is left as an exercise at the end of this chapter.

1.3.5   Reactions of General Order

There are no known examples of fourth-, fifth-, or higher order reactions in the chemical literature. The highest order which has been empirically encountered for chemical reactions is third order. Nevertheless, in this section we develop the general solution for a reaction which is nth order in one reactant. The rate expression for such a reaction is

$$R = -(1/n)\frac{d[A]}{dt} = k[A]^n \tag{1-61}$$

A simple integration of this expression yields the result

$$\frac{1}{(n-1)}\left(\frac{1}{[A]_t^{n-1}} - \frac{1}{[A]_0^{n-1}}\right) = nkt \tag{1-62}$$

which can be rewritten as

$$\frac{1}{[A]_t^{n-1}} - \frac{1}{[A]_0^{n-1}} = n(n-1)kt \tag{1-63}$$

Equation (1-63) is valid for any value of n except $n = 1$, in which case it is undefined and equation (1-32) must be used instead.

   In the general case, no simple plot can be constructed to test the order of the reaction, as can be done for the first- and second-order cases. When the order n is unknown, a van't Hoff plot can be constructed as an aid to deducing the order of the reaction. In a van't Hoff plot, the logarithm of the rate is plotted against the logarithm of the concentration of the reactant A. This is equivalent to making a plot of equation (1-63) on log-log graph paper. The slope of such a plot gives the order of the reaction n. Examples of van't Hoff plots for several reaction orders are shown in Figure 1-1. The van't Hoff plotting procedure is valid for integer and noninteger values of n.

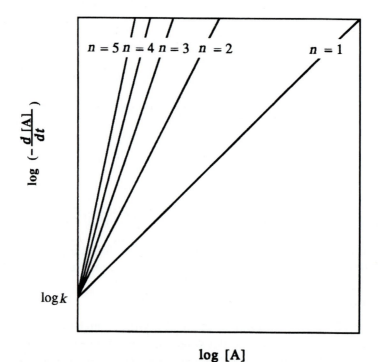

FIGURE 1-1   Van't Hoff plot of $\log -(d[A]/dt)$ versus $\log [A]$ for various reaction orders. Note that the slope of the line on this log-log plot is equal to n for each reaction order.

1.4   DETERMINATION OF REACTION ORDER: REACTION HALF-LIVES

Thus far, we have introduced the concept of a rate law and have shown that experimental rate laws can frequently be written as the product of concentrations of reacting species raised to some power; the exponents in such a rate law then define the order of the particular reaction. These reaction orders are empirically determined and may be nonintegral. We have also considered some simple rate laws and, by integrating them, have shown how experimental data can be plotted to enable the reaction order to be determined. The van't Hoff method has been introduced as a method for determining a general reaction order.

An alternative to the van't Hoff method, and one of the more popular methods for determining reaction order, is the *half-life method*. The reaction half-life $t_{1/2}$ as a function of initial reactant concentration can help establish the order with respect to that reactant. Consider the simple first-order reaction (1-11). The integrated rate equation is given by equation (1-32a),

$$\ln\left([A]_t/[A]_0\right) = -kt$$

By definition, at $t = t_{1/2}, [A]_t = [A]_0/2$; therefore, the rate equation can be rewritten as

$$-\ln\left(\frac{[A]_0/2}{[A]_0}\right) = kt_{1/2} \tag{1-64}$$

The half-life is then

$$t_{1/2} = \frac{\ln 2}{k} = \frac{0.693}{k} \tag{1-65}$$

Thus, for a first-order reaction, $t_{1/2}$ is independent of concentration. On the other hand, for a reaction of order $n \neq 1$ in a single reactant, the reaction half-life is

$$t_{1/2} = \frac{(2^{n-1} - 1)}{k(n - 1)[A]_0^{n-1}} \tag{1-66}$$

where k is the rate constant and $[A]_0$ is the initial concentration of the reactant. Thus, the half-life for orders $n \neq 1$ is a function of the initial concentration of the reactant; consequently, a plot of the logarithm of $t_{1/2}$ against the logarithm of $[A]_0$ should enable one to determine the reaction order. To illustrate this, let us take the logarithm of both sides of equation (1-66). We obtain

$$\log t_{1/2} = \log\frac{(2^{n-1} - 1)}{k(n - 1)} - (n - 1)\log[A]_0 \tag{1-67}$$

It is clear from this expression that the plot will be linear with a slope equal to $n - 1$, from which the order can be determined. With this information, and one or more absolute values of $t_{1/2}$, the rate constant can also be calculated. The reader is reminded that this procedure is valid only for reactions which are nth order in a single reactant.

Other methods of determining reaction orders are discussed in some earlier kinetics textbooks, including those by Hammes[3] and Benson.[4]

1.5 TEMPERATURE DEPENDENCE OF RATE CONSTANTS: THE ARRHENIUS EQUATION

We have seen that rate expressions are often simple functions of reactant concentrations with a characteristic rate coefficient k. If the rate expression is correctly formulated, the rate coefficient should indeed be a constant—that is, it should not depend on the concentrations of species appearing in the rate law, or of any other species which may be present in the reaction mixture.* The rate coefficient should also be independent of time. It does, however, depend strongly on temperature. This behavior was described by Svante Arrhenius in 1889[5] on the basis of numerous experimental rate measurements. Arrhenius found that rate constants varied as the negative exponential of the reciprocal absolute temperature, that is,

$$k(T) = A \exp(-E_{act}/RT) \qquad (1\text{-}68)$$

This relationship is now known as the *Arrhenius equation,* and a plot of $\ln k$ (or $\log_{10} k$) vs. $1/T$ is called an *Arrhenius plot.* In the Arrhenius equation, the temperature dependence comes primarily from the exponential term, although the quantity A, referred to as the *pre-exponential* or the *frequency factor,* may have a weak temperature dependence, no more than some fractional power of T. The units of A are the same as the units of the rate constant k, since the exponential term has no units. In the case of a first-order reaction, A has units of $\sec^{-1}$; for reactions of higher order, A has units (concentration)$^{1-p}\sec^{-1}$. The temperature T is, of course, in absolute or Kelvin units (degrees Celsius + 273.16)

The key quantity in the Arrhenius equation is the *activation energy* E_{act}. The activation energy can be thought of as the amount of energy which must be supplied to the reactants in order to get them to react with each other. Since this is a positive energy quantity, the majority of reactions have k increasing with temperature. For some reactions, however, the rate decreases with temperature, implying a negative activation energy. Such reactions are generally complex, involving the formation of a weakly bound intermediate species. An example is the recombination of iodine atoms in the presence of a molecular third body M, which proceeds via the following steps:[6]

$$I + M \rightarrow IM$$

$$IM + I \rightarrow I_2 + M$$

The IM species is a van der Waals complex whose stability decreases with increasing temperature.

The standard method for obtaining E_{act} is to graph experimental rate constant data on an Arrhenius plot, i.e., $\log_{10} k$ vs. $1/T$. The slope gives $E_{act}/2.303R$, where R = 8.3145 J mol^{-1}K^{-1}. The units of E_{act} are thus J mol^{-1}, but since the magnitudes of activation energies are typically in the range of a few to several hundred thousand J mol^{-1}, it is customary to report their values in kJ mol^{-1}. The older (non-SI) unit of calories of kcal mol^{-1} is still often encountered, but should be discouraged. The conversion factor[7] is 1 calorie = 4.184 J.

---

*The rate coefficient for reactions in liquid solutions may depend on pH, that is H$^+$ concentration; this will be discussed in chapter 4.

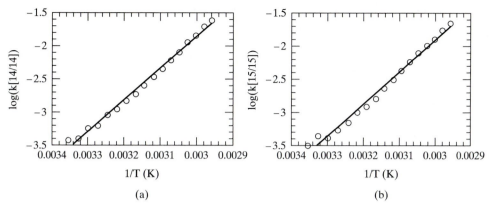

FIGURE 1-2   Arrhenius plots for the $(\mu\text{-}N_2)\{Mo[N(R)Ar]_3\} \rightarrow 2[Ar(R)N]_3Mo \equiv N$ reaction. (a) is for the $\mu\text{-}^{14}N_2$ isotopomer, and (b) is for the $\mu\text{-}^{15}N_2$ variety [From Ref. 8; reproduced with permission].

An illustration of the use of an Arrhenius plot to obtain activation energies is shown in Fig. 1-2, which shows data for the dinitrogen cleavage reaction of molybdenum (III) complexes first studied by Laplaza et al.[8] A key reaction in the overall mechanism is the conversion of a purple μ-dinitrogen complex into Mo (VI) terminal nitrido complexes, *viz.*,

$$\tag{1-69}$$

Purple μ-Dinitrogen Complex   **Molybdenum(VI) Terminal Nitrido Complexes**

The rate is monitored by following the decay of the absorption band at 547 nm due to the $(\mu\text{-}N_2)\{Mo[N(R)Ar]_3\}$ complex. Figure 1-2(a) shows the data for the $\mu\text{-}^{14}N_2$ complex, and Figure 1-2(b) for the $\mu\text{-}^{15}N_2$ complex. Note that the activation energies obtained for the two isotopic species are essentially the same, but that the pre-exponential factor is slightly smaller for the $^{15}N_2$ species. The reasons for this will be discussed in Chapter 10.

The origin of the activation energy is a barrier on the potential energy surface between the reactants and products; this is discussed in detail in chapters 7 and 10. For

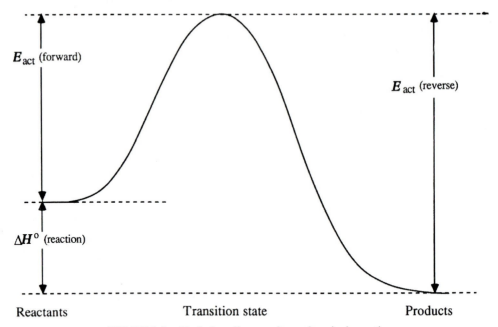

FIGURE 1-3   Enthalpy diagram for a chemical reaction.

the time being, the *enthalpy diagram* shown in Figure 1-3 may be instructive. This is simply a sketch of the thermodynamic energies associated with the reactants, the products, and a (for the moment) hypothetical *transition state* connecting the two. The energy difference between the reactants and products is the difference in their heats of formation and is given by

$$\Delta H^0_{reaction} = \Delta H^0_f(\text{products}) - \Delta H^0_f(\text{reactants}) \qquad (1\text{-}70)$$

A reaction which is highly *endothermic,* that is, which has a large positive $\Delta H^0_{reaction}$ is not likely to proceed spontaneously except at very high temperatures. A highly *exothermic* reaction, however, may do so unless the activation energy required to react to the transition state is very high; in that case, the reaction will be slow at other than very high temperatures. The calculation of reaction exothermicities is often a helpful guide in assessing the relative importance of individual reactions in a complex mechanism.

An important relationship which may be established by inspection of Figure 1-3 is that between forward and reverse activation energies for a given reaction. Clearly,

$$E_{act}(\text{forward}) - E_{act}(\text{reverse}) = \Delta H^0_{reaction} \qquad (1\text{-}71)$$

Thus, exothermic reactions have a larger activation energy for the reverse reaction, while the opposite is true for endothermic reactions.

For the great majority of elementary reactions, the rate coefficients exhibit Arrhenius behavior. However, there are some cases where the observed rate coefficient does not show the typical rapid increase with temperature. One such example is the reaction of OH radicals with CO, which is an important reaction in atmospheric and

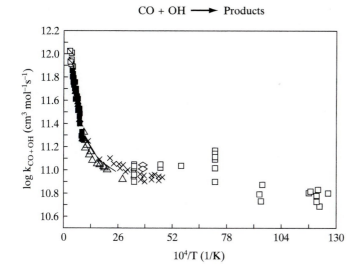

FIGURE 1-4 "Arrhenius plot" for the OH + CO → H + CO_2 reaction. The reaction rate is nearly temperature-independent between 100 K and 500 K but rises rapidly at T > 800 K. [Figure provided by D. M. Golden (1998); reproduced with permission.]

combustion chemistry for the conversion of CO into CO_2[9-11] (see Chapters 14 and 15). Remember that if a reaction obeys the Arrhenius equation, a plot of log k vs. $1/T$ should be a straight line with the slope and intercept determining the Arrhenius parameters E_{act} and A, respectively. A plot of log k vs. $1/T$ for the OH + CO reaction (Figure 1-4), clearly shows a marked deviation from linear behavior. This is a clear example of a non-Arrhenius temperature dependence. The explanation in this case is that the reaction is not a single elementary step, but involves a set of such steps in which OH radical reacts with CO to form a stable intermediate HOCO complex, which then decays either back to reactants or to H + CO_2 products. The HOCO complex (and its isotopic variant DOCO) has been experimentally observed by transient infrared spectroscopy.[12,13] Other sources of non-Arrhenius behavior will be discussed in Section 10.4.

1.6   REACTION MECHANISMS, MOLECULAR DYNAMICS, AND THE ROAD AHEAD

In this chapter we have concentrated mainly on elementary chemical reactions. We have examined ways in which the chemical kineticist can determine the reaction order from measurements of the time dependence of the reactant and product concentrations and the reaction activation energy from the temperature dependence of the rate constant. Unfortunately, with just this amount of information it is not possible to achieve a complete understanding of most chemical reactions. This is because reactions that are studied experimentally rarely occur in a single step. More often, there is a complex series of elementary reactions which take place during the transformation of reactants into products. Reactions that consist of several elementary reaction steps are called *complex,* or *composite,* reactions. The *mechanism* of a complex reaction is the assembly of individual steps that make it up. One of the principal aims of experimental studies in chemical kinetics is to elucidate mechanisms and to describe the overall observed reaction process in terms of the elementary reaction steps which constitute the mechanism. In chapter 2, we consider various types of complex reactions together with methods for analyzing their kinetics. Chapter 3 deals with some of the experimental methods that can be used to carry out rate measurements of complex reaction systems, with emphasis on gas-phase reactions. Chapter 4 then goes on to treat reaction in liquid solutions, and in chapter 5 we consider catalyzed reactions, including those which occur at the gas-solid interface.

Chapters 2 through 5 deal primarily with chemical *kinetics*, that is, the phenomenological behavior of reactions. In chapters 6 through 13, we turn our attention to chemical *dynamics*, the description of chemical reactions at the molecular level. The approach there will be to attempt to isolate elementary reaction steps experimentally, to examine them in great detail, and to relate the findings to microscopic molecular properties. Finally, chapters 14 and 15 present some examples of kinetic systems, that is, large coupled reactions sets which describe complex real-world phenomena.

REFERENCES

[1] K. J. Laidler, *Pure App. Chem. 53*, 753 (1981).

[2] G. Kornfeld and E. Klinger, *Z. physik. Chem. B4*, 37 (1929).

[3] G. G. Hammes, *Principles of Chemical Kinetics* (New York: Academic Press, 1978).

[4] S. W. Benson, *Foundations of Chemical Kinetics* (New York: McGraw-Hill, 1960).

[5] S. Arrhenius, *Z. physik. Chem. 4*, 226 (1889).

[6] G. Porter and J. A. Smith, *Proc. Roy. Soc. A261,* 28 (1961).

[7] I. Mills, ed., *Quantities, Units, and Symbols in Physical Chemistry* (Oxford: Blackwell, 1988).

[8] C. E. Laplaza, M. J. A. Johnson, J. C. Peters, A. L. Odom, E. Kim, C. C. Cummins, G. N. George, and I. J. Pickering, *J. Am. Chem. Soc. 118*, 8623 (1996).

[9] W. B. DeMore, S. P. Sander, D. M. Golden, R. F. Hampson, M. J. Kurylo, C. J. Howard, A. R. Ravishankara, C. E. Kolb, and M. J. Molina, *Chemical Kinetics and Photochemical Data for Use in Stratospheric Modeling*, JPL Publication 94-20, Jet Propulsion Laboratory, Pasadena, Calif. (1994).

[10] R. Atkinson, D. L. Baulch, R. A. Cox, R. F. Hampson, J. A. Kerr, and J. Troe, *J. Phys. Chem. Ref. Data 21,* 1125 (1993).

[11] D. M. Golden (to be published).

[12] J. T. Petty and C. B. Moore, *J. Chem. Phys. 99*, 47 (1993).

[13] T. Sears, W. M. Fawzy, and P. M. Johnson, *J. Chem. Phys. 97*, 3996 (1992).

BIBLIOGRAPHY

AMDUR, I., and HAMMES, G. G. *Chemical Kinetics: Principles and Selected Topics*. New York: McGraw-Hill, 1966.

CAPELLOS, C., and BIELSKI, B. *Kinetic Systems*. New York: Wiley, 1972.

ESPENSON, J. H. *Chemical Kinetics and Reaction Mechanisms*. New York: McGraw-Hill, 1981.

EYRING, H., and EYRING, E. M. *Modern Chemical Kinetics*. New York: Reinhold, 1963.

FLECK, G. M. *Chemical Reaction Mechanisms.* New York: Holt, Rinehart and Winston, 1971.

FROST, A. A., and PEARSON, R. G. *Kinetics and Mechanism.* 2d ed. New York: Wiley, 1961.

GARDINER, W. D., JR. *Rates and Mechanisms of Chemical Reactions.* Menlo Park, CA: W. A. Benjamin, 1969.

GLASSTONE, S., LAIDLER, K. J., and EYRING, H. *The Theory of Rate Processes.* New York: McGraw-Hill, 1941.

JOHNSTON, H. S. *Gas Phase Reaction Rate Theory.* New York: Ronald Press, 1961.

JORDAN, P. C. *Chemical Kinetics and Transport.* New York: Plenum, 1979.

KONDRAT'EV, V. N. *Chemical Kinetics of Gas Reactions.* Translated by J. M. Crabtree and S. N. Carnethes. Reading, MA: Addison-Wesley, 1964.

LAIDLER, K. J. *Chemical Kinetics.* 3d ed. New York: Harper and Row, 1987.

MOORE, J. W., and PEARSON, R. G. *Kinetics and Mechanism.* 3d ed. New York: Wiley, 1981.

NICHOLAS, J. *Chemical Kinetics.* New York: Wiley, 1976.

PILLING, M. J., and SEAKINS, P. W. *Reaction Kinetics.* Oxford: University Press, 1995.

PIMENTEL, G. C. (Chairman, Committee to Survey Opportunities in the Chemical Sciences). *Opportunities in Chemistry.* Washington, DC: National Academy Press, 1985.

WESTON, R. W., JR. and SCHWARZ, H. A. *Chemical Kinetics.* Englewood Cliffs, NJ: Prentice-Hall, 1972.

PROBLEMS

1.1 (a) What fractions of the molecules in H_2 and UF_6 have kinetic energies greater than 100 kJ/mole at 300 K and at 3,000 K (neglect dissociation)?

(b) In what proportion of binary collisions does the kinetic energy along the line of centers exceed 50 kJ/mole at 100 K, at 300 K, and at 2,000 K?

1.2 The units in the preceding problem were in kJ/mole, which is an example of SI units. Give values for the following quantities both in kJ/mole and in the indicated alternative units in parentheses:

(a) Heat capacity of liquid water at 15°C and 1 atm (calories/mole K).

(b) Vibrational fundamental frequency of $H^{35}Cl$ (cm^{-1}).

(c) Ionization potential of H atom (eV).

(d) R.M.S. average translational kinetic energy of Br_2 at 300 K (ergs/molecule).

(e) energy of CO_2 laser photons having a wavelength of 10.59 μm (Hz).

1.3 (a) Derive the integrated rate equation for a reaction of 3/2 order in a single reactant. Derive the expression of the half-life of such a reaction. Can you think of an example of such a reaction?

(b) Derive the integrated rate equation for a reaction of order n.

1.4 The first-order gas reaction $SO_2Cl_2 \rightarrow SO_2 + Cl_2$ has $k_1 = 2.20 \times 10^{-5} s^{-1}$ at 593 K. What percent of a sample of SO_2Cl_2 would be decomposed by heating at 593 K for 1 hour? For 3 hours? How long will it take for half the SO_2Cl_2 to decompose?

1.5 The kinetics of the formation of ethyl acetate from acetic acid and ethyl alcohol as homogeneously catalyzed by a constant amount of HCl has been studied by titrating 1-cc aliquots of the reaction mixture with 0.0612 N base at various times. The following data have been obtained at 25°C [O. Knoblauch, *Z. physik. Chem. 22,* 268 (1897)]:

Initial Concentrations

$[CH_3COOH] = 1.000$ M

$[C_2H_5OH] = 12.756$ M

$[H_2O] = 12.756$ M

$[CH_3COOC_2H_5] = 0$

t, min	Base, cc	t, min	Base, cc
0	24.37	148	18.29
44	22.20	313	14.14
62	21.35	384	13.40
108	19.50	442	13.09
117	19.26	∞	12.68

The overall reaction can be written as

$$CH_3COOH + C_2H_5OH \underset{k_{-1}}{\overset{k_1}{\rightleftharpoons}} CH_3COOC_2H_5 + H_2O$$

The reaction has been found to be first order with respect to each of the four reactants. Calculate the specific rate constants k_1 and k_{-1}. What is the equilibrium constant K_1 at 25°C?

1.6 Nitrogen pentoxide decomposes according to the reaction $N_2O_5 \rightarrow 2NO_2 + \frac{1}{2}O_2$ with rate constant k. The measured rates between 273 K and 338 K are as follows:

T(K)	273	298	308	318	328	338
$k(\times 10^5 \text{ sec}^{-1})$	0.0787	3.46	13.5	49.8	150	487

Make an Arrhenius plot of these data, and determine E_{act} and A for the first-order decomposition of nitrogen pentoxide [F. Daniels and F. H. Johnston, *J. Am. Chem. Soc. 43*, 53 (1921)].

1.7 The reaction

$$NO_3 + NO \rightarrow 2NO_2$$

is known to be an elementary process.

(a) Write the rate expression for the rate of disappearance of NO_3 and NO.

(b) Write the rate expression for the appearance of NO_2.

(c) Show how the rate constants in (a) and (b) are related.

1.8 T-butyl bromide is converted into t-butyl alcohol in a solvent containing 90 percent acetone and 10 percent water. The reaction is given by

$$(CH_3)_3CBr + H_2O \rightarrow (CH_3)_3COH + HBr$$

The following table gives the data for the concentration of t-butyl bromide versus time [L. C. Bateman, E. D. Hughes, and C. K. Ingold, *J. Chem. Soc. 960, 1940*]:

t (min)	$(CH_3)_3CBr(mol\ l^{-1})$
0	0.1056
9	0.0961
18	0.0856
24	0.0767
40	0.0645
54	0.0536
72	0.0432
105	0.0270

What is the order of the reaction? What is the rate constant of the reaction? What is the half-life of the reaction?

1.9 The reaction

$$A + B + C \rightarrow products$$

has the rate expression

$$R = -\frac{d[A]}{dt} = k[A][B][C]$$

Derive the integral form of the rate expression.

1.10 (a) What are the units of the rate constants of first-, second-, and third-order reactions if the concentrations are expressed in moles cm^{-3} and the time is given in seconds? What are the conversion functions that must be used to convert to concentration units of molecules/cm^3 in each case?

(b) If a reaction obeyed the rate law $R = k[A]^{1/2}[B]^{2/3}$, what would be the units of k?

1.11 Write the rate expression for the hypothetical reaction

$$2A + 3B \rightarrow X$$

and derive the integrated rate expression for the reaction.

CHAPTER 2

Complex Reactions

2.1 EXACT ANALYTIC SOLUTIONS FOR COMPLEX REACTIONS

2.1.1 Introduction

In chapter 1, we considered primarily reactions involving a single step; but most chemical processes are complex, i.e., they consist of a number of coupled elementary reactions. These complex reactions can be divided into several classes: (1) opposing or reversible reactions, (2) consecutive reactions, (3) parallel reactions, and (4) mixed reactions.

In this chapter we examine methods for determining exact analytic solutions for the time dependence of concentrations of species involved in complex reactions. Since it is difficult to obtain such exact solutions for many complex reactions, approximate treatments, such as the steady-state or pseudo-first-order methods, may need to be used to obtain analytic results. Accordingly, we discuss both of these methods in our treatment. Two other methods that are used in solving systems of kinetic equations for complex reactions are the Laplace transform method and determinant method. Using these methods, exact analytic solutions may be obtained for many complex reactions. However, when analytic solutions are impossible to obtain either directly or by using approximation methods, a numerical solution is generally possible.

In our discussion, we introduce the reader to the basic theory behind various numerical methods, discuss some of the limitations of these methods, and explain what must be considered in choosing a numerical method for solving kinetic equations for complex reactions. Finally, we introduce a more advanced topic, stochastic methods. This topic explores the origin of kinetic equations from a statistical view and gives a justification for the use of differential equations modeling the time evolution of species involved in complex reactions.

2.1.2 Reversible Reactions

Reversible or opposing reactions are reactions in which the product of the initial reaction can proceed to re-form the original substance. A chemical example of this is the cis-trans isomerization of 1,2-dichloroethylene:[1]

$$\text{(2-1)}$$

From this example, we can see that the simplest reversible reaction is of the form

$$A_1 \underset{k_r}{\overset{k_f}{\rightleftharpoons}} A_2 \tag{2-2}$$

and is first order in each direction. The differential equation for this mechanism is

$$-\frac{d[A_1]}{dt} = k_f[A_1] - k_r[A_2] \tag{2-3}$$

$$-\frac{d[A_2]}{dt} = k_r[A_2] - k_f[A_1] \tag{2-4}$$

If it is assumed that both A_1 and A_2 are present in the system at time $t = 0$, that is, $[A_1] = [A_1]_0$ and $[A_2] = [A_2]_0$ and, then at any time afterwards the total amount of reactant remaining and new product formed must equal the initial amount of the reactants before reaction. (This is the law of conservation of mass.) Hence,

$$[A_1]_0 + [A_2]_0 = [A_1] + [A_2] \tag{2-5}$$

Solving for $[A_2]$, we obtain

$$[A_2] = [A_1]_0 + [A_2]_0 - [A_1] \tag{2-6}$$

and substituting this into equation (2-3) yields

$$-\frac{d[A_1]}{dt} = +k_f[A_1] - k_r([A_1]_0 + [A_2]_0 + [A_1])$$

$$-\frac{d[A_1]}{dt} = -k_r([A_1]_0 + [A_2]_0) + (k_f + k_r)[A_1] \tag{2-7}$$

$$-\frac{d[A_1]}{dt} = (k_f + k_r)\left\{ \frac{-k_r([A_1]_0 + [A_2]_0)}{(k_f + k_r)} + [A_1] \right\}$$

To find the solution to equation (2-7) we introduce a parameter m, defined as

$$-m = \frac{-k_r([A_1]_0 + [A_2]_0)}{(k_f + k_r)} \tag{2-8}$$

This allows us to rewrite equation (2-7) as

$$-\frac{d[A_1]}{dt} = (k_f + k_r)\{-m + [A_1]\} \tag{2-9}$$

which we may then integrate

$$\int \frac{d[A_1]}{m - [A_1]} = (k_f + k_r)\int dt \tag{2-10}$$

The solution is

$$\ln\left(\frac{k_f[A_1] - k_r[A_2]}{k_f[A_1]_0 - k_r[A_2]_0}\right) = -(k_f + k_r)(t - t_0) \qquad (2\text{-}11)$$

If only $[A_1]$ is present in the system initially, at $t = 0$, then the solution reduces to

$$\ln\left(\frac{k_f[A_1] - k_r[A_2]}{k_f[A_1]_0}\right) = -(k_f + k_r)t \qquad (2\text{-}12)$$

which is just

$$[A_1] = \frac{[A_1]_0}{(k_f + k_r)}[k_r + k_f e^{-(k_f + k_r)t}] \qquad (2\text{-}13)$$

Using the mass conservation constraint $[A_1]_0 = [A_1] + [A_2]$, we can obtain the solution for $[A_2]$:

$$[A_2] = \frac{k_f[A_1]_0}{(k_f + k_r)}[1 - e^{-(k_f + k_r)t}] \qquad (2\text{-}14)$$

When equilibrium is reached, the individual reactions must be balanced; in other words, the reaction $A \rightarrow B$ must occur just as frequently as the reverse reaction. Consequently, detailed balancing is achieved by considering each reversible reaction to be independently able to bring about equilibrium between the reactant and the products. More formally, the principle of *detailed balance* (see Chapter 6), says that when a reaction system reaches equilibrium the forward and reverse reactions occur at the same rate. Consequently, for the reaction $A_1 \rightleftarrows A_2$ at equilibrium,

$$\frac{d[A_1]}{dt} = \frac{-d[A_2]}{dt} = 0 \qquad (2\text{-}15)$$

and

$$-k_f[A_1]_e + k_r[A_2]_e = 0 \qquad (2\text{-}16)$$

and we have the following definition of the equilibrium constant K_{eq} expressed in terms of rate constants:

$$\frac{k_f}{k_r} = \frac{[A_2]_e}{[A_1]_e} = K_{eq} \qquad (2\text{-}17)$$

The same argument can be extended to a reversible reaction that occurs in multiple stages.

A reversible reaction may also be of mixed order, such as

$$A_1 \underset{k_{-1}}{\overset{k_1}{\rightleftarrows}} A_2 + A_3$$

An example is $N_2O_4 \rightleftarrows 2NO_2$. The rate expression for this type of reaction is

$$\frac{d[A_1]}{dt} = -k_1[A_1] + k_{-1}[A_2][A_3] \qquad (2\text{-}18)$$

To find the solution, we introduce a progress variable,

$$x = [A_1]_0 - [A_1] \tag{2-19}$$

Introducing this variable into equation (2-18) transforms that equation into

$$\frac{dx}{dt} = k_1([A_1]_0 + x) - k_{-1}([A_2]_0 + x)([A_3]_0 + x)$$

or

$$\frac{dx}{dt} = k_1[A_1]_0 - k_{-1}[A_2]_0[A_3]_0 - (k_1 + k_{-1}[A_3]_0 + k_{-1}[A_2]_0)x - k_{-1}x^2 \tag{2-20}$$

We let

$$\alpha = k_1[A_1]_0 - k_{-1}[A_2]_0[A_3]_0$$
$$\beta = -(k_1 + k_{-1}[A_3]_0 + k_{-1}[A_2]_0) \tag{2-21}$$

and

$$\gamma = -k_{-1}$$

so that

$$\frac{dx}{dt} = \alpha + \beta x + \gamma x^2 \tag{2-22}$$

Then

$$\int \frac{dx}{\alpha + \beta x + \gamma x^2} = \int dt \tag{2-23}$$

The solution of equation (2-23) is

$$\ln\left[\frac{x + (\beta - q^{1/2})/2\gamma}{x + (\beta + q^{1/2})/2\gamma} \right] - \ln\left[\frac{x(t_0) + (\beta - q^{1/2})/2\gamma}{x(t_0) + (\beta + q^{1/2})/2\gamma} \right] = q^{1/2}(t - t_0) + \theta \tag{2-24}$$

where $q = \beta^2 - 4\alpha\gamma$ and $\theta = \ln((\beta - q^{1/2})/(\beta + q^{1/2}))$. At $t_0 = 0$, $x(0) = 0$, and the solution simplifies to

$$\ln\left[\frac{x + (\beta - q^{1/2})/2\gamma}{x + (\beta + q^{1/2})/2\gamma} \right] = q^{1/2}t + \theta \tag{2-25}$$

2.1.3   Consecutive Reactions

Irreversible reactions can be defined as those which start with an initial reactant and produce products or intermediates generally in only one direction. Consecutive reactions are sequential irreversible reactions. There are two classes of consecutive reactions: those which are first order and those which are mixed first order and second order. Let us consider first the simplest consecutive reaction of first order.

2.1.3.1   First-order consecutive reactions.   Consecutive reactions are further classified by the number of steps involved or the number of substances initially present, which define their order. We begin by considering the two-step (or -stage) first-order reaction.

2.1.3.1.1 First-order with two steps.   Consider the consecutive first-order reaction involving two stages:

$$A_1 \xrightarrow{k_1} A_2$$
$$A_2 \xrightarrow{k_2} A_3 \tag{2-26}$$

This mechanism can be described by the following set of rate expressions:

$$\frac{d[A_1]}{dt} = -k_1[A_1] \tag{2-27}$$

$$\frac{d[A_2]}{dt} = k_1[A_1] - k_2[A_2] \tag{2-28}$$

$$\frac{d[A_3]}{dt} = k_2[A_2] \tag{2-29}$$

Hence, the concentration of A_1 is obtained after integration as

$$[A_1] = [A_1]_0\, e^{-k_1 t} \tag{2-30}$$

The concentration of A_2 is computed from equation (2-28), which is a linear differential equation of first order. Substituting equation (2-30) into equation (2-28) we obtain

$$\frac{d[A_2]}{dt} + k_2[A_2] = k_1[A_1]_0\, e^{-k_1 t} \tag{2-31}$$

This general differential equation can be solved using standard methods. Doing so, we obtain the time dependence of $[A_2]$:

$$[A_2] = [A_2]_0 e^{-k_2 t} + \frac{k_1[A_1]_0}{k_2 - k_1}\left(e^{-k_1 t} - e^{-k_2 t}\right) \tag{2-32}$$

If $[A_2]_0 = 0$ at $t = 0$, then

$$[A_2] = \frac{k_1[A_1]_0}{k_2 - k_1}\left(e^{-k_1 t} - e^{-k_2 t}\right) \tag{2-33}$$

The concentration of $[A_3]$ can be determined from conservation of mass, i.e.,

$$[A_1]_0 = [A_1] + [A_2] + [A_3] \tag{2-34}$$

By substituting the expressions for $[A_1]$ and $[A_2]$ into equation (2-34) we obtain

$$[A_3] = [A_1]_0\left[1 - e^{-k_1 t} - \frac{k_1}{k_2 - k_1}\left(e^{-k_1 t} - e^{-k_2 t}\right)\right] \tag{2-35}$$

Simplifying this expression gives

$$[A_3] = [A_1]_0 \left[1 - \frac{k_2}{k_2 - k_1} e^{-k_1 t} + \frac{k_1}{k_2 - k_1} e^{-k_2 t} \right] \tag{2-36}$$

One obvious reason why the first-order consecutive reaction with two first order steps can be solved is that the depletion of A_1 does not depend upon A_2 and A_3. Thus even though there are three species here and only one conservation equation, it can be solved exactly.

Figure 2-1 shows typical kinetic curves of the accumulation of A_2 and A_3 for this case. As $[A_2]$ approaches its maximum value in the initial period, the rate of production of A_3 increases, but the rate of accumulation of A_3 is slow enough to allow an initial accumulation of A_2 to occur before it is converted to A_3, as shown in Fig. 2-1a. In reactions involving unstable intermediates such as free radicals, the production rate of the intermediate is usually slow; however, once the intermediate is formed, it rapidly transforms to products. In this case the rate of production of A_3 is faster than the rate of accumulation of the intermediate A_2, i.e., $k_2 \gg k_1$. In Fig. 2-1b, the concentration of the intermediate A_2 does not have sufficient time to build up. Consequently, its concentration remains small throughout the reaction.

Some caution must be exercised in interpreting kinetic plots such as those shown in Fig. 2-1. While it is natural to interpret the rising portion of the curve as the formation rate of the short-lived intermediate and the falling portion as its decay rate, that is not always the case. If $k_2 > k_1$, that is, the consumption rate of the unstable intermediate A_2 is faster than its formation rate (as is frequently the case), then for small times equation (2-33) approximates $\{k_1[A_1]/(k_2 - k_1)\} \times \{1 - \exp(-k_2 t)\}$, while the formation rate appears in the long-time behavior, which is proportional to $\exp(-k_1 t)$. If, on the other hand, $k_1 > k_2$, then we can rewrite equation (2-33) as

$$[A_2] = \frac{k_1[A_1]_0}{k_1 - k_2} (e^{-k_2 t} - e^{-k_1 t})$$

Now the rise time of $[A_2]$ goes as $(1 - e^{-k_1 t})$, while its decay time goes as $e^{-k_2 t}$; this is the case illustrated in Fig. 2-1a.

2.1.3.1.2 First order with three steps. The system of differential equations for a first-order consecutive reaction involving three steps, viz.:

$$A_1 \xrightarrow{k_1} A_2$$
$$A_2 \xrightarrow{k_2} A_3 \tag{2-37}$$
$$A_3 \xrightarrow{k_3} A_4$$

can be integrated in a way similar to the two-step case. The differential equations for $[A_1]$ and $[A_2]$ and their solutions do not differ from those obtained in the two-step case, while the differential equation for $[A_3]$ is

$$\frac{d[A_3]}{dt} = k_2[A_2] - k_3[A_3] \tag{2-38}$$

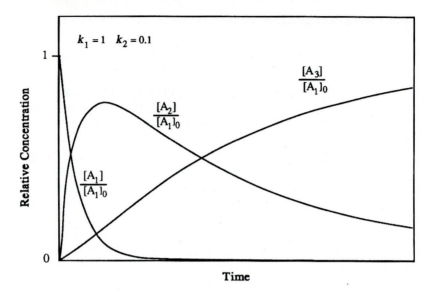

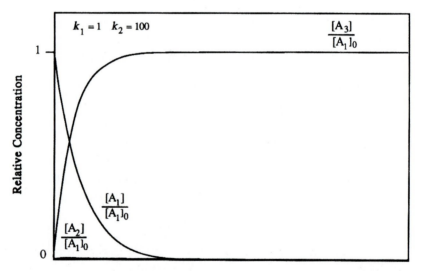

FIGURE 2-1   (a) Plot of the relative concentrations in the consecutive reaction $A_1 \rightarrow A_2 \rightarrow A_3$ for the case where $k_1/k_2 = 10$. (b) Plot of the relative concentrations for the case where $k_1/k_2 = 0.01$.

Substituting equation (2-33) for $[A_2]$ into equation (2-38) gives

$$\frac{d[A_3]}{dt} + k_3[A_3] = \left[\frac{k_1k_2[A_1]_0}{(k_2 - k_1)}\right][e^{-k_1t} - e^{-k_2t}] \qquad (2\text{-}39)$$

Integrating this linear equation subject to the initial condition that $[A_3] = 0$ at $t = 0$, we obtain

$$[A_3] = \left[\frac{k_1k_2[A_1]_0}{(k_2 - k_1)(k_3 - k_1)}\right]e^{-k_1t} + \left[\frac{k_1k_2[A_1]_0}{(k_1 - k_2)(k_3 - k_2)}\right]e^{-k_2t}$$

$$+ \left[\frac{k_1k_2[A_1]_0}{(k_1 - k_3)(k_2 - k_3)}\right]e^{-k_3t} \qquad (2\text{-}40)$$

By conservation of mass

$$[A_1]_0 = [A_1] + [A_2] + [A_3] + [A_4] \qquad (2\text{-}41)$$

Therefore,

$$[A_4] = [A_1]_0 - [A_1] - [A_2] - [A_3] \qquad (2\text{-}42)$$

Substituting the appropriate expressions for $[A_1]$, $[A_2]$, and $[A_3]$ into equation (2-42) we get

$$[A_4] = [A_1]_0 = \left[1 - \frac{k_2k_3e^{-k_1t}}{(k_2 - k_1)(k_3 - k_1)} - \frac{k_1k_3e^{-k_2t}}{(k_1 - k_2)(k_3 - k_2)} - \frac{k_1k_2e^{-k_3t}}{(k_1 - k_3)(k_2 - k_3)}\right]$$

$$(2\text{-}43)$$

In a similar manner the system of differential equations corresponding to $n - 1$ consecutive reactions

$$A_1 \rightarrow A_2$$
$$A_2 \rightarrow A_3$$
$$A_3 \rightarrow A_4 \qquad (2\text{-}44)$$
$$\vdots$$
$$A_{n-1} \rightarrow A_n$$

can be integrated. We leave this as an exercise in the problems at the end of this chapter.

2.1.3.2 Higher order consecutive reactions.

Some higher order consecutive reactions may be first order in one step and second order in the second, such as

$$A_1 \xrightarrow{k_1} A_2$$
$$A_1 + A_2 \xrightarrow{k_2} A_3 \qquad (2\text{-}45)$$

or both steps may be second order consecutive reactions, as in

$$A_1 + A_2 \rightarrow A_3$$
$$A_1 + A_3 \rightarrow A_4$$

$$(2\text{-}46)$$

The differential equations for most of these systems of reactions are nonlinear and generally they have no exact solutions. Nevertheless, analytic solutions for such systems can be obtained if time is eliminated as a variable.[2] To illustrate this method, consider the reaction sequence described by the system of equations (2-45):

$$\frac{d[A_1]}{dt} = -k_1[A_1] - k_2[A_1][A_2] \qquad (2\text{-}47)$$

$$\frac{d[A_2]}{dt} = +k_1[A_1] - k_2[A_1][A_2] \qquad (2\text{-}48)$$

$$\frac{d[A_3]}{dt} = k_2[A_1][A_2] \qquad (2\text{-}49)$$

To solve this system of equations we divide equations (2-48) by (2-47)

$$\frac{d[A_2]}{d[A_1]} = \frac{k_1[A_1] - k_2[A_1][A_2]}{- k_1[A_1] - k_2[A_1][A_2]} = -\left(\frac{k_1 - k_2[A_2]}{k_1 + k_2[A_2]}\right) \qquad (2\text{-}50)$$

If we let $K = k_1/k_2$, then

$$\frac{d[A_2]}{d[A_1]} = -\frac{K - [A_2]}{K + [A_2]} \qquad (2\text{-}51)$$

Equation (2-51) can be rearranged to

$$-\frac{(K - [A_2])d[A_2]}{K - [A_2]} + \frac{2Kd[A_2]}{(K - [A_2])} = -d[A_1] \qquad (2\text{-}52)$$

or

$$-d[A_2] + \frac{2K}{(K - [A_2])}d[A_2] = -d[A_1] \qquad (2\text{-}53)$$

Integrating equation (2-53) over the respective limits, i.e.,

$$-\int_{[A_2]=0}^{[A_2]} d[A_2] - 2K\int_{[A_2]=0}^{[A_2]} \frac{d(K - [A_2])}{(K - [A_2])} = -\int_{[A_1]=0}^{[A_1]} d[A_1] \qquad (2\text{-}54)$$

gives

$$\frac{[A_2]}{K} + 2\ln\left(1 - \frac{[A_2]}{K}\right) = \left(\frac{[A_1]_0}{K}\right)\left(\frac{[A_1]}{[A_1]_0} - 1\right) \qquad (2\text{-}55)$$

In the case of consecutive first- and second-order reactions, the only way that this system may be solved exactly is by eliminating time as an independent variable. In this manner, $[A_1]$ and $[A_2]$ are given at any time, as in equation (2-17); and by conservation of mass, the concentration relationship between $[A_3]$ and the other components is obtained. There are some special cases, however, of first- and second-order reactions that can be solved exactly without eliminating time as a variable.[3] One such reaction is

$$A_1 \xrightarrow{k_1} A_2$$
$$2\,A_2 \xrightarrow{k_2} A_3 \tag{2-56}$$

The rate equations for this reaction system are

$$\frac{d[A_1]}{dt} = -k_1[A_1] \tag{2-57}$$

$$\frac{d[A_2]}{dt} = k_1[A_1] - 2k_2[A_2]^2 \tag{2-58}$$

and

$$\frac{d[A_3]}{dt} = k_2[A_2]^2 \tag{2-59}$$

Integration of equation (2-57) yields

$$[A_1] = [A_1]_0\, e^{-k_1 t} \tag{2-60}$$

Substituting equation (2-60) into equation (2-58) yields

$$\frac{d[A_2]}{dt} = [A_1]_0\, k_1 e^{-k_1 t} - 2k_2[A_2]^2 \tag{2-61}$$

If we make the further substitution

$$[A_2] = \frac{1}{k_2 u(t)} \frac{du(t)}{dt} \tag{2-62}$$

we can transform the nonlinear first order equation (2-61) in $[A_2]$ into a linear second order equation in $u(t)$

$$\frac{d^2 u}{dt^2} - [A_1]_0 k_1 k_2 e^{-k_1 t} u = 0 \tag{2-63}$$

A further change of the independent variable t to $\tau = e^{-k_1 t}$ yields a Bessel equation

$$\frac{d}{d\tau}\left(\tau \frac{du}{d\tau}\right) - K u = 0 \tag{2-64}$$

the solution of which is

$$u = \alpha J_0(2i\sqrt{K\tau}) + \beta' i H_0^{(1)}(2i\sqrt{K\tau}) \tag{2-65}$$

where $K = [A_1]_0 k_2 / k_1$, α and β' are constants of integration, $i = \sqrt{-1}$ and J_0 and $H_0^{(1)}$ are Bessel functions of the first and third kind, respectively, of order zero. Then by the recurrence relations of Bessel functions, we have

$$\frac{dJ_0(2i\sqrt{K\tau})}{d\tau} = -\sqrt{K/\tau}\, iJ_1(2i\sqrt{K\tau}) \qquad (2\text{-}66)$$

and

$$i\frac{dH_0^{(1)}(2i\sqrt{K\tau})}{d\tau} = \sqrt{K/\tau}\, H_1^{(1)}(2i\sqrt{K\tau}) \qquad (2\text{-}67)$$

and the solution for $[A_2]$ is

$$[A_2] = [A_1]_0\sqrt{\tau/K}\,\frac{iJ_1(2i\sqrt{K\tau}) - \beta' H_1^{(1)}(2i\sqrt{K\tau})}{J_0(2i\sqrt{K\tau}) + \beta' H_1^{(1)}(2i\sqrt{K\tau})} \qquad (2\text{-}68)$$

The constant β' is determined by the initial condition $[A_2]_0 = 0$, which gives

$$\beta' = \frac{iJ_1(2i\sqrt{K})}{H_1^{(1)}(2i\sqrt{K})} \qquad (2\text{-}69)$$

The full solution to this problem is

$$[A_1] = [A_1]_0\tau$$

$$[A_1] = [A_1]_0\sqrt{\tau/K}\,\frac{iJ_1(2i\sqrt{K\tau}) - \beta' H_1^{(1)}(2i\sqrt{K\tau})}{J_0(2i\sqrt{K\tau}) + \beta' H_1^{(1)}(2i\sqrt{K\tau})} \qquad (2\text{-}70)$$

Finally, using conservation of mass, $[A_3]$ can be determined as:

$$[A_3] = \frac{1}{2}\left([A_1]_0 - [A_1] - [A_2]\right)$$

The definitions used in equation (2-70) are summarized below:

$$\tau = e^{-k_1 t}$$

$$K = [A_1]_0 k_2/k_1$$

$$\beta' = iJ_1(2i\sqrt{K})/H_1^{(1)}(2i\sqrt{K})$$

2.1.4 Parallel Reactions

Parallel reactions are defined as two or more processes in which the same species participate in each reaction step. The most common cases of parallel reactions are: (1) those in which the initial reactant decomposes into several products; (2) those in which the initial reactants are different, but yield the same products; and (3) those in which a substance reacts with two or more initial reactants.

2.1.4.1 First order decay to different products.   Consider the mechanism

$$
\begin{aligned}
A_1 &\xrightarrow{k_2} A_2\\
A_1 &\xrightarrow{k_3} A_3\\
A_1 &\xrightarrow{k_4} A_4
\end{aligned}
\qquad (2\text{-}71)
$$

which describes a kinetic system in which a single reactant decomposes into several products. At time $t = 0$ the initial concentrations of the four components are

$$[A_1] = [A_1]_0 \tag{2-72}$$

$$[A_2] = [A_3] = [A_4] = 0 \tag{2-73}$$

The differential equations for the system can be written, as exemplified for $[A_1]$,

$$-\frac{d[A_1]}{dt} = k_2[A_1] + k_3[A_1] + k_3[A_1]$$

$$= (k_2 + k_3 + k_4)[A_1] \tag{2-74}$$

$$= k_T[A_1]$$

where $k_T = (k_2 + k_3 + k_4)$ which is the sum of the rates for each of the independent parallel paths.

Integrating equation (2-74) yields

$$[A_1] = [A_1]_0 e^{-k_T t} = [A_1]_0 e^{-(k_2 + k_3 + k_4)t} \tag{2-75}$$

To solve for $[A_2]$ we substitute equation (2-75) into equation (2-76)

$$\frac{d[A_2]}{dt} = +k_2[A_1] \tag{2-76}$$

and upon integrating we obtain

$$[A_2] = [A_2]_0 + \frac{k_2[A_1]_0}{k_T} [1 - e^{-k_T t}]$$

or

$$[A_2] = \frac{k_2[A_1]_0}{k_T} [1 - e^{-k_T t}] \tag{2-77}$$

Expressions for $[A_3]$ and $[A_4]$, when derived in a similar way are, respectively

$$[A_3] = \frac{k_3[A_1]_0}{k_T} [1 - e^{-k_T t}] \tag{2-78}$$

and

$$[A_4] = \frac{k_4[A_1]_0}{k_T} [1 - e^{-k_T t}] \tag{2-79}$$

It follows that the relative rate constants can be determined by measuring the relative product yields:

$$\frac{[A_3]}{[A_4]} = \frac{k_3}{k_4} \tag{2-80}$$

This ratio defines the *branching ratio* for the reaction; note that this branching ratio is independent of time.

The above example can easily be generalized to reactions of a single reactant into n different products:

$$
\begin{aligned}
A_1 &\xrightarrow{k_2} A_2 \\
A_1 &\xrightarrow{k_3} A_3 \\
A_1 &\xrightarrow{k_4} A_4 \\
&\;\;\vdots \\
A_1 &\xrightarrow{k_n} A_n
\end{aligned}
\tag{2-81}
$$

The rate expression for the disappearance of $[A_1]$ is

$$
\frac{-d[A_1]}{dt} = k_T[A_1]_0
\tag{2-82}
$$

where $k_T = \Sigma_{i=2}^{n} k_i$; upon integration we find that

$$
[A_1] = [A_1]_0 e^{-k_T t}
\tag{2-83}
$$

To find the production of a given product A_n

$$
\frac{d[A_n]}{dt} = k_n[A_1]
\tag{2-84}
$$

we substitute equation (2-83) into (2-84) and integrate the resulting expression

$$
[A_n] = \frac{k_n[A_1]_0}{k_t}(1 - e^{k_T t})
\tag{2-85}
$$

2.1.4.2 First order decay to the same products. Consider the reaction sequence

$$
\begin{aligned}
A_1 &\xrightarrow{k_1} A_2 \\
A_3 &\xrightarrow{k_3} A_2
\end{aligned}
\tag{2-86}
$$

The rate expressions for the disappearance of $[A_1]$ and $[A_3]$ and the appearance of A_2 are

$$
-\frac{d[A_1]}{dt} = k_1[A_1]
\tag{2-87}
$$

$$
+\frac{d[A_2]}{dt} = k_1[A_1] + k_3[A_3]
\tag{2-88}
$$

$$
-\frac{d[A_3]}{dt} = k_3[A_3]
\tag{2-89}
$$

with the initial condition that $[A_1] = [A_1]_0$, $[A_3] = [A_3]_0$, and $[A_2] = 0$ at time $t = 0$. The equations describing the time dependence for each component are

$$[A_1] = [A_1]_0 e^{-k_1 t} \tag{2-90}$$

and

$$[A_3] = [A_3]_0 e^{-k_3 t} \tag{2-91}$$

Since

$$\frac{d[A_2]}{dt} = k_1 [A_1]_0 e^{-k_1 t} + k_3 [A_3]_0 e^{-k_3 t} \tag{2-92}$$

upon integration we obtain

$$[A_2] = [A_1]_0 - [A_1]_0 e^{-k_1 t} + [A_3]_0 - [A_3]_0 e^{-k_3 t} \tag{2-93}$$

Rearranging equation (2-93) yields

$$[A_2] = ([A_1]_0 + [A_3]_0) - ([A_1]_0 e^{-k_1 t} + [A_3]_0 e^{-k_3 t}) \tag{2-94}$$

2.1.4.3 Parallel second-order reactions.   In the case of parallel second-order reactions,

$$\begin{aligned} A_1 + A_2 &\xrightarrow{k_1} A_4 \\ A_1 + A_3 &\xrightarrow{k_2} A_5 \end{aligned} \tag{2-95}$$

two independent equations can be written:

$$-\frac{d[A_2]}{dt} = k_1 [A_1][A_2] \tag{2-96}$$

$$-\frac{d[A_3]}{dt} = k_2 [A_1][A_3] \tag{2-97}$$

If we consider conservation of mass, we obtain

$$[A_4] = [A_2]_0 - [A_2] \tag{2-98}$$

$$[A_5] = [A_3]_0 - [A_3] \tag{2-99}$$

and

$$\begin{aligned} [A_1]_0 - [A_1] &= [A_4] + [A_5] \\ &= [A_2]_0 - [A_2] + [A_3]_0 - [A_3] \end{aligned} \tag{2-100}$$

Eliminating time as a variable, as in Sec. 2.1.3.2, yields

$$\frac{d[A_3]}{d[A_2]} = \frac{k_2[A_3]}{k_1[A_2]} \tag{2-101}$$

which can be integrated under the initial condition that $[A_3] = [A_3]_0$ and $[A_2] = [A_2]_0$ at time $t = 0$. This gives

$$\frac{[A_3]}{[A_3]_0} = \left(\frac{[A_2]}{[A_2]_0}\right)^{k_2/k_1} \tag{2-102}$$

From the mass conservation relation we can determine $[A_5]$ by substituting equation (2-99) into equation (2-102). Rearranging then gives

$$[A_5] = [A_3]_0\left\{1 - \left(\frac{[A_2]}{[A_2]_0}\right)^{k_2/k_1}\right\} \tag{2-103}$$

or

$$[A_5] = [A_3]_0\left\{1 - \left(1 - \frac{[A_4]}{[A_2]_0}\right)^{k_2/k_1}\right\} \tag{2-104}$$

The expression for $[A_1]$ is obtained from equation (2-100), thus:

$$[A_1] = [A_1]_0 - [A_2]_0 - [A_3]_0 + [A_2] + [A_3] \tag{2-105}$$

Substituting equation (2-102) into equation (2-105), we obtain

$$[A_1] = [A_1]_0 - [A_2]_0 - [A_3]_0 + [A_2] + [A_3]_0\left(\frac{[A_2]}{[A_2]_0}\right)^{k_2/k_1} \tag{2-106}$$

Using this expression for $[A_1]$, we obtain a differential equation for $[A_2]$:

$$\frac{d[A_2]}{dt} = k_1[A_2]\left\{[A_1]_0 - [A_2]_0 - [A_3]_0 + [A_2] + [A_3]_0\left(\frac{[A_2]}{[A_2]_0}\right)^{k_2/k_1}\right\} \tag{2-107}$$

If we now let $\alpha = [A_1]_0 - [A_2]_0 - [A_3]_0$ and $\beta = [A_3]_0([A_2]_0)^{-k_2/k_1}$ we can rewrite equation (2-107) as

$$\frac{d[A_2]}{dt} = k_1[A_2]\{\alpha + [A_2] + \beta[A_2]^{k_2/k_1}\} \tag{2-108}$$

Integrating equation (2-108), we obtain

$$\int_{[A_2]_0}^{[A_2]_t} \frac{d[A_2]}{[A_2]\{\alpha + [A_2] + \beta[A_2]^{k_2/k_1}\}} = \int_0^t k_1 dt \tag{2-109}$$

There is no explicit solution to this integral, but we can consider some limiting cases.

Case I. $k_1 \gg k_2$

Under this condition the ratio k_2/k_1 is approximately zero, thus equation (2-109) reduces to the simple form

$$\int_{[A_2]_0}^{[A_2]_t} \frac{d[A_2]}{[A_2](\alpha + \beta + [A_2])} = \int_0^t k_1 dt \tag{2-110}$$

whose solution is

$$\ln\frac{(\alpha + \beta + [A_2]_t)[A_2]_0}{(\alpha + \beta + [A_2]_0)[A_2]_t} = -(\alpha + \beta)k_1 t \qquad (2\text{-}111)$$

Rearranging and solving for $[A_2]_t$ yields

$$[A_2]_t = [A_2]_0 \frac{\alpha + \beta}{(\alpha + \beta + [A_2]_0)e^{-(\alpha+\beta)k_1 t} - [A_2]_0} \qquad (2\text{-}112)$$

Case II. $k_1 = k_2$

When k_1 and k_2 are equal, the ratio of k_1 to k_2 is unity so the integral equation (2-109) simplifies once again to an expression which is integrable, viz.:

$$\int_{[A_2]_0}^{[A_2]_t} \frac{d[A_2]}{[A_2](\alpha + (1 + \beta)[A_2])} = \int_0^t k_1 dt \qquad (2\text{-}113)$$

The solution to equation (2-113) is

$$\ln\frac{(\alpha + (1 + \beta)[A_2]_t)[A_2]_0}{(\alpha + (1 + \beta)[A_2]_0)[A_2]_t} = -\alpha k_1 t \qquad (2\text{-}114)$$

Solving for $[A_2]_t$, we obtain

$$[A_2]_t = [A_2]_0 \frac{\alpha}{(\alpha + (1 + \beta)[A_2]_0)e^{-\alpha k_1 t} - (1 + \beta)[A_2]_0} \qquad (2\text{-}115)$$

Other cases can also be considered; however, when $k_1 = 2k_2$ and $k_1 \ll k_2$, it is difficult to obtain a solution for equation (2-109), and moreover the solutions that are obtained do not have a simple form.

2.2   APPROXIMATION METHODS

2.2.1   Steady-State Approximation

In section 2.1, we have derived analytic solutions for a few complex kinetic systems. As we have seen, some reaction sequences can lead to very cumbersome solutions which cannot be simplified to a tractable form. In some cases no analytic solution can be obtained using any of the conventional mathematical methods. Nevertheless, a simplified solution can often be obtained by application of the *steady-state approximation*. This method was originally proposed by Bodenstein[4] and further developed by Semenov.[5] It is particularly useful when intermediates which are present in small concentrations are involved in the reaction scheme. In that case, the rate of change of their concentration is negligible provided their concentration remains small compared to the major species in the kinetic scheme. Mathematically,

$$\frac{d[A_i]}{dt} \cong 0 \qquad (2\text{-}116)$$

where $[A_i]$ is the concentration of the intermediate A_i. This condition must be met to validate the steady-state method. In order to demonstrate the use of the steady-state method, let us consider the simple reaction

$$A_1 \rightarrow A_2 \rightarrow A_3 \tag{2-117}$$

which involves the conversion of species A_1 to an intermediate A_2 and finally to the product A_3. The kinetic equations for the species in this scheme are equations (2-27)–(2-29). Solving these equations exactly gives equations (2-30), (2-33), and (2-36), i.e.,

$$[A_1] = [A_1]_0 e^{-k_1 t}$$

$$[A_2] = \frac{k_1 [A_1]_0}{k_2 - k_1} (e^{-k_1 t} - e^{-k_2 t})$$

$$[A_3] = [A_1]_0 \left\{ 1 + \frac{1}{k_1 - k_2} (k_2 e^{-k_1 t} - k_1 e^{-k_2 t}) \right\},$$

under the initial condition that at $t = 0$, $[A_1] = [A_1]_0$ and $[A_2] = [A_3] = 0$.

Under steady-state conditions, the rate of change in the concentration of A_2 is zero, that is,

$$\frac{d[A_2]}{dt} = -k_2[A_2] + k_1[A_1] = 0 \tag{2-118}$$

In this case the steady-state condition yields the concentration of the intermediate as

$$[A_2]_{ss} = \frac{k_1}{k_2}[A_1] \tag{2-119}$$

which yields the following much simpler solutions for $[A_2]$ and $[A_3]$,

$$[A_2]_{ss} = \frac{k_1}{k_2}[A_1]_0 e^{-k_1 t} \tag{2-120}$$

$$[A_3]_{ss} = [A_1]_0 (1 - e^{-k_1 t}) \tag{2-121}$$

Comparison of equations (2-33) and (2-119) shows that $[A_2]$ and $[A_3]$ approximate closely the steady-state solution only if $k_2 \gg k_1$. In other words, the intermediate must be so reactive that it has little time to accumulate. This is illustrated quite clearly in Fig. 2-1(b). The difference between the two expressions provides information on the extent of deviation from the steady state approximation, and can be used to examine the validity of the method. Volk et al.[6] have developed another method for studying the validity of the steady-state approximation, which considers the effect of the deviation from the steady state for reactions of various orders.

One complex reaction mechanism which has wide application to many important kinetic processes is the first-order consecutive reaction with a reversible first step:

$$A_1 \underset{k_{-1}}{\overset{k_1}{\rightleftarrows}} A_2 \overset{k_2}{\longrightarrow} A_3 \tag{2-122}$$

The same sequence of reactions is used to describe enzyme-catalyzed reactions, which are discussed in detail in chapter 5, and thermally activated unimolecular reactions

(Chapter 11). We choose this reaction to illustrate how the steady-state treatment can simplify the mathematical complexity and yield an explicit solution of the rate equations for the reaction represented by equation (2-122). The rate equations are

$$\frac{d[A_1]}{dt} = -k_1[A_1] + k_{-1}[A_2] \tag{2-123}$$

$$\frac{d[A_2]}{dt} = k_1[A_1] - k_{-1}[A_2] - k_2[A_2] \tag{2-124}$$

$$\frac{d[A_3]}{dt} = k_2[A_2] \tag{2-125}$$

Unfortunately, it is not possible to obtain an explicit solution for these equations by straightforward integration. Using either the Laplace transform (Sec. 2-4) or matrix methods (Sec. 2-5), we can find an exact but very complex solution. Suppose that the steady-state assumption is valid for A_2 so that

$$\frac{d[A_2]}{dt} = k_1[A_1] - k_{-1}[A_2] - k_2[A_2] = 0 \tag{2-126}$$

Then the steady state concentration $[A_2]_{ss}$ is given by

$$[A_2]_{ss} = \frac{k_1}{(k_{-1} + k_2)} [A_1] \tag{2-127}$$

Substituting this expression into equations (2-123) and (2-125) gives a linear first-order differential equation for $[A_1]$ and $[A_3]$

$$\frac{d[A_1]}{dt} = \left(\frac{k_1 k_{-1} - k_1 k_2}{k_{-1} + k_2} \right) [A_1]$$

$$\frac{d[A_3]}{dt} = \left(\frac{k_1 k_2}{k_{-1} + k_2} \right) [A_1] \tag{2-128}$$

The solution to these equations is straightforward using standard methods. It is interesting that in the steady-state assumption case, $d[A_1]/dt = -d[A_3]/dt$. This is, of course, because A_2 is assumed to be small. Note that the validity of the steady state treatment requires either that A_2 is always small or that $[A_2]$ changes very slowly in time.

Depending on the relative magnitude of the rate constants in the reaction mechanism, either the reversible first step or the consecutive second step in equation (2-122) describing the mechanism can be *rate-determining;* in other words, the rate of one of the elementary reaction steps in the mechanism can limit the overall rate of product transformation. If the steady state assumption is valid and $k_2 \gg k_{-1}$, the intermediate is transformed to product A_3 much more rapidly than it is able to go back to being a reactant. In that case, the overall rate of product formation is given as

$$\frac{d[A_3]}{dt} = k_1[A_1] \tag{2-129}$$

Thus, the first step in the reaction mechanism is rate-determining.

Now suppose that the ratio k_{-1}/k_2 is very large, so that the reaction of the intermediate in the second step is too slow to perturb the equilibrium of $A_1 \rightleftarrows A_2$ in the first step. Then, the overall rate of product formation is

$$\frac{d[A_3]}{dt} = \frac{k_1 k_2}{k_{-1}}[A_1] \tag{2-130}$$

For this case, the rate-determining step is the reaction of the intermediate given by the second reaction step.

In applying the steady state method to more complex reaction mechanisms, care must be taken, since there are conditions in which departure from the steady state can lead to serious deviation from the exact solution.[7]

2.2.2 Pseudo-First-Order Method

For complex reactions with consecutive steps involving more than one reactant, finding an exact solution can become prohibitively difficult. Consider, for example, the following reaction sequence:

$$A_1 + A_2 \xrightarrow{k_1} \text{products}$$
$$A_1 + A_3 \xrightarrow{k_2} \text{products} \tag{2-131}$$

The second-order rate equations are given by

$$\frac{d[A_2]}{dt} = -k_1[A_1][A_2] \tag{2-132}$$

and

$$\frac{d[A_3]}{dt} = -k_2[A_1][A_3] \tag{2-133}$$

If the conditions of the experimental design are such that $[A_1] \gg [A_2]$ and $[A_3]$, then during the course of reaction the concentration of A_2 remains nearly constant; this implies that

$$\frac{d[A_2]}{dt} = -\kappa_1[A_2] \tag{2-134}$$

and

$$\frac{d[A_3]}{dt} = -\kappa_2[A_3] \tag{2-135}$$

where $\kappa = k_1[A_1]$ and $\kappa = k_2[A_1]$ are constants.

Equations (2-132) and (2-133) reduce to first order in the reactant; thus, the reaction follows pseudo-first-order kinetics. It should be pointed out that the pseudo-first-order method is not a mathematical approximation method so much as it is an experimental condition that must be adjusted so that second-order rate constants can be examined. The experimental technique is often referred to as *flooding*. In principle, even third-order consecutive reactions can be reduced to pseudo-second or -first order using this technique by merely having all reactants in considerable excess except one.

2.3 EXAMPLE OF A COMPLEX REACTION MECHANISM: THE HYDROGEN + HALOGEN REACTION

The reactions between molecular hydrogen and the halogens, first studied by Bodenstein[8–10] over a century ago, continue to engage the attention of chemical kineticists. Since these reactions illustrate several of the principles discussed in this chapter, we shall examine them in some detail. Moreover, since the $H_2 + F_2$ reaction, in particular, has been thoroughly studied from the point of view of microscopic, or state-to-state kinetics, we shall return to this system frequently throughout the text.

The overall reaction can be written as

$$H_2 + X_2 \rightleftarrows 2\,HX, \quad \text{where X = F, Cl, Br, or I}$$

With fluorine, the reaction can occur explosively and go to completion in a matter of seconds; with iodine, reaction times are typically an hour or so, eventually reaching an equilibrium mixture of H_2, I_2, and HI. This variety of behavior reflects both the range of stabilities of the diatomic halogens, and the varied kinetics which can occur in these systems.

It is now generally believed that the reaction does not occur between the diatomic molecules themselves, although in the case of $H_2 + I_2$, Bodenstein's original work seemed to imply that the reaction did indeed proceed in this way. Rather than this simple bimolecular reaction, the mechanism appears to require initial dissociation of the halogen, either thermally or photochemically. The halogen atom then reacts with the hydrogen molecule, initiating a *chain reaction* which proceeds until the reactants are consumed or the reactive species are deactivated, e.g., by atomic recombination.

2.3.1 The Hydrogen-Bromine Reaction

The rate of appearance of the HBr product during the initial stages of the reaction, when $[HBr] \ll [H_2]$ and $[Br_2]$, depends on concentration as follows:

$$\text{Rate} = \frac{d[HBr]}{dt} = (\text{const.})[H_2][Br_2]^{1/2} \tag{2-136}$$

At later times, there is inhibition of the rate by the HBr product:

$$\text{Rate} = \frac{d[HBr]}{dt} = (\text{const.})\frac{[H_2][Br_2]^{1/2}}{1 + k'[HBr]} \tag{2-137}$$

The appearance of a fractional order in equation (2-136) suggests the existence of a *free radical chain reaction* mechanism. The inhibition expressed in equation (2-137) further suggests that the product can react with one of the free radical species which carry the chain. Initiation of the reaction requires dissociation of one of the molecular species involved; under thermal (i.e., nonphotochemical) conditions, the species with the lowest bond dissociation energy will be the principal source of atoms. In the hydrogen-bromine system, the bond dissociation energies are as follows:

$$H_2: D_0^0 = 4.476 \text{ eV} = 430 \text{ kJ/mole}$$

$$HBr: D_0^0 = 3.75 \text{ eV} = 360 \text{ kJ/mole}$$

$$Br_2: D_0^0 = 1.97 \text{ eV} = 190 \text{ kJ/mole}$$

For reference, RT at 500 K is about 4 kJ/mole. Clearly, Br_2 will undergo the greatest extent of dissociation, although this will be only a small fraction at moderate temperatures. The reaction mechanism consists of this initial dissociation, plus the following sequence of reactions:

Initiation:	$Br_2 \xrightarrow{k_1} 2Br$	(2-138)
Propagation:	$Br + H_2 \xrightarrow{k_2} HBr + H$	
	$H + Br_2 \xrightarrow{k_3} HBr + Br$	(2-139)
Inhibition:	$H + HBr \xrightarrow{k_4} H_2 + Br$	(2-140)
Termination:	$2Br + M \xrightarrow{k_5} Br_2 + M$	(2-141)

The reader may wonder why the inhibition step involves a reaction of product with H and not with Br atoms. The reason is that the reaction $Br + HBr \rightarrow Br_2 + H$ is approximately 170 kJ/mole *endothermic,* as may readily be found by taking differences of the preceding bond dissociation energies. This reaction therefore proceeds with a negligible rate, particularly as compared with the H + HBr reaction, which is 67 kJ/mole *exothermic.* The termination reaction is written as a three-body combination, so that its rate coefficient k_5 is pressure-dependent. The termination step may also include contributions from wall recombination processes.

An analytic rate law may be derived from this mechanism by applying the steady-state approximation (see section 2.2.1) to the reactive intermediates, which in this case are the atomic free radicals H and Br. The steady-state conditions are

$$\frac{d[Br]}{dt} = 2k_1[Br_2] - k_2[Br][H_2] + k_3[H][Br_2] + k_4[H][HBr]$$
$$-2k_5[Br]^2 = 0 \tag{2-142}$$

and

$$\frac{d[H]}{dt} = k_2[Br][H_2] - k_3[H][Br_2] - k_4[H][HBr] = 0 \tag{2-143}$$

To solve this system of equations, note that equation (2-143) is just the negative of the middle three terms of equation (2-142) and is stipulated as equal to zero. Thus,

$$2k_1[Br_2] = 2k_5[Br]^2$$

and

$$[Br]_{SS} = \left(\frac{k_1}{k_5}[Br_2]\right)^{1/2} \tag{2-144}$$

Although the relation between [Br] and [Br_2] superficially resembles an equilibrium coefficient, these species are not necessarily in thermodynamic equilibrium, because k_1 and k_5 (which is pressure-dependent) are not related by detailed balancing. Similarly, the steady-state hydrogen atom concentration is found to be

$$[H]_{SS} = \frac{k_2(k_1/k_5)^{1/2}[H_2][Br_2]^{1/2}}{k_3[Br_2] + k_3[HBr]} \tag{2-145}$$

The rate of HBr formation is given by

$$\frac{d[HBr]}{dt} = k_2[Br][H_2] + k_3[H][Br] - k_4[H][HBr] \tag{2-146}$$

At this point, we could substitute the previous expressions for steady-state H and Br concentrations in equation (2-146) and grind away; but it is easier to note that, from equation (2-143), $k_2[Br][H_2] = k_3[H][Br_2] + k_4[H][HBr]$, so a simple substitution yields

$$\frac{d[HBr]}{dt} = 2k_3[H][Br_2]$$

$$= \frac{2k_2(k_1/k_5)^{1/2}[H_2][Br_2]^{1/2}}{1 + (k_4/k_3)[HBr][Br_2]^{-1}} \tag{2-147}$$

Note that equation (2-147) gives the correct concentration dependences, in accordance with equations (2-136) and (2-137), and provides explicit expressions for the empirical coefficients in terms of elementary rate processes.

The reaction between hydrogen and chlorine proceeds in a manner similar to that in the hydrogen-bromine system, except that the larger bond dissociation energy of chlorine (as compared to bromine) makes the initiation step more difficult, and the overall mechanism more susceptible to inhibition, especially by impurities such as oxygen. Dissociation of the chlorine molecule by photolysis, or at the surface of the reaction vessel, frequently dominates the kinetics. The higher exothermicity of the propagation steps makes the overall $H_2 + Cl_2$ reaction quite vigorous.

2.3.2   The Hydrogen-Iodine Reaction

A direct bimolecular reaction between H_2 and I_2,

$$H_2 + I_2 \underset{k_r}{\overset{k_f}{\rightleftharpoons}} 2\,HI \tag{2-148}$$

would have the rate law

$$\frac{d[HI]}{dt} = 2k_f[H_2][I_2] - 2k_r[HI]^2 \tag{2-149}$$

If we define the progress variable y as

$$[H_2]_t = H_0 - y(t)$$

and

$$[I_2]_t = X_0 - y(t)$$

where $H_0 = [H_2]_0$ and $X_0 = [I_2]_0$, then $[HI]_t = 2y(t)$, and the differential equation can be written as

$$2\frac{dy}{dt} = 2k_f[(H_0 - y)(X_0 - y)] - 2k_r(2y)^2$$

or

$$\frac{dy}{dt} = k_f[H_0X_0 - (H_0 + X_0)y + y^2] - 4k_ry^2 \tag{2-150}$$

$$= k_f[H_0X_0 - (H_0 + X_0)y + (1 - 4/K_{eq})]y^2$$

where $K_{eq} = k_f/k_r = [HI]^2_{eq}/[H_2]_{eq}[I_2]_{eq}$, by detailed balancing.

Integrating equation (2-150) as in Sec. 2.1.2.1 gives

$$\ln\left[\frac{y + (H_0 + X_0 - q^{1/2})/2(1 - 4/K_{eq})}{y + (H_0 + X_0 + q^{1/2})/2(1 - 4/K_{eq})}\right] - \ln\left[\frac{H_0 + X_0 - q^{1/2}}{H_0 + X_0 + q^{1/2}}\right] \tag{2-151}$$

$$= k_fq^{1/2}(t - t_0)$$

where $q = (H_0 + X_0)^2 - 4H_0X_0(1 - 4/K_{eq})$. Although mathematically a bit complicated, equation (2-151) predicts a straightforward dependence of the concentration of HI on both time and the initial concentrations of H_2 and I_2.

Bodenstein's measurements on this system[8-10] indicated that the concentrations of the various species did indeed appear to follow equation (2-151) accurately over a wide range of temperatures. A typical experimental result is shown in Fig. 2-2. The activation energy for the bimolecular reaction deduced from the data was approximately 170 kJ/mole. Thus, for the better part of a century, the $H_2 + I_2$ reaction was taken to be the paradigmatic example of a bimolecular reaction with a *"four-center"* transition state,

Further investigation of this reaction indicated that such a mechanism could not be supported. Early molecular beam experiments using supersonically accelerated I_2 (see chapter 9) with H_2 target gas failed to show any evidence for the production of HI. Such negative evidence was not conclusive, of course, because a reaction with a small probability would not be observed in such experiments. But in 1967, Sullivan[11] studied the photochemical reaction between I_2 and H_2 at low temperatures (0–200°C), at which the thermal reaction studied by Bodenstein would be too slow to be measurable. In this experiment, white light is used to efficiently dissociate excited iodine molecules

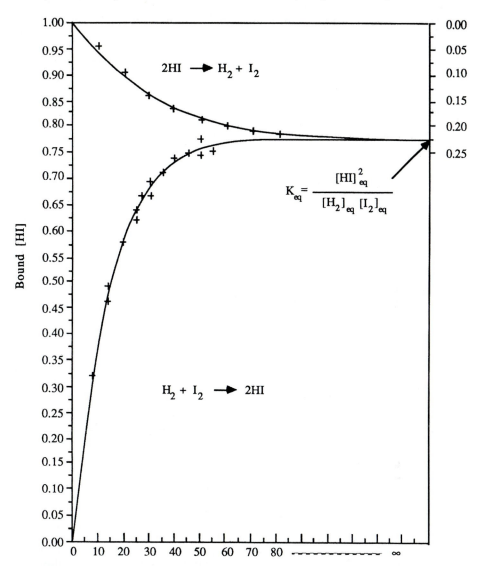

FIGURE 2-2   Plot of the composition of hydrogen-iodine mixtures with time. The left ordinate is the fraction of material present as HI; the right ordinate is the dissociated fraction, i.e., $H_2 + I_2$. The upper plot follows the dissociation of pure HI; the lower, the reaction of H_2 and I_2. The solid curves are the theoretical expression, equation (2-167). Both reactions reach a common equilibrium. The measurements shown were taken at a temperature of 448°C, maintained by a boiling sulfur bath [from M. Bodenstein, *Z. physik. Chem. 13,* 56 (1894)].

I_2^* via their A-X and B-X electronic transitions. This sets up a *photostationary* (steady-state, not thermal equilibrium) concentration of iodine atoms, according to the scheme:

$$I_2 + h\nu \rightarrow I_2^*$$

$$I_2^* \rightarrow 2I$$

$$2I + M \rightarrow I_2 + M \tag{2-152}$$

This is followed by the reaction of atomic iodine with hydrogen,

$$H_2 + 2I \rightarrow 2HI \tag{2-153}$$

The rate of production of HI is given by

$$\frac{d[HI]}{dt} = 2k_{atomic}[I]^2[H_2] \tag{2-154}$$

Sullivan measured the rate of HI formation, as well as the steady-state concentration of I atoms. His essential finding was that k_{atomic} had an activation energy of approximately 21 kJ/mole, which, when added to the bond dissociation energy of I_2 (148.5 kJ/mole), gave just the overall activation energy observed by Bodenstein in his thermal experiments. Furthermore, when this process was combined with the thermal equilibrium concentration of iodine atoms at the higher temperatures, the observed production rate of HI was reproduced. In other words, if we postulate that the total formation rate of HI includes both atomic and bimolecular contributions, and we then find that the total rate is equal to the atomic contribution, then the bimolecular contribution must be zero. The rate law of equation (2-149) arises simply from combining equation (2-154) with the equilibrium expression for [I], i.e.,

$$[I]^2 = K_{dissoc}[I_2] \tag{2-155}$$

to give

$$\frac{d[HI]}{dt} = 2k_{atomic} K_{dissoc}[H_2][I_2] \tag{2-156}$$

so that the "observed" bimolecular rate coefficient is simply the product of the rate coefficient for the atomic process and the thermal dissociation equilibrium coefficient.

Subsequent work on this reaction indicated additional complications. The atomic reaction (2-153) most likely consists of two steps, formation of a weakly bound IH_2 complex (which may be assisted by a third body), followed by

$$I + IH_2 \rightarrow 2HI \tag{2-157}$$

A straightforward steady-state treatment of the IH_2 complex produces the rate law 2-154. Hammes and Widom[12] further suggested that vibrationally excited I_2 molecules lying just below the molecular dissociation limit may play an important role in the reaction as well.

All of these proposed mechanisms lead to the same rate law, equation (2-149) in the differential form and equation (2-151) in the integrated form which may be compared with an experiment. This is a classic example of "Benson's Rule of Chemical Kinetics," which states that it is impossible to prove the correctness of a particular mechanism for a chemical reaction on the basis of experimental measurements; all that can be done is to disprove one or more of the postulated mechanisms on the basis of a discrepancy between predicted and experimental results.

We may summarize our state of knowledge about hydrogen-halogen kinetics as follows. The H_2-F_2, H_2-Br_2, and (with some qualification) the H_2-Cl_2 systems show rather complicated experimental rate laws, but these systems can be interpreted in terms of a rather straightforward free radical chain reaction mechanism. The H_2-I_2 system, on the other hand, has been studied longer than any of these other systems and possesses the simplest kinetic behavior; this system, however, is the one which possesses the most complicated mechanism. The hydrogen + halogen systems thus illustrate another dictum, due to Holmes rather than Benson, to the effect that " ... strange details, far from making the case more difficult, have really had the effect of making it less so"![13]

2.4   LAPLACE TRANSFORM METHOD

2.4.1   Introduction

The classical methods described in the preceding sections lead to the solutions of many kinetic linear differential equations, but as is plain, obtaining these solutions is not always straightforward. By contrast, using methods of *operational calculus* can lead directly to a solution satisfying given initial conditions without having to separately evaluate particular integrals or arbitrary constants. Through these methods, the calculation procedure can be both simplified and shortened considerably. The rules and operational procedures are introduced by means of a definite integral called a *transformation integral*, the general form of which is given by

$$F(p) = \int_a^b f(t)\mathcal{K}(p,t)\,dt \tag{2-158}$$

where $F(p)$ is the *transform* of $f(t)$ with respect to the *kernel*, $\mathcal{K}(p,t)$. For chemical kinetic systems the most commonly used transform is the Laplace transform, which is represented by the kernel $\mathcal{K}(p,t) = e^{-pt}$ and integrated over the limits 0 to ∞. Some properties of this integral and its use in connection with solving chemical kinetic problems will be discussed in this section and also in Appendix 2.1 to this chapter.

2.4.2   The Laplace Transform

The transform $F(p)$ of a function $f(t)$ subjected to the Laplace transformation is defined by the integral:

$$F(p) = \mathcal{L}[f(t)] = \int_0^\infty e^{-pt} f(t)\,dt \tag{2-159}$$

in which both factors, $f(t)$ and e^{-pt}, depend on t. After integration over the limits of $t = 0$ to $t = \infty$, the integral is a function of p only.

The Laplace transform of a given function may be determined by direct integration of (2-159). For example, the transform $F(p)$ of $f(t) = e^{-at}$ where a is a constant, is computed as follows:

$$F(p) = \mathscr{L}[e^{-at}] = \int_0^\infty e^{-at} e^{-pt} dt = \int_0^\infty e^{-(a+p)t} dt$$

$$= -\frac{1}{a+p} [e^{-(a+p)t}]_0^\infty \tag{2-160}$$

$$= \frac{1}{p+a} \quad (p > -a)$$

The condition $p > -a$ is required to make $e^{-(a+p)t}|_{t=\infty}$ equal to zero and to guarantee the convergence of the integral for $\mathscr{L}[e^{-at}]$.

2.4.3   Fundamental Properties of the Laplace Transform

The Laplace transform of a function has certain fundamental properties which are particularly useful in the evaluation of transforms and in the solution of chemical kinetic differential equations. Three of the most widely used properties will be mentioned here.

1. If $f(t)$ is a linear combination of known functions, i.e.,

$$f(t) = f_1(t) + f_2(t) + \cdots + f_n(t) \tag{2-161}$$

then its Laplace transform is

$$\mathscr{L}[f(t)] = \int_0^\infty [f_1(t) + f_2(t) + \cdots + f_n(t)] e^{-pt} dt$$

$$= \int_0^\infty f_1(t) e^{-pt} dt + \int_0^\infty f_2(t) e^{-pt} dt + \cdots + \int_0^\infty f_n(t) e^{-pt} dt$$

or

$$\mathscr{L}[f(t)] = F_1(p) + F_2(p) + \cdots + F_n(p) \tag{2-162}$$

This means that the Laplace transform of a linear combination of functions is the same linear combination of the transformed functions.

2. One of the most important features of the Laplace transform, especially in evaluating chemical kinetic equations, is that the Laplace transform of the derivative of $f(t)$, i.e., $f'(t)$, can be readily obtained. This is done by integrating equation (2-159):

$$\mathscr{L}[f'(t)] = \int_0^\infty f'(t) e^{-pt} dt$$

$$\mathscr{L}[f'(t)] = f(t) e^{-pt}]_0^\infty + p \int_0^\infty f(t) e^{-pt} dt \tag{2-163}$$

$$\mathscr{L}[f'(t)] = p\mathscr{L}[f(t)] - f(0)$$

For higher order derivatives, the Laplace transform can be found by applying equation (2-163) to the second derivative of $f(t)$, as follows:

$$\mathcal{L}[f''(t)] = p\mathcal{L}[f'(t)] - f'(0)$$

$$\mathcal{L}[f''(t)] = p(p\mathcal{L}[f(t)] - f(0)) - f'(0) \tag{2-164}$$

$$\mathcal{L}[f''(t)] = p^2\mathcal{L}[f(t)] - pf(0) - f'(0)$$

Higher order derivatives, $f^{(n)}(t)$, can be obtained by repeating this operation n times, i.e.,

$$\mathcal{L}[f^{(n)}(t)] = p^n\mathcal{L}[f(t)] - \sum_{i=1}^{n} f^{(i-1)}(0)p^{n-1} \tag{2-165}$$

3. The transform of an integral of a function $f(t)$ may be expressed as

$$\mathcal{L}\left[\int_0^t f(t)\,dt\right] = \frac{\mathcal{L}[f(t)]}{p} \tag{2-166}$$

Additional properties of the Laplace transform are given in Appendix 2.1. Some of the Laplace transforms most commonly encountered in kinetic problems appear in Table 2-1.2. These tables may be used to find both the Laplace transform and the inverse Laplace transform symbolized by

$$\mathcal{L}^{-1}[F(p)] = f(t) \tag{2-167}$$

2.4.4   Solving Chemical Kinetic Problems Using the Laplace Transform Method

As we can see from the general Laplace transform properties, the method allows differential and integral equations to be converted into linear algebraic equations that include initial conditions. The procedure for solving a set of kinetic equations is illustrated for the simplest example. Consider the opposing first-order reaction discussed in Sec. 2.1.2,

$$A_1 \underset{k_2}{\overset{k_1}{\rightleftharpoons}} A_2$$

The rate equations for this reaction are equations (2-3) and (2-4),

$$-\frac{d[A_1]}{dt} = k_1[A_1] - k_2[A_2]$$

and

$$-\frac{d[A_2]}{dt} = k_2[A_2] - k_1[A_1]$$

with initial conditions $[A_1] = A_0$ and $[A_2] = 0$ at $t = 0$. Laplace transformations can now be carried out on this set of differential equations to convert the equations into a set of algebraic equations.

$$(p + k_1)(\mathcal{L}[A_1]) - k_2\mathcal{L}[A_2] = [A_1]_0 \tag{2-168}$$

$$-k_1(\mathcal{L}[A_1]) + (p + k_2)\mathcal{L}[A_2] = [A_2]_0 \tag{2-169}$$

Solving for $\mathscr{L}[A_1]$ and $\mathscr{L}[A_2]$, and using Cramer's Rule for linear equations, we obtain

$$\mathscr{L}[A_1] = \frac{\begin{vmatrix} [A_1]_0 & -k_2 \\ [A_2]_0 & (p + k_2) \end{vmatrix}}{\begin{vmatrix} p + k_1 & -k_2 \\ -k_1 & p + k_2 \end{vmatrix}} \tag{2-170}$$

and

$$\mathscr{L}[A_2] = \frac{\begin{vmatrix} p + k_1 & [A_1]_0 \\ -k_1 & [A_2]_0 \end{vmatrix}}{\begin{vmatrix} p + k_1 & -k_2 \\ -k_1 & p + k_2 \end{vmatrix}} \tag{2-171}$$

The solution of this determinantal equation is

$$\mathscr{L}[A_1] = \frac{(p + k_2)[A_1]_0}{p(p + (k_1 + k_2))} + \frac{k_2[A_2]_0}{p(p + (k_1 + k_2))} \tag{2-172}$$

and

$$\mathscr{L}[A_2] = \frac{(p + k_1)[A_2]_0}{p(p + (k_1 + k_2))} + \frac{k_1[A_1]_0}{p(p + (k_1 + k_2))} \tag{2-173}$$

In order to find the time-dependent concentrations for A_1 and A_2, we carry out inverse Laplace transformations on equations (2-172) and (2-173). Using the Laplace tables, we see that the solution to equation (2-172) is

$$[A_1] = \frac{[A_1]_0}{(k_1 + k_2)}(k_2 + k_1 e^{-(k_1 + k_2)t}) + \frac{[A_2]_0 k_2}{(k_1 + k_2)}(1 - e^{-(k_1 + k_2)t}) \tag{2-174}$$

and similarly for equation (2-173). Then, since $[A_2]_0 = 0$, we find the previous solution, equation (2-13):

$$[A_1] = \frac{[A_1]_0}{(k_1 + k_2)}(k_2 + k_1 e^{-(k_1 + k_2)t})$$

Similarly for $[A_2]$, we recover equation (2-14):

$$[A_2] = \frac{k_1[A_1]_0}{(k_1 + k_2)}(1 - e^{-(k_1 + k_2)t})$$

As can be seen from this simple example, as long as the rate equations are linear with respect to the reactants, the Laplace method will yield solutions. Nonlinear rate equations can be made linear by making the concentrations of certain reactants large relative to others so that pseudo-first-order equations, which are linear, can be obtained.

A somewhat more complex example of this method is the infrared multiphoton dissociation (IRMPD, see Chapter 11, Section 13) of vinyl cyanide ($CH_2 = CHCN$) which yields electronically excited and ground state C_2 according to the reaction

$$CH_2 = CHCN \xrightarrow{nh\nu_{IR}} C_2(\tilde{X}^1\Sigma_g^+)$$
$$\longrightarrow C_2^*(\tilde{a}^3\Pi_u) \tag{2-175}$$

In the presence of a reactive scavenger such as O_2 the subsequent reactions are

$$C_2(\tilde{X}^1\Sigma_g^+) + O_2 \xrightarrow{k_1} \text{products} \tag{2-176}$$

$$C_2(\tilde{X}^1\Sigma_g^+) + O_2 \underset{k_e}{\overset{k_{e'}}{\rightleftharpoons}} C_2^*(\tilde{a}^3\Pi) + O_2 \tag{2-177}$$

and

$$C_2^*(\tilde{a}^3\Pi) + O_2 \xrightarrow{k_3} \text{products} \tag{2-178}$$

The reaction illustrated in equation (2-177) proceeds through collision induced inter-system crossing (see chapter 8). We can write equations (2-176) through (2-178) schematically as:

$$A + B \xrightarrow{k_1} D$$

$$A + B \underset{k_e}{\overset{k_e'}{\rightleftharpoons}} C + B \tag{2-179}$$

$$C + B \xrightarrow{k_3} D$$

This mechanism consists of a reversible and two concurrent reactions; the reversible step couples to the concurrent reactions. The rate equations for $[A] = [C_2(\tilde{X}^1\Sigma_g^+)]$ and $[C] = [C_2^*(\tilde{a}^3\Pi)]$ can be written as

$$\frac{d[A]}{dt} = -(k_1[B] + k_e'[B])[A] + k_e[B][C] \tag{2-180}$$

$$\frac{d[C]}{dt} = -(k_3[B] + k_e[B])[C] + k_e'[B][A] \tag{2-181}$$

Using a large excess of O_2, so that $[B]$ is approximately constant, we can rewrite these equations as

$$\frac{d[A]}{dt} = -K'[A] + K_e[C] \tag{2-182}$$

and

$$\frac{d[C]}{dt} = K_e'[A] - K[C] \tag{2-183}$$

where $K' = k_1[B] + k_e'[B]$, $K_e = k_e[B]$, $K_e' = k_e'[B]$, and $K = k_3[B] + k_e[B]$. The initial values of $[A]_0 = A_0$ and $[C]_0 = C_0$. The actual concentrations of $C_2(\tilde{X}^1\Sigma_g^+)$ and $C_2^*(\tilde{a}^3\Pi)$ may be determined by laser-induced fluorescence (see chapter 3). The $C_2^*(\tilde{a}^3\Pi)$ radicals are monitored by exciting the $d^3\Pi \leftarrow \tilde{a}^3\Pi(0,0)$ transition and detecting LIF in the $(0,1)$ vibronic band, and $C_2(\tilde{X}^1\Sigma_g^+)$ radicals are monitored by exciting the $\tilde{a}^1\Pi - \tilde{X}^1\Sigma_g^+(0,0)$ transition and detecting the $(4,1)$ vibronic band fluorescence.[14]

Taking the Laplace transforms of equations (2-182) and (2-183), we find

$$(p + K')\mathscr{L}[A] - K_e\mathscr{L}[C] = A_0 \tag{2-184}$$

$$-K'_e\mathscr{L}[A] + (p + K)\mathscr{L}[C] = C_0 \tag{2-185}$$

Solving equations (2-184) and (2-185) by using Cramer's Rule, we obtain

$$\mathscr{L}[A] = \frac{A_0(p + K)}{(p + K')(p + K) - K_e K'_e} + \frac{K_e C_0}{(p + K')(p + K) - K_e K'_e} \tag{2-186}$$

$$\mathscr{L}[C] = \frac{C_0(P + K')}{(p + K')(p + K) - K_e K'_e} + \frac{A_0 K'_e}{(p + K')(p + K) - K_e K'_e} \tag{2-187}$$

The roots of $(p + K')(p + K) - K_e K'_e = 0$, found by using the quadratic formula, are:

$$s_1 = 1/2\{-(K + K') + [(K + K')^2 - 4(KK' - K_e K'_e)]^{1/2}\} \tag{2-188}$$

and

$$s_2 = 1/2\{-(K + K') - [(K + K')^2 - 4(KK' - K_e K'_e)]^{1/2}\} \tag{2-189}$$

Hence, we may split $\mathscr{L}[A]$ and $\mathscr{L}[C]$ into the partial fractions:

$$\mathscr{L}[A] = \frac{A_0(p + K)}{(p - s_1)(p - s_2)} + \frac{C_0 K_e}{(p - s_1)(p - s_2)} \tag{2-190}$$

and

$$\mathscr{L}[A] = \frac{C_0(p + K')}{(p - s_1)(p - s_2)} + \frac{A_0 K'_e}{(p - s_1)(p - s_2)} \tag{2-191}$$

Taking the inverse Laplace Transform of equations (2-186) and (2-187) using nos. 4 and 9 of Table 2-1.2, we obtain

$$[A(t)] = [C_2(\tilde{X}^1\Sigma_g^+)]_t = \frac{1}{s_1 - s_2}\{(A_0 s_1 + A_0 K + C_0 K_e)e^{s_1 t} \\ -(A_0 s_2 + A_0 K + C_0 K_e)e^{s_2 t}\} \tag{2-192}$$

Similarly,

$$[C(t)] = [C_2^*(\tilde{a}^3\Pi)]_t = \frac{1}{s_1 - s_2}\{(C_0 s_1 + C_0 K' + A_0 K'_e)e^{s_1 t} \\ -(C_0 s_2 + C_0 K' + A_0 K'_e)e^{s_2 t}\} \tag{2-193}$$

2.5   DETERMINANT (MATRIX) METHODS

Another method of solving kinetic equations uses matrix algebra. We apply this to the chemical reaction system $A_1 \rightleftarrows A_2$, for which the differential equations are as before

$$-\frac{d[A_1]}{dt} = k_1[A_1] - k_2[A_2]$$

and

$$-\frac{d[A_2]}{dt} = -k_1[A_1] + k_2[A_2]$$

These equations can be expressed in matrix form as

$$\begin{pmatrix} \dfrac{d[A_1]}{dt} \\ \dfrac{d[A_2]}{dt} \end{pmatrix} = \begin{pmatrix} -k_1 & k_2 \\ k_1 & -k_2 \end{pmatrix} \begin{pmatrix} [A_1] \\ [A_2] \end{pmatrix} \tag{2-194}$$

Thus, the matrix of rate coefficients is given by

$$\mathbf{K} = \begin{pmatrix} -k_1 & k_2 \\ k_1 & -k_2 \end{pmatrix} \tag{2-195}$$

The matrix form of the set of differential equations may be expressed as

$$\frac{d\mathbf{A}}{dt} = \mathbf{KA} \tag{2-196}$$

where $\mathbf{K}$ is the rate coefficient matrix and $\mathbf{A}$ is the composition vector; because composition must be real and positive, $\mathbf{A}$ is a vector with real and positive components. $\mathbf{K}$ is an $n \times n$ matrix where n is the number of components, A_n. For this problem, $\mathbf{K}$ is diagonalizable and it has n linearly independent eigenvectors.[15]
    If we define an orthogonal matrix $\mathbf{P}$ which is invertible such that

$$\mathbf{A} = \mathbf{PB} \tag{2-197}$$

and $\mathbf{P}^{-1}\mathbf{KP} = \Lambda$, where Λ is the matrix of negative eigenvalues with $-\lambda_1, -\lambda_2, \ldots -\lambda_n$ as its successive diagonal elements and no off-diagonal elements, then equation (2-196) can be rewritten as

$$\frac{d\mathbf{B}}{dt} = \mathbf{P}^{-1}\mathbf{KPB} = \Lambda\mathbf{B} \tag{2-198}$$

Since equation (2-198) is uncoupled, its solution is

$$\mathbf{B} = e^{\Lambda t}\mathbf{B}_i \tag{2-199}$$

where $\mathbf{B}_i$ is the vector of initial values of $\mathbf{B}$ and is related to the initial conditions for $\mathbf{A}_i$ at $t = 0$ by

$$\mathbf{A}_i = \mathbf{PB}_i \tag{2-200}$$

If we multiply equation (2-197) by the inverse matrix of $\mathbf{P}$, from the left side we get

$$\mathbf{P}^{-1}\mathbf{A} = \mathbf{B} \tag{2-201}$$

Substituting $\mathbf{B}$ into equation (2-199) yields

$$\mathbf{P}^{-1}\mathbf{A} = e^{\Lambda t}\mathbf{B}_i = e^{\Lambda t}\mathbf{P}^{-1}\mathbf{A}_i \tag{2-202}$$

Multiplying through by $\mathbf{P}$ from the left we obtain the solution vector

$$\mathbf{A} = \mathbf{P}\, e^{\Lambda t}\, \mathbf{P}^{-1}\, \mathbf{A}_i \tag{2-203}$$

In order to evaluate the solution vector $\mathbf{A}$, the matrix $\mathbf{P}$ that diagonalizes $\mathbf{K}t$ must be found. To find the eigenvalues Λ, we must solve the characteristic equation of $\mathbf{K}$ satisfying the condition,

$$\det(\mathbf{K} - \Lambda\mathbf{1}) = 0 \tag{2-204}$$

The characteristic equation for the reaction $A_1 \rightleftarrows A_2$ can be written as

$$(-k_1 - \lambda)(-k_2 - \lambda) - k_1 k_2 = 0 \tag{2-205}$$

and the corresponding eigenvalues, which are roots of this equation, are

$$\lambda_1 = 0 \quad \text{and} \quad \lambda_2 = -(k_1 + k_2)$$

Thus the eigenvectors are

$$\mathbf{P}_1 = \begin{pmatrix} k_2 \\ k_1 \end{pmatrix} \quad \text{and} \quad \mathbf{P}_2 = \begin{pmatrix} 1 \\ -1 \end{pmatrix}$$

and they form a basis for the eigenspace corresponding to $\lambda_1 = 0$ and $\lambda_2 = -(k_1 + k_2)$. It is easy to check that $\mathbf{P}_1$ and $\mathbf{P}_2$ are linearly independent so that

$$\mathbf{P} = \begin{pmatrix} k_2 & 1 \\ k_1 & -1 \end{pmatrix} \tag{2-206}$$

diagonalizes $\mathbf{K}$. The reader should verify that $\mathbf{P}^{-1}\mathbf{K}\mathbf{P} = \Lambda$.

To find our solutions we use equation (2-203) and multiply the matrices. $\mathbf{A}_i$ is the initial value vector for the components given by

$$\mathbf{A} = \begin{pmatrix} [A_1] \\ [A_2] \end{pmatrix} = \begin{pmatrix} A_0 \\ 0 \end{pmatrix} \quad \text{at} \quad t = 0 \tag{2-207}$$

The inverse matrix, $\mathbf{P}^{-1}$, can be determined using standard matrix algebra.[15]

$$\mathbf{P}^{-1} = \begin{pmatrix} \dfrac{1}{(k_1 + k_2)} & \dfrac{1}{(k_1 + k_2)} \\ \dfrac{k_1}{(k_1 + k_2)} & \dfrac{-k_2}{(k_1 + k_2)} \end{pmatrix} \tag{2-208}$$

Therefore the solutions are given by

$$\mathbf{A} = \begin{pmatrix} [A_1] \\ [A_2] \end{pmatrix} = \frac{A_0}{(k_1 + k_2)} \begin{pmatrix} k_2 + k_1 e^{-(k_1 + k_2)t} \\ k_1 + k_1 e^{-(k_1 + k_2)t} \end{pmatrix} \tag{2-209}$$

and hence we recover equations (2-13) and (2-14),

$$[A_1] = \frac{A_0}{(k_1 + k_2)}(k_2 + k_1 e^{-(k_1 + k_2)t})$$

$$[A_2] = \frac{k_1 A_0}{(k_1 + k_2)}(1 - e^{-(k_1 + k_2)t})$$

The matrix method can be used to solve a set of first-order equations or pseudo-first-order equations. As is evident from the generalized treatment this procedure can also be used computationally to find the solution for n components (See Problem 2.5). One need only compute the eigenvalues and eigenvectors of **K** and assume a solution of the form given by equation (2-203). Computing eigenvalues for the rate coefficient matrix may be done without too much difficulty; calculating the eigenvectors can be done using matrix reduction algorithms such as the **QR** method.[16–18]

2.6   NUMERICAL METHODS

2.6.1   Introduction

It is usually best to solve kinetic equations exactly, provided that an exact solution does exist and also has a reasonable analytical form. As we have seen, for rather simple kinetic systems one can usually find an analytic solution, which makes it easy to obtain information about the concentration profiles and their time dependences. However, the search for an analytic solution has at least two limitations. The principal one is that such a solution may not be derivable using matrix or Laplace methods, or any other method for that matter. Secondly, even if an analytic solution does exist, it may be so complex that it may not yield the desired information easily nor in a useful form. An example of such a case is the reaction

$$A \xrightarrow{\ k_1\ } B+C \xrightarrow{\ k_2\ } D$$
$$\mathtt{\underline{\qquad\ \ k_3\qquad}}$$

The solution of this complex reaction system consists of Bessel and Henkel functions.

When analytic solutions are impossible to obtain directly, approximation methods may be employed. One such method, which has historically played a useful role in the modeling of chemical systems, is the steady-state approximation. Although it provides analytic solutions, the question of their validity is open. Questions of validity of the steady-state assumption have been raised in areas of kinetics that involve large sets of coupled differential equations, because it is difficult to verify experimentally the existence of steady-state transient species.[19–21]

In treating the kinetics of many-component systems, such as those typically encountered in atmospheric and combustion chemistry (see Chapters 14 and 15), a numerical solution is often required. In this section, we present some useful computational tools for solving differential equation models that arise in chemical kinetics.

2.6.2   Analyzing Complex Reaction Systems Numerically

2.6.2.1 Categorizing kinetic differential equations.   Before we can consider numerical methods that are suited to kinetic problems, we must be able to categorize the type of differential equation we have at hand. There are two distinct categories that are most commonly encountered. Equations involving derivatives of only one independent variable are called *ordinary differential equations*. If the equation contains more than one independent variable, then the equation becomes a *partial differential equation*. For most chemical kinetics problems, the equations are of the ordinary type, such as $y' = ky$. These may be further classified as either initial-value problems, in which the dependent variable and/or its derivative are required to have some specified value at the zero of the independent variable, i.e., $y(0)$ and $y'(0)$ are constants: or boundary-value problems, in which $y(x_1)$ and $y'(x_2)$ are specified constants, but x_1 and x_2 may be any values in the range of the independent variable.

Most coupled differential equations relevant to chemical kinetic systems are of the initial-value type and the problem is to solve a system of m equations involving n chemical species, each of which may be coupled to every other species by chemical reaction. In general, solving initial-value problems requires the problem to be in the form

$$\frac{dy}{dx} = f'(x,y) \qquad (2\text{-}210)$$

with initial condition $y(x_0) = y_0$. Assuming that $\partial f/\partial y$ is continuous over the interval $x_0 \le x \le x_1$ guarantees that equation (2-210) possesses a solution; $x_1 = x_0 + h$, where h is a small increment. Note that the numerical solution is only a set of points determined by the slope of the curve calculated from equation (2-210).

In the remainder of this section we describe various methods for obtaining a numerical solution that are applicable to systems of chemical kinetic equations. These methods fall into two categories: one-step methods and predictor-corrector methods. One-step methods are ones that use information about a single point or a previous step to find the next point. Some common methods employing this basic algorithm are Euler's,[22] the Modified Euler technique[22,23] and Runge-Kutta Methods.[24] Predictor-corrector methods are multistep methods which make use of information about the solution and its derivative at more than one point in order to extrapolate to the next point. Methods that make use of this algorithm are Milne's[25] and the Adams-Bashforth[26] and Gear[26,27] methods.

2.6.2.2 One-step methods.

Consider the first-order differential equation

$$y'(t) = f(t,y) \qquad (2\text{-}211)$$

where y' represents dy/dt subject to the initial condition $y(t_0) = y_0$. The purpose of the one-step method is to provide a means of calculating a set of points $[(t_i, v_i)]$ where $v_0 = y_0$ and each point (t_i, v_i) is an approximation to the corresponding point $(t_i, y(t_i))$.

2.6.2.2.1 Euler Methods.

Consider the first-order equation (2-211) and divide the interval $[t_0, t_N]$ into N equally spaced subintervals of step size,

$$h = \frac{t_N - t_0}{N} \qquad (2\text{-}212)$$

(Note that the step size does not have to be constant over the interval, but for the moment let us assume it is.) Each new increment in the independent variable can be approximated by

$$t_n = t_0 + nh \qquad (2\text{-}213)$$

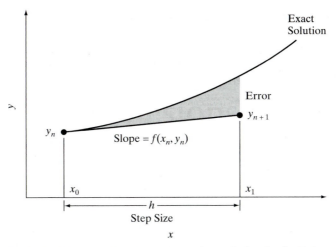

FIGURE 2-3   Graphical interpretation of the Basic Euler method.

where $n = 0, 1, 2, \ldots N$. If $y(t)$ is continuous and has a unique solution, then it can be expanded in a Taylor's series about the point t_0 thus:

$$y(t_0 + h) = y(t_0) + f(t_0, y_0) h + \frac{h^2}{2!} f'(t_0, y_0) + \cdots \qquad (2\text{-}214)$$

If the value of the step size is small, then higher order terms in h can be neglected. This is known as truncating and gives

$$y(t_0 + h) = y(t_0) + hy'(t_0) \qquad (2\text{-}215)$$

This technique is called the Euler method, and the process can be continued for many steps:

$$y_{n+1} = y_n + hy'(x_n, y_n), \qquad n = 0, 1, \ldots, N - 1 \qquad (2\text{-}216)$$

The Euler method is presented graphically in Fig. 2-3 and is the simplest of all methods for solving the initial-value problem.

By truncating after the second term of the Taylor's expansion, we introduce errors in the value of y_{n+1}, on the order of h^2. The truncation error in each step can be made small provided the step size chosen is sufficiently small. The question that arises is whether the Euler method is able to provide a sufficiently accurate approximation to equation (2-211).

To answer this question, consider the decomposition of the trifluoromethoxy radical, CF_3O. The reaction is first order, i.e.,

$$-\frac{d[CF_3O]}{dt} = k[CF_3O] \qquad (2\text{-}217)$$

and the reaction rate for its decomposition is[28–30]

$$k = 10^{15.2} e^{-15.65/T} \text{sec}^{-1} \qquad (2\text{-}218)$$

TABLE 2-1 Comparison of the Accuracy of Euler Integration of Equation (2-218) for 10- and 20-Step Integrations.

N = 10	t, sec	$[CF_3O]_{Euler}$	$[CF_3O]_{exact}$	Deviation
1	5E-9	−.33	.2644772613	−.5944772613
2	IE-8	.1089	.06994822174	.03895177826
3	1.5E-8	−.035937	.01849971412	−.05443671412
4	2E-8	.01185921	4.892753725E-3	6.966456275E-3
5	2.5E-8	−3.9135393E-3	1.294022105E-3	−5.207561405E-3
6	3E-8	1.291467969E-3	3.422394225E-4	9.492285465E-4
7	3.5E-8	−4.261844298E-4	9.051454518E-5	−5.166989749E-4
8	4E-8	1.406408618E-4	2.393903902E-5	1.167018228E-4
9	4.5E-8	−4.64114844E-5	6.331331477E-6	−5.274281588E-5
10	5E-8	1.531578985E-5	1.674493209E-6	1.364129664E-5

N = 20				
1	2.5E-9	.335	.5142735277	−.1792735277
2	5E-9	.112225	.2644772613	−.1522522613
3	7.5E-9	.037595375	.1360136542	−.09841827917
4	1E-8	.01259445062	.06994822174	−.05735377112
5	1.25E-8	4.219140959E-3	.03597251875	−.03175337779
6	1.5E-8	1.413412221E-3	.01849971412	−.0170863019
7	1.75E-8	4.734930942E-4	9.513913242E-3	−9.040420148E-3
8	2E-8	1.586201865E-4	4.892753725E-3	−4.734133539E-3
9	2.25E-8	5.313776249E-5	2.516213718E-3	−2.463075956E-3
10	2.5E-8	1.780115044E-5	1.294022105E-3	−1.276220955E-3
11	2.75E-8	5.963385396E-6	6.654813131E-4	−6.595179277E-4
12	3E-8	1.997734108E-6	3.422394225E-4	−3.402416884E-4
13	3.25E-8	6.69240926E-7	1.760046751E-4	−1.753354342E-4
14	3.5E-8	2.241957102E-7	9.051454518E-5	−9.029034947E-5
15	3.75E-8	7.510556292E-8	4.654923446E-5	−4.647412889E-5
16	4E-8	2.516036358E-8	2.393903902E-5	−2.391387865E-5
17	4.25E-8	8.428721799E-9	1.231121404E-5	−1.230278532E-5
18	4.5E-8	2.823621803E-9	6.331331477E-6	−6.328507855E-6
19	4.75E-8	9.459133039E-10	3.256036174E-6	−3.25509026E-6
20	5E-8	3.168809568E-10	1.674493209E-6	−1.674176328E-6

At 1000K, $k = 2.66 \times 10^8$ sec^{-1}, and at $t = 0$, $[CF_3O]_0 = 1$. The analytic solution is found by simple integration (Sec. 1.3.2) to be

$$[CF_3O]_t = [CF_3O]_0 \exp(-2.66 \times 10^8 t) \qquad (2\text{-}219)$$

Now, we solve equation (2-217) using the Euler method [equation (2-216)]

$$[CF_3O]_{n+1} = [CF_3O]_n - (k)h[CF_3O]_n \qquad (2\text{-}220)$$

Table 2-1 shows the results. Notice that for $N = 10$, the numerical results from the Euler calculation oscillate and also give negative concentrations. This is an example where the Euler method is *unstable*. By increasing N to 20, the concentration of $[CF_3O]$ is still underestimated compared to the concentration given by the exact expression equation (2-219). Furthermore, it is clear that the concentration of CF_3O calculated by the Euler method decreases more rapidly, by about 2 orders of magnitude, than the concentration given by the exact expression. Consequently, the error grows with increasing values of the independent variable. We define the global error as the difference between the computed value of the solution and the exact solution. It accounts for the total accumulated error. For the CF_3O example, the global error is about 35% in the first step, ($N = 1$). It continues to accumulate and by $N = 7$, the global error has reached 95%. Although the stability has been improved by increasing N, the method still remains unreliable. Results can be improved by increasing N further, which reduces the step size. This allows the function, which is rapidly decreasing, to be better estimated by the Euler method; but it increases the computation time since more steps are needed. To obtain more reliable and accurate numerical results for such problems, another method less susceptible to instability problems is required. Some higher order techniques, such as the Runge-Kutta methods, may be used.

2.6.2.2.2 Runge-Kutta Methods. Let us return to the Taylor expansion of the first-order equation (2-214). If the higher-order terms of the expansion are truncated at h^2, then

$$y(t_0 + h) = y(t_0) + hy'(t_0) + \frac{h^2}{2!} y''(t_0) + \cdots \qquad (2\text{-}221)$$

In order to calculate the new point, the second derivative $y''(t_0)$ must be known. This can be estimated as the finite difference

$$y''(t_0) = \frac{\Delta y'}{\Delta t} = \frac{y'(t_0 + h) - y'(t_0)}{h} \qquad (2\text{-}222)$$

Substituting this expression into the truncated Taylor expression results in

$$y(t_0 + h) = y(t_0) + hy'(t_0) + \frac{h^2}{2!} \left(\frac{y'(t_0 + h) - y'(t_0)}{h} \right) \qquad (2\text{-}223)$$

and combining terms in like powers of h gives

$$y(t_0 + h) = y(t_0) + \frac{h}{2} (y'(t_0 + h) + y'(t_0)) \qquad (2\text{-}224)$$

This is the second-order Runge-Kutta method because it utilizes the second term of the expansion, h^2. In order to derive other nth order Runge-Kutta methods, the nth order term of the Taylor expansion must be kept, and the nth order derivative calculated. This implies that more points are needed in order to evaluate the derivatives. The Runge-Kutta method thus provides a set of formulas, derived using the same procedure as for derivation of the second order method, for calculating the solutions of the first-order equation. The equations for the third-order method are

$$y_{n+1} = y_n + (\Delta Y_1 + 4\Delta Y_3 + \Delta Y_4)/6 \qquad (2\text{-}225)$$

where

$$\Delta Y_1 = hf(t_n, y_n)$$
$$\Delta Y_2 = hf(t_n + h, y_n + \Delta Y_1)$$
$$\Delta Y_3 = hf(t_n + h/2, y_n + \Delta Y_1/2)$$
$$\Delta Y_4 = hf(t_n + h, y_n + \Delta Y_2)$$

The fourth order Runge-Kutta method has found its widest application in chemical kinetics. This is because it retains all terms up through h^4 and consequently the error associated with any step is on the order of h^5. The equations for the fourth order method are

$$y_{n+1} = y_n + (\Delta Y_1 + 2\Delta Y_2 + 2\Delta Y_3 + \Delta Y_4)/6 \qquad (2\text{-}226)$$

where

$$\Delta Y_1 = hf(t_n, y_n)$$
$$\Delta Y_2 = hf(t_n + h/2, y_n + \Delta Y_1/2)$$
$$\Delta Y_3 = hf(t_n + h/2, y_n + \Delta Y_2/2)$$
$$\Delta Y_4 = hf(t_n + h, y_n + \Delta Y_3)$$

To illustrate the relative accuracy of the various one-step methods, let us model the reaction solved exactly in section 2.2.1 [equations (2-3) and (2-4)], *viz.*:

$$A \underset{k_2}{\overset{k_1}{\rightleftharpoons}} B$$

$$\frac{d[A]}{dt} = -k_1[A] + k_2[B]$$

$$\frac{d[B]}{dt} = k_1[A] - k_2[B]$$

where $[A] = [A]_0$ and $[B] = 0$ at $t = 0$. The exact solution is given by equations (2-13) and (2-14), namely,

$$[A] = \frac{[A_0]}{k_1 + k_2}[k_2 + k_1 e^{-(k_1 + k_2 t)}]$$

and from mass conservation

$$[B] = \frac{k_1[A]_0}{(k_1 + k_2)}[1 - e^{-(k_1 + k_2)t}]$$

We can use these equations to compare the relative accuracy of the various one-step methods for $k_1 = 1000$, $k_2 = 1$ and $[A]_0 = 1000$. Such a comparison is presented in Table 2-2. Clearly, the fourth-order Runge-Kutta method is better than the Euler method or the second-order method as shown also by the plot of the deviations given

TABLE 2-2 Comparison of the Accuracy of Various One-Step Methods for Integrating the Kinetic Equations for the A $\rightleftarrows$ B Reaction.

	Comparison of one-step methods for [A]			
Time	Euler	Modified Euler	Runge-Kutta-4	Exact
.00000	1000.00000	1000.00000	1000.00000	1000.00000
.00050	590.57145	607.16681	606.62113	606.62085
.00100	348.94227	368.80586	368.14394	368.14360
.00150	206.34191	224.17463	223.52745	223.57214
.00200	122.18458	136.41617	135.92920	135.92895
.00250	72.51813	83.16661	82.79743	82.79725
.00300	43.20687	50.85617	50.58748	50.58734
.00350	25.90849	31.25103	31.06091	31.06082
.00400	15.69963	19.35515	19.22337	19.22330
.00450	9.67476	12.13703	12.04712	12.04707
.00500	6.11910	7.75727	7.96998	7.69664
.00550	4.02069	5.09974	5.05932	5.05930
.00600	2.78229	3.48722	3.46048	3.46046
.00650	2.05143	2.50879	2.49122	2.49121
.00700	1.62010	1.91510	1.90362	1.90362
.00750	1.36555	1.55487	1.54741	1.54740
.00800	1.21533	1.33629	1.33146	1.33146
.00850	1.12667	1.20366	1.20055	1.20055
.00900	1.07434	1.12318	1.12118	1.12118
.00950	1.04347	1.07435	1.07307	1.07307
.01000	1.02524	1.04472	1.04390	1.04390

	Comparison of one step methods for [B]			
Time	Euler	Modified Euler	Runge-Kutta-4	Exact
.00000	.00000	.00000	.00000	.00000
.00500	409.42855	392.83319	393.37887	393.37915
.00100	651.05773	631.19414	631.85606	631.85640
.00150	793.65809	775.82537	776.42755	776.42786
.00200	877.81542	863.58383	864.07080	864.07105
.00250	927.48187	916.83339	917.20257	917.20275
.00300	956.79313	949.14383	949.41252	949.41266
.00350	974.09151	968.74897	968.93909	968.93918
.00400	984.30037	980.64485	980.77663	980.77670
.00450	990.32524	987.86297	987.95288	987.95293
.00500	993.88090	992.24273	992.30332	992.30336
.00550	995.97931	994.90026	994.94068	994.94070
.00600	997.21771	996.51278	996.53952	996.53954
.00650	997.94857	997.49121	997.50878	997.50879
.00700	998.37990	998.08490	998.09638	998.09638
.00750	998.63445	998.44513	998.45259	998.45260
.00800	998.78467	998.66371	998.66854	998.66854
.00850	998.87333	998.79634	998.79945	998.79945
.00900	998.92566	998.87682	998.87882	998.87882
.00950	998.95653	998.92565	998.92693	998.92693
.01000	998.97476	998.95528	998.95610	998.95610

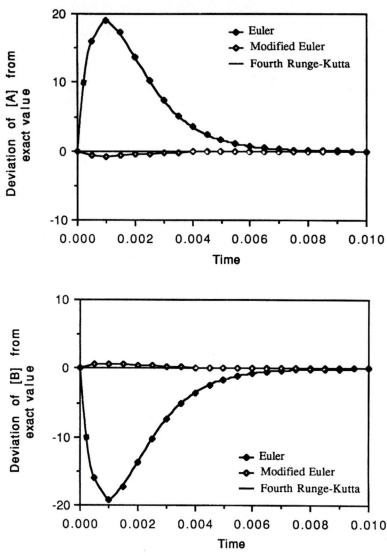

FIGURE 2-4 Plot of deviation of exact and numerical solution for the reaction A $\rightleftharpoons$ B for various one-step numerical methods. (a) Deviation for species A. (b) Deviation for species B.

in Fig. 2-4. Notice that at large values of t the Euler deviation tends to zero. To explore why this is so, recall that the Euler method is based on the Taylor expansion. If we use only the first two terms we get equation (2-215) and the deviation of the Euler method from the exact value is

$$\Delta y = y(t + \Delta t) - [y(t) + y'(t)\Delta t] \tag{2-227}$$

Thus,

$$\Delta y = \tfrac{1}{2} y''(t)\Delta t^2 + \tfrac{1}{6} y'''(t)\Delta t^3 + \cdots \tag{2-228}$$

In the case of equation (2-14) the higher derivatives are

$$\frac{d^n[A]}{dt^n} = (-1)^n \frac{k_1[A]_0}{(k_1 + k_2)^n} e^{-(k_1+k_2)t} \tag{2-229}$$

Consequently, as $t \to \infty$, the nth differential becomes close to zero, as does the deviation Δy, explaining why the Euler deviation tends to zero. Another way of looking at this physically (see Fig. 2-4) is that at large t, the curve becomes smoother and flatter, so that over the interval t_i and t_{i+1}, a linear extrapolation becomes a reasonable approximation. A computer program for the fourth-order Runge-Kutta integration of simultaneous differential equations representing the reaction $A \rightleftarrows B$ is presented in the Instructor's Manual. Other programs for numerical integration may be found in the sources cited in Appendix 2.1 and the Instructor's Manual.

2.6.2.3 Multistep Methods: Predictor-Corrector Method. The one-step methods discussed in the previous section all approximate the solution of the first-order differential equation [equation (2-211)] with initial condition that $y_0 = y(x_0)$. Using the Euler method, the solution is represented by the sum across a series of single intervals,

$$y_1 = y_0 + \int_{x_0}^{x_1} f(x_0, y_0)\, dx$$

$$y_2 = y_1 + \int_{x_1}^{x_2} f(x_1, y_1)\, dx \tag{2-230}$$

$$\vdots \qquad \vdots$$

$$y_{i+1} = y_i + \int_{x_i}^{x_{i+1}} f(x_i, y_i)\, dx$$

If the function $f(x_i, y_i)$ can be approximated by a polynomial that interpolates it at m points, where $x_i, x_{i-1}, \ldots, x_{i-m+1}$, and if the point at x_{i+1} is excluded, the method is known as an *explicit* type of formula. The *implicit* type is derived by basing the interpolating polynomial over the points x_i through x_{i-m+1}. All common multistep methods described by equation (2-230) that use the implicit and explicit formulas are known as *predictor-corrector* methods. In these methods, an approximate solution is obtained from the explicit formula predictor equation and refined iteratively by the implicit formula corrector equation. A flow chart of the predictor-corrector algorithm is shown in Fig. 2-5. The predictor-corrector procedure begins with a one-step method to start the iterative procedure from an initial value. Then, using the predictor equation, an approximation to $y_n = y(x_n)$ is obtained which uses the function $f(x,y)$ to evaluate an approximation to y_n'. From there, a corrector equation is employed, one evaluates again, corrects again, and so on until one decides to move to the next point x_{n+1}. The different predictor-corrector methods differ primarily in the choice of method used to

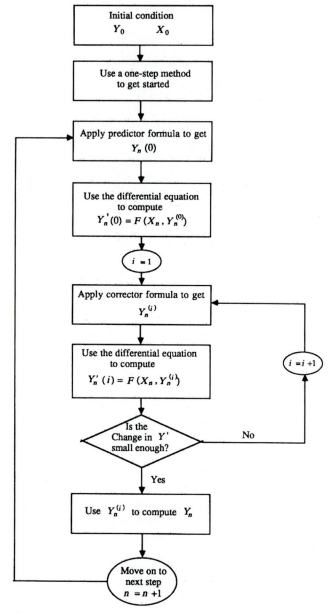

FIGURE 2-5   Flow chart of the predictor-corrector algorithm.

approximate the integral equation (2-230). Those that use Simpson's rule of integration form the basis of the Milne-Simpson predictor-corrector method.[25,31] The Adams-Bashforth method[25] uses the Newton backward interpolating formula of degree $k - 1$, where k refers to the number of previous steps used to interpolate $f(x,y)$. The order of the method is determined by the degree. The predictor equation for the fourth-order method is

$$y_{n+1} = y + \frac{h}{24}[55y_n' - 59y_{n-1}' + 37y_{n-2}' - 9y_{n-3}'] \qquad (2\text{-}231)$$

with an error $e_{i+1} = (251/720)h^5 y^{(v)}(\xi)$ where $x_{i-3} < \xi$. The corrector equation is

$$y_{n+1} = y_n + \frac{h}{24}[9y_{n+1}' + 19y_n' + 5y_{n-1}' + y_{n-2}'] \qquad (2\text{-}232)$$

with an error $e_{i+1}^* = -(19/720)h^5 y^{(v)}(\xi^*)$ and $\xi^* < x_{i+1}$.

2.6.3   Considerations in Selecting an Algorithm for Solving Differential Equations for Chemical Kinetic Systems

The first step in selecting a suitable algorithm for solving kinetic equations for a specific chemical application is to identify mathematically the type of differential equations, $y' = f(x,y)$ that describe the chemical system. If the differential equations are complicated, involving second- and/or third-order terms, a predictor-corrector method may be preferred since it requires only two evaluations of the function rather than the four required by the Runge-Kutta method. In specifying the problem the tolerance, $y' = f(x,y)$, must be considered. Usually the local error is bounded by the tolerance, that is, the local error per unit step

$$\frac{|y_i^{calc} - y_i^{exact}|}{h} \le \tau \qquad (2\text{-}233)$$

This specifies the accuracy required in the results of a differential equation, but it will depend on the step size. If the step size is too small, the computational process will consume vast amounts of computer time integrating through a large number of steps, which means that the number of operations increases and thus the global round-off errors increase. On the other hand, if the step size is too large, the local error due to truncation may be large, resulting in an accumulation of errors which will cause inaccurate results. Generally for a method of order n the local error will be on the order of a constant times the step size raised to the $n + 1$ power. Usually a differential equation of nth order will have n time constants. If two of them are of drastically different magnitudes, the numerical integration technique may become unstable. For systems of equations where the largest step-size is governed by the largest eigenvalue, the integration time for full evaluation of the solution is governed by the smallest eigenvalue. The system of kinetic equations which result in eigenvalues that differ in orders of magnitude is commonly referred to as *stiff equations*. These systems require that a step size be chosen small enough to account for the smallest eigenvalue. The difficulty with such a requirement is that the use of extremely small step sizes will cause the method to consume a large amount of time in reaching the end of the time interval. Also, round-off and truncation errors can accumulate through the large number of calculations that must be performed. However, these problems are not insurmountable, and the Gear routine is now commonly used for stiff chemical kinetic systems.[20]

2.7   STOCHASTIC METHOD

2.7.1   Introduction

Up to this point in the text, we have considered a number of complex kinetic processes and have gone through the mathematical theories used to predict the concentration of the reacting species as a function of time. The mathematical approach we have used thus far is *deterministic;* that is, once the initial values of the concentrations are speci-fied (the initial state of the system), we can express them at any later time in some mathematical form determined by the postulated mechanism. The deterministic approach regards the evolution of the system in time as a continuous process governed by a set of coupled differential equations. However, this approach has no theoretical justification, and its use is based solely on empiricism. The *stochastic approach,* on the other hand, is based on the probability that a single reaction occurs in an ensemble of molecules.[32] It is similar in spirit to the statistical approach to thermodynamics. The motivation for considering a probabilistic approach is that a chemically reacting sys-tem is not deterministic when its state is specified merely by the number of molecules of each chemical species. If the position and velocity of each molecule are specified, and the dynamics is governed by classical mechanics, then the system is indeed deter-ministic; this is the molecular-dynamic approach taken in Chapters 6, 8, and 9. But knowing *only* the average species concentrations removes essential information about the system; hence, molecular reactions are essentially random occurrences. By making some plausible assumptions, we can regard the stochastic process as a Markovian one, described by ensemble averages and fluctuations. If the fluctuations turn out to be neg-ligibly small compared to the average value, then we can say that the reactive process is effectively deterministic. For most simple reactions, the mean concentration will be essentially the same as the deterministic solution and the statistical fluctuations about the mean may be so small that they are negligible. However, a number of reactions are exceptions to this and are discussed in the review by McQuarrie.[33]

2.7.2   The Stochastic Approach

The stochastic approach to reaction kinetics stems from basic probability theory.[34] The probability of finding a number j of reactant molecules in the system at time t is repre-sented by $P_j(t)$. The aim is to find the probability that a later time $t + \Delta t$, the number of reactant molecules is equal to some other number n. The relationship between the two probabilities is given by

$$P_n(t + \Delta t) = \sum_j P_j(t)W_{j,n}(\Delta t) \tag{2-234}$$

where $W_{j,n}(\Delta t)$ is the conditional probability that n reactant molecules exist in the sys-tem at time $t + \Delta t$, given that there were j reactant molecules present at time t. The product of $P_j(t)$ and $W_{j,n}(\Delta t)$ is summed over all values of j to find the total probability P_n at $t + \Delta t$. Equation (2-234) is the fundamental equation for the application of the stochastic approach to chemical reactions.

Consider the irreversible reaction A $\rightarrow$ B.[35,36] As the reaction takes place, then some A molecules will change with time to B. At any given time then there should be some A and some B molecules in the system, but at $t = 0$ only A molecules are present. The probability that any one of the n reactant (or A) molecules will undergo a reaction during the time interval t and $t + \Delta t$ is equal to

$$W_{n,n-1}(\Delta t) = kn\Delta t + \mathcal{O}(\Delta t) \tag{2-235}$$

where k is a constant. Note that k is defined so that for sufficiently small Δt, $k\Delta t$ is the probability that any A molecule will react to yield a B molecule in the next interval Δt. If there are n A molecules, and Δt is so small that we can ignore the probability of more than one A molecule reacting in the next Δt, we may deduce that the term $kn\Delta t$ is the probability that any one of the A molecules will react in the next Δt. The term $\mathcal{O}(\Delta t)$ denotes the probability that more than one reaction occurs but is negligibly small by the preceding assumption. The probability that one of the $n + 1$ A molecules becomes a B molecule during Δt is given by

$$W_{n+1,n}(\Delta t) = k(n + 1)\Delta t + \mathcal{O}(\Delta t) \tag{2-236}$$

and the probability that the reactant stays in the same state and hence that an A molecule will remain an A molecule is

$$W_{n,n}(\Delta t) = 1 - W_{n,n-1}(\Delta t)$$

or

$$W_{n,n}(\Delta t) = 1 - (kn\Delta t + \mathcal{O}(\Delta t)) \tag{2-237}$$

According to equation (2-234), the probability that n molecules are present at the end of the time interval Δt is

$$P_n(t + \Delta t) = P_{n+1}(t)W_{n+1,n}(\Delta t) + P_n(t)W_{n,n}(\Delta t)$$

or

$$P_n(t + \Delta t) = k(n + 1)\Delta t P_{n+1}(t) + (1 - kn\Delta t)P_n(t) + \mathcal{O}(\Delta t) \tag{2-238}$$

The probability of finding n A molecules in the system at time $t + \Delta t$ is given by a sum of two possible ways of arriving at that state from the earlier configuration. The first term on the right-hand side of equation (2-238) is the probability that, given $n + 1$ A molecules at time t, one of those $n + 1$ A molecules will react in the next interval Δt. The second term is the probability that, given n A molecules at time t, none of those molecules will react in the next Δt. The third term is an error term, representing the probability that two or more A molecules might react in the interval Δt. Transposing $P_n(t)$ to the left-hand side and dividing by Δt, we get

$$\frac{P_n(t + \Delta t) - P_n(t)}{\Delta t} = k(n + 1)P_{n+1}(t) - knP_n(t) + \mathcal{O}(\Delta t) \tag{2-239}$$

If we now take the limit as $\Delta t \to 0$, we get the differential-difference equation,

$$\frac{dP_n}{dt} = k(n + 1)P_{n+1}(t) - knP_n(t) \quad (n = 1, \ldots, n_0 - 1) \tag{2-240}$$

which is a form of the master equation (see also Appendix 9.1). At $t = 0, n = n_0$; therefore, $P_{n_0+1}(0) = 0$, and we obtain

$$\frac{dP_n}{dt} = -knP_n(t) \tag{2-241}$$

The initial condition for equation (2-240) and (2-241) is:

$$P_n(0) = \begin{cases} 1, & \text{if } n = n_0 \\ 0, & \text{if } n \neq n_0 \end{cases}$$

Thus, the solution to equation (2-241) can easily be found. We first consider the solution that satisfies $P_{n_0}(0) = 1$, which is just

$$P_{n_0}(t) = e^{-kn_0 t} \tag{2-242}$$

Using this expression equation (2-240) becomes a relatively simple linear differential equation for $P_{n_0-1}(t)$:

$$\frac{dP_{n_0-1}}{dt} = k(n_0 - 1 + 1)e^{-kn_0 t} - k(n_0 - 1)P_{n_0-1} \tag{2-243}$$

The solution of this equation is subject to the initial condition $P_{n_0-1} = 0$, and is easily found. Iterating the procedure for decreasing n, to find $P_{n_0}(t), P_{n_0-1}(t), ..., P_0(t)$, we find the general result by induction to be

$$P_n(t) = \binom{n_0}{n}(e^{-kt})^n(1 - e^{-kt})^{n_0-n_x} \tag{2-244}$$

Equation (2-244) has a simple interpretation. The first factor is the number of ways of picking n specific A molecules from a total of n_0 molecules; the second factor is the probability that a specific set of n A molecules will not have decayed by time t, and the third factor is the probability that the other $n_0 - n$ A molecules will have decayed by time t. We can determine the average number of reactant molecules at time t by

$$\langle n \rangle_t = \sum_{n=0}^{n_0} nP_n(t) \tag{2-245}$$

The variance, which provides information on the fluctuations, is given by

$$\sigma_t^2 = \sum_{n=0}^{n_0} (n - \langle n \rangle_t)^2 P_n(t) \tag{2-246}$$

or

$$\sigma_t^2 = \sum_{n=0}^{n_0} n^2 P_n(t) - \langle n \rangle_t^2 \tag{2-247}$$

In the simple case of A $\rightarrow$ B the average number of reactant molecules at time t is given by

$$\langle n \rangle_t = n_0 e^{-kt} \tag{2-248}$$

and its variance is

$$\sigma_t^2 = n_0 e^{-kt}(1 - e^{-kt}) \qquad (2\text{-}249)$$

Now consider the more complex reaction

$$A \underset{k_2}{\overset{k_1}{\rightleftharpoons}} B$$

discussed earlier. We set up the master equation by first defining

$n_0 =$ the number of molecules of A at time zero;

$n\ =$ the number of molecules of A at time t;

$k_1 =$ the forward rate constant;

$k_2 =$ the backward rate constant.

At some time t, there are n A molecules and $(n_0 - n)$ B molecules. If we assume that there is no other type of reaction except A $\rightleftharpoons$ B, then the probability that one of the $(n + 1)$ A molecules will become a B molecule over the time interval $t + \Delta t$ is

$$W_{n+1,n}(\Delta t) = k_1(n + 1)\Delta t \qquad (2\text{-}250)$$

and the probability that one of the n A molecules will become a B molecule over the same time interval is

$$W_{n,n-1}(\Delta t) = k_1 n \Delta t \qquad (2\text{-}251)$$

Similarly, the probability that one of the $[n_0 - (n - 1)]$ B molecules will become an A molecule is

$$W_{n_0-n+1,n_0-n}(\Delta t) = k_2(n_0 - n + 1)\Delta t \qquad (2\text{-}252)$$

and the probability for one of the $(n_0 - n)$ B molecules to become an A molecule would be

$$W_{n_0-n,n_0-n-1}(\Delta t) = k_2(n_0 - n)\Delta t \qquad (2\text{-}253)$$

Finally, the probability that one of the n A molecules will remain an A molecule is

$$W_{n,n}(\Delta t) = 1 - W_{n,n-1}(\Delta t) - W_{n_0-n,n_0-n-1}(\Delta t) \qquad (2\text{-}254)$$

or

$$W_{n,n}(\Delta t) = 1 - k_1 n \Delta t - k_2(n_0 - n)\Delta t \qquad (2\text{-}255)$$

From the theorem of total probability, we have

$$P_n(t + \Delta t) = \sum_j P_j(t)W_{j,n}(\Delta t) \qquad (2\text{-}256)$$

where $P_j(t)$ is the probability of finding j reactant molecules at time t. Therefore,

$$P_n(t + \Delta t) = P_n(t)W_{n,n}(\Delta t) + P_{n+1}(t)W_{n+1,n}(\Delta t) + P_{n-1}(t)W_{n-1,n}(\Delta t) \qquad (2\text{-}257)$$

and substituting equations (2-250) through (2-255) into (2-257), we get

$$P_n(t + \Delta t) = P_n(t) - P_n(t)[k_1 n + k_2(n_0 - n)]\Delta t$$
$$+ P_{n+1}(t)k_1(n + 1)\Delta t + P_{n-1}(t)k_2(n_0 - n + 1)\Delta t \tag{2-258}$$

Dividing both sides of the equation by Δt and taking the limit as $\Delta t \to 0$, we obtain the master equation

$$\frac{dP_n}{dt} = k_2(n_0 - n + 1)P_{n-1}(t) + k_1(n + 1)P_{n+1}(t)$$
$$-[k_1 n + k_2(n_0 - n)]P_n(t) \tag{2-259}$$

To solve this equation for $P_n(t)$, we follow the approach previously used for equations (2-242) through (2-244). The average number of reactant molecules at time t is found to be

$$\langle n \rangle_t = \frac{n_0}{k_1 + k_2} [k_2 + k_1 e^{-(k_1 + k_2)t}] \tag{2-260}$$

which is just equivalent to the solution found by the deterministic approach, given by equation (2-14). A more generalized approach to setting up the master equation may be found in treatments by Gillespie.[37] An extension of Gillespie's stochastic method to bimolecular reactions of the type $A + B \to C + D$ has been presented by Piersall and Anderson.[38]

Since the stochastic and deterministic solutions are the same, why bother with a stochastic treatment? The beauty behind the formalism is that it provides a theoretical basis for the use of differential equations to model complex reaction systems. While solving the master equation for most complex reactions can be extremely difficult, there are still several good reasons for using the stochastic approach. First, its use may be appropriate either if σ_t is compatible to $\langle n \rangle_t$, or if the correlations present render $\langle n \rangle_t$ different from what you would get by solving the deterministic rate equations. One would never know for sure whether either of these conditions exist, unless the results of a stochastic analysis are examined. Second, one may want to try a stochastic rather than a deterministic approach, if one is forced to find a numerical solution, since the stochastic simulation is frequently less difficult to implement numerically than the integration of a set of reaction rate equations. Finally, the stochastic approach allows one to study fluctuations, and for systems involving a small number of molecules, fluctuations may be important. The stochastic approach can provide a measure of the fluctuations from the calculated variance.

It is clear from the simple examples considered in this section, the stochastic formulation can become mathematically intractable to solve analytically for more complex reaction systems. However, Gillespie[39-42] has developed a numerical method which simulates the stochastic evolution of any given chemical system in time. The basic stochastic numerical simulation algorithm and program are discussed in Appendix 2.3 and the Instructor's Manual. The method is used to analyze the problem we have just considered, i.e., solving the kinetic equations for the reaction $A \rightleftarrows B$.

As Fig. 2-6 shows, the evolution of the simple first-order reversible reaction shows significant fluctuations for the case where the initial reactant molecules are present in

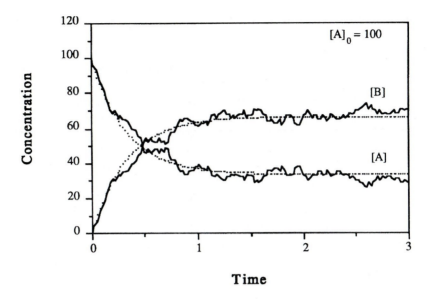

Time

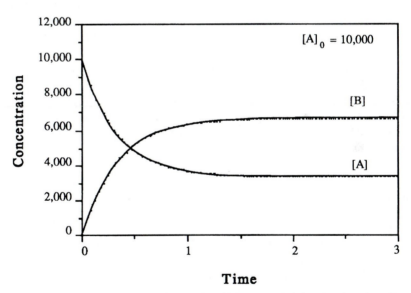

Time

FIGURE 2-6   Plot of two stochastic simulation runs for the reaction A $\rightleftarrows$ B. (a) Plot of the reaction with $[A]_0 = 100$ molecules. (b) Plot of the reaction with $[A]_0 = 10,000$ molecules. Dotted lines correspond to deterministic mean value, solid lines to stochastic values. In plot (b), the dotted line is coincident with the solid line.

small numbers. However, when the system has a large number of molecules, e.g., $n > 10,000$, the fluctuations in the system are averaged out. The solution one then obtains is no different from the deterministic solution. For reactions occurring in the gas or solution phase where the number of molecules typically ranges from 10^{10} to 10^{23} molecules per unit volume, it becomes quite clear why the deterministic approach considered in earlier sections is acceptable. The stochastic approach is particularly useful in analyzing *in vivo* biochemical reactions taking place in the individual cells of organisms.[36] Similarly, the study of enzyme amplifier kinetics, in which a few activated molecules initiate an avalanche type of reaction, can be studied in detail with the stochastic approach. In this case, the number of molecules participating in the activation process is too small to model deterministically.[43,44]

In the last decade, there has been considerable advances made in the detection of single molecules and atoms.[45–49] The spectroscopy and shape of a broad range of molecules have been investigated with single molecule detection methods. As research moves to explore the chemical reactivity of single molecules and atoms, the use of kinetic rate laws which give information on how fast the molecules react have to be viewed with some caution. The reason for this is that kinetic rate laws have all been derived from studies probing the ensemble of molecules. Consequently, information derived from such studies are quantities that represent average behavior of the ensemble of molecules (or they represent the first moment of the distribution). Single molecule studies of chemical reactivity offer intriguing direct insight into important fundamental issues of reactivity. As an example, Ertl and co-workers[50] have used scanning tunnelling microscopy to follow oxidation kinetics of individual CO molecules on a Pt(111) surface (see chapter 5). Parameters derived from statistics of individual reactions agree well with data from macroscopic kinetic measurements. But the question we need to ask is whether kinetic rate law formalisms provide a useful basis for interpretation of experimental information from single molecule or atom experiments. The stochastic method offers a fresh approach towards looking at the kinetics of single molecules and atoms.

REFERENCES

[1] P. M. Jeffers, *J. Phys. Chem. 76*, 2829 (1972).

[2] S. W. Benson, *Foundation of Chemical Kinetics* (New York: McGraw-Hill, 1960).

[3] J-Y. Chien, *J. Am. Chem. Soc. 70*, 2258 (1948).

[4] M. Bodenstein, *Z. physik. Chem. 85*, 329 (1913).

[5] N. N. Semenov, *Zhur. Fiz. Khim. 17*, 187 (1943).

[6] L. Volk, W. Richardson, K. H. Lau, M. Hall, and S. H. Lin, *J. Chem. Educ. 54*, 95 (1977).

[7] J. C. Gliddings and H. Y. Shin, *J. Chem. Phys. 36*, 640 (1962).

[8] M. Bodenstein, *Z. physik. Chem. 13*, 56 (1894).

[9] M. Bodenstein, *Z. physik. Chem. 22*, 1 (1897).

[10] M. Bodenstein, *Z. physik. Chem. 29*, 295 (1898).

[11] J. H. Sullivan, *J. Chem. Phys. 46*, 73 (1967).

[12] G. Hammes and B. Widom, *J. Chem. Phys. 96*, 7621 (1974).

[13] A. C. Doyle, "A Study in Scarlet," in *The Complete Sherlock Holmes,* edited by C. Morley (New York: Doubleday, 1930), p. 50.

[14] M. S. Mangir, H. Reisler, and C. Wittig, *J. Chem. Phys. 73,* 829 (1980).

[15] G. Strang, *Linear Algebra and Its Applications* (New York: Academic Press, 1976).

[16] J. G. F. Francis, *Computer J. 4,* 265, 332 (1961).

[17] R. S. Martin, P. Peters, and J. H. Wilkinson, *Numerische Math. 4,* 219 (1970).

[18] J. H. Wilkinson, *The Algebraic Eigenvalue Problem* (Oxford: University Press, 1965).

[19] L. A. Fanow and D. Edelson, *Int. J. Chem. Kinet. 6,* 787 (1974).

[20] D. Edelson, *J. Comp. Phys. 11,* 455 (1973).

[21] D. L. Allara and D. Edelson, *Int. J. Chem. Kinet. 4,* 345 (1972).

[22] D. F. deTar and C. E. deTar, *J. Phys. Chem. 70,* 3842 (1966).

[23] S. D. Conte and C. deBoor, *Elementary Numerical Methods,* 2d ed. (New York: McGraw-Hill, 1972).

[24] T. E. Shoup, *Applied Numerical Methods for the Microcomputer* (Englewood Cliffs, NJ: Prentice-Hall, 1984).

[25] R. W. Hamming, *J. Assoc. Comp. Mach. 6,* 37 (1959).

[26] C. W. Gear, *Commun. Am. Comput. Mach. 14,* 176 (1971).

[27] C. W. Gear, *Numerical Initial Value Problems in Ordinary Differential Equations* (Englewood Cliffs, NJ: Prentice-Hall, 1971).

[28] B. Descamps and W. Forst, *J. Phys. Chem. 80,* 933 (1976).

[29] R. C. Kennedy and J. B. Levy, *J. Phys. Chem. 76,* 3480 (1972).

[30] J. S. Francisco, Z. Li, and I. H. Williams, *Chem. Phys. Lett. 140,* 531 (1987).

[31] W. E. Milne, *Amer. Math. Month. 33,* 455 (1926).

[32] I. Oppenheim, K. Shuler, and G. H. Weiss, *Stochastic Processes in Chemical Physics: The Master Equation* (Cambridge, MA: M.I.T. Press, 1977).

[33] D. A. McQuarrie, *J. Appl. Prob. 4,* 413 (1967).

[34] W. Feller, *An Introduction to Probability Theory and Its Applications* (New York: Wiley, 1950).

[35] A. F. Bartholomay, *Bull. Math. Biophys. 20,* 175 (1958).

[36] E. A. Boucher, *J. Chem. Educ. 51,* 580 (1974).

[37] D. T. Gillespie, *J. Chem. Phys. 72,* 5363 (1980).

[38] S. P. Piersall and J. B. Anderson, *J. Chem. Phys. 95,* 971 (1991).

[39] D. T. Gillespie, *J. Phys. Chem. 81,* 2340 (1977).

[40] D. T. Gillespie, *J. Stat. Phys. 16,* 311 (1977).

[41] D. T. Gillespie, *J. Comp. Phys. 22,* 403 (1976).

[42] D. T. Gillespie, *J. Chem. Phys. 75,* 704 (1981).

[43] S. N. Levine, *Science 152,* 651 (1966).

[44] R. G. MacFarlane, *Nature 202,* 498 (1964).

[45] P. Avouris, *Acc. Chem. Res. 28,* 95 (1995).

[46] C. Bustamante and D. Keller, *Phys. Today 48,* 32 (1995).

[47] J. K. Trautman and J. J. Macklin, *J. Chem. Phys. 205,* 221 (1996).

[48] X. S. Xie, *Acc. Chem. Res. 29,* 598 (1996).

[49] M. M. Collinson and R.M. Wightman, *Science 268,* 1883 (1995).

[50] J. Wintterlin, S. Völkering, T. V. W. Jaussens, T. Zambelli, and G. Ertl, *Science 278,* 1931 (1995).

BIBLIOGRAPHY

BOWMAN, F. *Introduction to Bessel Functions.* New York: Dover, 1958.

BRADBURY, T. C. *Mathematical Methods with Applications to Problems in the Physical Sciences.* New York: Wiley, 1984.

CHURCHILL, R. V. *Operational Mathematics.* New York: McGraw-Hill, 1971.

COURANT, R., and HILBERT, D. *Methods of Mathematical Physics.* Vols. 1 and 2.

KAPLAN, W. *Ordinary Differential Equations.* Reading MA: Addison-Wesley, 1958.

MARGENAU, H., and MURPHY, G. M. *The Mathematics of Physics and Chemistry.* Vols. 1 and 2. New York: Van Nostrand, 1964.

MATHEWS, J., and WALKER, R. L. *Mathematical Methods of Physics.* 2d ed. New York: W. A. Benjamin, 1964.

RAINVILLE, E. D., and BEDIENT, P. E. *Elementary Differential Equations.* New York: Macmillan, 1981.

WILLIAMSON, R. E. *Introduction to Differential Equations.* Englewood Cliffs, NJ: Prentice-Hall, 1986.

APPENDIX 2.1   THE LAPLACE TRANSFORM

As discussed in section 2.4, the Laplace transform can be used to solve linear differential equations that describe kinetic processes occurring in reacting chemical systems. In

TABLE 2-1.1   Laplace Transform General Properties

No.	Name	$F(p)$	$f(t)$
1	Direct Laplace Transform	$\displaystyle\int_0^\infty f(t)\varepsilon^{-pt}\,dt$	$f(t)$
2	Inverse Laplace Transformation	$F(p)$	$\displaystyle\frac{1}{2\pi l}\int_{c-i\infty}^{c+i\infty} F(p)e^{pt}dp$
3	Linearity	$F_1(p) \pm F_2(p)$	$f_1(t) \pm f_2(t)$
4	Linearity	$aF(p)$ where a is independent of p and t	$af(t)$
5	Real Differentiation	$PF(p) - f(0+)$	$\dfrac{df(t)}{dt}$
6	Real Differentiation	$\displaystyle P^nF(p) - \sum_{\ell=1}^{n} p^{(\ell-1)} \cdot f^{(n-\ell)}(0+)$ where $f^{(n-\ell)}(0+) \equiv \left.\dfrac{d^{(n-\ell)}f}{dt^{(n-\ell)}}\right|_{t=0+}$	$\dfrac{d^n f(t)}{dt^n}$
7	Real Definite Integration	$\dfrac{1}{p}\,F(p)$	$\displaystyle\int_0^t f(t)\,dt$

section 2.4.3 we noted some properties which are particularly useful in carrying out Laplace transformations. Here, a few more useful properties are presented and a table of Laplace transforms most commonly encountered in kinetics problems is given.

For more comprehensive tables of Laplace transforms, see the following texts:

1. A. Erdelyi, W. Magnus, F. Oberhettinger, and F. Tricomi, *Tables of Integral Transform,* vols. 1 and 2 (New York: McGraw-Hill, 1954).

TABLE 2-1.2   Some Useful Laplace Transform Pairs

No.	$F(p)$	$f(t)$
1	$\dfrac{1}{p^2}$	t
2	$\dfrac{1}{p + a}$	e^{-at} Valid for complex a.
3	$\dfrac{1}{p(p + a)}$	$\dfrac{1}{a}(1 - e^{-at})$
4	$\dfrac{1}{(p + a)(p + b)}$	$\dfrac{1}{(b - a)}(e^{-at} - e^{-bt})$
5	$\dfrac{1}{p(p + a)(p + b)}$	$\dfrac{1}{ab}\left[1 + \dfrac{1}{(a - b)}(be^{-at} - ae^{-bt})\right]$
6	$\dfrac{1}{(p + a)(p + b)(p + c)}$	$\dfrac{1}{(b - a)(c - a)}e^{-at} + \dfrac{1}{(a - b)(c - b)}e^{-bt} + \dfrac{1}{(a - c)(b - c)}e^{-ct}$
7	$\dfrac{1}{(p + a)^2}$	te^{-at}
8	$\dfrac{1}{p(p + a)^2}$	$\dfrac{1}{a^2}[1 - e^{-at} - ate^{-at}]$
9	$\dfrac{p}{(p + a)(p + b)}$	$\dfrac{1}{(a - b)}[ae^{-at} - be^{-bt}]$
10	$\dfrac{p}{(p - a)(p - b)}$	$\dfrac{1}{(a - b)}[ae^{at} - be^{bt}]$
11	$\dfrac{p}{(p + a)(p + b)(p + c)}$	$\dfrac{-a}{(b - a)(c - a)}e^{-at} - \dfrac{b}{(a - b)(c - b)}e^{-bt} - \dfrac{c}{(a - c)(b - c)}e^{-ct}$
12	$\dfrac{p}{(p + a)^2}$	$(1 - at)e^{-at}$
13	$\dfrac{p + a}{(p + b)(p + c)}$	$\dfrac{1}{(c - b)}[(a - b)e^{-bt} - (a - c)e^{-ct}]$
14	$\dfrac{p^2}{(p + a)(p + b)(p + c)}$	$\dfrac{a^2}{(b - a)(c - a)}e^{-at} + \dfrac{b^2}{(a - b)(c - b)}e^{-bt} + \dfrac{c^2}{(a - c)(b - c)}e^{-ct}$

2. F. Oberhettinger, *Tabellen zur Fourier Transformation* (Berlin: Springer-Verlag, 1957).

3. P. A. McCollum and B. F. Brown, *Laplace Transform Tables and Theorems* (New York: Holt, Rinehart and Winston, 1965).

4. J. I. Steinfeld, J. S. Francisco, W. L. Hase, *Chemical Kinetics and Dynamics*, 1st Ed. (New York: Prentice Hall, 1989), pp. 77–91.

APPENDIX 2.2   NUMERICAL ALGORITHMS FOR DIFFERENTIAL EQUATIONS

There exists a variety of algorithms available for numerical solutions of coupled kinetic equations. The basic approach to solving a problem such as 2-12 or those in Chapter 15, involves the following steps:

1. Assume an initial concentration for each chemical species. Input step size and rate constants.

2. For each species calculate its incremental concentration change starting with the initial conditions for the time interval Δt.

3. Using the incremental concentration change, calculate the concentration of each species for the next interval.

4. Repeat until the desired time is reached.

Example: For the reaction

$$A \xrightarrow{k_1} \text{products}$$

The differential equation for the decay of A is

$$d[A]/dt = -k_1[A]$$

The incremental change in the concentration of A over the time interval Δt is

$$\Delta[A] = -k_1^*[A]_i^* \, \Delta t$$

$$[A]_{i+1} = [A]_i + \Delta[A]$$

A number of options are available for carrying out the numerical integration:

1. User-generated code in FORTRAN or some other computer language. The Instructor's Manual includes a sample FORTRAN computer program to simulate a set of rate equations using either Euler, modified Euler, or fourth-order Runge-Kutta integrations. In this approach, the user has to transcribe the kinetic mechanism into a set of coupled differential equations.

2. Set up finite-difference equations on a spreadsheet program such as EXCEL.

3. Use a commercial or public-domain numerical integration program. A good reference source is W. H. Press, B. P. Flannery, S. A. Teukolsky and W. T. Vetterling, *Numerical Recipes* (Cambridge University Press, 1986).

4. Commercial packages are also available which input a chemical mechanism, i.e., a set of chemical reactions and associated rate coefficients, along with initial concentrations. The program generates the numerical equations internally and outputs concentrations vs. time. Examples of such programs are CHEMKIN

(Reaction Design, San Diego, Calif.) and KINETICS (ARSofware, Landover, Md.). These software packages tend to be quite expensive, reflecting the programming effort that went into them.

APPENDIX 2.3   STOCHASTIC NUMERICAL SIMULATION OF CHEMICAL REACTIONS

Often the master equation (2-240) turns out to be intractable to solve analytically, and even numerically, for some cases, if attempted by finite-difference methods. However, Gillespie[39-41] has developed a numerical method which simulates the stochastic evolution in time of a given chemical system. To illustrate the method let us numerically analyze the problem previously considered in section 2.7.2 within the framework of the stochastic formulation by using Gillespie's stochastic simulation algorithm. The reaction is

$$A \underset{k_2}{\overset{k_1}{\rightleftharpoons}} B$$

We define a set of reaction events by R_i, so when $A \rightarrow B$ we say that an R_1 reaction has occurred, and when $B \rightarrow A$ we say that an R_2 reaction has occurred. We can write the set of reaction events as

$$R_i = \{R_1, R_2\}$$

We also define reaction rate coefficients for each reaction as

$$k_i = \{k_1, k_2\}$$

where k_1 is the forward rate constant and k_2 is the backward rate constant. Now, as mentioned earlier, k_1 is defined so that $k_1 \Delta t$ gives the probability that any given A molecule will turn into a B molecule in the next infinitesimal time interval Δt. With this definition, one can in principle calculate k_1 from a knowledge of the mechanism responsible for the reaction $A \rightarrow B$. Such a probabilistic formulation reflects either the random nature of collisions responsible for bimolecular reactions, or the random decay of molecules in unimolecular processes.

We consider a chemical system containing N molecular species denoted by the concentration variables $C_i = \{C_1, C_2\}$, where C_1 is the number of A molecules and C_2 is the number of B molecules. These species can interact through M chemical channels $R_i (i = 1, ..., M)$. We note that the number of chemical components need not be equal to the number of reaction channels. This particular problem can be regarded as a single-variable problem since $C_2 = N - C_1$.

The Gillespie algorithm is based on a reaction probability density function, $P(\tau, i)$, defined by

$P(\tau, i)d\tau$ = Probability at time t that the next reaction will occur in the differential time interval $(t + \tau, t + \tau + d\tau)$ and will be an R_i reaction

More formally,

$$P(\tau, i) = k_i C_i \exp\left\{-\sum_{i=1}^{2} k_i C_i \tau\right\} \qquad \text{(A3-1)}$$

where C_i is the number of molecules for species i and k_i is the set of rate constants. If we let

$$a_i = k_i C_i \qquad (i = 1,2)$$

and

$$a_t = \sum_{i=1}^{2} a_i = \sum_{i=1}^{2} k_i C_1$$

then equation (A3-1) becomes

$$P(\tau,i) = k_i C_i \exp[-(k_1 C_1 + k_2 C_2)\tau] \qquad (i = 1,2) \qquad \text{(A3-2)}$$

The probability density function, $P(\tau,i)$, is a function of two variables and can be written as the product of two probability density functions of one variable. In other words, $P(\tau,i)$, can be expressed as the product of the probability of a reaction occurring in time interval $d\tau$ and the probability of that reaction being an i reaction:

$$P(\tau,i) = P(\tau) \cdot P(i) \qquad \text{(A3-3)}$$

Here, $P(\tau)d\tau$ represents the probability that the next reaction will occur in the time interval $(t + \tau, t + \tau + d\tau)$ regardless of what the reaction is, and $P(i)$ is the probability that the next reaction will be an R_i reaction. Now by the addition theorem for probability, $P(\tau)$ is obtained from

$$P(\tau) = \sum_{i=1}^{2} P(\tau,i) \qquad \text{(A3-4)}$$

and $P(i)$ is obtained by substituting the above expression into equation (A3-3),

$$P(i) = P(\tau,i)/ \sum_{i=1}^{2} P(\tau,i) \qquad \text{(A3-5)}$$

Now, if we substitute equation (A3-2) into equation (A3-4)

$$P(\tau) = a_t \exp(-a_t\tau) \qquad \text{where } (0 \leq \tau < \infty) \qquad \text{(A3-6)}$$

and if we substitute equation (A3-2) into equation (A3-5) we obtain

$$P(i) = a_i/a_t \qquad (i = 1,2) \qquad \text{(A3-7)}$$

Notice that $P(\tau)$ and $P(i)$ are properly normalized in that they satisfy

$$\int_0^\infty P(\tau)d\tau = 1$$

and

$$\sum_{i=1}^{2} P(i) = 1 \qquad \text{(A3-8)}$$

The problem now is to generate a random number τ according to the probability function $P(\tau)$ in Eq. (A3-6), and a random number i according to the probability function $P(i)$ in Eq. (A3-7). The function $P(\tau)$ is known as *an exponential random variable* *with parameter a_t,* and it can be shown that one way to generate a sample value of such a variable is to first generate a random variable r_1 on the uniform interval between 0 and 1, and then take

$$\tau = (1/a_t)\ln(1/r_1) \tag{A3-9}$$

One can generate a sample value for the random number i according to eq. (A3-7) by first generating another random number r_2 on the uniform interval between 0 and 1, and then taking

$$i = \begin{cases} 1, & \text{if } r_2 < a_1/a_t \\ 2, & \text{if } r_2 > a_1/a_t \end{cases} \tag{A3-10}$$

The simulation time variable is then incremented by τ (i.e., we advance to the next occurrence of a reaction), and the concentration variables are changed as follows:

$$C_1 \rightarrow C_1 - 1, C_2 \rightarrow C_2 + 1, \quad \text{if } i = 1$$
$$C_1 \rightarrow C_1 + 1, C_2 \rightarrow C_2 - 1, \quad \text{if } i = 2$$

Summary of the Stochastic Algorithm Procedure

In general, Gillespie's procedure uses a Monte Carlo technique to simulate the stochastic process described by $P(\tau,i)$. The simulation algorithm may be outlined as follows:

1. Set the time variable $t = 0$. Then specify the initial values A(0) and B(0), the rate constants k_1 and k_2, and the time step size Δt and the "stopping time" t_{stop}.
2. Calculate the m equations $a_i = k_i C_i$ which collectively determine the reaction probability density function $P(\tau,i)$.
3. Using suitable Monte Carlo techniques generate one random pair (τ,i) according to the joint probability density function $P(\tau,i)$.
4. Using the numbers τ and i generated in step (c), advance t by τ, and change the C_i value of those species involved in reaction R_i to reflect the occurrence of one R_i whose reactant C_i values have just been changed.
5. If t has just been advanced through Δt, print out the current molecular population values. If $t > t_{\text{stop}}$, or if no more reactants remain (all $C_i = 0$), terminate the calculation; otherwise return to step (b).

A flow chart of this simulation algorithm is given in Fig. 2A-1. A simple Basic program that solves the reversible first-order reaction problem stochastically is provided in the Instructor's manual.

PROBLEMS

2.1 The reaction A + B → C takes place in two steps by the mechanism 2A ⇌ D followed by B + D $\xrightarrow{k_2}$ A + C. The first step comes to a rapid equilibrium (constant K_1). Derive an expression for the rate of formation of C in terms of $K_1, k_2, $ [A], and [B].

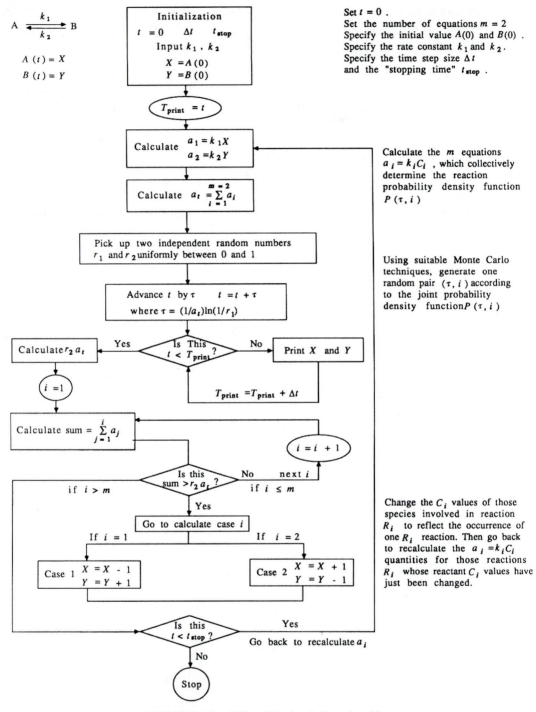

FIGURE 2A-1   Gillespie's simulation algorithm.

2.2 Set up the rate expressions for the following mechanisms:

$$A \xrightarrow{k_1} B \qquad B \xrightarrow{k_{-1}} A$$

$$B + C \xrightarrow{k_3} D$$

If the concentration of B is small compared with the concentrations of A, C, and D, the steady-state approximation may be used to derive the rate law. Show that this reaction may follow first-order kinetics at high pressures and second-order kinetics at low pressures.

2.3 Consider the reaction

$$A_1 \xrightarrow{k_1} A_2 \xrightarrow{k_2} A_3 \cdots A_{n-2} \xrightarrow{k_{n-2}} A_{n-1} \xrightarrow{k_{n-1}} A_n$$

$$\text{at } t = 0, [A_1] = [A_1]_0, [A_2] = [A_3] = \cdots = [A_{n-2}] = [A_{n-1}] = [A_n] = 0$$

Derive an expression that will describe the concentration of the intermediate existing at any time t during the reaction. [E. Abel, *Z. Phys. Chem. A56*, 558 (1906)].

2.4 Consider the reaction mechanism

$$A + B \underset{k_{-1}}{\overset{k_1}{\rightleftharpoons}} X + B$$

$$X \xrightarrow{k_2} C + D$$

(a) Write chemical rate equations for [A] and [X].

(b) Employing the steady-state approximation, show that an effective rate equation for [A] is

$$\frac{d[A]}{dt} = -k_{\text{eff}}[A][B].$$

(c) Give an expression for k_{eff} in terms of k_1, k_{-1}, k_2, and [B].

2.5 For each of the following reaction mechanisms, give the equilibrium concentration, if it exists, of all the chemical species for the initial concentrations stated.

(a) $A + B \underset{k_{-1}}{\overset{k_1}{\rightleftharpoons}} X \xrightarrow{k_2} C + D$

Initial concentrations

$$[A]_0 = 2, [B]_0 = 1, [X]_0 = [C]_0 = [D]_0 = 0.$$

(b) $A \underset{k}{\overset{k}{\underset{\nwarrow}{\overset{B}{\rightleftharpoons}}}} C$ All rate coefficients equal; initial concentrations $[A]_0 = 1, [B]_0 = [C]_0 = 0.$

(c) $X + X \underset{k_2}{\overset{k_1}{\rightleftharpoons}} 3X$ $X \xrightarrow{k_2} C$ Assume $k_1 = k_2$; initial concentrations $[X_0] = \frac{1}{2}, [C_0] = 0.$

2.6 Consider a reaction with two intermediates I_1 and I_2 which both decompose to the same product P by competing parallel steps [S. I. Miller, *J. Chem. Education 62*, 490 (1985)]:

$$A + M \underset{k_{-1}}{\overset{k_1}{\rightleftarrows}} I_1 \overset{k_3}{\searrow}$$

$$P$$

$$I_1 \underset{k_{-2}}{\overset{k_2}{\rightleftarrows}} I_2 \overset{k_4}{\nearrow}$$

(a) Find an expression for dP/dt by applying the steady state approximation with $[I_1]_{ss} \simeq [I_2]_{ss} \simeq 0$.

(b) Find an expression for dP/dt by applying the steady state approximation with $[A]_t = [A]_0 - [I_1] - [I_2] - [P]$.

(c) How do the expressions found in (a) and (b) differ, and what conclusions can be drawn from this difference?

2.7 Consider the first-order cyclic reversible reaction

$$A$$

$$B \rightleftarrows C$$

Assume that initially, at $t = 0$ $[A] = A_0$, $[B] = [C] = 0$, and that the amounts of A, B, and C which have reacted at a later time satisfy the following relation

$$[A]_0 = [A] + \{B\} + [C].$$

Using the detailed balance method, solve for $[A], [B]$, and $[C]$.

2.8 (a) The reaction $2NO + O_2 \rightarrow 2NO_2$ is third order. Assuming that a small amount of NO_3 exists in rapid reversible equilibrium with NO and O_2 and that the rate-determining step is the slow bimolecular reaction $NO_3 + NO \rightarrow 2NO_2$, derive the rate equation for this mechanism.

(b) Another possible mechanism for the reaction $2NO + O_2 \rightarrow 2NO_2$ is

(1) $NO + NO \rightarrow N_2O_2$ k_1
(2) $N_2O_2 \rightarrow 2NO$ k_2
(3) $N_2O_2 + O_2 \rightarrow 2NO_2$ k_3

Apply the steady state approximation to $[N_2O_2]$ to obtain the rate law for $d(NO_2)/dt$.

If only a very small fraction of the N_2O_2 formed in (1) goes to form products in reaction (3), while most of the N_2O_2 reverts to NO in reaction (2), and if the activation energies are $E_1 = 79.5$ kJ/mole, $E_2 = 205$ kJ/mole, and $E_3 = 84$ kJ/mole, what is the overall activation energy?

(c) How would you distinguish experimentally between the mechanisms suggested in parts (a) and (b)?

2.9 The mechanism for the decomposition of ozone into oxygen,

$$2 O_3 = 3 O_2$$

is stated to be as follows:

$$\text{Step 1, } O_3 \rightarrow O_2 + O, \qquad \text{rate constant, } k_1$$

$$\text{Step 2, } O + O_2 \rightarrow O_3, \qquad \text{rate constant, } k_2$$

$$\text{Step 3, } O + O_3 \rightarrow 2 O_2, \qquad \text{rate constant, } k_3$$

The activation energies, in kilojoules per mole, for each of the steps are as follows:

$$\text{Step 1, } E_{act} = 103 \text{ kJ/mole}$$

$$\text{Step 2, } E_{act} = 0$$

$$\text{Step 3, } E_{act} = 21 \text{ kJ/mole}$$

(a) Obtain the differential equation for the steady state rate of decomposition of ozone, $-d[O_3]/dt$, in terms of the constants, k_1, k_2 and k_3, the concentration of ozone, $[O_3]$, and the concentration of oxygen, $[O_2]$.

(b) On the basis of the values of E_{act} given above, simplify the expression for $-d[O_3]/dt$ obtained in (a) by eliminating any terms which can become negligible. State clearly the basis for the simplification.

(c) Calculate the energy of activation for the overall reaction.

2.10 As a final variation on the theme of Problems 8 and 9, let us consider the reaction between NO and O_3. In this experiment [E. Bar-Ziv, J. Moy, and R. J. Gordon, *J. Chem. Phys.* **68**, 1013 (1978)], it is necessary to distinguish between ozone in its ground (000), excited stretching (001), and excited bending (010) vibrational states.

A CO_2 laser is used to excite ozone from the (000) to the (001) state. The following reactions then ensue in the presence of nitric oxide:

(1) V-V equilibrium of ozone:

$$O_3(001) + NO \underset{k_{-1}}{\overset{k_1}{\rightleftharpoons}} O_3 (010) + NO$$

(2–4) reaction of ozone in specified vibrational states with NO to form NO_2:

$$O_3(000) + NO \xrightarrow{k_2} NO_2 + O_2$$

$$O_3(001) + NO \xrightarrow{k_3} NO_2 + O_2$$

$$O_3(010) + NO \xrightarrow{k_4} NO_2 + O_2$$

(5,6) V-T relaxation of ozone:

$$O_3(001) + NO \xrightarrow{k_5} O_3(000) + NO$$

$$O_3(010) + NO \xrightarrow{k_6} O_3(000) + NO$$

The ozone (000) could be followed by its Hartley-band absorption at $\lambda = 254$ nm, and the ozone (001) by its infrared emission at $\lambda = 9 \ \mu m$.

(a) Write the kinetic equations, from the above (simplified) mechanism, for the time derivatives of $[O_3(000)]$, $[O_3(010)]$, $[O_3(001)]$, and $[NO_2]$. If the experiment is carried out in a large excess of NO, then pseudo-first-order kinetics may be obtained; rewrite these four expressions in pseudo-first-order, using $K_i = k_i[NO]$.

(b) Find the Laplace transform of the four pseudo-first-order rate equations. (A table of Laplace transforms is included in the Appendix to Chapter 2.) Set up this set of transformed equations in determinantal form.

(c) Find $[O_3(000)]_t$ and $[O_3(010)]_t$ in terms of pseudo-first-order rate coefficients derived from k_1-k_6 and the initial concentrations of $O_3(000)$ and $O_3(001)$ following the laser pulse. Since the experiment is carried out at 300K, and $\nu_2 \simeq 700$ cm^{-1} for ozone, the initial concentration of $[O_3(010)]$ cannot be neglected.

2.11 Consider the following system of first order reversible reactions.

$$A \underset{k_1'}{\overset{k_1}{\rightleftharpoons}} A_2 \underset{k_2'}{\overset{k_2}{\rightleftharpoons}} A_3$$

Assume that initially, at $t = 0$ $[A_1] = [A_1]_0$ $[A_2]_0 = [A_3]_0 = 0$, and that amounts of A_1, A_2, and A_3, which have reacted at a later time, satisfy the following relation:

$$[A_1]_0 = [A_1] + [A_2] + [A_3]$$

(a) Solve for A_1, A_2, and A_3 using the detailed balance method.

(b) Solve for A_1, A_2, and A_3 using the Laplace transform method.

(c) Show that the two methods are equivalent as $t \to \infty$.

(d) Assume that $k_1 = k_1' = k_2 = k_2' = 1$ sec^{-1}, and $[A_1]_0 = 0.01$ M, provide a plot for A_1, A_2, and A_3 from $t = 0$ to $t = 6$ sec. Which of the two preceding methods is valid at shorter time periods? Indicate the time after the method which is invalid at shorter times becomes valid, and explain why it then becomes valid.

2.12 The free radical decomposition of methane occurs via

$$CH_4 \rightleftharpoons CH_3 + H \tag{1}$$

$$H + CH_4 \rightleftharpoons CH_3 + H_2 \tag{2}$$

$$2CH_3 \rightleftharpoons C_2H_6 \tag{3}$$

where $\quad k_1 = 14$ s^{-1}, $k_{-1} = 1.2 \times 10^{10}$ mole^{-1} 1 s^{-1}

$\quad\quad\quad k_2 = 1.5 \times 10^9$ mole^{-1} 1 s^{-1}, $k_{-2} = 2.9 \times 10^7$ mole^{-1} 1 s^{-1}

$\quad\quad\quad k_3 = 2.0 \times 10^{10}$ mole^{-1} 1 s^{-1}, $k_{-3} = 4.5 \times 10^4$ s^{-1}

with an initial methane concentration of $[CH_4]_0 = 1.0 \times 10^{-3}$ mole 1^{-1}

(a) Write the kinetic equations for this (simplified) mechanism for the time derivatives of the intermediates, reactant, and products.

(b) Assuming steady state concentrations for the free radicals, find the steady state expressions for $[H]_{ss}$ and $[CH_3]_{ss}$; derive an expression for the time dependence of $[C_2H_6]$ in terms of the concentration of $[CH_4]$.

(c) Plot the time dependence of $[CH_3]$ radicals and $[C_2H_6]$ over 30 μ sec.

(d) Using the Euler algorithm described in Appendix 2.2, numerically integrate the differential equations over 30 μsec using an integration step size of $\Delta t = 0.1$ μsec and 1.0 μsec. Plot the time dependence of $[CH_3]$ and $[C_2H_6]$.

(e) Compare the steady-state and numerical solutions. Over what times is it appropriate to use the steady-state assumption?

(f) What percentage increase in the concentration of C_2H_6 is observed if k_2 is increased by a factor of 5?

2.13 Consider the reaction system

$$A_1 + A_2 + A_3 + \cdots + A_{n-1} + A_n \underset{k_2}{\overset{k_1}{\rightleftharpoons}} B_1 + B_2 + B_3 + \cdots + B_{n-1} + B_n$$

Let $A_1(t), A_2(t), A_3(t), ..., A_{n-1}(t), A_n(t)$ and $B_1(t), B_2(t), B_3(t), ..., B_{n-1}(t), B_n(t)$ be random variables which represent the number of molecules of species $A_1, A_2, A_3, ..., A_{n-1}, A_n$ and $B_1, B_2, B_3, ..., B_{n-1}, B_n$, respectively, present at time t. We let $a_1, a_2, a_3, ..., a_{n-1}, a_n$ and $b_1, b_2, b_3, ..., b_n$ be the respective (integer) values which these random variables can achieve. The possible states of the system at time t which could lead to the state specified by $a_1, a_2, a_3, ..., a_{n-1}, a_n$ and $b_1, b_2, b_3, ..., b_{n-1}, b_n$ at time $t + \Delta t$ involving not more than one molecular transformation in the interval Δt are

$$\{a_1 + 1, a_2 + 1, a_3 + 1, ..., a_{n-1} + 1, a_n + 1, b_1 - 1, b_2 - 1, b_3 - 1, ..., b_{n-1} - 1, b_n - 1\},$$

$$\{a_1 - 1, a_2 - 1, a_3 - 1, ..., a_{n-1} - 1, a_n - 1, b_1 + 1, b_2 + 1, b_3 + 1, ..., b_{n-1} + 1, b_n + 1\},$$

and

$$\{a_1, a_2, a_3, ..., a_{n-1}, a_n, b_1, b_2, b_3, ..., b_{n-1}, b_n\}$$

(a) Write an expression $P(a_1, a_2, a_3, ..., a_{n-1}, a_n; b_1, b_2, b_3, ..., b_{n-1}, b_n; t + \Delta t)$ for probability vs. time during the course of this reaction. At time t, $A_1(t) = a_1$, $A_2(t) = a_2$, ..., $A_n(t) = a_n$, $B_1(t) = b_1$, $B_2(t) = b_2$, ..., $B_n(t) = b_n$.

(b) Since the system is conserved, the random variables $a_1, a_2, a_3, ..., a_{n-1}, a_n$ and $b_1, b_2, b_3, ..., b_{n-1}, b_n$ can be related through the initial concentrations which are taken to be $A_1(0) = \alpha_1$, $A_1(0) = \alpha_2$, ..., $A_n(0) = \alpha_n$, $B_1(0) = \beta_1$, $B_2(0) = b_2$, ..., $B_n(0) = \beta_n$. Clearly $\alpha_1 - a_1 = \alpha_2 - a_2 = \cdots = \alpha_n - a_n = b_1 - \beta_1 = b_2 - \beta_2 = \cdots = b_n - \beta_n$. Write an expression for probability vs. time during the course of the reaction for a single random variable $A_1(t)$.

2.14 In the reaction

$$X + Y \xrightarrow{k_1} 2Y$$

$$2Y \xrightarrow{k_2} Z$$

there are two steady-state solutions [M. Malek-Mansour and G. Nicolis, *J. Stat. Phys.* **13**, 197 (1975)].

(a) Using the steady-state approximation, find these two solutions.

(b) Using the stochastic method, set up the master equation for this problem.

(c) Simulate the reaction using the numerical stochastic method described in Appendix 2-3. Can you obtain the two steady-state solutions?

(d) What does the simulation imply about the physical significance of one of the steady-state solutions which you found?

2.15 A somewhat different problem from the fully reversible cyclic mechanism in Problem 2.7 is the following system of reactions*:

$$d[A]/dt = -k_1[A] - k_3[A]$$

$$d[B]/dt = k_1[A] - k_2[B]$$

$$d[C]/dt = k_2[B] + k_3[A]$$

---

*We are indebted to Prof. D. E. Hansen, Department of Chemistry, Amherst College, for bringing this interesting problem to our attention. In the system he studied, A is N-(phenylacetyl)glycyl-D-valine, B is glycyl-D-valine, and C is D-valine.

which can be represented as

Using Laplace transforms and/or a symbolic manipulation program such as MACSYMA or MATHEMATICA, find explicit expressions for $[A]_t$, $[B]_t$, and $[C]_t$ in terms of an initial concentration $[A]_0 = A_0$. Assume $[B]_0 = [C]_0 = 0$.

CHAPTER 3

Kinetic Measurements

3.1 INTRODUCTION

Experiments in macroscopic chemical kinetics have two broadly defined goals. The first aim is to characterize and confirm details of the mechanism involved in complex chemical processes. In these studies, the experimenter attempts to measure as many properties as possible of the reaction, such as the dependence of the reaction rate on concentration and temperature, the identity of all species present in the system and their concentrations and time dependences, or the extent of progress of the reaction. Such measurements provide the criteria for testing the proposed model of the reaction mechanism. With the experimental data obtained, the validity of the proposed mechanism can be tested by transforming the postulated mechanism into a mathematical model from which predictions can be made to see whether the model is reasonable and consistent with prior and new experimental data.

The next stage involves fitting the proposed mechanism to the experimental data. Once new rate parameters are obtained, the adequacy of the mechanism can be tested by investigating the deviations of the predictions from the experimental results. One procedure is to measure parameters such as rate constants at a variety of initial concentrations. If the model is correct, the values of the rate constants will be invariant under such changes. If the model is inadequate, a new mechanism must be proposed which is consistent with new and existing experimental data. Changes are usually made in species that are involved in the mechanisms, for which evidence is provided by the experimental detection and identification of such species in the reacting system. Then the cycle is repeated until the model explains the set of experimental data within the limits of experimental error. (But, as noted in chapter 2, sec. 2.3.2, no mechanism can be proven to be correct; it can only be disproved.)

The basic strategy for carrying out the preceding process is shown in Figure 3-1 and is the essence of applied macroscopic chemical kinetics.

A second goal of experiments in macroscopic chemical kinetics is to study the kinetics of individual elementary reactions. These studies provide rate data for elementary steps that are needed in assessing complex reaction mechanisms as well as examine the relation between the measured macroscopic rate constant and details of the

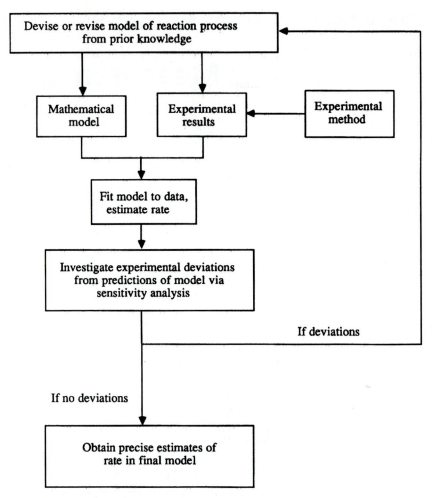

FIGURE 3-1 The basic approach to applied macroscopic chemical kinetics.

molecular dynamics (i.e., the microscopic kinetics), allowing us to test various microscopic rate theories. Thus, the experimental design is one that allows a single reaction step to be studied in complex reaction mechanisms. Such studies are now possible largely because of the advances that have been made in direct time-resolved detection of short-lived intermediates by spectroscopic means.

It is our aim in this chapter to introduce the basic concepts of kinetic measurement techniques and show how information on rate constants may be determined from such measurements. In the last portion of the chapter, we consider how a postulated mechanism is fitted to a set of experimental data. In discussing the evaluation of rate constants for elementary steps by fitting observations on complex systems, one cannot overemphasize the importance of sensitivity analysis, because knowing the experimental errors allows a realistic appraisal of the data. This information is needed not only by experimentalists to assess the sensitivities of calculations from models, e.g., in atmospheric chemistry, combustion, or pollution studies, but also by theorists who need to know the reliability of results from such measurements in guiding and assessing various microscopic theories of reaction kinetics. Many of the techniques discussed in this chapter can be used for either gas phase or solution kinetics studies. However, the major emphasis in the chapter is on gas phase kinetics. Techniques specific to solution kinetics are described in chapter 4.

3.2   TECHNIQUES FOR KINETIC MEASUREMENTS

3.2.1   General Features

Experimental methods in chemical kinetics are characterized by three features: a method of initiation, a suitable medium in which a reaction may take place, and a method of detection. Let us first consider some methods which are used to initiate chemical reactions.

3.2.1.1 Initiation methods.   Before a reaction can take place, the reactant molecules must be activated in some way to a minimum total energy necessary to induce the reaction. Various methods can be used to initiate reactions, including thermal activation (heating), electrical discharge, chemical activation, and photoactivation. Perhaps the most conventional means of starting a reaction is by thermal activation. Here, a reactant molecule is activated by collision with another molecule or with the walls of a reaction vessel heated by a furnace. In the collision, energy is transferred to the reactant molecule and distributed internally throughout it. (We shall go into further details of this process in subsequent chapters.) If the energy content of the activated molecule is above its minimum energy for reaction, reaction may occur. Typically the system is characterized by a single temperature, which for most experiments ranges from 300 to 3,000 K.

Electrical discharges are also effective in initiating reactions. The two types most often used in kinetics studies are microwave ($\approx$450 MHz) and radio-frequency ($\approx$20 MHz) discharges. In both cases, free electrons which are present in the gaseous reaction mixture are accelerated to high kinetic energies by the rf or microwave fields. These electrons in turn produce additional charged species by electron-impact ionization, as well as by molecular fragmentation and electronic excitation. Because the light electrons are inefficient at transferring momentum to heavier molecular particles, the kinetic temperature of the gas remains relatively cool. Unlike thermal excitation, which tends to break only the weakest bond in the molecule, electrical discharges cause extensive fragmentation and produce atoms, radicals, and ions. Consequently, such discharges are not clean sources of free radicals, but are an excellent source of atoms.

Photoactivation sources usually initiate a reaction by photodissociating the reactant molecule following its absorption of light. The typical wavelengths used to initiate photochemical reaction range from 120 nm to 700 nm. If the light is of low intensity, then usually one photon of light will be absorbed by the molecule, in accordance with Einstein's law of photochemical equivalence. If the light is ultraviolet, say, 300 nm in wavelength, a photon will have an energy of about 400 kJ/mole. [This figure is obtained by converting the wavelength λ (nm) to energy units E(kJ/mol) by the conversion expression $E = 1.1962 \times 10^5/\lambda$ kJ/mol.] With this amount of energy, the molecule may be promoted to electronically excited states and dissociate. One conventional light source that can provide these energies is the mercury lamp, which is basically a high-intensity lamp with output up to 1,000 watts covering a broad range of wavelengths that extend from the ultraviolet to the visible region. In order to do accurate and quantitative kinetic studies using photoactivation, it is necessary to use monochromatic light. Unfortunately, atomic lamp sources frequently have a broad wavelength distribution, and the output is not tunable over the range of their operation. Some degree of selectivity in wavelength can be obtained by using a monochromator, but this reduces the available intensity. These limitations may be overcome through the use of lasers, which have the unique advantage of being monochromatic, tunable, and highly intense. The energy available in laser pulses, with pulse widths that can range from femtoseconds to milliseconds, makes the laser an extremely useful initiation source, since species can be produced for kinetic study over a broad range of time scales. This expands the scope of kinetic problems that can be studied in both solution and state-to-state chemistry. Furthermore, with high-intensity laser sources it is possible to induce dissociation by absorption of multiple photons rather than just one photon. One difficulty with inducing dissociation by absorption of a single ultraviolet photon is that the process can leave the resulting photofragments highly excited. Dissociation by multiple infrared photons can overcome this problem. Details of the process are discussed in chapter 11, section 11.13. Lasers have become major tools for chemical kinetic studies and are sure to increase in importance as their frequency, pulse length, and power characteristics continue to improve.[1]

3.2.1.2 Kinetic systems (measurement media)

3.2.1.2.1 Static systems.   Chemical reactions are generally observed taking place in some container—a gas bulb, test tube, living cell, etc.—which holds very large numbers of each of the reactive species. Such a closed, constant-volume container, which serves as the measurement medium, constitutes a static system. For gas phase reaction, the reaction vessel, often cylindrically or spherically shaped and made of glass, is situated in a temperature-controlled furnace and connected to a vacuum system. After evacuation, the vessel is filled to some measured pressure with an appropriate reactant or mixture of reactants and possibly an inert buffer gas, and is then closed off from the system. If a mixture of reagents is used, the reaction vessel in the static system must be designed so that mixing can take place during admission of the reagents into the vessel. In a static system there is usually a finite time required for the reactant to fill the vessel, mix with the other reagents, and reach an equilibrated temperature. This time often limits the rates of some reactions that can be measured using conventional measurement techniques. For a cylindrical reaction vessel, the mixing time can be estimated[2] as

$$t = \frac{3\ell^2}{4\mathscr{D}} \tag{3-1}$$

where ℓ_t is the length of the vessel in cm and $\mathscr{D}$ is the diffusion coefficient in $(cm^2 \ sec^{-1})$. The diffusion coefficient depends on the total pressure in the reaction vessel according to

$$\mathscr{D} = \frac{3\bar{v}}{32n_i\sigma^2} \qquad (3\text{-}2)$$

where n_i is the total density of molecules (cm^{-3}), σ is the collision diameter for the molecular pair, and the mean relative velocity

$$\bar{v} = \left(\frac{8k_BT}{\pi\mu}\right)^{1/2} \qquad (3\text{-}3)$$

where, in turn, k_B is the Boltzmann constant $(1.38 \times 10^{-23}$ J K$^{-1})$, T is the absolute (Kelvin) temperature, and μ is the reduced mass of the collision pair. As a typical example, $\ell = 25$ cm, $T = 343$ K, $\sigma = 3 \times 10^{-8}$ cm^2, and $\bar{v} = 5 \times 10^4$ cm sec^{-1}, giving $\mathscr{D} = 3 \times 10^{-1}$ cm^2 sec^{-1} at a pressure of 5 Torr, so that the diffusion time $t \approx 15$ sec. Clearly, the reaction time to be measured must be longer than a few seconds, or serious errors in rate measurements will result.

A way to overcome this limitation of mixing time is to initiate the reaction by electric discharge or pulsed light and to observe the products immediately following initiation by time-resolved spectroscopic detection. It is then possible to study many reactions in premixed, static systems. Consequently, kinetic experiments which would otherwise be limited to intermittent measurements in static systems can be monitored continuously.

3.2.1.2.2 Flow systems.   Although laser initiation techniques have greatly expanded the range of reactant species and time scales that can be studied in static systems, there are a number of reactions that cannot be conveniently studied in this way. These include reactions of atoms and ions formed in a continuous discharge. For such reactions, flow systems are convenient measurement media. Unlike the closed, static systems discussed thus far, a flow system is an open system, in that reactants are constantly entering and products leaving the reaction zone. A typical discharge-flow apparatus is shown in Figure 3-2.

The analysis of a flow system is particularly simple if mixing is complete in a direction transverse to the flow but does not take place at all along the direction of flow. The conditions required for this to hold true will be discussed later in this section. In this case, the net change in the number N of species i in some volume dV a distance z down the flow is given by

$$dN_1 = r_idV - udc_i$$

On the right, the first term represents the chemical rate of change r_i taking place in volume element $dV = Adz$, where A is the cross-sectional area of the flow tube, while the second term represents the transport of species i through volume element dV with volume flow rate u. At steady-state flow conditions $dN_i = 0$ at a fixed z; therefore,

$$r_idV = udc_i$$

Integration gives

$$\frac{1}{u} = \int dV = \frac{A}{u}\int dz = \frac{A}{u}z = \int_{c(0)}^{c(z)} \frac{dc_i}{r_i} \qquad (3\text{-}4)$$

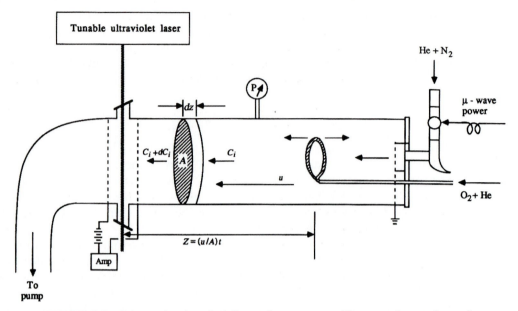

FIGURE 3-2   Schematic of typical flow tube apparatus. The experiment shown is the measurement of the rate of the $N + O_2 \rightarrow NO + O$ reaction. N is produced by microwave discharge dissociation of N_2 and is mixed with O_2 via the movable "shower-head" injector. The detection system illustrated works through multi-photon ionization of the product NO molecules (see section 3.2.2.5 and chapter 9) [From I. C. Winkler, R. A. Stachnik, J. I. Steinfeld, and S. M. Miller, *J. Chem. Phys. 85,* 890 (1986)].

For a simple first-order reaction $X \rightarrow$ products with rate constant k_1,

$$r_x = \frac{dc_x}{dt} = -k_1 c_x$$

Substituting this expression into equation 3-4 gives

$$\frac{Az}{u} = -\int_{c(0)}^{c(z)} \frac{dc_x}{k_{1c_x}} = -\frac{1}{k_1} \ln \frac{c(z)}{c(0)} \tag{3-5}$$

If the concentration c of species X is measured at intervals along the flow tube, and if $(u/A)\ln(c/c(0))$ is plotted against z, the slope of the plot will be $-1/k_1$. The same result would be obtained by measuring the concentration decay with time and plotting $\ln(c(t)/c(0))$ vs. t; therefore, the flow method is equivalent to replacing the reaction time t by the flow distance $z = (u/A)t$. For a second-order reaction we may simply use a large excess of one of the reactants to establish *pseudo-first-order* kinetics, as discussed in chapter 2, sec. 2.2.2.

The principal advantage of the flow system is that processes that proceed at a moderately fast rate can be measured. For example, linear flow velocities of several thousand cm sec^{-1} can be achieved with Roots blowers or other high-speed pumps, and, if steady flow conditions are maintained, so that the concentration-distance profile is stable, a series of measurements may be made along the flow tube with sufficient time to afford a good signal-to-noise ratio. Of course, many of the detection techniques employed in flow-tube measurement, such as mass spectrometric sampling, laser-induced fluorescence, and multiphoton ionization, are not simple to move along the tube; but in such cases the distance z is varied by moving the reactant injector along the flow direction.

The principal requirement for successful flow measurements is nonturbulent flow, the condition for which is that the Reynolds number

$$\mathrm{Re} = \frac{vd\rho}{\eta} \tag{3-6}$$

be less than 2,300. In equation (3-6) v is the linear flow rate u/A, d is the tube diameter, ρ is the gas density, and η is the gas specific viscosity. In typical experiments a 100-scfm (160 m³/h) pump is used, giving $u \approx 5 \times 10^4$ cm³ sec⁻¹; in a tube of 10 cm² cross section, $v = u/A \approx 5 \times 10^3$ cm sec⁻¹. The tube diameter $d = (4A/\pi)^{1/2} = 3.6$ cm, the density of atmospheric gases at $p = 1$ torr is about 2×10^{-6} g cm⁻³, and $\eta(\mathrm{air}) \approx 200$ μpoise $= 2 \times 10^{-4}$ dyne sec cm⁻². Inserting these values in equation 3-6 gives Re ≈ 180, safely under 2,300. Since measurements of $\ln(c/c(0))$ over one or two decades can be made over a distance of 1 to 10 cm, this means that rate constants on the order of 10^3 sec⁻¹ can be measured using flow techniques. Faster reactions can be measured by using larger pumps to get faster flow speeds and also lower partial pressures (for bimolecular reactions); but the limitation then becomes the mixing time following injection of the reagent into the flow, which may be several msec. Other factors which limit flow system measurements, such as concentration gradients and nonlaminar flow corrections, are discussed in detail in several excellent treatments[3]. Flow systems have also proven very useful in studies of ion-molecule reactions over a wide temperature range.[4] A turbulent flow technique, which circumvents many of these difficulties, and permits measurements at higher pressures, has recently been developed by Molina and coworkers.[5]

3.2.1.2.3 Shock tubes.   Another method of initiating reactions which is especially useful in attaining high temperatures is the shock tube method. In this method, the reactants are pre-mixed in a chamber which is coupled to a second chamber containing a high pressure inert gas (N_2 or Ar) by a thin metal diaphragm. When the diaphragm is burst (by failure at a specified pressure or a small explosive charge), the high pressure "driver" gas creates a shock wave in the sample which in turn results in a transient high temperature of several thousand K. The reaction may be followed by the optical or mass spectrometric techniques described below. The fluid dynamics of shock tubes are complicated, but the method has found considerable use for reactions which are otherwise difficult to initiate. The shock tube technique and some recent applications are described in a review by Tsang and Lifshitz[6] and the references cited therein.

3.2.2   Description of Techniques

As noted earlier, the typical apparatus for kinetic studies includes a suitable initiation source, measurement medium, and detection method. Since the essence of kinetic studies is measuring changes in composition with time, the detection method used must be sensitive to the magnitude of the expected concentration changes on the time scale at which they are changing. Analytical techniques used in classical kinetic studies to make such measurements include pressure measurements, nuclear magnetic and electron paramagnetic resonance spectroscopy, ultraviolet and visible spectroscopy, gas titration, chemical trapping, ultrasonic measurement, polarography, gas chromatography, and mass spectroscopy. Many of these techniques are incapable of monitoring reactions that occur on fast time scales. A general requirement for any technique is that the analysis time must be short compared to the time during which the reaction occurs. There are several ways of achieving this requirement, one of which is simply to slow the reaction down. Since the reaction rate can be expressed as $k = A \exp(-E_{act}/RT)$ (see chapter 1, sec. 1.4), lowering the temperature will reduce the reaction rate to a point at which the reaction can be monitored. For example, a reaction with a temperature-independent activation energy of 84 kJ/mol will be about 10^7 times slower at 157 K than at room temperature. This method has found good use in low-temperature matrix isolation studies, but it is hardly an adequate solution, of course, for very fast reactions.

A limitation of any detection system is electronic processing of the signal; the best state-of-the-art devices have 10-GHz bandwidths (about 0.1 nsec response time). The most important technological advance in detection is the laser, for many of the same reasons that make it a superb device for initiating reactions (see section 3.2.1.1). Its narrow-bandwidth, tunable output, together with its short pulse duration and high peak power, make it possible to monitor specific species on short time scales. For example, mode-locked laser systems are capable of producing femtosecond (10^{-15} sec) pulses, which allow details of molecular motion in condensed phases to be monitored in real time (see chapter 10, section 10.9).

Figure 3-3 indicates the time scales which can be accessed by a number of fast-reaction techniques. In the sections that follow, we introduce several of these techniques, with an emphasis on methods that employ lasers for detection. An additional set of techniques, namely, relaxation methods, is discussed in chapter 4, section 7.

3.2.2.1 Flash photolysis and laser flash photolysis.

One of the early techniques which overcame many of the limitations of classical techniques is flash photolysis. Originally developed by Norrish and Porter in the 1950s,[7,8] flash photolysis has established itself as a powerful technique for directly studying the rates of elementary reactions both in the gas phase and in solution. In flash photolysis, a reactant is irradiated with an intense flash of ultraviolet or visible light. The intensity of the light is sufficiently high to initiate a chemical reaction in the mixture, and the duration of the flash is short compared with the time scale of the reaction to be studied. This allows the unstable intermediates that are generated to be followed in real time using a suitable monitoring technique, such as absorption spectroscopy. Consequently, the concentration of the intermediates can be measured as a function of time, and their physical and chemical properties can be determined.

The typical flash photolysis apparatus is shown in Figure 3-4. The reaction vessel consists of a quartz cylinder up to 1 m in length with plane windows at each end. One or more photoflash tubes are situated parallel to the cell, and the complete assembly is surrounded by suitable reflectors, usually coated with magnesium oxide.

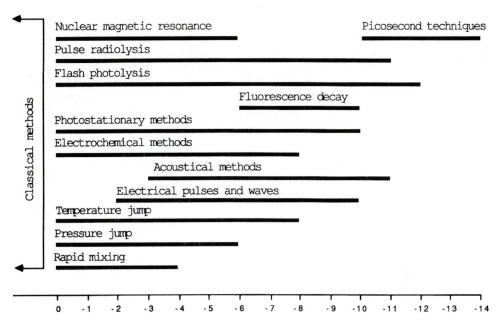

FIGURE 3-3   Summary of various experimental techniques and their time ranges [adapted from G. G. Hammes, *Principles of Chemical Kinetics* (New York: Academic Press, 1978)].

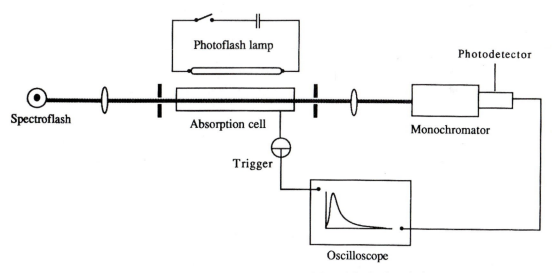

FIGURE 3-4   Schematic diagram illustrating principles of flash photolysis.

The photoflash lamp consists of a long quartz tube fitted at each end with metal electrodes, usually made of magnesium. The tube contains an inert gas at low pressure. The light flash is generated by discharging a large capacitor, previously charged to a high potential, across the two electrodes. If the inductance of the circuit is kept low and the conductivity of the leads is high, the duration of the flash (the fall time to half the maximum intensity) is about 20 μs. In general, the duration depends on the energy dissipated and varies from nanoseconds for microjoule flashes to milliseconds for flashes of several hundred joules. An empirical relation that exists between the duration of the flash and the dissipated electrical energy is that a tenfold reduction in duration requires a hundredfold reduction in energy. Since scattered light from the flash interferes with spectroscopic measurements, the time resolution of the technique is limited by the duration of the flash. Although the technique is necessarily restricted to molecules with suitable absorption characteristics, the light output is essentially a continuum down to the quartz cutoff, so that the range of reactions which can be investigated is clearly very large.

Ensuing reactions must be followed in real time. Ideally, one would like to scan the absorption spectrum repeatedly after the flash, but this is ruled out by signal-to-noise problems. The usual procedure is either to monitor a narrow spectral band continuously or to obtain a complete spectrum at a predetermined interval after the flash. The latter has proved the more profitable, at least until the reaction mechanism has been fully analyzed. The experiment is repeated with different time delays until the complete pattern of events has been resolved.

The absorption spectrum is obtained by utilizing a second flash lamp situated at one end of the reaction vessel. The spectroflash energy and duration are chosen such that the detector situated at the opposite end of the reaction vessel receives light that is adequate for detection but insufficient to interfere with ongoing photochemistry.

Once the transient spectrum has been established, kinetic data may be obtained by measuring time-resolved absorption at a fixed wavelength. A typical flash photolysis kinetic curve is shown in Figure 3-5. For an atomic recombination reaction such as $2A + M \rightarrow A_2 + M$, the instantaneous concentration of atoms can be found[9] from such a curve to be

$$[A]_t = 2[A_2]_\infty \frac{\ln(x_t/x_\infty)}{\ln(x_0/x_\infty)}$$

Using $x_t = x_\infty + \Delta x$, $\ln(x_t/x_\infty) = \ln(1 + \Delta x/x_\infty)$. For optically thin samples, this is approximately equal to $\Delta x/x_\infty$, so that

$$[A_t] = \gamma \Delta x$$

where

$$\gamma = \frac{2[A_2]_\infty}{x_\infty \ln(x_0/x_\infty)}$$

One difficulty with the flash photolysis technique has been the inability to generate a concentration of transient species sufficiently high to permit the recording of an absorption spectrum from conventional flash lamps of short duration and low energy. However, pulsed lasers provide a source of excitation which is not subject to the energy-duration characteristics of flash lamps. The peak power of these lasers can be enormous (10^9 to 10^{12} watts), and the emitted light is coherent, so the beam can be focused on very small areas. The beam can also be monochromatic and of very short

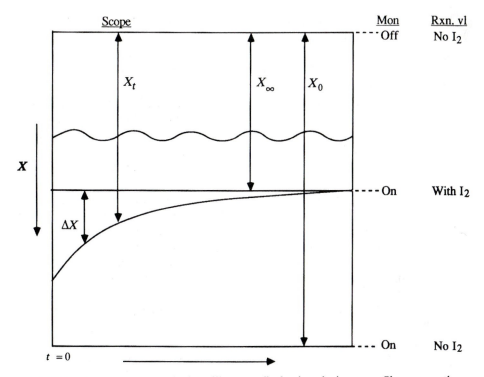

FIGURE 3-5 Plot of a typical oscilloscope flash photolysis curve. Shown are the scope deflections x_0, x_∞, x_t, and $\Delta x = x_t - x_\infty$. Mon = monitor light, Rxn.vl = reaction vessel. The flash was fired at $t = 0$. [Adapted from Yamanashi and Nowak, *J. Chem. Ed. 45,* 705 (1968).]

(nanosecond to femtosecond) pulse duration. The first laser flash-photolysis measurement was carried out by Porter and Steinfeld in 1966.[10] Further details may be found in reference 1 and chapter 9 of this text.

Flash photolysis and its laser variant are extremely versatile methods. They have been used to study a variety of reactions: radical-radical recombination, unimolecular decay, and radical-molecule, atom-atom, and atom-molecule bimolecular reactions. They also have proved particularly useful in the identification of transient species such as radicals, atoms, and molecules in excited states. A similar technique, which uses transient infrared absorption spectroscopy, is described in the next section.

3.2.2.2 Absorption spectroscopy.   Absorption spectroscopy encompasses all those techniques that measure a change in the intensity of some light source due to absorption that occurs as the light passes through a gas along some specified path length. The absorption measurement approach is in general quite versatile, and the technique can use light sources which range from broadband to monochromatic laser; in principle, it can cover the entire electromagnetic spectral region. Absorption spectroscopy is used in kinetic measurements as a means of quantitatively detecting species whose reaction rate is of interest. Consequently, a requirement for using this technique for kinetic studies is that there exists a unique spectral signature associated with the species of interest which allows it to be monitored and is such that the isolated absorption line or feature must be free from interference from other species produced in the reaction.

The infrared region has traditionally been considered the "fingerprint" for polyatomic molecules, since most such molecules have vibrational-rotational bands in that region. A technique which is very versatile as well as highly species-specific is the time-resolved diode laser absorption technique, abbreviated TRDLA. Because of its great specificity and selectivity, this technique is used to study gas phase reactions involving transient species.[11–13]

The selectivity of TRDLA is based on the fact that diode lasers have a very high spectral resolution, usually 10^{-3} cm^{-1}, so that the real spectral resolution for molecules in the gas phase can easily achieve the Doppler limit and the individual vibrational-rotational absorption lines of small molecules with fewer than ten atoms can be resolved. Because of the high spectral resolution a particular species of interest can be easily identified, even in a very complicated mixture, provided that the spectroscopy of the species is well established. The high sensitivity of TRDLA is related to its high selectivity: since individual vibrational-rotational lines are usually resolved, the background of the signal is the intensity profile of the diode laser beam instead of contours of unresolved clusters of lines as in low-resolution spectroscopy. The sensitivity of the diode laser absorption can be characterized[14] by the minimum detectable concentration $n_{\min}$. For a path length of 100 cm,

$$n_{\min} = \frac{6.3 \times 10^{14} \gamma_D}{s(n_j/n)} \text{ molecule cm}^{-3} \tag{3-7}$$

where s is the absorption band strength in cm^{-2} atm^{-1}, γ_D is the Doppler half-width at half maximum, and n_j/n is the fraction of molecules in a particular vibrational-rotational state. For most small molecules in gas phase, $n_{\min} \approx 10^{12}$ molecule cm^{-3}, which corresponds to a partial pressure of about 10^{-3} torr at room temperature.

The principle of TRDLA is that the absorbance at the diode laser frequency is approximately proportional to the concentration of the absorbing species. The measured absorption signal is therefore a direct measure of the species of interest, and the evolution with time of its concentration can be monitored in real time. Because of the direct detection of the species under investigation, kinetic quantities such as reaction rate constants are likely to be more reliable than those obtained with indirect methods such as gas chromatography or pressure measurements, and the errors involved in the experiment can be more easily analyzed and estimated.

This is not to say that there are no limitations to the TRDLA technique. First, the technique is limited to species which have reasonable absorption band strengths within the operating region of the diode laser. Second, the spectral appearance may be very complicated due to the high resolution of the diode laser. Therefore, the spectroscopy of the species to be monitored must be well known for positive identification of that species. Furthermore, calibrations in the infrared region are inherently more difficult than in the visible region. This is because in a frequency region where calibration lines are sparse, one may have to transfer a calibration frequency from a known line several wavenumbers away from the wanted line, and the practice will introduce errors in frequency measurements of the line. In that case it will be very difficult to identify the wanted line from a group of closely spaced lines of different species. Finally, even though the sensitivity of the diode laser absorption is high, it is in some cases not high enough to be used in kinetic measurements if the absorption of the species is too small.

In principle, for kinetic measurements, any reaction involving a transient species which absorbs infrared laser radiation can be investigated by TRDLA. To illustrate how the technique is used in chemical kinetic measurement, consider the rate constant for methyl (CH_3) recombination, i.e.,

$$CH_3 + CH_3 \xrightarrow{k_R} C_2H_6$$

A diagram of the basic components of the TRDLA system for these measurements is shown in Figure 3-6. First, a KrF laser is used to photolyze CH_3I to generate CH_3 radicals, the concentration of which is measured by the transient absorption of the diode laser output. The KrF laser and the diode laser beams are made colinear by a combination of lithium fluoride plates which serve as 90% reflectors to the diode laser while passing the KrF laser beam. Absorption of CH_3 radicals from the diode radiation is detected by a HgCdTe infrared detector, and the signal from this detector is recorded and averaged by a transient digitizer controlled by a computer. An example of the signal from methyl radical recombination is shown in Figure 3-7.

From chapter 2, the kinetic rate equation for the bimolecular recombination of methyl is

$$\frac{d(CH_3)}{dt} = -2k_R [CH_3]^2 \tag{3-8}$$

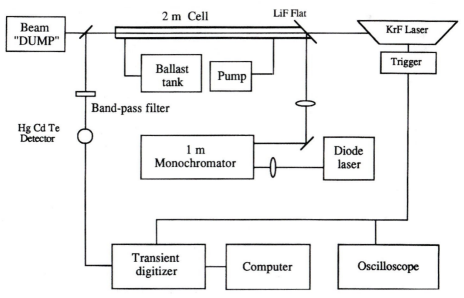

FIGURE 3-6 Schematic arrangement for detection of species using the time-resolved diode laser absorption technique.

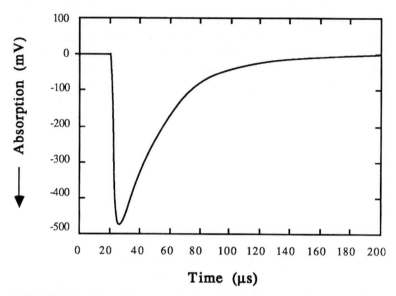

FIGURE 3-7   Absorption signal of methyl radicals as a function of the time after the photolysis pulse. The radicals are monitored by the time-resolved diode laser absorption technique. [Adapted from G. A. Laguna and S. L. Baughcum, *Chem. Phys. Lett. 88,* 568 (1982).]

which yields

$$\frac{[CH_3]_0}{[CH_3]_t} = 1 + 2k_R [CH_3]_0 (t - t_0) \tag{3-9}$$

A least-squares fit of the signal shown in Figure 3-6 to equation (3-9) is made to obtain the recombination rate constant k_R using the initial concentration of CH_3 calculated from the CH_3I pressure, photolysis laser power, and absorption of the gas. We shall examine this reaction in more detail in a later section.

3.2.2.3 Laser photolysis/chemiluminescence.

The chemiluminescence technique uses the emission from excited species as a means for detecting and following the kinetics of these species.[15] The emission from the excited species is one of the several paths by which excess energy may be lost by the molecule, the general phenomenon of light emission from an excited state being known as luminescence. Emission following excitation by chemical reaction is referred to as chemiluminescence. For example, the reaction between NO and O_3 yields the reaction product NO_2, some of which is generated in an excited electronic state ($NO + O_3 \rightarrow NO_2^*$) and is subsequently relaxed by photon emission. Detection of the NO_2^* provides a measure of the reaction rate of $NO + O_3$.

In the laser photolysis/chemiluminescence technique, the role of the laser is primarily that of initiating the reaction. For example, in studying the reaction of Cl atoms with hydrogen halides (HX), the laser is used to photolyze molecular chlorine (Cl_2) and yield $Cl(^2P_{3/2})$ atoms according to the reaction[16]

$$Cl_2 \xrightarrow{h\nu(347.1 \text{ nm})} 2Cl(^2P_{3/2})$$

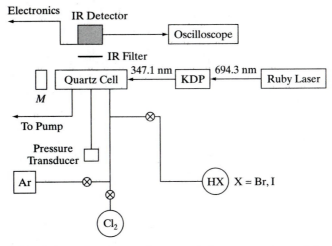

FIGURE 3-8 Schematic diagram of the apparatus used for measurement of the Cl + HX rate constant [adapted from F. J. Wodarczyk and C. B. Moore, *Chem. Phys. Lett. 26*, 484 (1974).]

which is followed by the reactions

$$Cl + HX \xrightarrow{\ k_1\ } HCl^* + X \qquad (X = Br, I)$$

and

$$HCl^* + M \xrightarrow{\ k_2\ } HCl + M$$

The rate of the reaction under investigation is followed by monitoring the time-resolved infrared chemiluminescence of the excited product HCl molecules (see Figure 3-8). If the reaction conditions are such that $[HX] \gg [Cl]_0$, where $[Cl]_0$ is the initial concentration of chlorine atoms produced by the photolyzing laser (pseudo-first-order conditions), then the chemiluminescence intensity signal is related to the rate by the expression

$$I \propto \exp(-k_2[M]t) - \exp(-k_1[HX]_0 t) \tag{3-10}$$

By fitting the signal to a functional form

$$S(t) = A[\exp(-t/t_d) - \exp(-t/t_r)] \tag{3-11}$$

where A is a constant and t_d and t_r are the decay and rise times, respectively, we can obtain the rise time, which is related to the rate constant of interest by

$$\frac{1}{t_r} = k_1[HX]_0 \tag{3-12}$$

Thus, we need know only the initial concentration of HX and the rise time in order to determine rate constants using this technique.

The detection limit of the chemiluminescence technique is determined by the background against which the measurements are made. Nevertheless, the sensitivity of the technique is about 10^{12} molecules cm^{-3}. For greater sensitivity, the fluorescence technique (see next section) may be used.

3.2.2.4 Laser-induced fluorescence.   The basic laser-induced fluorescence (LIF) approach typically employs one or more pulsed lasers to selectively produce an excited state of the species being monitored.[17,18] This is done by varying the laser frequency until the laser output overlaps an absorption line of the molecule. If the lifetime of the excited state is short compared to the collisional deactivation process, the excited state will decay to a more stable state via fluorescence. This fluorescence generally occurs at ultraviolet or visible wavelengths and can be detected using high-efficiency photomultiplier tubes and photon-counting techniques. To further enhance the signal-to-noise ratio of the detection process, gated photon-counting synchronized to the pulsed laser is generally employed. The basic experimental arrangement is shown in Figure 3-9. In order to use this technique for kinetic measurements, three major requirements must be fulfilled. First of all, the molecule must fluoresce. Although this requirement is well satisfied in general for small molecules, particularly

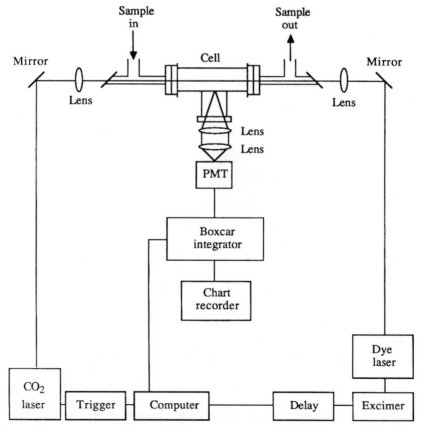

FIGURE 3-9   Schematic diagram of a typical laser photolysis using a CO$_2$ laser followed by laser-induced fluorescence detection of the fragments produced.

diatomic molecules, there are many polyatomic molecules whose excited states are known to undergo radiationless transitions instead of fluorescence. The second requirement is that the band system must be well characterized spectroscopically. The third requirement is that the molecular band system should be accessible with available lasers or frequency-multiplication techniques.

Despite its limitation to molecules which satisfy these requirements, LIF does have an advantage over typical absorption techniques such as time-resolved diode laser absorption spectroscopy. In the absorption technique, the sensitivity of the technique is limited by the smallest $\Delta I/I_0$ that can be measured. In LIF, the fluorescence is measured against a zero background, and the intensity of the fluorescence is proportional to the intensity of the irradiating light. While single atoms or molecules can be detected by LIF, which makes it far more sensitive than the TRDLA technique, in practice this limit is rarely reached because of scattered light and saturation problems. For practical purposes, LIF can detect species at about 10^{10} particles/cm^3. However, with great effort, OH radicals, for example, have been detected at concentration levels of 10^6 molecules/cm^3. With these sensitivity limits, a variety of reactions can be studied with the LIF method: atom-molecule, radical-molecule, and unimolecular. Zare and co-workers[19,20] have studied the family of reactions between Ba atoms and hydrogen halides, i.e.,

$$Ba + HX(X = F, Cl, Br, I) \rightarrow BaX + H$$

using the LIF technique. Other species which can be detected by LIF include NO, CN, NO_2, CF_2, NO_3, and CH_3O. Since the intensity of the fluorescence is proportional to the population of the state or states from which fluorescence occurs,[21] by measuring the intensity of the fluorescence during the reaction and plotting intensity versus time, a rate constant can be obtained.

3.2.2.5 Photoionization Techniques.   Photoionization spectroscopy is a sensitive laser technique which is well suited for kinetic measurements. Its main advantage is that the ionization process is highly selective, so that one can ionize and identify transient or stable species at low concentrations in the presence of a large excess of other species. Furthermore, molecules that cannot be detected using LIF because fluorescence quantum yields are too low can be studied using photoionization spectroscopy. The basic concept of the technique utilizes one or more pulsed visible- or ultraviolet-light lasers focused between two biased ion collection plates housed inside a sample cell. If the photon energy absorbed by the molecule exceeds its ionization threshold, ions will be generated and collected by collection plates. The resulting signal is a measure of the number of molecules in the cell. Molecules can be ionized using the multiple photon ionization process (MPI). This is achieved by first exciting the molecule from its ground electronic state to an intermediate state, and then ionizing the molecule from the intermediate state by one photon excitation. The technique is referred to as the $n + 1$ MPI process, where n photons are used to reach the intermediate state and one additional photon is used to ionize that excited state.[22] The 2 + 1 MPI process is illustrated[23] for Br atoms in Figure 3-10. In general, the MPI process has high sensitivity. For example, detection limits of Br atoms using 2 + 1 MPI is reported to be about 10^{10} atoms/cm^3. Not only is the technique highly sensitive, but also it is highly

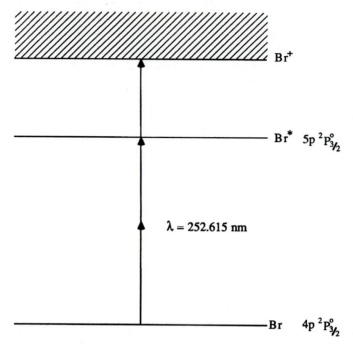

FIGURE 3-10 Multiple-photon ionization of bromine atoms. Two-photon excitation at 252.615 nm excites the $5p^2P^0_{3/2}$ $\leftarrow 4p^2P^0_{3/2}$ transition, and an additional photon of the same wavelength ionizes the atom.

specific. The net result of the absorption of $n + 1$ photons by the molecule results in the selective production of the ionized molecule in a complex product mixture with the advantage of not ionizing interfering compounds of the reaction mixture. The sensitivity and selectivity of the method make MPI well suited for kinetic studies.

Coupling photoionization spectroscopy with mass spectrometry produces a powerful technique which allows one to identify molecules not only by their mass but also by their characteristic optical ionization spectrum. A typical experimental setup is shown in Figure 3-11. Since the ions produced in the photoionization process are generated in a very short time interval, usually a few nanoseconds during the laser pulse, and in a small volume of about 10^{-5} cm^3 (the focal volume of the laser), it is most convenient to use time-of-flight mass spectroscopic detection. Slagle et al.[24] provide a good illustration of the use of the photoionization mass spectroscopic technique for kinetic rate constant measurement for the reaction of oxygen atoms with ethylene, i.e.,

$$O + C_2H_4 \xrightarrow{k_1} CH_3 + HCO$$

The complete mechanism includes the additional reactions

$$O + CH_3 \xrightarrow{k_2} H_2CO + H$$
$$C_2H_4 + CH_3 \xrightarrow{k_3} C_3H_7$$
$$CH_3 + O_2 \xrightarrow{k_4} CH_3O_2$$

(3-13)

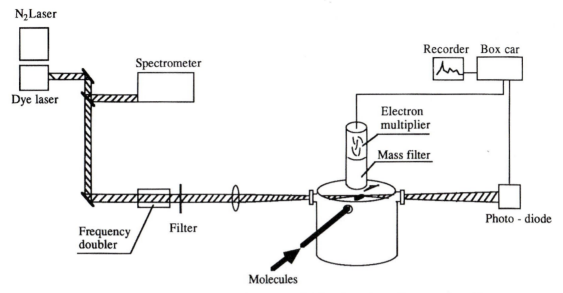

FIGURE 3-11   Experimental setup for mass-selective detection of ions produced in a multiple-photon ionization process.

and

$$2CH_3 \xrightarrow{\;k_5\;} C_2H_6$$

The desired conditions of the model are achieved by mixing O atoms and C_2H_4 in a flow system in which the concentrations of the reactants are held constant to ensure a constant rate of CH_3 production. Under these conditions, $[CH_3]$ remains constant and equal to the steady-state concentration $[CH_3]_{ss}$. The second and third reactions in the mechanism given by equations (3-13) are the only significant reactions consuming CH_3. Since the CH_3 concentration is proportional to the CH_3 ion signal, k_1 can be determined by fitting the half-life of the CH_3 ion-count growth to its steady-state value, as shown in Figure 3-12. Additional aspects of the photoionization technique are discussed in references 25 and 26 and in chapter 9.

3.3   TREATMENT OF KINETIC DATA

3.3.1   General Analysis of Kinetic Data

3.3.1.1 Introduction.   Two questions usually arise when rate data are obtained experimentally. The first deals with the inherent errors in the data as contributed by instrumental factors and experimental procedures; the second pertains to how these errors determine the uncertainty in the fitted rate constants. Correct estimates of kinetic parameters and their uncertainties are fundamentally important for chemical kinetics, because such data are needed for assessing kinetic models and microscopic rate theories in chemical kinetics. These two questions are also important in assessing

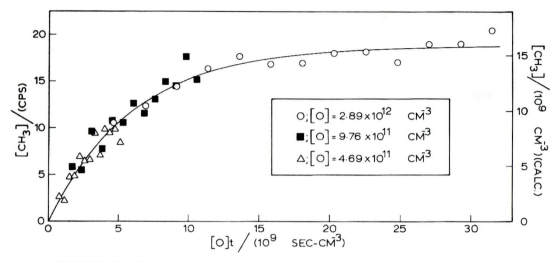

FIGURE 3-12   Plot of CH_3 ion signal versus [O], in a kinetic experiment employing multi-photon ionization with mass spectrometric detection [From I. R. Slagle, F. J. Pruss, and D. Gutman, *Int. J. Chem. Kinet. 6*, 111 (1974)].

just how well correlated a postulated reaction mechanism is to the raw data. Since errors are inherent in any measurement and correlations are never exact, a method for judging whether a particular correlation is significant is needed to assess the model. This section is given over to considering briefly some of these points.

3.3.1.2 Types of measurement errors.   No measurement can be made perfectly and with zero error; thus, it becomes important to know the uncertainty of a measurement and how different errors that arise from the experiment or the instrument enter into it. Furthermore, once sources of errors are identified, ways of reducing them may be undertaken. It is the duty of every experimental kineticist to take every precaution to minimize errors and to report the accuracy of the final result. The accuracy of kinetic results involves estimating the systematic errors in each observation in the experimental apparatus or procedure. Although this step is difficult to quantify, it is possible to correct for the errors incurred. The precision of the result depends on random or statistical errors, which are fluctuations in repeated measurements and are usually beyond the control of the experimentalist. To estimate such errors, we treat the experimental data statistically using standard methods to obtain the standard deviation in the results from the mean value. Knowing these uncertainties in the experimental data permits the uncertainty in the measured rate constants to be estimated and also enables us to further understand how these sources of errors contribute to the overall uncertainty in the measured rate. In this section our purpose is not to provide the reader with an extensive review of statistical methods of treating data (indeed, this has already been done by others),[27-29] but rather to acquaint the would-be kineticist with the procedures for analyzing and reporting kinetic data together with appropriate error limits.

3.3.1.2.1 Systematic errors.   Systematic errors in kinetic measurements have been discussed in some detail by Cvetanovic et al.[30-33] In kinetic measurements there are four kinds of systematic errors: (1) instrumental (shortcomings of the instruments or the effects of the environment on the instrument or its user); (2) operational (the personal judgment of the kineticist, which enters into readings of the instrument, and errors in instrument calibration); (3) methodological (inaccuracies in the modeling equations, which are imperfect approximations to the true solutions); and (4) mechanistic (errors introduced by inaccurate representations of the underlying chemical mechanism or some unrecognized chemical interference from secondary reactions or impurities in reagents). There are several ways of searching for systematic errors in measured kinetic parameters:

1. *Comparison of measured kinetic parameters under a variety of experimental conditions.* Inconsistencies between measured rate coefficients greater than the cumulative random error indicate the presence of systematic errors. Caution must be exercised since apparently consistent results can conceal systematic errors.

2. *Comparison of the absolute value of rate parameters obtained by different techniques.* Here, the importance of determining the same kinetic parameters by vastly different techniques cannot be overemphasized.

3. Through *insight and guidance from theoretical and semi-empirical methods* such as those to be presented in later chapters.

In chemical kinetics, an inadequate mechanism is frequently the largest source of systematic errors. If a mechanistic complication is the cause of systematic error, its magnitude and sign may be assessed by computer simulation. In this case, the reaction or reactions that may be significant are included. Once upper and lower bounds of the systematic error are estimated, it is possible to evaluate their contribution to the overall uncertainty in the results of the measurements. When upper (e_s^u) and lower (e_s^L) bounds of the systematic error are given, a correction for the error is given by the mean,

$$\bar{e}_s = \frac{(e_s^u + e_s^L)}{2} \tag{3-14}$$

In this case the remaining error becomes

$$\pm (e_s)_{max} = \frac{\pm (e_s^u + e_s^L)}{2} \tag{3-15}$$

An excellent discussion of other ways of estimating systematic errors is given by Eisenhart.[34,35]

3.3.1.2.2 Random errors.   Random errors are usually due to unknown causes and may occur even when all systematic errors have been accounted for. They also contribute to the uncertainty of the measured value. Because of their randomness, there is no known method of controlling or correcting these errors. The only way to offset them is to increase the number of repetitive measurements and use statistical means to obtain the estimate of the true value of the parameter under measurement. For a large number of observations, the random errors have a Gaussian distribution;[27,36] that is, the probability of their occurrence is given by

$$f(\varepsilon) = \left(\frac{1}{\sigma\sqrt{2\pi}}\right) \exp\left(\frac{-(x - \mu)^2}{2\sigma^2}\right) \tag{3-16}$$

where ε is the magnitude of the random error given by $x - \mu$, and μ and σ are moments of the distribution. The quantity μ is the *mean* of the population, and σ is the *standard deviation* of the population—it measures true spread of the individual observation about the mean. The term $x - \mu$ represents the magnitude of the random error in an individual measurement.

3.3.1.2.3 Propagation of errors.   Rate constants of chemical reactions are calculated from experimental measurements of quantities such as gas pressure, intensity alteration, intensity of chemiluminescence or fluorescence, etc. Consequently, the uncertainty in the rate constant is dependent on the individual uncertainties in these measured parameters. For example, consider a resonance fluorescence experiment done in a flow system. The critical parameters affecting the results of measurement are (1) gas flow, (2) pressure, (3) temperature, (4) distance, and (5) light intensity. Table 3-1

TABLE 3-1   Random Error Sources and Magnitudes in Experimental Flow-Tube Measurements

Measurement (units)	Device/method used	Range used	Uncertainty range[a]
Gas flow, u	Volume displacement	0.05–250	0.01–15
(mliter (STP)sec^{-1})	Critical flow orifices	5–500	1.0–5.0
Pressure, P (torr)	Closed-end manometers	1–1000	0.1–5.0
	Bourdon gauges	1–50	0.1–2.5
	U-tube manometer	0.5–20	0.05–1.0
	Capacitance manometer	0.001-100	0.0005–1.0
Temperature, T (°C)	Chromel-alumel thermocouple	−80–200	0.1
	Pt vs. Pt-Rh thermocouple	300–2000	10–60
Distance, z (cm)	Centimeter scale	0–100	0.05
Light intensity			
I(mV)	Absorption-monochromator/PMT	10–100	0.5–2.0
F(mV)	Fluorescence-light filter/PMT	1–20	0.1–1.0
I_{CL}(pA)	Chemiluminescence-calibrated monochromator/PMT	0.1–500	0.02–50[b]

[a]Values correspond to uncertainties at extremes of range; units are the same as measurement units.
[b]Includes contribution of scatter in detection system calibration ($\approx\pm5\%$) used to obtain absolute intensities.
Source: A. Fontijn and W. Felder in *Reactive Intermediates in the Gas Phase: Generation and Monitoring,* edited by D. W. Setser (New York: Academic Press, 1979), pp. 59–151.

provides a list of typical uncertainty ranges for these parameters.[37] Uncertainties in these values together with scatter in the kinetic data ultimately determine the precision of reported results. Using the propagation-of-errors method,[27,28,36] the overall uncertainty in the measured parameter is given by

$$\sigma_k = \left[\sum_i \left(\frac{\partial f}{\partial \phi_i} \sigma_i \right)^2 \right]^{1/2} \tag{3-17}$$

where the σ_i are the uncertainties in the measurement parameters and the partial derivatives $(\delta f / \delta \phi_i)$ are measured at the sample mean of the measurement parameters.

Let us determine the precision of a simple first-order rate constant from a flow reaction system (see section 3.2.1.2.2). The specific rate constant k is given by

$$k = \frac{-v(\ln[A]_1 - \ln[A]_2)}{(z_1 - z_2)} \tag{3-18}$$

where $[A]_1$ and $[A]_2$ are the concentrations of A present at the distances z_1 and z_2, respectively, and $v = u/A$ is the linear gas flow velocity. If we assume that errors in the measured quantities are independent of each other, then we can use equation 3-17 to estimate the uncertainty Δk in k as follows:

$$(\Delta k)^2 = \left(\frac{\partial k}{\partial z_1}\right)^2 (\Delta z_1)^2 + \left(\frac{\partial k}{\partial z_2}\right)^2 (\Delta z_2)^2 + \left(\frac{\partial k}{\partial \ln[A]_1}\right)^2 + (\Delta \ln[A]_1)^2$$
$$+ \left(\frac{\partial k}{\partial \ln[A]_2}\right)^2 + (\Delta \ln[A]_2)^2 + \left(\frac{\partial k}{\partial v}\right)^2 (\Delta v)^2 \tag{3-19}$$

The relative error can be found by dividing through by k^2 and then multiplying and dividing each term by the square of the independent variable:

$$\left(\frac{\Delta k}{k}\right)^2 = \left(\frac{z_1}{k}\frac{\partial k}{\partial z_1}\right)^2 \left(\frac{\Delta z_1}{z}\right)^2 + \left(\frac{z_2}{k}\frac{\partial k}{\partial z_2}\right)^2 \left(\frac{\Delta z_2}{z_2}\right)^2 + \left(\frac{\ln[A]_1}{k}\frac{\partial k}{\partial \ln[A]_1}\right)^2 \left(\frac{\Delta\ln[A]_1}{\ln[A]_1}\right)^2$$
$$+ \left(\frac{\ln[A]_2}{k}\frac{\partial k}{\partial \ln[A]_2}\right)^2 \left(\frac{\Delta\ln[A]_2}{\ln[A]_2}\right)^2 + \left(\frac{v}{k}\frac{\partial k}{\partial v}\right)^2 \left(\frac{\Delta v}{v}\right)^2 \tag{3-20}$$

We use equation (3-18) to calculate the partial derivatives, and if we substitute the values in equation (3-20), we obtain

$$\left(\frac{\Delta k}{k}\right)^2 = \left(\frac{z_1}{z_1 - z_2}\right)^2 \left(\frac{\Delta z_1}{z}\right)^2 + \left(\frac{z_2}{z_1 - z_2}\right)^2 \left(\frac{\Delta z_2}{z_2}\right)^2 + \left(\frac{\ln[A]_1}{\ln[A]_1 - \ln[A]_2}\right)^2 \left(\frac{\Delta\ln[A]_1}{\ln[A]_1}\right)^2$$
$$+ \left(\frac{\ln[A]_2}{\ln[A]_1 - \ln[A]_2}\right)^2 \left(\frac{\Delta\ln[A]_2}{\ln[A]_2}\right)^2 + \left(\frac{\Delta v}{v}\right)^2 \tag{3-21}$$

To show how we use equation (3-21), suppose that for a particular reaction $z_1 - z_2 = 100$ cm, $[A]_2 = 0.80[A]_1$, the relative error in determining $[A]_1$ and $[A]_2$ is about $\pm 1\%$, the flow velocity is 400 cm/sec, its uncertainty is ± 5 cm/sec, and the uncertainty in the distance is ± 0.1 cm. Substituting these values into equation 3-21, we obtain, for the precision in k,

$$\frac{\Delta k}{k} = \pm \{(0.005)^2 + (0.001)^2 + (0.0448)^2 + (0.0448)^2 + (0.0125)^2\}^{0.5} \tag{3-22}$$

$$= \pm 0.0645$$

so the expected error is $\pm 6.5\%$.

3.3.1.3 Evaluation of rate constants

3.3.1.3.1 Least-squares analysis.   Once the uncertainties in each measurement are known, how does one determine the uncertainties in the rate constant? Here, we turn to the question of relating the mechanistic model expressed in some mathematical form to the actual experimental rate data. Since there are errors in the measurements and perhaps some inaccuracies in the model, it is often difficult to obtain an exact fit. Consequently, it becomes difficult to judge the appropriateness of the model to the experimental data by inspection. Assuming that the errors, excluding systematic errors, are random, a suitable technique typically used in kinetic data treatment is the least-squares method.[27,36,38] To estimate values of parameters in the model, we must minimize some measure of the errors. In the least-squares method the measure of error is defined as the sum of squares of the errors of the individual measurements. If we have some model to be fitted to the form

$$\mathbf{y} = f(\mathbf{p},t) \tag{3-23}$$

where $\mathbf{y}$ is the set of experimental data points and $\mathbf{p}$ are parameters we wish to find, the least-squares criterion gives

$$Q = \sum_{j=1}^{M} \omega_j [\, y_j - f(\mathbf{p},t_j)]^2 = \sum_{j=1}^{M} \omega_j \varepsilon_j^2 \tag{3-24}$$

In this equation $\varepsilon_j = y_j - f(\mathbf{p},t_j)$ represents the residual error given by the difference between the observed value y_j and the calculated value $f(\mathbf{p},t_j)$, which is then weighted by ω_j for each observation. It is the quantity Q that is minimized with respect to estimated parameters.

Two cases are associated with the least-squares procedure. In the linear least-squares case, the functional form of the values of $\mathbf{y}$ are presumed to follow some general linear equation of the form

$$y = p_1 + p_2 t \tag{3-25}$$

In the second case, the function y has a nonlinear form. For many cases in kinetics, the general linear functional form is more commonly used. If the y values have unequal weights, equation (3-24) can be rewritten as

$$Q = \sum_{j=1}^{M} \omega_j [y_j - (p_1 + p_2 t_j)]^2 \tag{3-26}$$

The derivatives of Q with respect to both p_1 and p_2 must be taken and equated to zero in order to find the optimal values. Taking the derivative of equation 3-26 with respect to p_1 yields

$$\frac{\partial Q}{\partial p_1} = -2 \sum_{j=1}^{M} \omega_j (y_j - p_1 - p_2 t_j) = 0 \tag{3-27}$$

so that

$$p_1 = \frac{\displaystyle\sum_{j=1}^{M} \omega_j y_j - p_2 \sum_{j=1}^{M} \omega_j t_j}{\displaystyle\sum_{j=1}^{M} \omega_j} \tag{3-28}$$

Thus, we now have an equation which relates the optimal value of p_1 to the optimal value of p_2 and the means of the y and t values. Taking the derivative of Q with respect to p_2 yields

$$\frac{\partial Q}{\partial p_2} = -2 \sum_{j=1}^{M} t_j \omega_j (y_j - p_1 - p_2 t_j) = 0 \tag{3-29}$$

Rewriting this expression, we obtain

$$\sum_{j=1}^{M} \omega_j t_j y_j - p_1 \sum_{j=1}^{M} \omega_j t_j - p_2 \sum_{j=1}^{M} \omega_j t_j^2 = 0 \tag{3-30}$$

Substituting equation (3-27) into equation (3-29) yields

$$p_2 = \frac{\left(\sum \omega_j\right)\left(\sum \omega_j t_j y_j\right) - \left(\sum \omega_j t_j\right)\left(\sum \omega_j y_j\right)}{\left(\sum \omega_j\right)\left(\sum \omega_j t_j^2\right) - \left(\sum \omega_j t_j\right)^2} \tag{3-31a}$$

and substituting this into equation (3-28) yields

$$p_1 = \frac{\sum \omega_j t_j^2 \sum \omega_j y_j - \sum \omega_j t_j \sum \omega_j t_j y_j}{\left(\sum \omega_j\right)\left(\sum \omega_j t_j^2\right) - \left(\sum \omega_j t_j\right)^2} \tag{3-31b}$$

The standard deviations of p_1 and p_2 are useful indicators of the goodness of fit. They are given by

$$\sigma(p_1) = \sigma \left(\frac{\sum \omega_j t_j^2}{\left(\sum \omega_j\right)\left(\sum \omega_j t_j^2\right) - \left(\sum \omega_j t_j\right)^2} \right)^{1/2} \tag{3-32}$$

and

$$\sigma(p_2) = \sigma \left(\frac{\sum \omega_j t_j}{\left(\sum \omega_j\right)\left(\sum \omega_j t_j^2\right) - \left(\sum \omega_j t_j\right)^2} \right)^{1/2} \tag{3-33}$$

where

$$\sigma = \left(\frac{\displaystyle\sum_{j=1}^{M} \omega_j (y_j - y_j^*)^2}{(M-1)} \right)^{1/2} \tag{3-34}$$

In this expression, y_j^* are the computed values for y_j from $y = p_1 + p_2 t$ using the computed values of p_1 and p_2.

There are other measures of the quality of fit of the data. The simplest is the root mean square (rms) deviation, which is

$$\text{rms error} = \left(\frac{\displaystyle\sum_{j=1}^{M} (\omega_j \varepsilon_j)^2}{n} \right)^{1/2} \tag{3-35}$$

where n is the number of data points. Another commonly used quantity is the correlation coefficient, which allows the significance of correlation to be tested. This quantity is given by

$$r = \frac{\sum \omega_j t_j \sum \omega_j y_j - n \sum \omega_j t_j y_j}{\left[\left(\sum \omega_j y_j \right)^2 - n \left(\sum \omega_j y_j^2 \right) \right] \left[\left(\sum \omega_j t_j \right)^2 - n \left(\sum \omega_j t_j^2 \right) \right]} \tag{3-36}$$

When $r = 1$, there is a perfect correlation between the calculated line and experimental points; when $r = 0$, the data cannot be correlated linearly. If kinetic data are reasonably fitted, the correlation is generally $r > 0.95$, while poor correlations have $r < 0.90$.

Although the linear least-squares method is widely used in the treatment of kinetic data, it does have its limitations. One such limitation is that the method applies only to cases that can be expressed in linear form. Thus, for most complex kinetic models, the linear least-squares method is not applicable. Sometimes the equation can be linearized with respect to the constants by simple rearrangement. For example, the expression for the concentration-vs.-time dependence of the second-order reaction, $A + A \xrightarrow{\kappa} $ products, is

$$[A] = \frac{[A]_0}{1 + \kappa [A]_0 t} \tag{3-37}$$

This form is nonlinear, but it can be rewritten linearly as

$$\frac{1}{[A]} = \frac{1}{[A]_0} + \kappa t \tag{3-38}$$

This form is identical to equation (3-25) with $1/[A] = y$, $1/[A]_0 = p_1$, and $\kappa = p_2$. When an equation is linearized by either rearrangement or transformation, the original errors are often not preserved. It may then be the case that statistical weights have to be adjusted appropriately in order to obtain the correct estimates of p_1 and p_2 by the least-squares method.[30–33] Furthermore, for models that involve several dependent variables, linearization is not feasible. In such cases we turn to the nonlinear least-squares method,[39] which does not require that the model equations be linear in the unknown parameters. Nonlinear least-square methods are beyond the scope of this treatment.

3.3.1.3.2 Evaluation of rate constants at a given temperature from direct measurements.   A simple illustration of the use of the least-squares method to evaluate rate constants from experimental determinations is illuminating. Let us consider how the rate constant for the recombination of methyl radicals is measured using the laser photolysis time-resolved diode laser absorption technique (see section 3.2.2.2). The rate equation for the recombination reaction

$$CH_3 + CH_3 \xrightarrow{\kappa_1} C_2H_6$$

is given by

$$\frac{d[CH_3]}{dt} = -2\kappa_1 [CH_3]^2 \tag{3-39}$$

which yields, upon integration,

$$\frac{[CH_3]_0}{[CH_3]} = 1 + 2\kappa_1[CH_3]_0(t - t_0) \tag{3-40}$$

In the experiment the measured quantity is the voltage (V) on a voltage meter corresponding to the signal from a HgCdTe infrared detector. According to Beer's law

$$I = I_0 \exp(-\varepsilon c \ell) \tag{3-41}$$

where ε is the molar extinction, c the molar concentration, and ℓ the absorption path length. If $\varepsilon c \ell$ is very small (weak absorption), we can expand equation (3-41) via a Taylor series and truncate the expansion beyond the linear term to obtain

$$I = I_0(1 - \varepsilon c \ell) \tag{3-42}$$

Consequently, the concentration c is given as

$$c = \frac{(\varepsilon \ell)^{-1}(I_0 - I)}{I_0} \tag{3-43}$$

From equation (3-43), we see that the concentration of the absorbing molecule is proportional to the relative change in the transmittance I. Since the transmittance is proportional to the voltage, we have

$$c \propto \frac{V_0 - V}{V_0} = \frac{S}{V_0} \tag{3-44}$$

where S is the strength of the signal which is the difference between the voltage for $c = 0$ and $c \neq 0$. Now, we can rewrite equation (3-40) as

$$\frac{S(t_0)}{S(t)} = 1 + 2\kappa_1 [CH_3]_0 (t - t_0) \tag{3-45}$$

and if we let

$$y = \frac{S(t_0)}{S(t)} - 1 \tag{3-46}$$

$\kappa' = 2\kappa_1[CH_3]_0$, and $x = t - t_0$, we get a simple expression of the form

$$y = \kappa' x \tag{3-47}$$

Using the formula of linear least-squares fitting described in section 3.3.1.3.1, we get

$$\sigma_{\kappa'} = \frac{\sigma}{\left[\sum_i \omega_i x_i^2\right]^{1/2}} \tag{3-48}$$

where

$$\sigma = \frac{\left(\sum \omega_j (y_j - \kappa_1 x_j)^2\right)^{1/2}}{n - 1} \tag{3-49}$$

$$\omega_i = \frac{c^2}{\sigma_i^2} \tag{3-50}$$

and where n is the number of y_i measurements and c is an arbitrary proportionality constant. To evaluate σ_i, we apply the propagation-of-errors formula, equation (3-17), as follows:

$$\sigma_i = \left[\left(\frac{\partial y}{\partial S(t_0)} \sigma_s(t_0)\right)^2 + \left(\frac{\partial y}{\partial S(t_i)} \sigma_s(t_i)\right)^2\right]^{1/2} \tag{3-51}$$

This is no more than

$$\sigma_i = \left[\left(\frac{\sigma_s(t_0)}{S(t_0)}\right)^2 + \left(\frac{\sigma_s(t_i)}{S(t_i)}\right)^2\right]^{1/2} \frac{S(t_0)}{S(t_i)} \tag{3-52}$$

where $\sigma_s(t_0)$ and $\sigma_s(t_i)$ are the standard deviations of measured signal at $t = t_0$ and $t = t_i$, respectively. To evaluate the standard deviation of κ', we substitute equations (3-51) and (3-52) into equations (3-49) and (3-50). To evaluate the standard deviation of κ_1, we again apply equation (3-17)

$$\kappa_1 = \frac{\kappa'}{2[CH_3]_0} \tag{3-53}$$

to obtain

$$\sigma_{\kappa_1} = \left[\left(\frac{\partial \kappa_1}{\partial \kappa'} \sigma_{\kappa'}\right)^2 + \left(\left(\frac{\partial \kappa_1}{\partial [CH_3]_0}\right) \sigma_{[CH_3]_0}\right)^2\right]^{1/2} \tag{3-54}$$

giving

$$\sigma_{\kappa_1} = \kappa_1 \left[\left(\frac{\sigma_{\kappa'}}{\kappa_1}\right)^2 + \left(\frac{\sigma_{[CH_3]_0}}{[CH_3]_0}\right)^2\right]^{1/2} \tag{3-55}$$

Now, $[CH_3]_0$ is proportional to the pressure P_0 of the initial reactant (CH_3I) and the photolysis laser fluence Φ, i.e.,

$$[CH_3]_0 \propto P_0\,(CH_3)\Phi \qquad (3\text{-}56)$$

So again, using the equation (3-17) with equation (3-56), we obtain

$$\sigma\,[CH_3]_0 = [CH_3]_0 \left\{ \left(\frac{\sigma_{P_0}}{P_0}\right)^2 + \left(\frac{\sigma_\Phi}{\Phi}\right)^2 \right\}^{1/2} \qquad (3\text{-}57)$$

Finally,

$$\sigma_{\kappa_1} = \kappa_1 \left[\left(\frac{\sigma_{\kappa'}}{\kappa'}\right) + \left(\frac{\sigma_{P_0}}{P_0}\right) + \left(\frac{\sigma_\Phi}{\Phi}\right)^2 \right\}^{1/2} \qquad (3\text{-}58)$$

The quantity σ_κ' is the standard deviation of the sample mean value $\kappa' = 2\kappa_1[CH_3]_0$ obtained by evaluating equations (3-48) through (3-52), σ_{P_0} is the standard deviation of the sample mean value of the reactant $(CH_3I)_0$ pressure, and σ_Φ is the standard deviation of the sample mean value of photolysis laser fluence.

3.3.1.3.3 Evaluation of temperature-dependent rate constants.   Perhaps the simplest illustration of the use of the least-squares method to analyze kinetic data is in evaluating Arrhenius parameters from temperature-dependent rate data. In the form

$$\kappa(T) = A \exp\left(\frac{-E_{act}}{RT}\right) \qquad (3\text{-}59)$$

the temperature rate expression is nonlinear. Accordingly, if the linear least-squares method is to be used to evaluate the coefficients, this expression must be linearized. The common practice is to transform equation (3-59) to the logarithmic form

$$\ln \kappa(T) = \ln A - \frac{B}{T} \qquad (3\text{-}60)$$

where $B = E_{act}/R$. Then, to obtain E_{act} and A (the preexponential factor), a linear least-squares fit can be performed on equation (3-60) rather than using a nonlinear fit to equation (3-59). However, when equation (3-59) is thus linearized and a least-square fit performed, the weights of the $\kappa(T)$ values must be adjusted to the weight of $\ln \kappa(T)$ to ensure that the least-squares procedure does indeed minimize the residuals of the measured rate constant $\kappa(T)$.

3.3.1.3.4 Reporting kinetic data.   Results of kinetic measurements have great value for anyone who uses them, particularly for those performing atmospheric and combustion modeling. However, the results are useful only when they are accompanied by realistic estimates of their overall uncertainty. In reporting kinetic data, it is recommended[34] that information regarding both random and systematic errors be given. In particular, one should report known and even suspected systematic errors and, when available, estimates of their bounds and how the estimates were obtained. The value of the random error of the sample mean should be reported at a confidence level of at least $\pm 2\sigma$. In reporting kinetic data, the form $a \pm b$ is used, but only when the exact meaning of the uncertainty $\pm b$ is explicitly and clearly stated.

3.3.2   Sensitivity Analysis

In chemical kinetics it is often the case that the concentrations of chemical species are known more precisely than the rate constants of the kinetic mechanism. Consequently, in analyzing results from kinetic experiments, it is important to assess the effects of such parameter uncertainties on the predicted concentrations of the various intermediate and product species. Such information can be useful in identifying those rate constants which need to be determined more precisely, and may also help the experimentalist decide which concentration to monitor in order to improve the estimate of a specific rate constant. An analysis of this type is known as a *sensitivity analysis*.[40-49] Its purpose is to assess the sensitivity of the system's solutions to changes in input parameters, which in turn can be of enormous help in specifying the reaction mechanism and in designing experimental measurements.

How does one carry out a sensitivity analysis? Suppose we have a kinetic mechanism described by a set of coupled first-order differential equations having the following general form:

$$\frac{dy_i}{dt} = f_i(y_1, y_2, \ldots, y_N, t, \kappa_1, \kappa_2, \ldots, \kappa_M) \tag{3-61}$$

where there are N chemical components with concentrations y_i and M reactions in the mechanism with rate parameters κ_j. The solution of equation (3-61) will be a function of the two variables κ and t, that is $y_i(\kappa, t)$. A rate constant is called sensitive if a small change in its value produces large changes in the solution. Likewise, if changes in rate constants produce little or no change in the solution, then the rate constant is said to be insensitive. Therefore, to measure the sensitivity of the system of equations given by equation (3-61) with respect to a particular rate constant parameter, we need to know what small change in y_i is produced as a result of a small change in the rate constant κ_j, i.e.,

$$\Delta y_i = y_i(\ldots, \kappa_j + \Delta\kappa_j, \ldots) - y_i(\ldots, \kappa_j, \ldots) \tag{3-62}$$

If this expression has a limiting value as $\Delta\kappa_j$ approaches zero, i.e., if

$$\lim_{\Delta\kappa_j \to 0} \frac{y_i(y_i, t, \kappa_j + \Delta\kappa_j) - y_i(y_i, t, \kappa_j)}{\Delta\kappa_j} = \frac{\partial y_1}{\partial \kappa_j} \tag{3-63}$$

then we can define the sensitivity matrix element Z_{ij} as

$$\frac{\partial y_i}{\partial \kappa_j} = Z_{ij} \tag{3-64}$$

The coefficients of the sensitivity matrix are referred to as *sensitivity coefficients*. In order to calculate Z_{ij} we must relate it to the set of differential equations of the system, i.e., equation (3-61). We do this by interchanging the order of differentiation

$$\frac{d(Z_{ij})}{dt} = \frac{d\partial y_i}{dt\partial \kappa_j} = \frac{\partial}{\partial \kappa_j}\frac{dy_i}{dt} \tag{3-65}$$

Since y_i is a function of κ_j the right-hand expression is expanded to give

$$\frac{d(Z_{ij})}{dt} = \frac{\partial f_i}{\partial \kappa_j} + \sum_{\ell=1}^{N} \frac{\partial f_i}{\partial y_\ell} \frac{\partial y_\ell}{\partial \kappa_j} \tag{3-66}$$

where $i = 1, 2, ..., N$ and $j = 1, 2, ..., M$.

Finally, substituting equation (3-64) into equation (3-66), we obtain

$$\frac{d(Z_{ij})}{dt} = \frac{\partial f_i}{\partial \kappa_j} + \sum_{\ell=1}^{N} \frac{\partial f_i}{\partial y_\ell} Z_{\ell j} \tag{3-67}$$

Equation (3-67) is the fundamental sensitivity equation.

In deriving the sensitivity equation (3-67), we did not speak of the initial condition. If in equation (3-61) the initial condition is

$$f_i(y_i, 0, \kappa_j) = 0 \tag{3-68}$$

then there is no deviation between the two systems. The initial condition for equation (3-67) is

$$Z_{ij}(0) = 0 \tag{3-69}$$

This is sometimes referred to as the constant parameter case. Indeed, for most chemical kinetic systems one needs to consider only this case. The solution of equation (3-67), then, provides us with information about the sensitivity of the system to variations in each rate constant parameter. To obtain the sensitivity coefficient we must solve the differential equation (3-67) along with equation (3-61), since both of these two sets of equations are usually coupled. First, equation (3-61) must be solved to yield $y_i(\kappa, t)$ in order to evaluate $\partial f_i / \partial \kappa_i$ and $\partial f_i / \partial y_i$; then equation (3-67) must be solved. To measure the total sensitivity of the system we use the sensitivity coefficients to calculate the total change contributed by each of the N parameters from the expression

$$\Delta y_i = \sum_{j=1}^{M} Z_{ij} \Delta \kappa_j \tag{3-70}$$

As a simple illustration of the "direct" method of sensitivity analysis[40-42], we shall calculate sensitivity coefficients for the temperature-dependent rate constant of a first-order reaction

$$A_1 \xrightarrow{k} A_2 \tag{3-71}$$

which has the form $k = \kappa_1 \exp(-\kappa_2/T)$. We wish to examine the sensitivity of the concentration of A_1 to variations in the Arrhenius parameters κ_1 and κ_2 at $T = 298$ K. The rate equation for the decay of A_1 is

$$\frac{d[A_1]}{dt} = -k[A_1] = -[A_1]\kappa_1 e^{-\kappa_2/T} \tag{3-72}$$

subject to the initial condition that

$$[A_1]_t = [A_1]_0 \text{ at } t = 0.$$

Integration of equation (3-72) yields the expression for the time and temperature dependence of $[A_1]$ as

$$[A_1]_t = [A_1]_0 \exp(-K_1 t e^{-K_2/t}) \tag{3-73}$$

The sensitivity coefficients of the sensitivity equations are determined simply by differentiating equation (3-73) with respect to κ_1 and κ_2

$$\frac{\partial [A_1]}{\partial \kappa_1} = Z_{11} = -[A_1]_0 t e^{-\kappa_2/T} \left[\exp(-\kappa_1 t e^{-\kappa_2/T})\right] \tag{3-74}$$

and

$$\frac{\partial [A_1]}{\partial \kappa_2} = Z_{12} = \frac{\kappa_1}{T} [A_1]_0 t e^{-\kappa_2/T} \left[\exp(-\kappa_1 t e^{-\kappa_2/T})\right] \tag{3-75}$$

The remaining terms of the sensitivity equations are determined by

$$\frac{\partial f_1}{\partial \kappa_1} = -[A_1] e^{-\kappa_2/T} \tag{3-76}$$

$$\frac{\partial f_1}{\partial \kappa_1} = \frac{\kappa_1 [A_1]}{T} e^{-\kappa_2/T} \tag{3-77}$$

and

$$\frac{\partial f_1}{\partial [A_1]} = \kappa_1 e^{-\kappa_2/T} \tag{3-78}$$

With these terms we can write out the sensitivity differential equations as

$$\frac{dZ_{11}}{dt} = -[A_1] e^{-\kappa_2/T} - \kappa_1 e^{-\kappa_2/T} Z_{11} \tag{3-79}$$

$$\frac{dZ_{12}}{dt} = +\frac{\kappa_1 [A_1] e^{-\kappa_2/T}}{T} - \kappa_1 e^{-\kappa_2/T} Z_{12} \tag{3-80}$$

The sensitivity coefficients are

$$\frac{\partial [A_1]}{\partial \kappa_1} = Z_{11} = -[A_1]_0 t e^{-\kappa_2/T} \left[\exp(-\kappa_1 t e^{-\kappa_2/T})\right] \tag{3-81}$$

and

$$\frac{\partial [A_1]}{\partial \kappa_2} = Z_{12} = \frac{\kappa_1}{T} [A_1]_0 t e^{-\kappa_2/T} \left[\exp(-\kappa_1 t e^{-\kappa_2/T})\right] \tag{3-82}$$

To see what the sensitivity coefficients tell us about the effect of variations in κ_1 and κ_2 on the concentrations of A_1, let us choose $\kappa_1 = 1.79 \times 10^6$ sec^{-1} and $\kappa_2 = 500$ K. Also, let us suppose that $[A_1]_0 = 1$ mole ℓ^{-1} at $T = 298$ K and that the uncertainties in κ_1 and κ_2 are

$$8.97 \times 10^5 \le \kappa_1 \le 3.59 \times 10^6 \text{ sec}^{-1} \quad \text{and} \quad 0 \le \kappa_2 \le 1000 \text{ K}$$

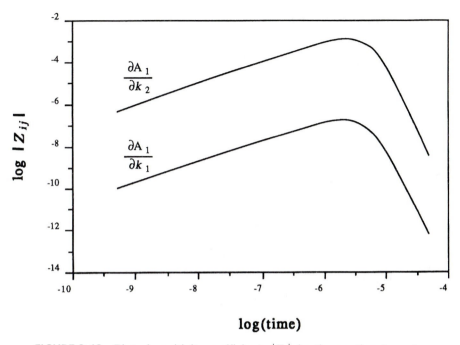

FIGURE 3-13 Plot of sensitivity coefficients $|Z_{ij}|$ for the reaction $A_1 \rightarrow A_2$.

Figure 3-13 shows the sensitivity function at various times for these values of the constants; it can be seen that the most sensitive parameter is κ_2. Note that the sensitivity function varies as the reaction proceeds in time. The relative sensitivities are similar to those in Figure 3-13. For the parameter range we have studied, all the sensitivity measures, including the analytical sensitivity coefficients, indicate that the concentration of A_1 is more sensitive to changes in κ_2 than to changes in κ_1. If we increase the uncertainty in κ_1 by two more orders of magnitude, then the concentration of A_1 becomes more sensitive to κ_1 than to κ_2.

The method we have just illustrated for the simple case above is often referred to as the direct method of sensitivity analysis.[40-44] If we wish to explore reactions of larger systems that include more parameters, we have to resort to numerical means of solving equations (3-61) and (3-67). There are several strategies for carrying this out:[41,42,49]

(1) Brute-force finite-difference method. At each time step t, calculate

$$\frac{\partial c_i(t)}{\partial k_j} \approx \frac{c_i(t, k_j + \delta k_j) - c_i(t, k_j)}{\delta k_j}$$

for all $i = 1, ..., N$ and $j = 1, ..., M$. For a large coupled system, this is obviously impractical because of the mammoth amount of computation required.

(2) Using the analytic form of dc/dt to find differential equations for Z [see equation (3-67)] and either solving (3-61) and (3-67) simultaneously, or solving equation (3-61) to yield $y_i(\kappa, t)$, then interpolating two values of $y_i(\kappa, t)$ to evaluate the terms in equation (3-67). The latter technique amounts to solving N differential equation $(M + 1)$ times.

(3) A Green's function method, which transforms the differential equation into an integral equation, has also been applied.[43-45]

Other methods such as a polynomial approximation which transforms the differential equation into a set of linear algebraic equations, and a stochastic or Monte Carlo method, are described in the review article by Turanyi.[49] With these methods, it becomes computationally feasible to explore the importance of parameter variations in large systems, such as those pertinent to atmospheric and combustion systems. An example of an application to combustion kinetics will be given in chapter 14.

In this section, we have introduced a technique which can be of great utility in analyzing complex kinetic models. Although the emphasis here has been on analysis of phenomenological kinetic equations, the same principles can be applied to elucidating the relationships between microscopic quantities such as details of the potential energy surface, discussed in chapter 7, and properties such as scattering cross sections (chapter 8) and statistical rate coefficients (chapters 10 and 11). This use of sensitivity techniques[47-49] can answer such basic questions as which characteristic molecular parameters are most relevant for determining the rates and products of chemical reactions.

REFERENCES

[1] L. J. Kovalenko and S. R. Leone, *J. Chem. Ed. 65*, 681 (1988).

[2] L. Batt and F. R. Cruickshank, *J. Chem. Soc. A*, 261 (1967).

[3] C. J. Howard, *J. Phys. Chem. 83*, 3 (1976); K. H. Hoyermann, "Interaction of Chemical Reactions, Transport Processes, and Flow" in *Physical Chemistry, An Advanced Treatise: Kinetics of Gas Reactions,* edited by H. Eyring, D. Henderson, and W. Jost (New York: Academic Press, 1975). Vol. 6B, pp. 931–1006; F. Kaufman, *Prog. React. Kinet. 1*, 3 (1961).

[4] V. M. Bierbaum, G. B. Ellison, and S. R. Leone, in *Gas Phase Ion Chemistry,* Vol. 3, edited by M. T. Bowers, (New York: Academic Press, 1984), p. 1.

[5] J. V. Seeley, J. T. Jayne, and M. J. Molina, *Intl. J. Chem. Kinetics, 25*, 571 (1993).

[6] W. Tsang and A. Lifshitz, *Ann. Rev. Phys. Chem. 41*, 599 (1990).

[7] G. Porter, *Proc. Roy. Soc. (London) A200*, 284 (1950).

[8] M. I. Christie, R. G. W. Norrish, and G. Porter, *Proc. Roy. Soc. (London) A216*, 152 (1952).

[9] B. S. Yamanashi and A. V. Nowak, *J. Chem. Ed. 45*, 705 (1968).

[10] G. Porter and J. I. Steinfeld, *J. Chem. Phys. 45*, 3456 (1966).

[11] G. A. Laguna and S. L. Baughcum, *Chem. Phys. Lett. 88*, 568 (1982).

[12] A. Wahner and C. Zetzsch, *Berg. Bunsenges. Phys. Chem. 89*, 323 (1985).

[13] R. J. Balla and L. Pasternack, *J. Phys. Chem. 91*, 73 (1987).

[14] J. Wormhoudt, A. C. Stanton, and J. A. Silver, *Proc. SPIE 452*, 88 (1984).

[15] P. L. Houston, *Adv. Chem. Phys. 47*, 615 (1981).

[16] F. J. Wodarczyk and C. B. Moore, *Chem. Phys. Lett. 26*, 484 (1974).

[17] R. N. Zare and P. J. Dagdigian, *Science 185*, 739 (1974).

[18] J. L. Kinsey, *Ann. Rev. Phys. Chem. 28*, 349 (1977).

[19] A. Schultz, H. W. Cruse, and R. N. Zare, *J. Chem. Phys. 57*, 1354 (1972).

[20] H. W. Cruse, P. Dagdigian, and R. N. Zare, *Farad. Disc. Chem. Soc. 55*, 277 (1973).

[21] R. Altkorn and R. N. Zare, *Ann. Rev. Phys. Chem. 35*, 265 (1984).

[22] P. M. Johnson and C. E. Otis, *Ann. Rev. Phys. Chem. 32,* 139 (1981).

[23] S. Arepalli, N. Presser, D. Robie, and R. J. Gordon, *Chem. Phys. Lett. 117,* 64 (1985).

[24] I. R. Slagle, F. J. Pruss, and D. Gutman, *Int. J. Chem. Kinet. 6,* 111 (1974).

[25] D. C. Jacobs and R. N. Zare, *J. Chem. Phys. 85*, 5457 (1986).

[26] D. C. Jacobs, R. J. Madix, and R. N. Zare, *J.Chem. Phys. 85*, 5469 (1986).

[27] P. R. Bevington, *Data Reduction and Error Analysis for the Physical Sciences* (New York: McGraw-Hill, 1969).

[28] H. D. Young, *Statistical Treatment of Experimental Data* (New York: McGraw-Hill, 1962).

[29] N. R. Draper and H. Smith, *Applied Regression Analysis* (New York: Wiley, 1966).

[30] R. J. Cvetanovic, R. P. Overend, and G. Paraskevopoulos, *Int. J. Chem. Kinet. S1,* 249 (1975).

[31] R. J. Cvetanovic and D. L. Singleton, *Int. J. Chem. Kinet. 9,* 481 (1977).

[32] R. J. Cvetanovic and D. L. Singleton, *Int. J. Chem. Kinet. 9,* 1007 (1977).

[33] R. J. Cvetanovic, and R. P. Overend, and G. Paraskevopoulos, *J. Phys. Chem. 83,* 50 (1979).

[34] C. Eisenhart, *Science 160,* 1201 (1968).

[35] C. Eisenhart, *J. Res. Natl. Bur. Stand., Sec. C 67,* 161 (1963).

[36] E. M. Pugh and G. H. Winslow, *The Analysis of Physical Measurements* (Reading: Addison-Wesley, 1966).

[37] A. Fontijn and W. Felder in *Reactive Intermediates in the Gas Phase: Generation and Monitoring,* edited by D. W. Setser (New York: Academic Press, 1979), pp. 59–151.

[38] C. Daniel and F. S. Wood, *Fitting Equations to Data,* 2d ed. (New York: Wiley, 1980).

[39] D. W. Marquardt, *J. Soc. Indust. Appl. Math. 11,* 431 (1963).

[40] R. W. Atherton, R. B. Schainker, and E. R. Ducot, *AIChE J. 21,* 441 (1975).

[41] R. P. Dickinson and R. J. Gelinas, *J. Comput. Phys. 21,* 123 (1976).

[42] J. W. Tilden, V. Costanza, G. J. McRae, and J. H. Seinfeld in *Modelling of Chemical Reaction Systems,* edited by K. H. Ebert, P. Deuflhard, and W. Jüger (Berlin: Springer-Verlag, 1981) pp. 69–91.

[43] E. P. Dougherty, J.-T. Hwang, and H. Rabitz, *J. Chem. Phys. 71,* 1794 (1979).

[44] J.-T. Hwang, E. P. Dougherty, S. Rabitz, and H. Rabitz, *J. Chem. Phys. 69,* 5180 (1978).

[45] E. P. Dougherty and H. Rabitz, *Int. J. Chem. Kinet. 11,* 1237 (1979).

[46] A. A. Boni and R. C. Penner, *Combust. Sci. Tech. 15,* 99 (1976).

[47] H. Rabitz, *Comput. Chem. 5,* 167 (1981).

[48] L. Eno and H. Rabitz, *Adv. Chem. Phys. 51,* 177 (1982).

[49] T. Turanyi, *J. Math. Chem. 5*, 203 (1990).

PROBLEMS

3.1 Glucose is formed by the reaction of cellulose in the presence of HCl at various temperatures. The rate is found to be first order in the cellulose concentration, but depends on the HCl concentration [E. E. Harris and A. A. Kline, *J. Phys. Colloid Chem. 53,* 344 (1949)]. The following data are measured for k_1, the rate of hydrolysis of Douglas-fir wood cellulose at various temperatures:

[HCl] moles/ℓ	k_1, sec^{-1}		
	$T = 160°C$	$T = 180°C$	$T = 190°C$
0.055	0.00203	0.0190	0.0627
0.11	0.00486	0.0455	0.149
0.22	0.01075	0.011	0.357
0.44	0.0261	0.284	—
0.88	0.0672	—	—

Find the best least-squares values of k_1' and n that fit the relationship

$$k_1 = k_1' \, [\text{HC1}]^n$$

each temperature.

3.2 An experiment is designed in which the reaction rate between gaseous H_2 and Br_2 will be measured in a static photolysis vessel of 50 cm in length at a total pressure of 1 torr. How much time must be allowed for the sample to be well mixed before the reaction is initiated?

3.3 This problem examines the sensitivity of the concentration of a substance A to variations in Arrhenius parameters at $T = 208$ K.

(a) Calculate analytic sensitivity coefficients for the second-order reaction

$$A + A \xrightarrow{K} \text{products}$$

where $K = k_1 \exp(k_2/T)$. Suppose that

$$[A]_0 = 2.0 \text{ mole } \ell^{-1}$$

$$k_1 = 1.79 \times 10^{10} \, \ell \text{ mole}^{-1} \text{ sec}^{-1}$$

$$k_2 = 500 \ K$$

and the uncertainty range for the parameters is

$$8.9 \times 10^9 < k_1 < 3.59 \times 10^{10}$$

and

$$0 < k_2 < 1,000$$

(b) Plot the sensitivity coefficients as a function of time. What can you conclude from your plot?

(c) Plot the relative sensitivities $\Delta k_i (\partial A_j / \partial k_i)$ as a function of time. What can you conclude about the importance of the effects of parameter uncertainties on the concentration of A?

(d) Suppose that the range of uncertainty for the parameter k_1 is

$$0 < k_1 < 3.6 \times 10^{10}$$

Plot the sensitivity coefficients and the relative sensitivities as a function of time. What can you conclude?

3.4 Consider the following set of four reactions in three components.

$$C_1 \underset{k_2}{\overset{k_1}{\rightleftarrows}} C_2$$

$$k_3 \seararrow \swarrow k_4$$

$$C_3$$

Numerical values are assigned to the rate constants and initial conditions as follows:

$$k_1 = 1 \qquad\qquad [C_1]_0 = 2$$
$$k_2 = 499.5 \qquad\qquad [C_2]_0 = 1$$
$$k_3 = 499.5 \qquad\qquad [C_3]_0 = 0$$
$$k_4 = 1$$

(a) Using the Laplace transform method or the matrix method solve for $[C_1]$, $[C_2]$, and $[C_3]$. Plot the evolution with time of these concentrations.

(b) Calculate analytic sensitivity coefficients for this reaction system.

(c) Plot the sensitivity coefficients as a function of time. When will the sensitivity function attain a maximum?

(d) Suppose that the uncertainty range for the parameters is

$$0.1k_i < k_i < 10k_i \qquad\qquad (i = 1, \ldots, 4)$$

Plot the relative sensitivity as a function of time.

C H A P T E R   4

Reactions in Solution

Although the principal emphasis in this text is on reactions that take place in the gas phase, many important chemical transformations occur in liquid solutions. The synthesis of organic chemicals, the commercial production of many chemical commodities, and the metabolic processes which enable a person to be reading this book all occur in liquids, often in aqueous solutions. It is therefore essential to have an understanding of the nature of chemical reactions in solutions.

4.1   GENERAL PROPERTIES OF REACTIONS IN SOLUTION

There are significant differences between reactions in solution and reactions in the gas phase. In solution, the concept of isolated collisions between reactant molecules is practically meaningless. The reactant molecules interact continuously with solvent molecules. The distribution of collisions is also quite different in the two cases. In the gas phase, there is a random distribution of collisions between reactant molecules, so that the occurrence of a collision does not affect the likelihood of a subsequent collision. Such processes are said to be *Markovian* in nature. By contrast, the mean collision frequency per unit density may be the same in liquids as in gases, but the distribution of the collisions in time is quite different. If two solute molecules collide in solution, it is highly probable that they will undergo numerous successive collisions before they separate. In effect, the liquid structure forms a "cage" of solvent molecules around the reactants, holding them in close proximity.[1] Such a set of consecutive collisions in solution is referred to as an *encounter*. The different distributions of collisions for the gas phase and in solution are illustrated in Figure 4-1.

The effects of solvent on reaction rate may be either chemical or physical, i.e., environmental. If the solvent molecules participate directly in the reaction mechanism, the effect is chemical. In this case, the solvent will appear explicitly in the stoichiometric equation as a reactant or product. If the effect of the solvent is merely to modify the interaction between reacting species, as in the alteration of the collision distribution just described, its effect may be termed physical. Solvent molecules may also play a catalytic role (see chapter 5), in which case they appear in the explicit kinetic equations but not in the stoichiometric equation for the reaction.

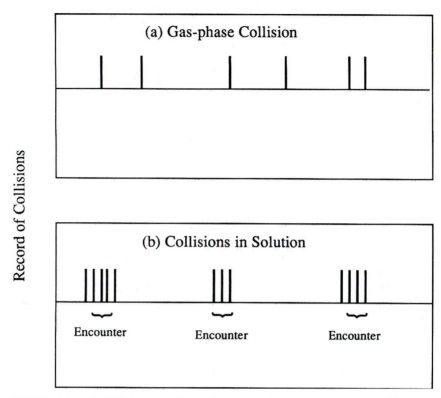

FIGURE 4-1   Distribution of collisions between solute molecules. Record (a) shows schematically the gas-phase collisions, while record (b) shows the distribution of collisions when a solvent is present.

The solvent may play a variety of roles during the course of the reaction. For example, solvation stabilizes the formation of ions: $NaCl \rightarrow Na^+ + Cl^-$ is highly endothermic in the gas phase, but exothermic in aqueous solution. Thus, ionic reactions are quite common in solution mechanisms, but relatively rare in the gas phase. Rapid energy transfer between reactant and solvent molecules maintains thermal equilibrium in solution reactions, while thermal disequilibrium is often important in gas-phase reactions. Also, as discussed in the sections that follow, interactions between reacting ions are influenced by the dielectric character of the solvent and by the presence of other ions in solution.

4.2 PHENOMENOLOGICAL THEORY OF REACTION RATES

There is an upper limit to the rate of a bimolecular reaction in solution which is analogous to the hard-sphere encounter frequency in the kinetic theory of gases (see Sec. 8.1). The rate with which two reactant molecules A and B can come together is governed by the phenomenological laws of diffusion and conduction. Once the reactants have come together, they do not simply collide and separate; instead, they are constrained by the solvent molecules, and typically remain trapped for approximately 10^{-11} sec undergoing numerous collisions with each other, as shown in Figure 4-1(b). If the activation energy for the chemical reaction is low, so that the reaction between A and B is facile, the encounter frequency sets an upper bound to the bimolecular reaction rate. This frequency, in turn, is controlled by the processes of diffusion and conduction, which determine the rate at which the reactive molecules can get into the same solvent cage.

4.2.1 Diffusion and Conduction

The two factors which control the movement of reactant molecules in solution are random diffusive motion and any conductive forces that are present if a potential energy gradient exists in the solution. The former process describes the tendency of solute molecules to move from regions of high to low concentration in the solution. This is expressed by Fick's first law of diffusion,

$$\phi_i = -D_i \nabla c_i \tag{4-1}$$

where ϕ_i is the flux of the solute i in units of molecules $cm^{-2}sec^{-1}$, ∇c_i is the concentration gradient in units of molecules $cm^{-3}sec^{-1}$, and D_i is the diffusion coefficient for solute i relative to the solvent. Both ϕ_i and ∇c_i are vectors with $x, y,$ and z components. However, if the concentration gradient is spherically symmetric, without loss of generality, one of the components may be placed along r, the distance from an origin, so that equation (4-1) becomes

$$\phi_i = -D_i \left(\frac{dc_i}{dr} \right) \tag{4-2}$$

If ϕ_i is multiplied by $4\pi r^2$, the area of a spherical shell around the origin, the resulting term $4\pi r^2 \phi_i$ has units of molecules/sec and is the rate at which the solute i passes through the spherical shell; i.e.

$$\frac{dn_i}{dt} = 4\pi r^2 \phi_i = -4\pi r^2 D_i \left(\frac{dc_i}{dr} \right) \tag{4-3}$$

When c_i decreases with increase in r, dc_i/dr is negative and molecules pass through the spherical shell in the direction of increasing r. The solute moves in the opposite direction for positive dc_i/dr.

   If there is a difference in the potential energy of the solute in different parts of the solution, a potential energy gradient will arise which can also cause a net motion of the solute in a definite direction. The solute flux arising from a potential gradient may be written as

$$\phi_i = c_i \mathbf{v}_i \tag{4-4}$$

where $\mathbf{v}_i$ and c_i are the velocity vector and concentration, respectively, of the solute i. Potential energy gradients, arising from an electric field E, are particularly important for ions in solution. An ion i with a charge $z_i e$ moving in an electric potential gradient ∇E has a mobility tensor μ_i given by

$$\mathbf{v}_i = \mu_i \nabla E \tag{4-5}$$

Following the above discussion, for a spherically symmetric potential gradient ∇E becomes dE/dr. If the concentration is spherically symmetric, we have, from equations (4-4) and (4-5),

$$\phi_i = \frac{-z_i}{|z_i|} \mu_i c_i \frac{dE}{dr} \tag{4-6}$$

where the factor $-z_i/|z_i|$ is introduced to fix the sign. A negative ion has a positive flux in the direction of increasing r with increasing E. Standard units for dE/dr and μ_i are volts cm^{-1} and cm^2 volt^{-1}sec^{-1}. In equation (4-6) ϕ_i is the flux through the spherical shell of radius r about the origin, c_i is the concentration of the solute on this shell, and dE/dr is the potential gradient on the shell. To simplify the notation, μ_i in equation (4-6) is not a tensor but the ion mobility for the spherically symmetric potential gradient.

The total flux of a solute is the sum of the diffusion term in equation (4-2) and the term for the potential energy gradient in equation (4-6). Thus, for an ion, we have

$$\phi_i = -\left(D_i \frac{dc_i}{dr} + \frac{z_i}{|z_i|} \mu_i c_i \frac{dE}{dr}\right) \tag{4-7}$$

The first term in equation (4-7) represents diffusion, the second conduction. In a chemical reaction both terms are important. For example, in a reaction between two positively charged ions conduction will cause the ions to move apart. Diffusion must then bring the reactants together.

At equilibrium the flux must be zero everywhere in the solution, so that for any value of r

$$\phi_i = 0 = -D_i \frac{dc_i}{dr} - \frac{z_i}{|z_i|} \mu_i c_i \frac{dE}{dr}$$

and

$$\frac{dc_i}{c_i} = \frac{-z_i}{|z_i|} \frac{\mu_i}{D_i} dE \tag{4-8}$$

Integrating equation (4-8) gives

$$\ln \frac{c_i}{c_i^0} = \frac{-z_i}{|z_i|} \frac{\mu_i}{D_i} E \tag{4-9}$$

where c_i^0 is the concentration of ions at a reference value of r (e.g. $r = \infty$) for which $E = 0$, and c_i and E are the concentration and potential at r. It can be assumed that at equilibrium the relative concentration of the solute at different values of r must obey the Boltzmann law so that

$$c_i = c_i^0 \exp\left[\frac{-V(r)}{k_B T}\right] \tag{4-10}$$

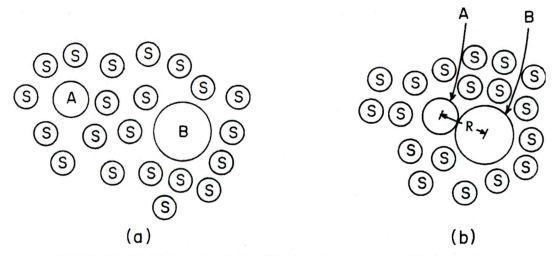

FIGURE 4-2 Model for AB relative diffusion through solvent. (a) A and B separated by solvent sheath. (b) A and B have diffused into contact when reaction occurs; the contact distance is R. [Adapted from P. C. Jordan, *Chemical Kinetics and Transport* (New York: Plenum, 1979).]

with the potential energy $V(r)$ of the ions related to the electric field by $V(r) = z_i eE$. Replacing $V(r)$ by $z_i eE$, equations (4-9) and (4-10) can be combined to obtain the relation between the ion's mobility and its diffusion coefficient, viz.,

$$\frac{\mu_i}{D_i} = \frac{|z_i|e}{k_B T} \qquad (4\text{-}11)$$

This equation is important for determining an ion's diffusion coefficient, since its mobility can be determined directly from a conductance measurement. Finally, if equations (4-10) and (4-11) are inserted into equation (4-6), we have

$$\phi_i = -D_i \left[\frac{dc_i}{dr} + \frac{1}{k_B T} c_i \frac{dV(r)}{dr} \right] \qquad (4\text{-}12)$$

4.2.2 Solution of the Conduction-Diffusion Equation

The relative motion of A and B molecules in a solvent is depicted in Figure 4-2. At a microscopic level this motion is related to local concentration and potential energy gradients. It will thus be assumed that on the average these gradients are described by the preceding equations. Consequently, from equation (4-12), the flux of B across a sphere of radius r around each A is given by

$$\phi_B = -(D_A + D_B) \left[\frac{d[B]_r}{dr} + \frac{1}{k_B T} [B]_r \frac{dV(r)}{dr} \right] \qquad (4\text{-}13)$$

where the diffusion coefficient is $D_A + D_B$, since A and B are diffusing simultaneously, and $[B]_r$ is the local concentration of B at a distance r from A. Reaction may occur whenever the A and B molecules come into contact, which is stipulated by the contact radius R. An effect of reaction is to decrease $[B]_r$ with decrease in r. To show this, consider the case where $V(r)$ in equation (4-13) is zero. From equation (4-3), the rate of B diffusing through the spherical surface of radius r towards A, i.e. J_B, is

$$J_B = \frac{-d[B]_r}{dt} = 4\pi r^2 (D_A + D_B) \frac{d[B]_r}{dr} \tag{4-14}$$

Integrating this equation, using the boundary condition $[B]_r$ equals the bulk concentration $[B]$ for $r = \infty$, gives

$$[B]_r = \frac{-J_B}{(D_A + D_B)4\pi r} + [B] \tag{4-15}$$

This equation shows that $[B]_r$ increases to the bulk concentration as $r \to \infty$.

If A and B react whenever they come in contact, the rate of loss of B as a result of collisions with a single A molecule is given by equation (4-3) and equals $-4\pi R^2 \phi_B$. The total reaction rate is the rate at which the B molecules come into contact with all the A molecules and is given by

$$\frac{-d[B]}{dt} = k[A][B] = -[A]4\pi R^2 \phi_B \tag{4-16}$$

where $[A]$ and $[B]$ represent the bulk concentrations of A and B. At steady state the concentration of B at any value of r, i.e. $[B]_r$, does not change with time. As a result, the total rate of B passing through a sphere of radius r about A, which is $4\pi r^2 \phi_B(r)$, is constant for all values of r (Note the flux term $\phi_B(r)$ depends on r). Thus, R in equation (4-16) can be replaced by r and ϕ_B can be replaced with the right-hand side of equation (4-13) to give

$$k[B] = 4\pi r^2 [D_A + D_B] \left[\frac{d[B]_r}{dr} + \frac{[B]_r}{k_B T} \frac{dV(r)}{dr} \right] \tag{4-17}$$

since k and $[B]$ on the left-hand side are independent of r. To integrate equation (4-17), write it in the form

$$k[B] = 4\pi r^2 [D_A + D_B] e^{-V(r)/k_b T} \frac{d}{dr} \{[B]_r e^{V(r)/k_B T}\} \tag{4-18}$$

Dividing both sides of the equation by $r^2 \exp[-V(r)/k_B T]$ and integrating over r gives

$$k[B] \int_R^\infty \frac{e^{V(r)/k_B T}}{r^2} dr = 4\pi (D_A + D_B) \int_R^\infty \left(\frac{d}{dr} \{[B]_r e^{V(r)/k_B T}\} \right) dr \tag{4-19}$$

To complete the integration first define

$$\beta^{-1} \equiv \int_R^\infty \frac{e^{V(r)/k_B T}}{r^2} dr \tag{4-20}$$

and then change the limits of integration from distance to concentration. At the limits $r = R$ and $r = \infty$, the concentration of B is $[B]_R$ and the bulk concentration, $[B]$, respectively. Thus equation (4-19) becomes

$$k[B] = 4\pi (D_A + D_B)\beta \int_{[B]_R}^{[B]} d\{[B]_r e^{V(r)/k_B T}\} \tag{4-21}$$

As $r \to \infty$, $V(r) \approx 0$, so one has, from equation (4-21),

$$k = \frac{4\pi (D_A + D_B)\beta}{[B]} \{[B] - [B]_R \, e^{V(R)/k_B T}\} \tag{4-22}$$

For the general case there is a finite rate constant k_R for reaction during the encounter, where

$$\frac{-d[B]}{dt} = k[A][B] = k_R[A][B]_R \tag{4-23}$$

Substituting $[B]_R + (k/k_R)[B]$ into equation (4-22) and rearranging gives

$$k = \frac{4\pi (D_A + D_B)\beta}{1 + [4\pi(D_A + D_B)\beta/k_R \exp\{-V(R)/k_B T\}]} \tag{4-24}$$

This equation is the general phenomenological rate constant for a bimolecular reaction in solution. Before illustrating its uses, it is worthwhile to review the model assumed in deriving the equation. Equation (4-12), which describes bulk diffusion and conduction, is used to describe the relative motion between molecules of A and molecules of B at the microscopic level. Note that to integrate equation (4-18) the variation of the diffusion coefficient with concentration is neglected. However, if the reactant concentrations are low and if a constant solution environment is maintained, the diffusion coefficient does remain approximately constant during the reaction. To obtain the general diffusion equation (4-12), a steady state has been assumed; this assumption breaks down for reactions with short half-lives. Finally, spherical symmetry has been assumed for the reactants, but it holds only for simple reactions. Introducing nonspherical symmetry results in steric and alignment effects for reaction and reduces the solid angle in equation (4-24) from 4π to a smaller value. [See discussion preceding equation (4-16)].

4.3 DIFFUSION-LIMITED RATE CONSTANT

The diffusion-limited rate constant results when each encounter between A and B molecules leads to reaction. There are two approaches for finding this rate constant from the foregoing equations. If A and B do indeed react at each encounter, $[B]_R$ is zero since no B molecules exist at the critical radius. Thus, from equation (4-22), the diffusion-limited rate constant is

$$k_D = 4\pi(D_A + D_B)\beta \tag{4-25}$$

Alternatively, for the rate constant to be diffusion-limited, the rate constant k_R during the encounter is very large, so that the denominator of equation (4-24) becomes unity. For uncharged reactants $V(r) \approx 0$, so that equation (4-20) gives $\beta = R$. The diffusion-limited rate constant is then

$$k_D = 4\pi(D_A + D_B)R \tag{4-26}$$

This expression may also be obtained from equations (4-14) and (4-15). J_B in equation (4-14) equals $k[B]$, since as discussed above $d[B]_r/dt$ is the same for all values of r.

For a diffusion limited reaction $[B]_R = 0$, so that J_B in equation (4-15) equals $4\pi R(D_A + D_B)\,[B]$. Equating these two expressions for J_B gives equation (4-26).

For the spherical isotropic particle model used in the derivation of equation (4-24), the diffusion coefficient is related to the viscosity via the Einstein-Stokes relationship

$$D = \frac{k_B T}{6\pi r\eta(T)} \tag{4-27}$$

where $\eta(T)$ is the temperature-dependent viscosity of the solution, and r is the radius of the particle. Writing the critical radius R as $(r_A + r_B)$ and substituting equation (4-27) into equation (4-26) gives

$$k_D = \frac{2k_B T (r_A + r_B)^2}{3\eta(T) r_A r_B} \tag{4-28}$$

Thus, the rate constant is proportional to temperature and inversely proportional to viscosity.

It is of interest to estimate k_D for neutral reactants. A representative value for r_A and r_B is 2×10^{-8} cm, so that the critical radius R is 4×10^{-8} cm. A typical aqueous solution diffusion coefficient at 300 K is 1×10^{-5} cm^2/sec. Thus, from equation (4-26), the value for k_D in liters/mole-sec is

$$k_D = \frac{4\pi(4 \times 10^{-8})\,(2 \times 10^{-5})\,(6 \times 10^{23})}{10^3}$$

$$= 6 \times 10^9$$

This may be compared with 3×10^{11} liters/mole-sec for a normal binary gas collision rate constant. Accordingly, liquid-phase reactions are slowed down by a factor of about 50 due to diffusion, with the rate dependent on the viscosity of the solvent.

A diffusion-limited rate constant can also be evaluated for reactions between ions. The potential energy between two ions in solution is (see Chapter 7, Section 7.1)

$$V(r) = \frac{z_A z_B e^2}{\varepsilon r} \tag{4-29}$$

where ε is the dielectric constant. Inserting this expression for $V(r)$ into equation (4-20) yields

$$\beta = \frac{-z_A z_B r_0}{1 - \exp(z_A z_B r_0 / R)} \tag{4-30}$$

The parameter r_0 has dimensions of length and is given by

$$r_0 = \frac{e^2}{\varepsilon k_B T} \tag{4-31}$$

For water at 25°C, r_0 equals 7×10^{-8} cm. The term β is positive for any combination of ion charges. For ions of the same charge the exponential in equation (4-30) is larger than unity, since r_0 is ordinarily greater than R. If the ions are oppositely charged, the

TABLE 4-1   Effect of Reactant Charge on the
Diffusion-Limited Rate Constant[a]

$z_A z_B$	β/R
0	1
−1	2.14
−2	3.65
−3	5.30
1	0.362
2	0.105
3	0.029

[a]β/R is for an aqueous solution at 25°C; r_0 in
equation (4-31) equals 7.1×10^{-8} cm and R
equals 4.0×10^{-8} cm.

exponential is small and β is approximately $-z_A z_B r_0$. The diffusion-limited rate constant for reactions between ions is found by inserting the expression for β in equation (4-30) into equation (4-25). The rate constant here differs from that for reactions between neutrals, given by equation (4-26), by the ratio β/R. Representative values of this ratio are listed in Table 4-1 for a water solution with R equal to 4×10^{-8} cm. As expected, ions of opposite charge react more rapidly than neutral molecules, while ions of the same charge react more slowly.

4.4 SLOW REACTIONS

The preceding solution reactions may all be depicted by the mechanism

$$A + B \underset{k_{uni}}{\overset{k_D}{\rightleftharpoons}} AB \xrightarrow{k_2} \text{products} \tag{4-32}$$

where AB is the encounter complex and k_{uni} is the unimolecular rate constant for dissociation of the encounter complex (i.e., diffusion of the reactants apart from the solvent cage). A steady-state treatment of this mechanism yields the rate law

$$\frac{-d[B]}{dt} = k[A][B] = \frac{k_2 k_D}{k_{uni} + k_2} [A][B] \tag{4-33}$$

Only if k_2 is much larger than k_{uni} is the overall reaction rate diffusion-limited. For a slow reaction where $k_{uni} \gg k_2$, the first step in equation (4-32) is in equilibrium and

$$k = K k_2 \tag{4-34}$$

where $K = k_D/k_{uni}$.

The approach used in section 4.2 for calculating the bimolecular rate constant may also be used to calculate the maximum rate constant for dissociation of the encounter complex. The only difference is in the boundary conditions which become $[AB] = 0$ and $V = 0$ at $r = \infty$, and $[AB] = 1/\Delta V$ and $V = V(R)$ at $r = R$. In other words, at infinite separation the complex does not exist. On the other hand, the volume

ΔV contains the encounter complex AB and is approximately $4/3\pi R^3$. Beginning with equation (4-13), and using the same reasoning as before, we obtain

$$k_{uni} = 4\pi(D_A + D_B)\,\frac{\beta\exp[V(R)/k_BT]}{\Delta V} \tag{4-35}$$

The equilibrium constant in equation (4-34) is then

$$K = \frac{k_D}{k_{uni}} = \frac{4\pi R^3}{3}\,10^{-3}\,N_0\exp[-V(R)/k_BT] \tag{4-36}$$

for concentration in units of M^{-1}.

The phenomenological treatment developed in section 4.2 can be used to interpret slow reactions. In these cases there is a substantial barrier for the encounter complex to form reaction products, so that $k_D \gg k_R\exp[-V(R)/k_BT]$. Thus, equation (4-24) reduces to

$$k = k_R\exp[-V(R)/k_BT] \tag{4-37}$$

This equation can be combined with equations (4-34) through (4-36) to find the relationship between k_2 in equation (4-32) and k_R.

Equation (4-37) has been used most frequently for interpreting reactions between ions. Substituting the potential given by equation (4-29) gives a rate constant dependent on the dielectric constant and contact radius; i.e.,

$$k = k_R\exp(-z_Az_Be^2/\varepsilon k_BTR) \tag{4-38}$$

By varying the composition of the solvent, the dielectric constant for the reaction can be adjusted. A plot of the logarithm of k versus $1/\varepsilon$ should be linear with the slope giving the value of R. Such an analysis has been performed for slow ionic reactions, and physically realistic values of R have been obtained.

4.5   EFFECT OF IONIC STRENGTH ON REACTIONS BETWEEN IONS

If the ionic strength of a solution is sufficiently high ($>10^{-4}$ M) so that the ion-atmosphere affects conductance and equilibrium properties, the derivation of the diffusion-controlled rate constant using equations (4-29) through (4-31) is no longer valid. Instead, the simple coulombic potential given by equation (4-29) must be modified to account for the shielding which results from the ion atmosphere. The approach taken here is to use a form for the potential developed by Debye and Hückel.[2] Using that approach, a general expression for the phenomenological rate constant can be derived from equation (4-24).

To calculate the potential $V(r)$ between two ions in an ionic atmosphere, we begin with the Poisson differential equation. In the case of a spherically symmetric distribution of charges about a central ion, the electrical potential $E(r)$ is a function only of r, the distance from the central ion. For this situation the Poisson equation is

$$\frac{1}{r^2}\frac{d}{dr}\left[r^2\frac{dE(r)}{dr}\right] = -\frac{r\pi\rho}{\varepsilon} \tag{4-39}$$

where ρ is the charge density at radius r. The charge density is positive if positive charges predominate, negative if negative charges predominate, and zero if they balance. The electrostatic potential $E(r)$ is due to the central ion and its surrounding atmosphere. The solutions to equation (4-39) give this potential at any r as a function of ρ. To solve equation (4-39) requires evaluating ρ for the distribution of ions.

In an ionic solution ρ will not be zero in the region surrounding an ion A, since there is a tendency for ions of opposite sign to concentrate around A. The Boltzmann law, equation (4-10), gives the concentration of a particular ion i at a radius r from A. Since $V(r)$ is related to the electric potential via $V(r) = z_i e E$, we have

$$c_i = c_i^0 \exp[-z_i e E(r)/k_B T] \tag{4-40}$$

The charge density in a region of the solution is the summation of $c_i z_i e$ over the different kinds of ions (positive and negative) that may be present; i.e.,

$$\rho = \sum_i c_i z_i e$$

or

$$\rho = \sum_i c_i^0 z_i e \exp[-z_i e E(r)/k_B T] \tag{4-41}$$

The Debye-Hückel approximation, based on the assumption that the solution is so dilute that ions will rarely be close together, is now made. For this situation $z_i e E(r) \ll k_B T$, since the interionic potential is much less than the average thermal energy. The exponential factor in equation (4-41) may then be expanded as

$$\exp[-z_i e E(r)/k_B T] = 1 - \frac{z_i e E(r)}{k_B T} + \cdots \tag{4-42}$$

where the higher-order terms are negligible. With this expansion equation (4-41) becomes

$$\rho = \sum_i c_i^0 z_i e - \frac{E(r)e^2}{k_B T} \sum_i c_i^0 z_i^2 \tag{4-43}$$

The first term is zero because there is electrical neutrality at large r, where $V(r)$ equals zero. Thus,

$$\rho = \frac{-e^2 E(r)}{k_B T} \sum_i c_i^0 z_i^2 \tag{4-44}$$

Substituting equation (4-44) into equation (4-39) gives the Poisson-Boltzmann equation

$$\frac{1}{r^2} \frac{d}{dr} \left[r^2 \frac{dE(r)}{dr} \right] = \frac{4\pi e^2 E(r)}{\varepsilon k_B T} \sum_i c_i^0 z_i^2$$

or

$$\frac{d}{dr} \left[r^2 \frac{dE(r)}{dr} \right] = b^2 r^2 E(r) \tag{4-45}$$

where

$$b^2 = \frac{4\pi e^2}{\varepsilon k_B T} \sum_i c_i^0 z_i^2 \tag{4-46}$$

The quantity $1/b$ has the dimensions of length and is often called the *Debye screening length*. It may be regarded as an approximate measure of the thickness of the ionic atmosphere surrounding an ion.

The term $\Sigma c_i^0 z_i^2$ is closely related to the ionic strength, defined as $I = 1/2 \, \Sigma m_i z_i^2$, where m_i is the molality of ionic species i in the solution. Since c_i^0 is the number of ions per cm^3, we have

$$\sum_i c_i^0 z_i^2 = \frac{2 N_0 d}{1000} I \tag{4-47}$$

where N_0 is Avogadro's number and d is the density of the solution. Thus b^2 can be expressed as

$$b^2 = \frac{8\pi e^2 N_0 d}{1,000 \varepsilon k_B T} I \tag{4-48}$$

The solution to equation (4-45) is obtained by substituting $x = rE$ from which $d^2 x / dr^2 = b^2 x$, and therefore,

$$x = C_1 e^{-br} + C_2 e^{br}$$

or

$$E(r) = \frac{C_1}{r} e^{-br} + \frac{C_2}{r} e^{br} \tag{4-49}$$

The integration constants C_1 and C_2 are determined by the boundary conditions. Clearly, C_2 is zero, since $E(r)$ approaches zero as r goes to infinity. C_1 can be determined by noting that when $b = 0$, i.e., the ionic strength is zero, the electrical potential is simply that of a single ion A in the absence of any other ions and is given by $E(r) = z_A e / \varepsilon r$. Thus, the final solution is

$$E(r) = \frac{z_A e}{\varepsilon r} \exp(-br) \tag{4-50}$$

Since, as previously discussed, a dilute ionic solution with small b is assumed in deriving this expression, the exponential may be expanded without further approximation to give

$$E(r) = \frac{z_A e}{\varepsilon r} - \frac{z_E e}{\varepsilon} b \tag{4-51}$$

The first term is simply the electrical potential of an ion in a medium with dielectric constant ε. The second term is the potential resulting from the ions which form the ionic atmosphere about the ion A.

The interaction potential $V(r)$ between two ions A and B in an ionic solution can now be written. For B with charge z_B in the electric field of A, the potential energy is

$$V(r) = z_B e E(r) = \frac{z_A z_B e^2}{\varepsilon r} - \frac{z_A z_B e^2}{\varepsilon} b \tag{4-52}$$

The first term is the potential at infinite dilution, denoted by $V^0(r)$. Dividing by $k_B T$ gives

$$\frac{V(r)}{k_B T} = \frac{V^0(r)}{k_B T} - z_A z_B r_0 b \tag{4-53}$$

where r_0 is as defined in equation (4-31). To obtain the rate constant, β in equation (4-20) is evaluated using equation (4-53). The result is

$$\beta = \beta^0 \exp(z_A z_B r_0 b) \tag{4-54}$$

where β^0, given by equation (4-30), is the value of β at infinite dilution. Inserting this expression for β and equation (4-53) for $V(r)$ into equation (4-24) yields

$$k = \frac{k_D^0 \exp(z_A z_B r_0 b)}{1 + k_D^0 / k_R \exp[-V^0(R)/k_B T]} \tag{4-55}$$

which may be written as

$$k = k^0 \exp(z_A z_B r_0 b) \tag{4-56}$$

The rate constants k_D^0 and k^0 are for infinite dilution.

Equation (4-56) becomes

$$\log_{10} k = \log_{10} k^0 + 1.02 z_A z_B I^{1/2} \tag{4-57}$$

at 25°C in water. Thus, a plot of log k vs. $I^{1/2}$ is linear with a slope of $1.02 z_A z_B$. A few examples are illustrated in Figure 4-3. The effect of the ionic strength is to increase the rate for ions with charges of the same sign and to decrease it when the charges are opposite in sign; the increased ionic atmosphere counteracts the simple electrostatic interaction between A and B. This variation of rate constants with ionic strength is often called the *primary salt effect*.

4.6   LINEAR FREE-ENERGY RELATIONSHIPS

Empirical correlations have been very useful in systematizing a large number of rate coefficients and equilibrium constants for organic reactions in solution. They provide a predictive method for estimating rate data when no such data are available. Many of these correlations are of a general type called *linear free-energy relationships*. Consider a general solution reaction denoted by

$$RX + A \rightarrow products \tag{4-58}$$

where R contains the reactive site and X is a substituent not directly involved in the reaction. The empirical correlation which is made is to plot log k vs. log K for reactions

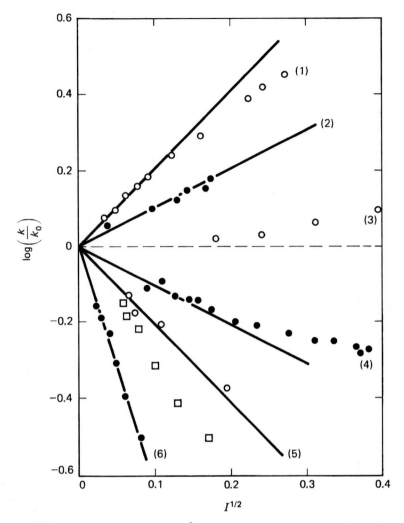

FIGURE 4-3   Plots of $\log_{10}(k/k^0)$ versus the square root of the ionic strength for ionic reactions of various types. The reactants are:
(1) $BrCH_2COO^- + S_2O_3^{2-}$
(2) $e_{aq} + NO_2^-$
(3) $H_3O^+ + C_{12}H_{22}O_{11}$ (inversion of sucrose)
(4) $H_3O^+ + Br^- + H_2O_2$
(5) $OH^- + Co(NH_3)_5Br^{2+}$ [ionic strength varies with NaBr (circles) and Na$_2$SO$_4$ (squares)]
(6) $Fe(H_2O)_6^{2+} + Co(C_2O_4)_3^{3-}$
[Adapted from R. E. Weston, Jr., and H. A. Schwarz, *Chemical Kinetics* (Englewood Cliffs, NJ: Prentice-Hall, Inc., 1972).]

of this type which simply differ by the substituent X, where k is the rate constant and K the equilibrium constant for the particular reaction. What is observed for many cases is that the plot is linear, i.e., $\log k = \alpha \log K + \text{constant}$, so that

$$\frac{k'}{k} = \left(\frac{K'}{K} \right)^{\alpha} \tag{4-59}$$

where k' and K' are for the reaction with the substituent X′.

Such a relationship between k and K is characteristic of an overall reaction mechanism controlled by one elementary reaction, which is apparently the case for many solution reactions. To derive equation (4-59), two assumptions must be made. The first is that the elementary rate constant is proportional to $\exp(-\Delta G^{a}/RT)$, where ΔG^{a} is the free energy of activation. In Chapter 10 it is shown that this proportionality is predicted by transition state theory. The second assumption is that, for a specific class of reactions like those in equation (4-58), ΔG^{a} is related to the free energy of the overall reaction ΔG^{0} according to

$$\Delta G^{a} = \alpha \Delta G^{0} + \text{constant} \tag{4-60}$$

This is the linear free-energy relationship.

One of the best known linear free-energy relationships is the Hammett equation,[3]

$$\log_{10} K = \log_{10} K_0 + \sigma \rho \tag{4-61}$$

which correlates the effects of *meta-* or *para-*substituents on equilibrium constants for reactions involving aromatic compounds. K is the equilibrium constant for the reaction involving the substituted compound, K_0 is for the same reaction of the unsubstituted compound, σ is an empirical parameter that varies only with the substituent and its position, and ρ is an empirical parameter that is characteristic of the particular reaction series, including the solvent, that is being studied.

The ionization of benzoic acid in water at 25°C is the reference reaction used to establish values for σ and ρ. The definitions are $\rho = 1$ and $\sigma = \log(K_i/K_{i0})$, where K_i and K_{i0} are the ionization constants for the substituted and unsubstituted benzoic acid, respectively. Representative values of σ are listed in Table 4-2. These values can be inserted into equation (4-61) to determine values of ρ for benzoic acid in other solvents, or other sets of equilibria in various solvents.

By combining equations (4-59) and (4-61), a free-energy relationship is found for correlating rates of substituted aromatic compounds. This relationship is given by

$$\log_{10} k = \log_{10} k_0 + \sigma \rho^{a} \tag{4-62}$$

where ρ^{a} is used to differentiate it from ρ in the equilibrium free-energy equation (4-61). The σ values are those for substituted benzoic acids given in the previous paragraph. As stated above, the parameter ρ depends on the particular reaction.

Table 4-3 lists some values of ρ for acid dissociation equilibria. There are also extensive lists of ρ-values for rates of various reactions in different solvents. Essentially, the parameter ρ gives the sensitivity of the reaction (or equilibrium) to a change in substituent. A positive value of ρ means that a reaction (or particular equi-

TABLE 4-2 σ-Values of Common Substituents

Substituent	σ_{meta}	σ_{para}
H	0	0
NH$_2$	-0.16	-0.66
CH$_3$	-0.07	-0.17
C$_6$H$_5$	0.06	-0.01
OH	0.12	-0.37
OCH$_3$	0.12	-0.27
F	0.34	0.06
Cl	0.37	0.23
Br	0.39	0.23
I	0.35	0.18

TABLE 4-3   Values of the Reaction Constant ρ for Acid Dissociation Equilibria

Acid[a]	Solvent	ρ
Ar-COOH	H$_2$O	1.00
Ar-COOH	C$_2$H$_5$OH	1.96
Ar-OH	H$_2$O	2.26
Ar-CH$_2$COOH	H$_2$O	0.56
Ar-NH$_3$	H$_2$O	2.94
Ar(NO$_2$)-COOH	H$_2$O	0.91

[a] Ar = substituted phenyl group

librium) responds to the substituent in the same ways as does benzoic acid ionization equilibrium. If $\rho > 1$ the reaction (equilibrium) is more sensitive to the substituent, while the reaction is less sensitive if $0 < \rho < 1$. If ρ is negative, the substituents affect the reaction (equilibrium) in a manner opposite to what is found for benzoic acid dissociation.

There are several ways to identify the Hammett equations (4-61) and (4-62) as linear free-energy relationships. Since the equilibrium constant is related to the free energy via $\Delta G^0 = -RT\ln K$, equation (4-61) can be written as

$$\Delta G^0 = \Delta G_0^0 - 2.303\ RT\sigma\rho \tag{4-63}$$

The same equation holds for another class of reactions, except that ΔG^0, ΔG_0^0, and ρ are different and identified by a ($'$). Dividing the two equations by ρ and ρ', respectively, and subtracting gives

$$\frac{\Delta G^0}{\rho} - \frac{\Delta G^{0'}}{\rho} = \frac{\Delta G_0^0}{\rho} - \frac{\Delta G_0^{0'}}{\rho} \tag{4-64}$$

where it is assumed that ΔG^0 and $\Delta G^{0'}$ are for the same substituent and position. This equation can be rewritten as

$$\Delta G_0^{0'} - \Delta G^{0'} = \frac{\rho'}{\rho}\ (\Delta G_0^0 - \Delta G^0) \tag{4-65}$$

Thus, in the new reaction series the difference between the free-energy changes for the unsubstituted and substituted reactions is proportional to the analogous difference for the initial reaction class. An equation similar to equation (4-65) can be derived for the free energies of activation.

A relationship can also be written between the free energy of activation ΔG^a and the overall free energy change ΔG^0 for a chemical reaction. From equation (4-62) and the relation $k \propto \exp(-\Delta G^a/RT)$, we have

$$\Delta G^a = \Delta G_0^a - 2.303RT\sigma\rho^a \tag{4-66}$$

where ρ^a is used to differentiate it from ρ in the equilibrium free-energy equation (4-63). Combining equations (4-63) and (4-66) yields

$$\Delta G^a = \frac{\rho^a}{\rho}\,\Delta G^0 + \left[\Delta G_0^a - \frac{\rho^a}{\rho}\,\Delta G_0^0\right] \tag{4-67}$$

This is the relationship between ΔG^a and ΔG^0 predicted by the Hammett equations and may be compared with equation (4-60). Although the Hammett correlations are extremely successful, they do not apply well to the reactions of aliphatic and some aromatic compounds. Related correlations have been suggested for these cases. Another widely used free-energy relationship is the Brönsted correlation of acid-base catalysis.[4]

4.7   RELAXATION METHODS FOR FAST REACTIONS

Most solution reactions involve a complex multistep mechanism. In analyzing the mechanism, it is often necessary to determine rate constants for the individual elementary reactions. The time scales for these reactions may range from minutes to 10^{-12} sec. Different experimental techniques have been developed for measuring rate constants over such a broad time range. Flow techniques may be used if the reaction time is less than 10^{-4} sec, while ESR and NMR spectral line shape methods are applicable for reaction times between 10^{-1} and 10^{-10} sec. One of the most successful techniques, which spans the 1—10^{-11}-sec time scale, is the *relaxation method*.

As a typical example, consider the formation of transition-metal complexes given by the reactions

$$M^{+2}(H_2O)_6 + L \underset{k_{-1}}{\overset{k_1}{\rightleftharpoons}} ML^{+2}(H_2O)_5 + H_2O$$

$$ML^{+2}(H_2O)_5 + L \underset{k_{-2}}{\overset{k_2}{\rightleftharpoons}} ML_2^{+2}(H_2O)_4 + H_2O \tag{4-68}$$

$$ML_2^{+2}(H_2O)_4 + L \underset{k_{-3}}{\overset{k_3}{\rightleftharpoons}} ML_3^{+2}(H_2O)_3 + H_2O$$

$$\vdots \qquad \vdots \qquad \vdots$$

where M^{+2} is a cation such as Ni^{+2}, Co^{+2}, or Cu^{+2} and L is a unidentate ligand like NH_3. The mechanism can be easily generalized to include charged ligands or a multidentate ligand such as $NH_2(CH_2)_3NH_2$. If all six reactions are occurring at once, analysis of the system is quite complex. However, by choosing the proper relative concentrations, it is often possible to isolate one or two of the reactions. For example, with $[M^{+2}] \gg [L]$, the major species are M^{+2} and ML^{+2}. At a higher concentration of L there may be predominantly ML_2^{+2} and ML_3^{+2}. Thus, for the correct choice of initial concentrations the complex mechanism may be dominated by only one equilibrium, denoted by

$$ML_{n-1} + L \underset{k_{-n}}{\overset{k_n}{\rightleftharpoons}} ML_n$$

with equilibrium constant $K_n = k_n/k_{-n}$. The generic reaction under investigation is then

$$A + B \underset{k_{-1}}{\overset{k_1}{\rightleftharpoons}} C \tag{4-69}$$

with concentrations [A], [B], and [C]. The rate constants can be measured by the relaxation method developed by Eigen[5] in the 1950s. In this technique, the position of the equilibrium is shifted and the time required for equilibrium to be reestablished is measured. If the displacement from equilibrium is small, the relaxation to the new equilibrium will be first order.

Equation (4-69) is described by the usual rate equation,

$$\frac{-d[A]}{dt} = \frac{-d[B]}{dt} = \frac{d[C]}{dt} = k_1[A][B] - k_{-1}[C] \tag{4-70}$$

If $a, b,$ and c are the new equilibrium concentrations resulting from the perturbation of the system, we have

$$x = a - [A] = b - [B] = [C] - c \tag{4-71}$$

where x is the difference between the instantaneous concentrations and the concentrations for the new equilibrium. Equation (4-70) may then be written as

$$\frac{dx}{dt} = k_1 (a - x) (b - x) - k_{-1}(c + x)$$

or

$$\frac{dx}{dt} = k_1ab - k_1(a + b)x + k_1x^2 - k_{-1}c - k_{-1}x \tag{4-72}$$

By the definition of equilibrium, $k_1ab - k_{-1}c = 0$. The equation can then be *linearized* by making the assumption that the displacement is small, so that $x^2 \ll x$. When we do so, we have

$$\frac{dx}{dt} = -[k_1(a + b) + k_{-1}]x$$

$$= -kx = \frac{-x}{\tau} \tag{4-73}$$

where τ is the relaxation time. By measuring τ as a function of $a + b$, and knowing the equilibrium constant, both k_n and k_{-n} can be obtained. Note that in deriving equation (4-73) we have identified a general principle, viz., that "all rate equations near equilibrium turn into first-order differential equations which have simple exponential solutions."

The relaxation time τ in equation (4-73) is the average "lifetime" of x. To show this, consider the integrated solution of equation (4-73), which is $x = x_0 \exp(-kt)$. The probability that x has a lifetime in the range $t \to t + dt$ is then

$$P(t)dt = \frac{-dx}{x_0} = k \exp(-kt) \, dt \tag{4-74}$$

The average lifetime is

$$\langle t \rangle = \int_0^\infty tP(t) \, dt = k \int_0^\infty t \exp(-kt) \, dt \tag{4-75}$$

which, when integrated, gives $\langle t \rangle = k^{-1} = \tau$.

Often it is impossible to isolate one reaction in a complex mechanism. In that case, a set of coupled reactions as depicted in equation (4-68) must be solved. By using the procedure just described, we can derive, for the coupled reactions, a system of linear differential equations which may be solved by the matrix method described in chapter 2, section 2-5. To illustrate this method, consider the two coupled equations

$$A + B \underset{k_{-1}}{\overset{k_1}{\rightleftharpoons}} C$$

and

$$C + B \underset{k_{-2}}{\overset{k_2}{\rightleftharpoons}} D \tag{4-76}$$

If the rate equations are linearized and the material balance and equilibrium relations used, one obtains the linear differential equations

$$\frac{-d\Delta[A]}{dt} = \{k_1(a + b) + k_{-1}\}\Delta[A] + (k_{-1} - k_1 a)\Delta[D]$$

$$= c_{11}\Delta[A] + c_{12}\Delta[D] \tag{4-77}$$

and

$$\frac{d\Delta[D]}{dt} = k_2(b - c)\Delta[A] + \{k_2(b + c) + k_{-2}\} \Delta[D]$$

$$= c_{21}\Delta[A] + c_{22}\Delta[D] \tag{4-78}$$

where the Δ's are the displacement in the concentrations from equilibrium; i.e., $[A] = a + \Delta[A]$.

These equations are solved as outlined in chapter 2, section 2-5, and the relaxation times are found from the secular determinantal equation

$$\begin{vmatrix} c_{11} - 1/\tau & c_{12} \\ c_{21} & c_{22} - 1/\tau \end{vmatrix} = 0 \tag{4-79}$$

The two solutions to the resulting quadratic equation are given by

$$\frac{1}{\tau_{1,2}} = \frac{1}{2}\{(c_{11} + c_{22}) \pm [(c_{11} + c_{22})^2 - 4(c_{11}c_{22} - c_{12}c_{21})]^{1/2}\} \tag{4-80}$$

and the displacements $\Delta[A]$ and $\Delta[D]$ from equilibrium are sums of exponentials; e.g.,

$$\Delta[A] = a_1 \exp(-t/\tau_1) + a_2 \exp(-t/\tau_2) \tag{4-81}$$

The coefficients a_1 and a_2 are found from the secular equations as described in chapter 2, section 2.5.

Several experimental methods are available for performing relaxation kinetics measurements on microsecond time scales. One of the most versatile, pioneered by Eigen and coworkers, is the *temperature-jump* method. In this method, the electrical charge on a storage capacitor is released through a small volume of electrolytic solution containing the reactants. The energy heats the solution, which has a small but finite electrical resistance. In a typical apparatus, 45 J of electrical energy may be discharged in 1 μsec through 10 cm^3 of solution. The specific heat of aqueous solutions is essentially that of water, viz., 4.18 J/g. The rise in temperature of the solution is thus on the order of 1°C, which is enough to shift the equilibrium by van't Hoff's law,

$$\left(\frac{\partial \log_{10} K}{\partial T} \right)_P = \frac{\Delta H^\circ}{2.303 \, RT^2} \tag{4-82}$$

A variation of this technique is the *pressure-jump* method, in which a volume of compressed gas is suddenly released into a chamber containing the solution. In this case, the equilibrium shift is given by

$$\left(\frac{\partial \log_{10} K}{\partial P} \right)_T = - \frac{\Delta \tilde{V}}{2.303 RT} \tag{4-83}$$

Properties of excited electronic states can be exploited in this technique. Since acid dissociation constants of organic species are often quite different in the ground and excited electronic states, excitation may release protons into the solution, resulting in a *pH-jump*. The excitation may be carried out by a pulsed laser, thus affording nanosecond or even picosecond time resolution.

Of course, the method used for following the adjustment of the reaction to the new equilibrium must be fast enough to detect whatever changes occur. The two most commonly used methods are optical absorption (frequently with a pH indicator added to amplify the absorbance changes) and electrical conductance for ionic reactions. In addition to acid-base and metal-ligand reactions, the kinetics of many enzyme-catalyzed reactions have been elucidated by the use of relaxation techniques.

REFERENCES

[1] E. Rabinowitch and W. C. Wood, *Trans. Faraday Soc. 32,* 1381 (1936).

[2] P. Debye and E. Hückel, *Physik. Z. 24,* 185 (1923).

[3] L. P. Hammett, *Physical Organic Chemistry* (New York: McGraw-Hill, 1940), chapter 7.

[4] J. N. Brönsted and K. Pederson, *Z. Phys. Chem.* (*Leipzig*) *108,* 185 (1924).

[5] M. Eigen, *Disc. Faraday Soc. 17,* 194 (1954).

BIBLIOGRAPHY

BERRY, R. S., RICE, S. A., and ROSS, J. *Physical Chemistry.* New York: Wiley, 1980.

GARDINER, W. C., JR., *Rates and Mechanisms of Chemical Reactions.* Menlo Park, CA: W. A. Benjamin, 1972.

HAMMES, G. G. *Principles of Chemical Kinetics.* New York: Academic Press, 1978.

JORDAN, P. C. *Chemical Kinetics and Transport.* New York: Plenum Press, 1979.

LAIDLER, K. J. *Chemical Kinetics.* New York: Harper and Row, 1987.

LEFFLER, J. E., and GRUNWALD, E., "*Rates and Equilibria of Organic Reactions as treated by Statistical, Thermodynamic, and Extrathermodynamic Methods.*" New York: Wiley, 1963.

LOWRY, T. H., and RICHARDSON, K. S. *Mechanism and Theory in Organic Chemistry.* New York: Harper and Row, 1981.

MOORE, W. J. *Physical Chemistry*. Englewood Cliffs, NJ: Prentice-Hall, 1962.

WESTON, R. E., Jr., and SCHWARZ, H. A. *Chemical Kinetics*. Englewood Cliffs, NJ: Prentice-Hall, 1972.

PROBLEMS

4.1 The results of the alkaline hydrolysis of ethyl nitrobenzoate at various times are reported below. Determine the order of the reaction and find the rate constant.

t(sec)	0	100	200	300	400	500	600	700	800
[A] (10^{-2}mol/dm³)	5.00	3.55	2.75	2.25	1.85	1.60	1.48	1.40	1.38

4.2 Consider the following mechanism for a chemical reaction in solution.

$$A + B \underset{k_{-d}}{\overset{k_d}{\rightleftarrows}} (AB) \overset{k}{\longrightarrow} \text{products, P}$$

(a) Show that the overall rate law is second-order if the steady-state assumption is applied to AB.

(b) What is the expression for the second-order rate constant k_2?

(c) Show what relations between the rate constants give the diffusion-controlled limit and the activation-controlled limit.

4.3 The substitution constant for OCH_3 is +0.12 in the meta position and -0.27 in the para position. What is the rate constant ratio for the ionization of

meta   and   *para*

in H_2O at 25°C?

4.4 The bimolecular rate constant for a reaction in solution is

$$k = \frac{4\pi(D_A + D_B)\beta}{1 + [4\pi(D_A + D_B)\beta/k_R e^{-V(R)/k_B T}]}$$

(a) What are the units for the rate constant?

(b) What is the expression for the diffusion limited rate constant?

(c) Express the diffusion limited rate constant in units of liters/mole-sec.

(d) What is the rate constant expression for a slow reaction?

4.5 How does increasing the ionic strength affect the rates of the following reactions?

(a) $S_2O_8^{-2} + 2I^- \rightarrow I_2 + 2SO_4^{-2}$

(b) $H_2O_2 + 2H^+ + 2Br^- \rightarrow 2H_2O + Br_2$

(c) $Co(NH_3)_5 Br^{+2} + OH^- \rightarrow Co(NH_3)_5 OH^{+2} + Br^-$

(d) $O_2NNCOOC_2H_5^- \rightarrow N_2O + CO_3^{-2} + C_2H_5OH$

For each of these cases, what ionic strength will cause a change in rate of 50%?

4.6 Following the linearizing of rate expressions presented in Section 4.7, determine the relationships between the relaxation time and the rate constants for the equilibria

$$A \rightleftarrows B, 2A \rightleftarrows A_2, \text{ and } A + B \rightleftarrows C + D$$

4.7 What is the temperature dependence of a slow solution reaction? What is the temperature dependence of the rate constant for the diffusion-controlled reaction $A^+ + B^- \rightarrow AB$?

4.8 The complexation reaction between $Rh_2(OAc)_4 \cdot 2H_2O$ and 5'–AMP (adenosine monophosphate) has been investigated using temperature-jump techniques. The reaction provides an interesting example of relaxation kinetics because two equilibria are maintained. Denoting the dirhodium complex by M and the 5'–AMP by L, we find that the important equilibria are

$$M \cdot 2H_2O + L \underset{k_{-1}}{\overset{k_1}{\rightleftarrows}} M \cdot L(H_2O) + (H_2O), \quad K_1 = 1,893 \ M^{-1}$$

$$M \cdot L(H_2O) + L \underset{k_{-2}}{\overset{k_2}{\rightleftarrows}} M \cdot 2L + (H_2O), \quad K_2 = 202 \ M^{-1}$$

where the equilibrium constants were determined at 25°C. The approach to equilibrium was followed by a spectrophotometric technique which was sensitive to the concentration of the dirhodium complex $M \cdot 2H_2O$. The decay constant was determined at 25°C for a variety of *total* concentrations of both M and L, c_M, and c_L. The data are given in the following table:

Dependence of the Decay Constant in the $Rh_2(OAc)_4$
$2H_2O$ − (5'-AMP) System on Concentration

c_M (mM)	c_L (mM)	$1/\tau \times 10^{-3}$ (sec^{-1})	[M] (mM)	[L] (mM)
1.94	1.96	5.88	0.79	0.67
1.92	2.91	6.25	0.50	1.21
1.92	3.38	7.14	0.41	1.51
1.91	3.85	7.69	0.33	1.84
1.90	4.31	8.70	0.27	2.19
1.89	4.76	9.09	0.23	2.54

Source: K. Das, E. L. Simmons, and J. L. Bear, *Inorg. Chem. 16,* 1268 (1977); L. Rainen, R. A. Howard, A. O. Kimball, and J. L. Bear, *Inorg. Chem. 14,* 2752 (1975).

(a) Show that if the reaction $ML + L \rightleftarrows ML_2$ were being monitored for this system, the decay constant would be

$$\frac{1}{\tau} = k_2\left([ML] + [L] + \frac{1}{K_2}\right)$$

(b) Use the data in the table and the values of K_1 and K_2 to compute [ML].

(c) Test the hypothesis of part (a) both graphically and via a least-squares procedure.

4.9 The following bimolecular rate coefficients relate to a substitution reaction of bromacetic acid. Deduce the product of the formal charges of the reactants.

Rate coef/dm^{3}mol^{-1}min^{-1}:	0.288	0.301	0.334	0.371	0.429
10^3 × ionic strength/mol dm^{-3}	1	2	4	10	20

4.10 The diffusivity of I_2 in CCl_4 is 1.5×10^{-5}cm^{2}s^{-1} at 320 K. At the same temperature the rate coefficient for combination of atomic iodine in CCl_4 was recorded as 7×10^{12} cm^3 mol^{-1}s^{-1} with a flash method. If the reaction is diffusion controlled, and the diffusivity of the atom is the same as for the diatomic molecule, do the data give an acceptable model? The equilibrium nuclear separation of I_2 is 2.66×10^{-8} cm.

4.11 Verify equation (4-80) for relaxation times τ_1 and τ_2 using equations (4-77) through (4-79). The material balance relations are

$$\Delta[A] = -\Delta[C] - \Delta[D] \quad \text{and} \quad \Delta[A] - \Delta[B] = -\Delta[D]$$

C H A P T E R 5

Catalysis

Many of the mechanisms discussed in the preceding chapters are characterized by a *rate-limiting step,* that is, one elementary reaction whose rate is significantly slower than that of the other steps with which it is coupled. The overall rate of the chemical transformation is constrained by the slow step through which the reactants must pass. (A toll booth across a busy highway is a familiar analogy!) Very often, in practical situations, the greatest possible net rate of chemical conversion, or *throughput,* is desirable. While the rate coefficient of any reaction possessing typical Arrhenius behavior (see chapter 1, section 1.5) can be increased by raising the temperature, it is often impractical to do so because the equilibrium concentration of the desired product may decrease with temperature, or undesirable side reactions which consume product or yield impurities may occur at the high temperatures. In such situations the use of a *catalyst** is usually helpful. A catalyst is defined as a chemical substance which increases the rate of a chemical reaction without itself being consumed in the reaction.

If the catalyst is uniformly dispersed in a gas or liquid phase solution, we may speak of *homogeneous* catalysis (see section 5.2). Enzyme-catalyzed biochemical reactions (see section 5.4) are a special case of homogeneous catalysis. Very often, however, the catalytic process occurs at the surface of a solid particle that is in contact with the gaseous or liquid solution. This is called *heterogeneous* catalysis (see section 5.5). Many important industrial catalysts are of the heterogeneous variety. Before dealing with these specific types of catalysis in detail, we consider, in the next section, some general principles of catalysis.

5.1 CATALYSIS AND EQUILIBRIUM

The effect of a catalyst C on a slow reaction A $\rightarrow$ B can be written formally as

$$A + C \rightarrow B + C$$

---

* The word "catalyst" is derived from the Greek $\kappa\alpha\tau\alpha$ (down) and $\lambda\upsilon\varepsilon\iota\nu$ (to loosen), evoking how a catalyst *lowers* the activation energy and *loosens* chemical bonds.

If the overall reaction is reversible, i.e., if

$$A \underset{k_{-1}}{\overset{k_1}{\rightleftharpoons}} B$$

with

$$\frac{k_1}{k_{-1}} = K_{eq} = \frac{[B]_{eq}}{[A]_{eq}} \tag{5-1}$$

then the catalyzed path is

$$A + C \underset{k_{-2}}{\overset{k_2}{\rightleftharpoons}} B + C$$

with

$$\frac{k_2}{k_{-2}} = K'_{eq} = \frac{[B]_{eq}[C]}{[A]_{eq}[C]} = \frac{[B]_{eq}}{[A]_{eq}} = K_{eq} \tag{5-2}$$

That is, the equilibrium coefficient K'_{eq} for the catalyzed path must be equal to the equilibrium coefficient K_{eq} for the uncatalyzed path. The catalyst concentration $[C]$ appears in the rate expression, but not in the equilibrium ratio. Another way of saying this is that for the catalyzed reaction, the forward rate coefficient k_2 must be accelerated by exactly the same factor as the reverse rate coefficient k_{-2} with respect to the uncatalyzed rate coefficients k_1 and k_{-1}. This can be interpreted in terms of an enthalpy diagram similar to that discussed in chapter 1; such a diagram is shown in Figure 5-1.

5.2   HOMOGENEOUS CATALYSIS

Homogeneous catalysis occurs when the catalyst is uniformly dispersed in the reaction mixture, be it a gaseous or a liquid solution. An example of each kind will be given.

Consider the reaction

$$2SO_2 + O_2 = 2SO_3$$

at 300 K. The free energy change for this reaction, as calculated from standard free energies of formation, is

$$\Delta G° = 2(-395) - (0) - 2(-297) = -196 \text{ kJ/mole}$$

so that the equilibrium coefficient for the reaction as written is

$$K_p = e^{-\Delta G°/RT} = 2.3 \times 10^{34}$$

for the reactants in their standard states. Despite this enormously favorable equilibrium coefficient, mixtures of SO_2 and O_2 are stable at room temperature: the reaction does not proceed spontaneously.

As a step in the production of sulfuric acid, the oxidation of SO_2 is a very important industrial process toward which much effort has been expended in finding an efficient catalyst. A solid catalyst containing vanadium is most widely used. In the gas

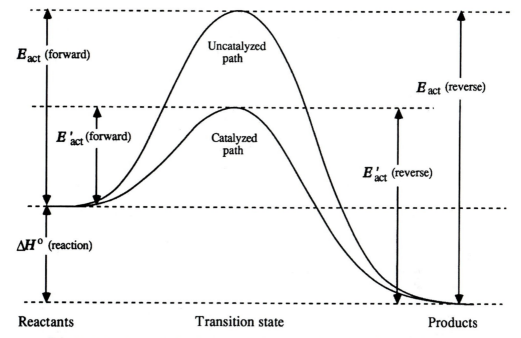

FIGURE 5-1 The enthalpy diagram shown here is based on the potential energy surface for a chemical reaction, which will be discussed in chapter 7. From this figure, we can see that the catalyst reduces the enthalpy of activation for the forward reaction by exactly the same amount as it reduces the enthalpy of activation for the reverse reaction. Thus, the net enthalpy change in the reaction, $\Delta H° = H^{\ddagger}_{fwd} - \Delta H^{\ddagger}_{rev}$, remains unchanged. The equilibrium constant can be written as $K_{eq} = \exp(-\Delta G°/RT)$, where $\Delta G° = \Delta H° - T\Delta S°$; since the entropy change in the reaction is also unaffected by the presence of the catalyst, the position of equilibrium remains the same for the catalyzed and uncatalyzed processes.

phase, the principle of catalysis can be illustrated by the effect of adding nitric oxide. The following reactions occur:

$$2NO + O_2 \rightarrow 2NO_2 \tag{5-3}$$

$$NO_2 + SO_2 \rightarrow NO + SO_3 \tag{5-4}$$

At elevated temperatures, the second reaction proceeds much faster than the first. Taking the net reaction to be one of step 1 combined with two of step 2, we have the following conditions:

$$\frac{d[NO]}{dt} = 0 \tag{5-5a}$$

and

$$\frac{d[NO_2]}{dt} = 0 \tag{5-5b}$$

Equation (5-5a) is just the necessary condition on a catalyst, viz., that its concentration does not change during the reaction. Equation (5-5b) can be regarded as a steady-state condition on NO_2, which, since (5-4) is stipulated to proceed much faster than (5-3), remains at low concentration during the course of the reaction. The rate of formation of SO_3 is

$$\frac{d[SO_3]}{dt} = k_2[NO_2][SO_2]$$

Applying Equation (5-5b) to the kinetic equation for $[NO_2]$, we have

$$\frac{d[NO_2]}{dt} = 2k_1[NO]^2[O_2] - k_2[NO_2][SO_2] = 0$$

so that

$$[NO_2]_{ss} = \frac{2k_1[NO]^2[O_2]}{k_2[SO_2]}$$

and

$$\frac{d[SO_2]}{dt} = k_2[SO_2]\frac{2k_2[NO]^2[O_2]}{k_2[SO_2]} = 2k_1[NO]^2[O_2] \qquad (5\text{-}6)$$

Notice that the concentration of SO_2 does not appear in the expression for the rate of formation of SO_3 (equation 5-6). This is because the rate-limiting step (1) does not include SO_2 as a reactant. This will be true until the reaction proceeds so far that most of the SO_2 is used up. Reaction (5-3) will then no longer be rate-limiting, and equation (5-6) will no longer be valid.

One of the most celebrated and well-studied examples of gas-phase catalysis is the destruction of ozone in the Earth's stratosphere by reaction cycles involving OH, nitrogen oxides, and halogens. These will be discussed in considerable detail in Chapter 15.

A simple example of a catalyzed process in a liquid solution is the hydrolysis of an ester, such as ethyl acetate:

$$CH_3COOCH_2CH_3 + H_2O \rightarrow CH_3COOH + C_2H_5OH$$

This reaction proceeds immeasurably slowly in a neutral solution, but quite readily in the presence of added acid or base. The mechanism of the acid-catalyzed hydrolysis is the following:

$$CH_3COOCH_2CH_3 + H_3O^+ \rightleftarrows \overset{\overset{\textstyle H}{|}}{\underset{\underset{\textstyle O}{\|}}{CH_3C}}\!\!-\!\!\overset{\oplus}{O}CH_2CH_3 + H_2O$$

$$\uparrow \qquad\qquad \downarrow$$

$$+\,H_2O \qquad CH_3\overset{\oplus}{COOH_2} + CH_3CH_2OH$$

$$\downarrow\uparrow$$

$$CH_3COOH + H^+$$

In this mechanism, two rapid protonation equilibria are coupled to the rate-determining step which liberates the free alcohol. The overall rate of disappearance of ester, given by

$$-\frac{d[\text{ester}]}{dt} = k_{\text{hyd}}[\text{H}_3\text{O}^+][\text{ester}]$$

is proportional to the concentration of acid. The overall equilibrium coefficient,

$$K_{\text{eq}} = \frac{a_{\text{acid}} a_{\text{alcohol}}}{a_{\text{ester}} a_{\text{water}}}$$

is expressed in terms of molar activities rather than concentrations, as is appropriate for a concentrated solution. K_{eq} is independent of acidity, except as the pH may affect the activity coefficients.

In the absence of added acid, the only H^+ that would be present results from the self-hydrolysis of water, viz.,

$$2\text{H}_2\text{O} \rightleftharpoons \text{H}_3\text{O}^+ + \text{OH}^-$$

with

$$K_W = \frac{(\text{H}_3\text{O}^+)a(\text{OH}^-)}{a^2(\text{H}_2\text{O})} = \frac{[\text{H}_3\text{O}^+][\text{OH}^-]}{[\text{H}_2\text{O}]^2} \cdot \frac{\gamma_{\text{H}^+}\gamma_{\text{OH}^-}}{\gamma_W^2} \tag{5-7}$$

where the γ's are activity coefficients for H^+, OH^-, and water. The hydrolysis rate would thus be inversely proportional to the square root of the activity coefficient product in equation (5-7), which, according to the Debye-Hückel theory,[1] is related to the ionic strength $I = 1/2 \sum_i c_i z_i^2$. This means that adding an inert electrolyte to a solution, which increases the value of $c_i z_i^2$, will increase the hydrolysis rate even though the reactants themselves are electrically neutral but the catalytic species is charged. This is often referred to as the *secondary salt effect*.

5.3   AUTOCATALYSIS AND OSCILLATING REACTIONS

An interesting new feature is introduced by the presence of *autocatalysis*, which is just the appearance of one of the products of the reaction as a reactant in the same or a coupled reaction. The simplest possible example is the reaction

$$\text{A} + \text{B} \xrightarrow{k} 2\text{B}$$

whose rate law is

$$-\frac{d[\text{A}]}{dt} = k[\text{A}][\text{B}]$$

To find a solution, let $x = [\text{A}(t)]$ and since $[\text{A}]_0 - [\text{A}(t)] = [\text{B}(t)] - [\text{B}]_0$, we have

$$[\text{B}(t)] = [\text{A}]_0 + [\text{B}]_0 = [\text{A}(t)] = A_0 + B_0 - x$$

Then

$$-\frac{dx}{dt} = kx(A_0 + B_0 - x)$$

which is simply integrated:

$$\int \frac{dx}{(A_0 + B_0)x - x^2} = -\int k\, dt$$

The solution is

$$\frac{1}{A_0 + B_0} \ln\left(\frac{A_0}{B_0} \cdot \frac{[B]}{[A]}\right) = -kt$$

or

$$[B(t)] = \frac{A_0 + B_0}{1 + (A_0/B_0)e^{-k(A_0 + B_0)t}} \tag{5-8}$$

The time dependence of [B] is shown in Figure 5-2. Note that at $t = 0$, [B] just has its initial value B_0. If $B_0 = 0$, then the only solution is [B] $= 0$ for all t. If any small amount $B_0 > 0$ is initially present, then there is first a slow increase in [B] called an *induction period;* the smaller B_0 is, the longer is this period. The rate of increase of [B] then increases (the autocatalytic effect), goes through an *inflection point* (maximum $d[B]/dt$) at some time t^*, and then decreases to zero, while [B] levels off at a value equal to $A_0 + B_0$ at long times.

The concentration-vs.-time dependence shown in Figure 5-2 is the classic "S" curve associated with increasing populations. For example, if A is food and B is bacte-

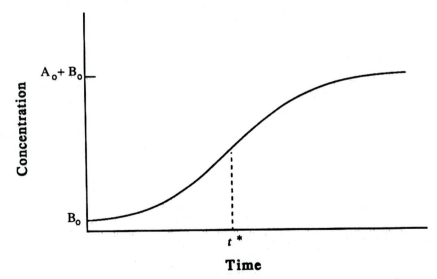

FIGURE 5-2 Concentration vs. time for product of an autocatalytic reaction, as given by equation (5-8). The inflection point occurs at t*.

ria (or, perhaps, people), the organisms eat the food and in turn produce more organisms. When the food supply runs out, the population can no longer increase. The reaction is intrinsically irreversible unless cannibalism is included in the mechanism. An actual population model for humans or other organisms that reproduce sexually would necessarily be more complex:

$$A + X_m \rightarrow X_m^*$$
$$A + X_f \rightarrow X_f^*$$
$$X_m^* + X_f^* \rightarrow X_m + X_f + B$$
$$A + B \rightarrow X, \text{ etc.}$$

In this model the organisms are divided into subsets of m (males) and f (females), an initial activation step is included, and the activated species then react to produce B (babies), which consume food and in turn become reproductive species. Another kind of population model is explored in problem 2 at the end of the chapter.

5.3.1 Autocatalytic Formation of Ozone

An interesting example of autocatalysis related to atmospheric chemistry has been found by Slanger and coworkers.[2] The mechanism for production of ozone is, in simplified form,

$$O_2 + h\nu(\lambda < 240 \text{ nm}) \rightarrow 2O$$
$$O + O_2 + M \rightarrow O_3 + M$$
$$O + O_3 \rightarrow 2O_2$$
$$O_3 + h\nu\ (\lambda < 320 \text{ nm}) \rightarrow O_2 + O$$

The net effect is a photochemical stationary state, $3O_2 \rightleftarrows 2O_3$. This is discussed in much greater detail, in its atmospheric context, in Chapter 15, Section 15.4.

The wavelength limits specified above result from the spectroscopic properties of the oxygen and ozone molecules; for example, molecular oxygen cannot be photodissociated by light with wavelengths longer than 240 nm. However, when oxygen is irradiated with 248-nm light from a KrF laser, or 254-nm light from a Hg lamp, a slow buildup of ozone is observed. This is attributed to an autocatalytic reaction sequence involving very small amounts of ozone initially present, which dissociate to form vibrationally "hot" oxygen, $O_2^\#$:

$$O_3 + h\nu\ (\lambda = 248, 254 \text{ nm}) \rightarrow O_2^\# + O$$

The $O_2^\#$ can, in turn, be dissociated by the u.v. light or react with another O_2 molecule to produce ozone:[3]

$$O_2^\# + h\nu \rightarrow 2O, \ O + O_2 + M \rightarrow O_3 + M$$

or $\qquad \dfrac{O_2^\# + O_2 \rightarrow O_3 + O}{}$

Net: $\quad O_3 + 3O_2 \rightarrow 3O_3$

In either case, an autocatalytic mechanism is set up which results in the "characteristic S-shaped curve of O_3 production."[2] As it turns out, this mechanism probably has little to do with ozone production in the atmosphere, because of the rapid removal of the autocatalytic species, $O_2^{\#}$, by deactivating collisions.[3]

5.3.2   Chemical Oscillations

The presence of one or more autocatalytic steps in a complex chemical mechanism can lead to the phenomenon known as *chemical oscillation*. Normally, chemical systems approach equilibrium in a smooth, frequently exponential relaxation; under special circumstances, however, the concentrations can oscillate between several stationary states. By analogy with electrical or electro-optical systems that display oscillation or bistability, such systems are sometimes described as examples of *chemical feedback*. In addition to temporal oscillation, these systems can also produce nonuniform spatial structures which appear to arise spontaneously from a homogeneous medium. While chemical oscillation is rather unusual, the phenomenon has been the subject of a considerable amount of study, in part because of its possible significance in biological processes. Further background may be found in references 4-6.

In order to illustrate how chemical oscillation may arise, consider the reaction system

$$A + X \xrightarrow{k_1} 2X$$
$$X + Y \xrightarrow{k_2} 2Y \qquad (5\text{-}9)$$
$$Y \xrightarrow{k_3} Z$$

This "Lotka" mechanism was specifically invented to display oscillatory behavior; it does not correspond to any chemical system that is known. All its steps are irreversible, and steps (1) and (2) are autocatalytic, as in the reaction $A + B \rightarrow 2B$ considered previously. We also postulate that the system is open and that [A] is maintained equal to A_0 at all times. This can be realized, for example, in a flow reactor, in which species A is introduced and Z is drawn off continuously, with intermediates X and Y remaining at a steady state. The kinetic equations for these intermediates are then

$$\frac{d[X]}{dt} = +k_1 A_0[X] - k_2[X][Y] \qquad (5\text{-}10a)$$

and

$$\frac{d[Y]}{dt} = +k_2[X][Y] - k_3[Y] \qquad (5\text{-}10b)$$

A steady-state solution will occur when $d[X]/dt = d[Y]/dt = 0$, or

$$k_2[Y]_{ss} = k_2[Y]_0 = k_1[A]_0 \qquad (5\text{-}11a)$$

and

$$k_2[X]_{ss} = k_3 \qquad (5\text{-}11b)$$

Note that, since [A] is to be maintained at a constant value A_0, the steady-state concentrations given by equations (5-11a) and (5-11b) are completely independent of time.

We now wish to see what will happen when the concentrations are temporarily displaced from their steady-state values. In order to treat the problem mathematically, we introduce *displacement variables* as in relaxation kinetics (see chapter 4, section 4.7). In this case these are

$$[X] = x(t) + X_0$$

and

$$[Y] = y(t) + Y_0$$

In terms of these displacement variables, equation (5-10a) becomes

$$\frac{dx}{dt} = k_1 A_0 (x + X_0) - k_2 (x + X_0)(y + Y_0)$$

$$= k_1 A_0 X_0 + k_1 A_0 x - k_2 xy - k_2 X_0 y - k_2 Y_0 x - k_2 Y_0 X_0$$

Using equation (5-11a), $k_2 Y_0 = k_1 A_0$, to cancel terms, we have

$$\frac{dx}{dt} = -k_2 X_0 y - k_2 xy \tag{5-12a}$$

Similarly, from equations (5-10b) and (5-11b), we obtain

$$\frac{dy}{dy} = k_2 (x + X_0)(y + Y_0) - k_3 (y + Y_0) = k_2 Y_0 x + k_2 xy \tag{5-12b}$$

If the displacements from the steady state are small, we can neglect the xy terms in equations (5-12) and obtain a simple set of coupled linear first-order differential equations:

$$\frac{dx}{dt} = -k_2 X_0 y$$
$$\frac{dy}{dt} = +k_2 Y_0 x \tag{5-13}$$

By writing $\dot{x} = -ay$, $\dot{y} = +bx$, we can see that equations (5-13) are equivalent to a single second-order differential equation,

$$\ddot{x} = -a\dot{y} = -(ab)x \tag{5-14}$$

The solution of equations (5-13) or equation (5-14) can be written down by inspection; but let us proceed systematically, as was done previously, and find an eigenvalue solution. We write the solution as

$$x(t) = [X]_t - X_0 = (\text{const})e^{-t/\tau}$$

The eigenvalue τ^{-1} is found from the determinant

$$\begin{vmatrix} +\dfrac{1}{\tau} & -k_2 X_0 \\ +k_2 Y_0 & +\dfrac{1}{\tau} \end{vmatrix} = 0$$

or

$$\left(\frac{1}{\tau}\right)^2 + k_2^2 X_0 Y_0 = 0$$

$$\left(\frac{1}{\tau}\right)^2 = -k_2^2(X_0 Y_0)$$

(5-15)

Since the rate coefficient k_2 and the concentrations X_0 and Y_0 are all positive quantities, the eigenvalues given by equation (5-15) must be purely imaginary and thus can be written as

$$\frac{1}{\tau} = \pm\, i(k_2^2 X_0 Y_0)^{1/2} = \pm\, i(k_1 k_3 A_0)^{1/2}$$

The time-varying displacement is then

$$x(t) = c_1 e^{i\omega t} + c_2 e^{-i\omega t}$$

(5-16)

The functional form of equation (5-16) is an oscillation with angular frequency $\omega = |1/\tau| = |k_1 k_3 A_0|^{1/2}$, since oscillating functions such as $\cos(\omega t)$ can be written as $e^{i\omega t} + e^{-i\omega t}$. Physically, this means that if the system is displaced from its steady state, the species concentrations, rather than simply returning to their previous values, will oscillate in time, as shown in Figure 5-3a. If additional non-linear terms are present, as in the Higgins Model,[7] small displacements may even grow in time, as shown in Figure 5-3b.

Note that if all steps are reversible, that is, if the mechanism is actually

$$A + X \underset{k_{-1}}{\overset{k_1}{\rightleftarrows}} 2X \qquad X + Y \underset{k_{-2}}{\overset{k_2}{\rightleftarrows}} 2Y \qquad Y \underset{k_{-3}}{\overset{k_3}{\rightleftarrows}} Z$$

then simple exponential relaxation, and not oscillation, occurs. This is explored further in problem 2.

5.3.3 Belousov-Zhabotinsky Reaction

While no chemical system is known which exactly follows the Lotka mechanism given in the reactions (5-9), several "chemical oscillators" have been discovered. The one which has been most extensively studied[4-9] is the Belousov-Zhabotinsky reaction, which is the oxidation of malonic acid by bromate, catalyzed by cerium ions. There are at least nine coupled reactions in this system. Two of these are

$$2H^+ + BrO_3^- + Br^- \xrightarrow{k_1} HBrO_2 + HOBr$$

(1)

and

$$H^+ + HBrO_2 + Br^- \xrightarrow{k_2} 2HOBr$$

(2)

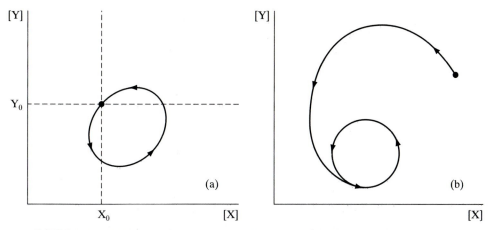

FIGURE 5-3 X-Y phase plane representations of oscillating concentrations.
(a) Lotka-Volterra reaction. (b) Stable limit cycle following initial displacement.

These are coupled with the two rapid reactions,

$$Br^- + HOBr + H^+ \rightarrow Br_2 + H_2O \tag{3}$$

and

$$Br_2 + CH_2(COOH)_2 \rightarrow BrCH(COOH)_2 + H^+ + Br^- \tag{4}$$

An overall reaction α, leading to bromination of malonic acid, is obtained by adding to steps 1 and 2 three times each of reactions 3 and 4:

$$BrO_3^- + 2Br^- + 3CH_2(COOH)_2 + 3H^+ \rightarrow 3BrCH(COOH)_2 + 3H_2O \tag{α}$$

Reactions leading to the oxidation of cerium ions are

$$H^+ + BrO_3^- + HBrO_2 \xrightarrow{k_5} 2BrO_2 + H_2O \tag{5}$$

$$H^+ + BrO_2 + Ce^{III} \xrightarrow{k_6} HBrO_2 + Ce^{IV} \tag{6}$$

and

$$2HBrO_2 \xrightarrow{k_7} BrO_3^- + HOBr + H^+ \tag{7}$$

Adding two of reaction 5 to four of reaction 6 and one of reaction 7 gives an overall reaction β, given by

$$BrO_3^- + 4Ce^{III} + 5H^+ \rightarrow HOBr + 4Ce^{IV} + 2H_2O \tag{β}$$

Finally, the oxidation of bromomalonic acid by cerium is

$$BrCH(COOH)_2 + 4Ce^{IV} + 2H_2O \rightarrow Br^- + HCOOH + 2CO_2 + 4Ce^{III} + 6H^+ \tag{8}$$

coupled with a fast reaction

$$HOBr + HCOOH \rightarrow Br^- + CO_2 + H^+ + H_2O \tag{9}$$

to give a net reaction γ of

$$BrCH(COOH)_2 + 4Ce^{IV} + HOBr + H_2O \rightarrow 2Br^- + 3CO_2 + 4Ce^{III} + 6H^+ \tag{γ}$$

The overall Belousov-Zhabotinsky reaction, obtained by adding together reactions α, β, and γ, is

$$2BrO_3^- + 3CH_2(COOH)_2 + 2H^+ \xrightarrow{Ce^{III}, Br^-} 2BrCH(COOH)_2 + 3CO_2 + 4H_2O$$

Note that in this reaction the concentrations of cerium ion and bromide ion are canceled out, so that these species play a true catalytic role.

In order to analyze such a complicated set of reactions, simplifying assumptions are of course necessary. The most appropriate such assumptions are steady-state conditions on intermediate-valence and odd-electron bromine species, so that

$$\frac{d[HBrO_2]}{dt} = \frac{d[BrO_2]}{dt} = 0$$

From the kinetic equations, this gives

$$\frac{d[HBrO_2]}{dt} = k_1[H^+]^2[BrO_3^-][Br^-] - k_2[H^+][HBrO_2][Br^-]$$
$$-k_5[H^+][BrO_3^-][HBrO_2] + k_6[H^+][BrO_2][Ce^{III}] \tag{5-17}$$
$$-2k_7[HBrO_2]^2 = 0$$

and

$$\frac{d[BrO_2]}{dt} = 2k_5[H^+][BrO_3^-][HBrO_2] - k_6[H^+][BrO_3^-][Ce^{III}] = 0 \tag{5-18}$$

Equation (5-18) allows us to replace $k_6[H^+][BrO_2][Ce^{III}]$, the fourth term in equation (5-17), with $2k_5[H^+][BrO_3^-][HBrO_2]$, to give the relatively simple steady-state condition,

$$k_1[H^+]^2[BrO_3^-][Br^-] - k_2[H^+][HBrO_2]_{SS}[Br^-]$$
$$+ k_5[H^+][BrO_3^-][HBrO_2]_{SS} - 2k_7[HBrO_2]_{SS}^2 = 0 \tag{5-19}$$

Two limiting cases can be distinguished:

1. If the steady-state bromide ion concentration is much greater than the steady-state $HBrO_2$ concentration we may neglect the last term in equation (5-19) and solve for

$$[HBrO_2]_{SS(1)} = \frac{k_1[BrO_3^-][Br^-][H^+]}{k_2[Br^-] - k_5[BrO_3^-]}$$

2. If, on the other hand, the steady-state bromide ion concentration is much less than the steady-state $HBrO_2$ concentration, the first and second terms in equation (5-19) may be neglected, and we obtain

$$[HBrO_2]_{SS(2)} = \frac{k_5[H^+][BrO_3^-]}{2k_7}$$

So there are two possible steady-state solutions for at least one of the key intermediates, and the system may be *nonstationary,* that is, it may switch back and forth between these two steady states. Another term for such switching behavior is *bistability.* We can see qualitatively how such bistability may arise. If initially the bromide ion concentration is small, then reaction cycle α is shut off, cerous ion is oxidized to the ceric form, and reaction cycle γ proceeds to generate Br^-. As the latter builds up, reaction cycle α turns back on, Ce^{IV} is no longer produced, and the Br^- is consumed. The concentrations of these species may oscillate back and forth between the two steady states, as shown in Figure 5-4.

In addition to temporal oscillations, reactions of this type may generate spatial inhomogeneities known as *dissipative structures* or *Turing structures.* Additional examples may be found in Ref. 10, and an example involving heterogeneous catalysis will be described later in this chapter.

5.4   ENZYME-CATALYZED REACTIONS

A very important class of catalysts is the *biological enzymes:* without them, you would not be reading this book, nor indeed would it ever have been written. Enzymes are protein molecules which, through a process of chemical evolution, have become exquisitely tailored to carry out highly selective and efficient biochemical reactions in living organisms. The biochemistry of enzymes is a subject of great depth and complexity; in this section we can only touch on the simplest aspects of the kinetics of these systems. For greater detail, the reader should consult the texts by Hammes[9] and Walsh.[11]

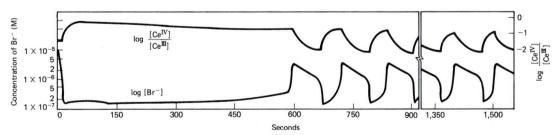

FIGURE 5-4   Oscillatory behavior of bromide ion concentration and ratio $[Ce^{III}]/[Ce^{IV}]$ in the Belousov-Zhabotinsky reaction. [From R. J. Field, *Am. Scientist 73*, 142 (1985).]

A generic enzyme-catalyzed mechanism may be illustrated as follows:

S (Substrate)   E (Enzyme)   X or E · S   P₁   P₂   E
                                     (Enzyme–Substrate   (Products)
                                      Complex)

The substrate may be anything from a protein to a simple sugar molecule to a strand of RNA or even a complete cell; the enzyme, typically a protein of molecular weight 10,000 to 100,000 atomic mass units, may in fact be much bigger than the substrate, or it may attach to a specific site such as a part of a cell membrane. The enzyme-substrate complex is a true chemical intermediate, but is typically present in such low concentrations that it cannot be isolated. The product may be separated molecules, or the protein with one bond cleaved or one site oxidized. A simple example of this mechanism is the hydrolysis of a complex sugar, sucrose, to two smaller and more readily metabolized sugars, with the assistance of an enzyme, called invertase,* that is derived from yeast:

Sucrose   Glucose   Fructose

Kinetic equations can readily be derived from the general mechanism described earlier:

$$E + S \underset{k_{-1}}{\overset{k_1}{\rightleftarrows}} X \underset{k_{-2}}{\overset{k_2}{\rightleftarrows}} E + P$$

$$-\frac{d[X]}{dt} = (k_2 + k_{-1})[X] - k_1[E][S] - k_{-2}[E][P]$$

(5-20)

In a reaction such as the hydrolysis of sucrose, in which the products are separated molecules, the route from complex to products is essentially irreversible, so that k_{-2} may safely be set equal to zero. Also, because the enzyme-substrate complex is rapidly destroyed and thus is present at very low concentration, the steady-state condition $d[X]/dt = 0$ may be safely invoked. Further, we have conservation conditions on the total enzyme, which is present either as free enzyme or bound in the complex, thus

---

* The suffix "-ase" is a general designator for enzymes, with the remainder of the name suggesting the specific functionality. The yeast enzyme is named invertase because the hydrolysis converts a dextrorotary sugar ($[\alpha]_D^{20} = +66.53$) into an overall levorotary mixture of glucose ($[\alpha]_D^{20} = +52.7$) and fructose ($[\alpha]_D^{20} = -92.4$). The first measurements on this system were indeed performed using quantitative polarimetry, but the current technique employs a colorimetric assay.[12]

$$[E]_0 = [E] + [X] \tag{5-21a}$$

and on the total substrate-cum-product concentration, viz.,

$$[S]_0 = [S] + [P] \tag{5-21b}$$

since $[X] \approx [E] \ll [S]$. Hence,

$$-\frac{d[S]}{dt} = +\frac{d[P]}{dt} = k_1[E][S] - k_{-1}[X]$$

Setting equation (5-20) equal to zero, solving for $[X]_{SS}$, and substituting in the preceding expression gives

$$-\frac{d[S]}{dt} = +\frac{\{k_1 k_2[S] - k_{-1} k_2[P]\}[E]_0}{k_1[S] + k_{-2}[P] + k_{-1} + k_2} \tag{5-22}$$

Measurements of the reaction velocity v are typically carried out at initial times, when $[S] \gg [P]$. This permits us to neglect the terms containing the concentration of product in equation 5-22, and so obtain

$$v = -\frac{d[S]}{dt}\bigg|_{t \approx 0} = +\frac{k_1 k_2[S][E]_0}{k_1[S] + k_1 + k_2}$$

or

$$v = \frac{k_2[E]_0}{1 + \dfrac{k_{-1} + k_2}{k_1[S]}} \tag{5-23}$$

If we define $v_s = k_2[E]_0$ and $K_M = (k_{-1} + k_2)/k_1$, then equation (5-23) takes the form

$$v = \frac{v_s}{1 + K_M/[S]} \tag{5-24}$$

which is known as the *Michaelis-Menten* equation. The ratio K_M is known as the *Michaelis Constant.* Although a ratio of rate coefficients, it is not a true equilibrium constant, since its definition does not satisfy detailed balancing (see chapter 2). A plot of the Michaelis-Menten equation is shown in Figure 5-5(a).

Because of the curvature of the Michaelis-Menten plot, it is often difficult to determine the limiting value v_s, the reaction velocity when all of the enzyme is tied up in the enzyme-substrate complex. Similarly, K_M, which is numerically equal to the substrate concentration when v is exactly equal to one-half v_s, is not precisely determined. To obtain better precision, it is customary to recast the kinetic data into a linear form. Two such forms are commonly used, the *Lineweaver-Burk* plot,

$$\frac{1}{v} = \frac{K_M}{v_s}\frac{1}{[S]} + \frac{1}{v_s} \tag{5-25}$$

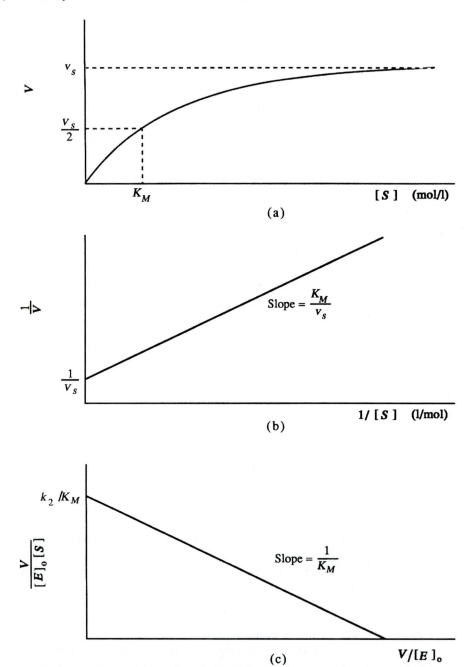

FIGURE 5-5   Kinetic plots for a simple enzyme-catalyzed reaction. (a) Michaelis-Menten plot, equation (5-24). (b) Lineweaver-Burk plot, equation (5-25). (c) Eadie-Hofstee plot, equation (5-26).

and the *Eadie-Hofstee* plot, one form of which is

$$\frac{v}{[E]_0[S]} = \frac{k_2}{K_M} - \frac{v}{K_M[E]_0} \tag{5-26}$$

Lineweaver-Burk (LB) and Eadie-Hofstee (EH) plots are shown in Figures 5-5(b) and (c), respectively. In the LB plot, $1/v_s$ is obtained as the intercept when $1/[S] \rightarrow 0$, and the slope of the straight line is equal to K_M/v_s. In the EH plot, the negative of the slope gives $1/K_M$ directly, and the intercept as $v/[E]_0 \rightarrow 0$ is k_2/K_M.

No matter what method of analysis is used, at best two parameters of the enzyme reaction can be extracted using initial-velocity data. If we know $[E]_0$, then we can find k_2 from v_s. In many cases, however, the molecular weight, and thus the exact concentration of the enzyme, is not precisely known, since the enzyme is obtained only as a complex mixture of proteins from some biological extraction. In such a case, only the *specific activity* (moles substrate converted per mg of enzyme preparation) can be reported. If k_2 can be determined, then combining that with the measurement of K_M allows us to find the ratio k_1/k_{-1}. To obtain all of the rate coefficients for a particular enzyme-catalyzed reaction, additional measurements are necessary. If the reaction is intrinsically reversible, then it can be run backwards, starting with products, and the corresponding v_p and $K_M(p)$ values, combined with those of v_s and $K_M(s)$, would permit determination of all of the k's. When this is not possible, as is often the case, fast-reaction techniques such as stopped-flow or relaxation measurements (see chapter 4, section 4.7) must be used. Additional information may be found in Hammes[9] and Walsh.[11]

5.5   HETEROGENEOUS CATALYSIS AND GAS-SURFACE REACTIONS

An extremely important aspect of chemical kinetics, from the points of view of both fundamental processes and practical applications, is *heterogeneous catalysis,* i.e., reactions occurring at the gas-solid or liquid-solid interface. The discussion here touches on only a few of the principal features of these processes; further details are available in References 13–17.

5.5.1   General Description of Heterogeneous Reactions

The generalized surface-catalyzed reaction bears a formal similarity to an enzyme-catalyzed system:

$$A \xrightarrow[\text{on surface S}]{} P(\text{roducts})$$

As we shall see, the resemblance is more than superficial.

The basic mechanism for a surface-catalyzed reaction begins with *adsorption* of the reactant at a localized site on the surface:

$$A + S \rightleftharpoons A \cdot S \tag{5-27}$$

The magnitude of the adsorption coefficient $K_{ads}(A)$ may depend on the nature of the surface S and the ambient temperature, as well as on the identity of species A. As discussed in chapter 12, this is because the interaction between the surface and the adsorbate can range from the rather weak binding of a valence-saturated species due to van der Waals attraction (physisorption) to the formation of a strong chemical bond between the adsorbate and a surface atom (chemisorption).

In order to express K_{ads} quantitatively, we must be able to define the concentration of the adsorbed species, A · S. This is best done in terms of a dimensionless quantity, the *coverage* θ. For a given solid catalyst, there is a finite number S_0 of binding sites available. The maximum number is the number of surface atoms, on the order of 10^{14} to 10^{15} per cm^2 of surface area, but typically the number of active binding sites is a small fraction of that. The coverage is defined as

$$\theta = \frac{\text{number of A} \cdot \text{S}}{S_0} \tag{5-28}$$

The numerator and denominator of equation (5-28) are expressed in terms of either the total number of adsorbed species and binding sites for a given piece of catalyst or the site density per unit area, but must be consistent in either case.

The physical resemblance to enzyme catalysis may be apparent at this point. An enzyme is typically a large protein whose "surface" includes one or more active sites. The enzyme-catalyzed reaction begins when the substrate binds to one of these active sites; the resulting enzyme-substrate complex is formally analogous to the adsorbed molecule. If we define the equilibrium coefficient for the adsorption process given by equation (5-27) as K_{ads}, the coverage is

$$\theta = \frac{K_{ads}[\text{A}]}{1 + K_{ads}[\text{A}]} \tag{5-29}$$

If the adsorption is from a low-pressure gas which obeys the ideal gas law $p = (N/V)k_B T$, equation (5-29) can be expressed as

$$\theta = \frac{bp}{1 + pb} \tag{5-30}$$

where $b = K_{ads}/k_B T$. Equation (5-30) is the familiar *Langmuir adsorption isotherm;* notice its similarity to the Michaelis-Menten expression, equation (5-24).

If the ensuing surface reaction is simply conversion of the adsorbed species to product, i.e.,

$$\text{A} \cdot \text{S} \xrightarrow{k_2} \text{products}$$

then the rate of conversion is given by

$$R = -\frac{d[\text{A}]}{dt} = k_2 \theta S_0 = \frac{k_2 K_{ads} S_0 [\text{A}]}{1 + K_{ads}[\text{A}]} \tag{5-31}$$

Two limiting cases of equation (5-31) should be noted. At low [A], $K_{ads}[\text{A}] \ll 1$, so that the rate is proportional to the product of the partial pressure of A and the total amount of catalyst. By contrast, at large [A], $K_{ads}[\text{A}] \gg 1$, and the rate becomes independent of the pressure of A. The catalyst is then said to be *saturated.*

For a surface-catalyzed bimolecular reaction, two mechanisms are possible. In the first, both species must be preadsorbed, and the reaction occurs between adsorbed

A and B species at neighboring sites on the surface. This mechanism can be written as follows:

$$A + S \rightleftharpoons A \cdot S \qquad K_{ads}(A)$$

$$B + S \rightleftharpoons B \cdot S \qquad K_{ads}(B)$$

$$A \cdot S + B \cdot S \longrightarrow products \qquad k_2'$$

The rate of product formation associated with this *Langmuir-Hinshelwood mechanism* is

$$R = k_2' \theta_A \theta_B S_0^2 = \frac{k_2' K_{ads}^{(A)}[A] K_{ads}^{(B)}[B] S_0^2}{\{1 + K_{ads}^{(A)}[A] + K_{ads}^{(B)}[B]\}^2} \qquad (5\text{-}32)$$

Note that there is a competition for available surface sites between A and B molecules; one consequence of this is that the rate equation (5-32) at constant p_A may go through a maximum as p_B is varied.

The other possible mechanism, often referred to as an *Eley-Rideal* process, involves reaction of an impinging gas-phase species with an adsorbate. This can be written

$$A + S \rightleftharpoons A \cdot S \qquad K_{ads}^{(A)}$$

$$A \cdot S + B \longrightarrow products \qquad k_2''$$

The rate expression for this mechanism is

$$R = k_2'' \theta_A S_0 p_B = k_2'' \frac{K_{ads}^{(A)}[A] p_B S_0}{1 + K_{ads}^{(A)}[A]} \qquad (5\text{-}33)$$

This expression is linear in [B] and saturates at high [A]; it does not possess a maximum with variation of p_B.

5.5.2 Some Examples of Heterogeneous Reactions

A prototypical example of heterogeneous catalysis is the formation of hydrocarbons from hydrogen and carbon monoxide over a suitable solid catalyst. The methanation reaction

$$3H_2 + CO \rightarrow CH_4 + H_2O$$

is the simplest case of such a reaction. The free-energy change in this reaction at 500 K is

$$\Delta G^\circ = -219.05 - 32.74 - (3(0) - 155.41) = -96.38 \text{ kJ/mole}$$

giving a very favorable equilibrium ratio

$$K_{eq} = \exp(-\Delta G^\circ/RT) = \exp(-(-96.38/0.0083 \times 500)$$

$$= 1.22 \times 10^{10} \text{ at 500 K}$$

However, the homogeneous reaction rate between H_2 and CO in the gas phase is vanishingly small. The reaction is carried out efficiently at moderate temperatures (250°C–450°C) by contacting the gas mixture with finely divided transition metals, such as "Raney nickel" (60% Ni supported on Al_2O_3). Variation of the reaction conditions produces mixtures of higher molecular weight paraffins and olefins, a process known as *Fischer-Tropsch synthesis*. This process is of technical importance because it provides a route from the products of coal gasification (CO and H_2) to products which must otherwise be obtained by refining petroleum.[18,19] Use of a different catalyst, such as copper-zinc oxide on Al_2O_3, produces methanol and higher alcohols rather than hydrocarbons via the reaction

$$2H_2 + CO \rightarrow CH_3OH$$

even though the equilibrium for this reaction is much less favorable than for methane production.

The detailed mechanism of the Fischer-Tropsch synthesis has been the subject of much debate. Over a dozen mechanisms have been proposed, ranging from complete dissociation of the reactants on the surface, e.g.,

$$CO \rightarrow C(ads) + O(ads)$$

$$H_2 \rightarrow 2H(ads)$$

$$4H(ads) + C(ads) \rightarrow CH_4(g)$$

to a concerted, super-Eley-Rideal mechanism in which CO and H_2 are conveniently bound at the same site, e.g.,

$$CO(g) + H_2(g) \rightarrow H_2CO(ads)$$

$$H_2CO(ads) + 2H_2(ads) \rightarrow CH_4(g) + H_2O(g)$$

(The surface S is implicit in all of these reactions.)

A mechanism which plausibly accounts for the variety of products observed would include the adsorption of CO on neighboring surface sites and dissociative chemisorption of H_2. This is followed by reaction of chemisorbed H with chemisorbed CO. Subsequent hydrogenation reactions on the surface will yield stable products such as CH_4 and methanol, followed by desorption. In this mechanism, higher molecular weight olefins and alcohols can arise from polymerization of the adsorbed H_2CO species prior to desorption of the gaseous products. Rate laws for the production of hydrocarbons, derived from such mechanisms, are complex, to say the least. Typically, they involve irrational powers of the partial pressures of CO and hydrogen.

A seemingly simpler reaction, namely, oxidation of CO on the surface of expensive metals such as Pt or Pd, turns out to have unexpectedly complex features. Adsorption of both CO and O_2 on Pt is highly exothermic:

$$CO(g) + S \rightarrow CO\ (ads) \qquad \Delta H_{ads} = -146 \text{ kJ mol}^{-1}$$

$$O_2 + S \rightarrow O_2\ (ads) \rightarrow 2O\ (ads) \qquad \Delta H_{ads} = 2 \times (-115 \text{ kJ mol}^{-1})$$

The adsorbed CO and dissociated O can react on the surface:

$$CO(ads) + O(ads) \rightarrow CO_2(ads)$$

followed by desorption of CO_2 which is 22 kJ/mole endothermic but occurs spontaneously under vacuum. The overall reaction exothermicity, -283 kJ/mole, is exactly the same as in the gas phase, but the gas-phase activation energy (ca. $+200$ kJ/mole) is entirely absent in the presence of the Pt catalyst.

While this seems to be a straightforward mechanism, Ertl[20,21] has found that, in certain ranges of reactant pressures and surface temperatures, the oxidation rate can show oscillatory behavior of the kind discussed earlier in connection with the Belousov-Zhabotinsky reaction (Section 5.3). The oscillations result from switching between two surface phases of Pt, having different reactivities, which is triggered by the heat of adsorption of CO. Spatio-temporal ("dissipative") structures can also appear on the surface, associated with these differences in coverage and reactivity.

Often, the phenomenological rate laws are insufficient for deciding among the various mechanisms postulated for surface-catalyzed reactions. Their proof (or, more likely, disproof) rests on experimental observations of the postulated surface-adsorbed species, using sensitive techniques such as photoelectron spectroscopy, Auger electron spectroscopy, electron energy-loss spectroscopy, and thermal desorption mass spectrometry. These techniques are discussed in books on surface science, such as Gasser.[15] However, such measurements must be carried out under ultra-high-vacuum conditions, typically 10^{-13} atm. Extrapolating from these conditions to those under which catalytic conversions are frequently carried out, which may be tens or hundreds of atmospheres, is by no means straightforward. Other techniques, which are suitable for use under higher pressure conditions, include Fourier-Transform Infrared Spectroscopy (e.g., ref. 22) and imaging techniques such as Scanning Tunneling Microscopy or Atomic Force Microscopy.

There has been an enormous amount of work investigating the reactions of gaseous species such as CO, H_2, and NH_3 on metal or metal-oxide surfaces, because of their importance in industrial processes as well as their intrinsic scientific interest. Recently, a new class of heterogeneous reactions has been found to play a key role in atmospheric chemistry.[17,23] Halogen-containing species such as HCl, HOCl, or $ClONO_2$ (as well as N_2O_5) can adsorb and react on the surfaces of naturally occurring stratospheric aerosols, generating catalytically active species which can destroy stratospheric ozone. These aerosols consist of water and nitric acid in definite proportions, often in combination with sulfuric acid which acts as a condensation nucleus. Further details about these reactions, and their profound effect on stratospheric chemistry, will be given in chapter 15.

REFERENCES

[1] R. S. Berry, S. A. Rice, and J. Ross, *Physical Chemistry* (New York: Wiley, 1980), section 26.5.

[2] T. G. Slanger, L. E. Jusinski, G. Black, and G. E. Gadd, *Science 241*, 945 (1988).

[3] R. L. Miller, A. G. Suits, P. L. Houston, R. Toumi, J. A. Mack, and A. M. Wodtke, *Science 265, 1831* (1994).

[4] P. C. Jordan, *Chemical Kinetics and Transport* (New York: Plenum Press, 1979), section 7.4–7.6.

[5] R. J. Field, *Amer. Scientist 73,* 142 (1985).

[6] K. L. C. Hunt, P. M. Hunt, and J. Ross, *Ann. Rev. Phys. Chem. 41,* 409 (1990).

[7] K. L. Queeney, E. P. Marin, C. M. Campbell, and E. Peacock-Lopez, *The Chemical Educator* (1996); available on-line at http://journals.springer.ny.com/chedr/S1430-4171(96)03035-X.

[8] J. H. Espenson, *Chemical Kinetics and Reaction Mechanisms* (New York: McGraw-Hill, 1981), section 7.8.

[9] G. G. Hammes, *Principles of Chemical Kinetics* (New York: Academic Press, 1978), chapter 9.

[10] I. Lengyel, S. Kadar, and I. R. Epstein, *Science 259,* 493 (1993); M. Dolnik and I. R. Epstein, *J. Chem. Phys. 98,* 1149 (1993).

[11] C. T. Walsh, *Enzymatic Reaction Mechanisms* (San Francisco: W. H. Freeman Co., 1979).

[12] D. P. Shoemaker, C. W. Garland, J. I. Steinfeld, and J. W. Nibler, *Experiments for Physical Chemistry,* 4th ed. (New York: McGraw-Hill, 1981), pp. 271–281.

[13] P. C. Jordan, *op. cit.,* pp. 147–152.

[14] R. J. Madix, *Science 233,* 1159 (1986).

[15] R. P. H. Gasser, *An Introduction to Chemisorption and Catalysis by Metals* (Oxford, England: Clarendon Press, 1985).

[16] K. J. Laidler, *Chemical Kinetics,* 3rd ed. (New York: Harper and Row, 1987), chapter 7.

[17] M. J. Molina, L. T. Molina, and C. E. Kolb, *Ann. Rev. Phys. Chem. 47,* 327 (1996).

[18] M. A. Vannice, *Catal. Rev. Sci. Eng. 14,* 153 (1976).

[19] J. Haggin, *Chem. and Eng. News* (Oct. 26, 1981), pp. 22–32.

[20] G. Ertl, *Science 254,* 1750 (1991).

[21] R. Imbihl and G. Ertl, *Chem. Revs. 95,* 697 (1995).

[22] N.-Y. Topsøe, *Science 265,* 1217 (1994).

[23] A. R. Ravishankara, *Science 276,* 1058 (1997).

PROBLEMS

5.1 **(a)** Using integration by partial fractions, derive the autocatalytic rate law, equation (5-8). That is, integrate

$$\int \frac{dx}{(A_0 + B_0)x - x^2} = -k \int dt$$

(b) Find the time at which the curve of $[B(t)]$ undergoes a point of inflection.

5.2 Consider the hypothetical mechanism

$$A + X \xrightarrow{k_1} 2X, \quad X + Y \xrightarrow{k_2} 2Y, \quad Y \xrightarrow{k_3} Z$$

which does not describe any known chemical process, but has the following crude ecological analog. If A represents cabbage, X rabbits, Y foxes, and Z dead foxes (or fertilizer), then rabbits eat cabbage and reproduce, foxes eat rabbits and reproduce, and foxes die. In wilderness management it is well known that animal populations oscillate rather than settle into a steady state. As we have seen, the preceding mechanism can account for such behavior.

The related problem, where reverse reaction is possible, is

$$A + X \underset{k_{-1}}{\overset{k_1}{\rightleftharpoons}} 2X, \quad X + Y \underset{k_{-2}}{\overset{k_2}{\rightleftharpoons}} 2Y, \quad Y \underset{k_{-3}}{\overset{k_3}{\rightleftharpoons}} Z$$

For simplicity, assume that [A] has a constant value A_0 which may be arranged if A is initially in great excess or if the system is open and A is continually replenished. Since reverse reaction is possible, this mechanism admits of a true equilibrium state.

(a) Using the principle of detailed balance, find the equilibrium constraints on the system. Write the coupled rate equation for the system, i.e., $d[X]/dt$, $d[Y]/dt$, and $d[Z]/dt$.

(b) Define displacement variables $\xi = [X] - X_0$, $\eta = [Y] - Y_0$, and $\zeta = [Z] - Z_0$, and use the result of part (a) to show that, near equilibrium,

$$\dot{\xi} = -(k_{-1}X_0 + k_2Y_0)\xi + k_{-2}Y_0\eta$$

$$\dot{\eta} = k_2Y_0\xi - (k_{-2}Y_0 + k_3)\eta + k_{-3}\zeta$$

$$\dot{\zeta} = k_3\eta - k_{-3}\zeta$$

(c) Assume that $[Z] = [Z_0]$, i.e., that product is always drawn off so that $\zeta = 0$. Show that the decay constants are solutions of the equation

$$\lambda^2 - [k_{-1}X_0 + (k_{-2} + k_2)Y_0 + k_3]\lambda + k_{-1}X_0 + (k_3 + k_{-2}Y_0) + k_2k_3Y_0 = 0$$

(d) Show that both roots of the preceding equation are positive and therefore that the system relaxes to equilibrium without oscillation. (The roots of the equation $\lambda^2 - B\lambda + C = 0$ satisfy the relation $\lambda_1\lambda_2 = C$.)

5.3 Using the stochastic method described in chapter 2, section 2.7, and appendix 2.3,

(a) Set up the stochastic master equation for problem 2.

(b) Show by stochastic numerical simulation what happens to the fox and rabbit populations.

(c) If there is a short-term fluctuation in the supply of cabbage, what effect does this have on the fox and rabbit populations? Choose reasonable numerical values for the rate constants and initial "concentrations". The following values give interesting results:

	Set I	Set II	Set III	Set IV
k_1	2×10^{-4}	2×10^{-3}	2×10^{-3}	1.5×10^{-3}
k_{-1}	0.0	0.0	4×10^{-3}	0
k_2	1×10^{-4}	1×10^{-3}	1×10^{-3}	1×10^{-3}
k_{-2}	0.0	0.0	2×10^{-3}	0
k_3	1×10^{-2}	1×10^{-1}	1×10^{-1}	1.1×10^{-1}
k_{-3}	0.0	0.0	2×10^{-1}	0

5.4. The reaction catalyzed by the enzyme hexokinase is

$$MgATP + G \rightleftharpoons MgADP + G6P$$

where MgATP is magnesium adenosine 5'-triphosphate, G is glucose, MgADP is magnesium adenosine 5'-diphosphate, and G6P is glucose 6-phosphate. What mechanisms are consistent with the initial-rate steady-state data given in the following table for yeast hexokinase at pH 8, 25°C, and 0.3 M $(CH_3)_4NCl$ [G. G. Hammes and D. Kochavi, *J. Am. Chem. Soc.* **84**, 2069, 2073, 2076 (1962)].

[MgATP], $M \times 10^4$	[G], $M \times 10^4$	$E_0/v \times 10^3$, sec
4.73	10	2.55
	5	2.90
	2	4.54
	1	7.48
9.20	10	2.21
	5	2.51
	2	3.90
	1	6.10
19.4	10	1.83
	5	2.17
	2	3.17
	1	4.87
40.0	10	1.72
	5	2.03
	2	2.93
	1	4.44

CHAPTER 6

The Transition from the Macroscopic to the Microscopic Level

Thus far, we have considered chemical kinetics from the macroscopic, or molar, point of view. Although informed throughout by the microscopic, or molecular, viewpoint, concepts such as the rate coefficient, the reaction order, mechanism, and even activation energy could be interpreted without reference to the discrete, atomic nature of matter. We now turn to the deeper insight afforded by a description of chemical reactions in terms of ensembles of molecular collisions.[1,2]

6.1 RELATION BETWEEN CROSS SECTION AND RATE COEFFICIENT

Let us return to the prototypical bimolecular chemical reaction with which we began our discussion in chapter 1. Rather than dealing with molar concentrations in a bulk system, we shall instead consider a scattering process in laboratory coordinates; this will be an idealized version of the molecular-beam experiments to be discussed in chapter 9. In lieu of a thermodynamic temperature, we shall characterize the reacting A and B particles by their velocities. In the following section, we shall also take explicit account of the rotational, vibrational, and electronic states of the reactant and product molecules, but at this point we will consider only the chemical identity and kinematic velocity of each species. The reaction is then described as

$$A(v_A) + B(v_B) \rightarrow C(v_C) + D(v_D)$$

As shown in Figure 6-1, we measure the number dN_c of C molecules collected per unit time and the solid angle $d\Omega$ emerging in a direction specified by angles θ and ϕ from the scattering center:

$$\frac{dN_C(\theta,\phi)d\Omega}{dt} = I_R(\theta,\phi;v)vn_A f_A(v_A)n_B f_B(v_B)V d\Omega \qquad (6\text{-}1)$$

Let us take a closer look at the variables in this equation, paying particular attention to their dimensional units:

n_A and n_B are the densities of A and B particles, respectively, with units of number/cm^3 (dimension ℓ^{-3})

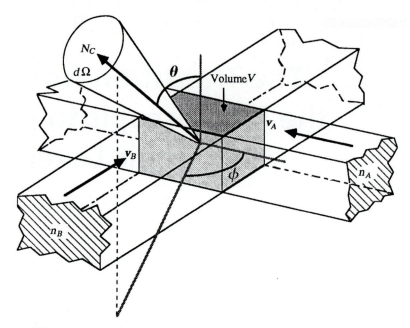

FIGURE 6-1   Definition of the scattering process $A + B \rightarrow C + D$ in laboratory coordinates, with collection of product C.

dN_C/dt is the number of C molecules collected per second at scattering angles θ and ϕ (dimension t^{-1})

$v = |v_A - v_B|$ is the relative velocity between A and B particles (the transformation from laboratory to center-of-mass coordinates is discussed in greater detail in section 9.1)   (dimension ℓt^{-1})

$d\Omega = \sin\theta\ d\theta d\phi$ is the element of solid angle   (dimensionless)

$f_i(v_i)$ is the dimensionless normalized probability density for species i having a velocity between v_i and $v_i + dv_i$

V is the volume defined by the intersection of the beams of A and B molecules   (dimension ℓ^3)

A straightforward dimensional analysis allows us to determine the units of the coefficient $I_R(\theta,\phi;v)$:

$$(t^{-1}) = (?)(\ell t^{-1})(\ell^{-3})(\ell^{-3})(\ell^3) = (?)(\ell^{-2}t^{-1})$$

Therefore, the units of I_R must be ℓ^2 which corresponds to a *cross section*. I_R is termed a *differential* cross section, since it depends on the angular variables θ and ϕ. We can also define a *total* cross section by integrating over the angular variables, thus:

$$\sigma_R(v) = \int_0^{2\pi}\int_0^\pi I_R(\theta,\phi;v)\ \sin\theta\ d\theta d\phi \qquad (6\text{-}2)$$

The cross section $I(\theta,\phi)$ or σ is a property of a *pair* of atoms or molecules which measures the interaction between them. The larger the cross section, the higher the frequency of collisions and thus the more likely that a reaction will occur. Cross sections for some simple forms of molecular interactions are given in sections 8.1 and 8.2.

Now we wish to see how the cross section, as just defined, connects with the phenomenological rate coefficient which has formed the basis for our discussion thus far. To do this, we expand the beam intersection volume V to be the entire reaction vessel containing A and B molecules and integrate over all velocities and angular variables:

$$\frac{dn_C}{dt} = \frac{d}{dt}\left(\frac{N_C}{\text{volume}}\right) = \frac{1}{V}\frac{dn_C}{dt}$$

$$= \int_0^{2\pi}\int_0^{\pi}\int_0^{\infty}\int_0^{\infty} vI_R(\theta,\phi;v)n_A f_A(v_A)n_B f_B(v_B) \times 4\pi v_A^2\, dv_A 4\pi v_B^2\, dv_B \sin\theta\, d\theta\, d\phi$$

or

$$\frac{dn_C}{dt} = \int\int_0^{\infty} v\sigma_R(v)n_A f_A(v_A)n_B f_B(v_B)4\pi v_A^2\, dv_A\, 4\pi v_B^2\, dv_B \tag{6-3}$$

Recalling the definition of the phenomenological rate coefficient given in chapter 1, i.e.,

$$\frac{dn_C}{dt} = -\frac{dn_A}{dt} = kn_A n_B$$

we can identify the right-hand side of this equation with that of equation (6-3) and take the species densities n_A and n_B outside of the integral to give an expression for the rate coefficient as the velocity-weighted, velocity-averaged reactive cross section:

$$k = \int\int_0^{\infty} v\sigma_R(v)f_A(v_A)f_B(v_B)4\pi v_A^2 4\pi v_B^2 dv_A\, dv_B \tag{6-4}$$

We now assume that the molecular velocities are described by a Maxwell-Boltzmann distribution

$$f_A(v_A;T) = \left(\frac{m_A}{2\pi k_B T}\right)^{3/2} e^{-m_A v_A^2/2k_B T}$$

We shall use the symbol k_B for Boltzmann's constant, which has the value 1.381×10^{-23} J per degree K. This is the molecular analogue of the gas constant $R = k_B N = (1.381 \times 10^{-23})(6.022 \times 10^{23}) = 8.3145$ J per degree K per mole. Using the Maxwell-Boltzmann distribution in equation (6-4) gives the following expression for the thermal rate coefficient:

$$k(T) = \left(\frac{\mu}{2\pi k_B T}\right)^{3/2}\int_0^{\infty} v\sigma_R(v)e^{-\mu v^2/2k_B T} 4\pi v^2\, dv \tag{6-5a}$$

Substituting the relative kinetic energy $E = 1/2\,\mu v^2$ results in the form

$$k(T) = \frac{1}{k_B T}\left(\frac{8}{\pi\mu k_B T}\right)^{1/2}\int_0^{\infty} E\sigma_R(E)e^{-E/k_B T}\, dE \tag{6-5b}$$

In equations (6-5a) and (6-5b), the reduced mass μ is $m_A m_B/(m_A + m_B)$.

6.2   INTERNAL STATES OF THE REACTANTS AND PRODUCTS

In reality, molecules are not structureless blobs possessing only a kinematic velocity. They also possess internal states which, as we shall see in chapter 9, can have a profound influence on the course of the reaction. Molecules may react preferentially from specified initial states (selective energy consumption), or a reaction may preferentially populate specific product states (specific energy release). The variables which are most relevant for reaction dynamics are the vibrational and rotational states of the molecules; excited electronic states may also play a role if sufficient energy is available in the reaction. Nuclear and electronic spin energies are very small compared to these other terms, but spin states tend to be conserved in chemical reactions and thus may determine their outcome (cf. section 7.10). We shall denote these internal states collectively by the symbol Γ.

The basic procedure for including reactant and product internal states in the calculation of the rate coefficient (sometimes referred to as the canon of molecular dynamics)[3] is to average over initial (reactant) states and sum over final (product) states. The result is

$$\frac{dN_C(\theta,\phi;\Gamma')d\Omega}{dt} = \sum_{\Gamma} I_R(\theta, \phi; v\Gamma\Gamma')vn_A f_A(v_A)f_A(\Gamma_A)n_B f_B(v_B)f_B(\Gamma_B)V\,d\Omega \qquad (6\text{-}6)$$

This expression gives us the rate of appearance of product C in a specified state Γ' at scattering angles θ and ϕ. The distribution over reactant states $f(\Gamma)$ may be Maxwell-Boltzmann, or may be some nonequilibrium distribution prepared in the experiment, e.g., by laser pumping. Summing over all product states gives the total rate of appearance of species C:

$$\frac{dN_C(\theta,\phi)d\Omega}{dt} = \sum_{\Gamma'} I_R(\theta, \phi; v\Gamma')vn_A f_A(v_A)n_B f_B(v_B)V\,d\Omega \qquad (6\text{-}7)$$

If internal state specification is retained in the rate coefficient expressions (6-5a) or (6-5b), we obtain *state-to-state rate coefficients.* Measurement and interpretation of state-to-state rates will be discussed in chapter 9.

6.3   MICROSCOPIC REVERSIBILITY AND DETAILED BALANCING

In addition to providing us with a formula for transforming from scattering cross section to rate coefficient, and thus forming the basis for subsequent developments of microscopic rate theory, equation (6-4) affords insight into the real origin of detailed balancing (see chapter 2, section 2.1.2). Consider the collision process

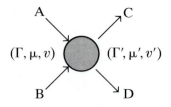

which takes A and B, having internal states Γ, reduced mass μ, and relative velocity v, into C and D having corresponding internal states Γ', μ', and v'. The process is described by the differential cross section $I_R(\Gamma \to \Gamma', v, \theta, \phi)$. The fundamental physical principle of invariance under time reversal gives us the relationship between the forward and the reverse processes, i.e., $C + D[\Gamma', \mu', -v'] \to A + B[\Gamma, \mu, -v]$. This relationship,[4] called *microscopic reversibility,* is

$$(\mu v)^2 g_\Gamma I_R(\Gamma \to \Gamma'; v\theta\phi) = (\mu' v')^2 g_{\Gamma'} I_R(\Gamma' \to \Gamma; v'\theta\phi) \tag{6-8}$$

The ratio $(\mu v)^2/(\mu' v')^2$ is just the ratio of the translational states available to the reactants and the products; from conservation of energy, v' and v are related by

$$\frac{1}{2}\mu' v'^2 + E_{\Gamma'} = \frac{1}{2}\mu v^2 + E_\Gamma - \Delta E_{int}(\text{chemical}) \tag{6-9}$$

In equation (6-9), E_Γ is the energy associated with internal state Γ, and ΔE_{int} is the amount of energy consumed or liberated in the reaction—essentially the reaction enthalpy, see equation (1-69). In an *exothermic* reaction, $\Delta E_{int} < 0$, and energy is available to be partitioned among vibrational, rotational, and translational (and possibly even electronic) states of the products. In an endothermic reaction, $\Delta E_{int} > 0$, and energy must be supplied as kinetic and/or internal energy of the reactants. The ratio $g_\Gamma/g_{\Gamma'}$ is the ratio of statistical weights for the internal states of the molecules; e.g., $g_J = 2J + 1$ for rotational state J.

Substituting the microscopic reversibility relationship (6-8) into equation (6-4) gives the rigorous form of the detailed balancing principle, viz.,

$$\frac{k[AB(\Gamma) \to CD(\Gamma'); T]}{k[CD(\Gamma) \to AB(\Gamma'); T]} = \left(\frac{\mu'}{\mu}\right)^{3/2}\left(\frac{g_{\Gamma'}}{g_\Gamma}\right)e^{-\Delta E_{int}/k_B T} \tag{6-10}$$

Since the entropy change ΔS is just the logarithm of the ratio of the statistical weights, the right-hand side of equation (6-10) can be written as $\exp(\Delta S/k_B - \Delta E_{int}/k_B T) = \exp(-\Delta G°/k_B T) = K_{eq}(T)$, which recovers the form of detailed balancing given earlier in chapter 2. Thus, detailed balancing is contingent upon the existence of Boltzmann velocity distributions, which define a kinetic temperature for the system; if in addition, averaging over internal states Γ is carried out, then a Boltzmann distribution among these variables is required as well. The principle of microscopic reversibility always holds, however, no matter what the degree of disequilibrium in the system, since it is based solely on the invariance of the collision trajectories under time reversal.

6.4 THE MICROSCOPIC-MACROSCOPIC CONNECTION

The hierarchy of relationships between microscopic quantities such as cross sections and state-to-state probabilities and macroscopic quantities such as rate coefficients is summarized in Figure 6-2.[5] Several comments on the figure, as well as an indication of its optimum pathways, are in order. First, the multiple levels of averaging make clear the difficulty of attempting to go "up the ladder", i.e., from rate coefficients to details of the potential energy surface U. Even reversing a single stage of averaging is fraught

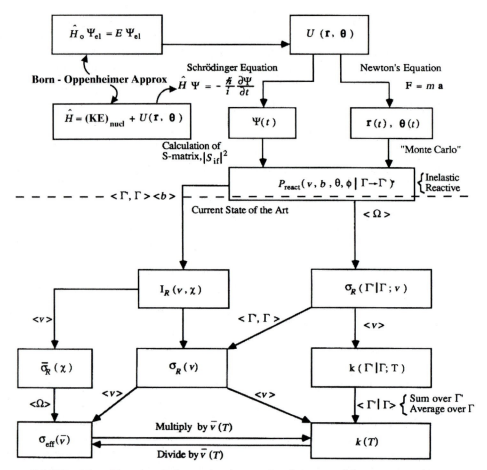

FIGURE 6-2 "Flowchart" from the intermolecular potential energy surface to thermally averaged rate coefficients. The bracket notation is used to indicate averaging over a variable; i.e., $\langle \odot \rangle$ denotes $\int \dots f(\odot)d\odot$. [Adapted from I. W. M. Smith, *Kinetics and Dynamics of Elementary Gas Reactions* (London: Butterworths, 1980).]

with difficulties. For example, the deconvolution of $k(T)$ to yield $\sigma_R(v)$ would be determinate only if rate data were available with essentially infinitely good precision over an infinite temperature range. Indeed, attempting to sample the reaction cross section with a thermal equilibrium distribution of reactants has been likened to measuring an absorption spectrum without using a monochromator:[1] it would be as if one varied the temperature of a black-body light source and attempted to extract the wavelength dependence of the extinction coefficient from the change in total transmittance at different color temperatures! Since rate data do not possess the requisite range or precision, it is generally not possible to extract microscopic details such as energy-dependent cross sections from conventional kinetics experiments. The preferable approach is to calculate microscopic quantities from some model and then perform the downward averaging for comparison with measured quantities.

Two theoretical paths are available for navigating Fig. 6-2. The first step in either case is to obtain a *potential energy surface* (PES), which is essential for any calculation of cross sections and reaction probabilities. Methods for generating the PES are discussed in Chapter 7. We may then solve the classical equations of motion on the PES using Newtonian mechanics, and obtain a classical trajectory describing the motion of the constituent atoms in space. These trajectories can then be analyzed to determine the likelihood of a reactive encounter which transforms A and B into C and D. Alternatively, we may recognize that phenomena at the atomic level are intrinsically quantum-mechanical, and attempt to solve the time-dependent Schrodinger equation, using the PES as the potential function in the Hamiltonian operator, to obtain the probability of product molecules appearing in the collision. Both of these methods are discussed in Section 8.3. In either case, the objective is to establish connections between microscopic properties and macroscopic phenomena, which is one of the basic goals of chemical kinetics and dynamics.

The current state of the art in experimental techniques, some of which are described in greater detail in chapter 9, permits measurements of state-to-state rate coefficients $k(\Gamma' | \Gamma; T)$ and of angle- and velocity-dependent cross sections $I_R(v, \theta)$. Measurements in which relative velocity, scattering angle, and product and reactant states may all be simultaneously determined are now possible using combinations of laser and molecular-beam techniques. Such measurements are essential for stringent tests of computed potential energy surfaces.

A final point worth considering is whether all of the intermediate details are really necessary. In many theoretical calculations, such details are averaged out if all one is interested in is, say, the collision energy dependence of the cross section. Considerations such as these are the province of *information theory,* which is considered in chapter 13. A second simplifying approach is *statistical rate theory* (transition state theory), which seeks to go from the potential U to the rate coefficient $k(T)$ in a single jump. This approach is described in chapter 10. First, however, we shall examine how microscopic quantities such as scattering cross sections and transition probabilities may be calculated from intermolecular potential energy surfaces. Although simple model potentials are frequently used in such calculations, modern computational techniques permit the use of accurate, *ab initio* potential functions. In the next chapter, we shall learn how these functions may be obtained.

REFERENCES

[1] J. I. Steinfeld and J. L. Kinsey, *Progr. Reaction Kinetics 5,* 1 (1970).

[2] K. Shuler, J. C. Light, and J. Ross, *Equilibrium, Transport, and Collision Processes in Gases and Plasmas,* IDA-ARPA Re-entry Physics Series (New York: Academic Press, 1967).

[3] A. Ben-Shaul, Y. Haas, K. L. Kompa, and R. D. Levine, *Lasers and Chemical Change* (Berlin: Springer-Verlag, 1981).

[4] J. C. Polanyi and J. L. Schreiber, "The Dynamics of Bimolecular Reactions" in *Physical Chemistry, an Advanced Treatise: Vol. 6A, Kinetics of Gas Reactions,* edited by H. Eyring, D. Henderson, and W. Jost (New York: Academic Press, 1974).

[5] I. W. M. Smith, *Kinetics and Dynamics of Elementary Gas Reactions* (London: Butterworths, 1980).

BIBLIOGRAPHY

BERNSTEIN, R. B. (ed.), *Atom-Molecule Collision Theory: A Guide for the Experimentalist*. New York: Plenum Press (1979).

BILLING, G. D., and MIKKELSEN, K. V. *Introduction to Molecular Dynamics and Chemical Kinetics*. New York: John Wiley and Sons (1996).

LEVINE, R. D., and BERNSTEIN, R. B. *Molecular Reaction Dynamics and Chemical Reactivity*. Oxford: University Press (1987).

MCCOURT, F. R. W., BEENAKKER, J. J. M., KOHLER, W. E., and KUSCER, I. *Nonequilibrium Phenomena in Polyatomic Gases. Vol. 1: Dilute Gases*. Oxford: Clarendon Press (1990).

PROBLEMS

6.1 Use equation (6-5a) or (6-5b) to calculate $k(T)$ given the following forms for the reactive cross section:

(a) $\sigma_R(v) = \pi d^2$, i.e., a constant. If d is the molecular diameter, this represents an "encounter limited rate coefficient".

(b) $\sigma_R(E) = \sigma_0(1 - e^{-aE})$

(c) $\sigma_R(E) = \sigma_0[1 - e^{-a(E-E_0)}]$ where $E_0 > 0$ represents a minimum energy threshhold for the reaction. Take $\sigma_R(E) = 0$ for $E < E_0$.

(d) $\sigma_R(v) = \sigma_0 \exp(-A/v)$

Do any of these calculated k's have an Arrhenius-like temperature dependence?

C H A P T E R   7

Potential Energy Surfaces

The potential energy surfaces considered in this chapter are based on the Born-Oppenheimer separation of nuclear and electronic motion.[1] A justification for this separation is the disparity in the electron and nuclear masses, which results in very slow nuclear motion compared to electronic motion. With the Born-Oppenheimer separation, each electronic state of the chemical reactive system has a potential energy surface. The primary focus of this chapter will be the *ground electronic state* potential energy surface. In the last section potential energy surfaces will be considered for excited electronic states.

For a nonlinear molecule, consisting of N atoms, the potential energy surface depends on 3N-6 independent coordinates, and depicts how the potential energy changes as relative coordinates of the atomic nuclei involved in the chemical reaction are varied. An analytic function which represents a potential energy surface is called a *potential energy function*. Understanding the relationship between properties of the potential energy surface and the behavior of the chemical reaction is a central issue in chemical kinetics. Both the macroscopic thermal rate constant for a chemical reaction and its microscopic counterpart such as a quantum mechanical state-to-state reaction probability may be interpreted in terms of a potential energy surface.

One of the simplest possible chemical reactions is rupture of a diatomic bond as in $HF \rightarrow H\cdot + F\cdot$. In this case, the dependence of the potential energy on the internuclear bond length r is a one-dimensional potential energy curve $V(r)$ illustrated in Figure 7-1. If three or more atoms participate in the chemical reaction, the potential energy will depend on more than one coordinate and a multidimensional surface results instead of a curve.

The forces between atoms which participate in a chemical reaction may be determined from the potential energy function. An interatomic force is defined by the negative of the derivative of the potential with respect to the internuclear coordinate. For the diatomic potential the force is given by

$$F_r = -\frac{dV(r)}{dr}.$$

(7-1)

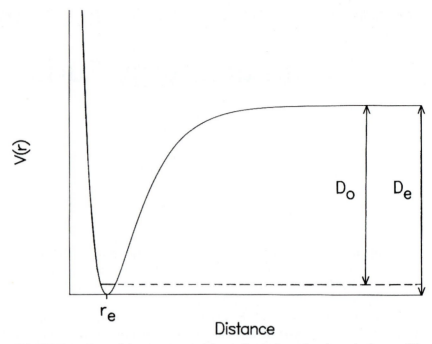

FIGURE 7-1   Potential energy curve for a diatomic molecule. r_e is the equilibrium bond length, D_e the classical dissociation energy.

From the plot in Figure 7-1, it is clear that the force depends on r and goes through zero at the potential minimum at r_e.

7.1   LONG-RANGE POTENTIALS

Long-range classical electrostatic interactions between two particles depend on the inverse of the interparticle distance. A familiar example is the Coulombic potential between two charges Z_1e and Z_2e separated by a distance r; i.e.,

$$V(r) = \frac{Z_1 Z_2 e^2}{r} \tag{7-2}$$

For the interaction of a charge Z_1e and a dipole μ_2, the potential is

$$V(r) = -\frac{Z_1 e \mu_2 \cos \theta}{r^2} \tag{7-3}$$

where $\cos \theta$ is the cosine of the angle between $\boldsymbol{\mu}_2$ and $\mathbf{r}$. Similar to the above charge-charge and charge-dipole interactions, dipole-dipole interactions vary as r^{-3}, dipole-quadrupole as r^{-4}, and quadrupole-quadrupole as r^{-5}. These latter three interactions depend on the orientation of the multipoles as well as the interparticle distances.

Even if a molecule does not have a permanent dipole, a dipole may be induced by the electric field of an ion. The induced dipole and the electric field are related by

$$\boldsymbol{\mu} = \alpha_2 \mathbf{E} \tag{7-4}$$

where α_2 is the polarizability of the nonpolar molecule and has units of volume. The magnitude of the ion's electric field is given by

$$E = -d[Z_1 e/r]/dr = Z_1 e/r^2 \tag{7-5}$$

so that the magnitude of the induced dipole is $\mu = \alpha_2 Z_1 e/r^2$. As a result of the ion's electric field the magnitude of the Coulomb force on the induced dipole is

$$F = \boldsymbol{\mu} \cdot \frac{d\mathbf{E}}{dr} \tag{7-6}$$

To find the ion-induced dipole interaction energy at a given r, the differential $dV(r) = -\mathbf{F} \cdot \mathbf{dr}$ must be integrated from a separation of infinity, where the potential is zero to r;

$$V(r) = -\int_{\infty}^{r} \mathbf{F} \cdot \mathbf{dr} = -\int_{\infty}^{r} F \, dr \tag{7-7}$$

Combining equations (7-4) through (7-6) and inserting them into equation (7-7) gives

$$V(r) = -\alpha_2 (Z_1 e)^2 / 2r^4 \tag{7-8}$$

A dipole as well as an ion can induce a dipole in a nonpolar molecule. For this case, the attractive potential varies by $1/r^6$, and a detailed calculation gives

$$V(r, \theta) = \frac{-\alpha_2 \mu_1^2 (3\cos^2\theta + 1)}{2r^6} \tag{7-9}$$

where θ is the angle between the permanent dipole and the inter-particle axis, α_2 is the polarizability of the nonpolar molecule, and μ_1 is the permanent dipole.

Dipole moments and polarizabilities for various molecules are listed in Tables 7-1 and 7-2. It is useful to estimate the magnitudes of the ion-ion, ion-dipole, ion-induced dipole, and dipole-induced dipole interactions discussed above. For illustration we will use the representative values for μ and α of 1.5D and 3 Å^3, respectively, and $Z = 1$ and θ equal to zero so that the dipoles are aligned along the interparticle axis. With these parameters and r of 5 and 10 Å, we obtain the interaction energies in Table 7-3.[2] Only the ion-ion interaction energy is comparable to a typical covalent single bond energy of 380–420 kJ/mol. The dipole-dipole interaction, which can be present in the interaction of two neutral molecules, is comparable to the thermal energy $3/2 \, k_B T$ at 300 K.

Equations (7-3) and (7-9) give the instantaneous interaction energy at a particular orientation of two particles. For a large number of particles there is an average interaction energy $\langle V(r) \rangle$, which is obtained by averaging the interaction potential at a fixed r, $V(\theta_1, \theta_2; r)$, over all possible orientations θ_1 and θ_2 of the two interacting particles.

TABLE 7-1 Experimental Dipole Moments.	
Molecule	Dipole moment (debyes)[a]
C_3H_8	0.08
H_2O_2	2.2
H_2O	1.85
H_2S	0.97
NaCl	9.0
KCl	10.3
HCl	1.12
CH_3D	0.0006
CH_2Cl_2	1.60
SO_2	1.63
HCN	2.98

[a]1 debye $= 10^{-18}$ esu-cm $= 3.336 \times 10^{-30}$ Coulomb-meter.
Source: Natl. Bur. Stand. U.S. Publ. NSRDS-NBS 10, 1967;
A. L. McClellan, *Tables of Experimental Dipole Moments,*
Vol. 1 (Freeman, New York, 1963) and Vol. 2 (Rahara
Enterprises, 1974).

TABLE 7-2 Experimental Polarizabilities.	
Molecule	Polarizability (Å^3)
Ar	1.63
N_2	1.76
F_2	1.3
I_2	10.2
CH_4	2.6
C_3H_8	6.29
CO	1.98
O_2	1.60
HCl	2.60
CO_2	2.63
NH_3	2.22
C_6H_6	10.4
CH_3Cl	4.53
$CHCl_3$	8.50

Source: N. J. Bridge and A. D. Buckingham, *Proc. Roy. Soc. (London) A295,* 334 (1966).

At each orientation the potential must be weighted by the Boltzmann factor given by $\exp[-V(\theta_1, \theta_2; r)/k_BT)]$. This averaging may alter n in the r^{-n} term of the potential. As discussed above, the instantaneous dipole-dipole interaction is proportional to r^{-3}. However, the average dipole-dipole interaction energy with the proper Boltzmann weighting depends on r^{-6} and is

$$\langle V(r) \rangle = -\frac{2}{3} \frac{\mu_1^2 \mu_2^2}{k_BT r^6} \tag{7-10}$$

Equation (7-9) can be generalized to represent the total dipole-induced dipole interaction potential between two polar molecules. Averaging this potential over all orientations of the two dipoles μ_1 and μ_2 and Boltzmann weighting gives

$$\langle V(r) \rangle = -\frac{(\mu_1^2 \alpha_2 + \mu_2^2 \alpha_1)}{r^6} \tag{7-11}$$

For this interaction, averaging over all orientations does not affect the r^{-n} term in the potential. The average potential in equation (7-11) is half the value in equation (7-9) for $\theta = 0$.

Even when a net charge or permanent dipole moment is absent from the molecule, there remain London or dispersion forces. In contrast to those above, these forces cannot be understood in terms of classical electrostatics. They are quantum mechanical in origin and arise from the correlated motion of the electrons in the two molecules. Although the time-averaged electron distribution of a nonpolar molecule is symmetric, fluctuations in the distribution can lead to a transient dipole moment, which can induce a dipole in a neighboring molecule. These fluctuations become correlated and an attractive interaction results. A useful approximate expression for estimating the

TABLE 7-3   Long-Range Interaction Energies.[a]

	$r = 5$ Å	$r = 10$ Å
Ion-ion, Eq. (7-2)	278	139
Ion-dipole, Eq. (7-3)	17.4	4.35
Ion-induced dipole, Eq. (7-8)	3.34	0.21
Dipole-induced dipole, Eq. (7-9)	0.052	0.00081

[a]Interaction energies are in units of kJ/mol, and are taken from R. S. Berry, S. A. Rice, and J. Ross, *Physical Chemistry* (New York: Wiley, 1980).

interaction energy due to dispersion forces, derived from quantum mechanical perturbation theory, is

$$V(r) = \frac{-3I_1 I_2}{2(I_1 + I_2)} \frac{\alpha_1 \alpha_2}{r^6} \tag{7-12}$$

where the I's are first ionization potentials and the α's are the polarizabilities of the two interacting particles.

If the particles are neutral and only dipoles are considered, the long-range potential is a sum of the average dipole-dipole interaction in equation (7-10), the average dipole-induced dipole interaction in equation (7-11), and the dispersion forces, each of which depends on r^{-6}. For most cases the dispersion contribution is the largest and is the only interaction when the particles are nonpolar. Including average dipole-quadrupole and quadrupole-quadrupole interactions gives rise to terms in the long-range potential which depend on r^{-8} and r^{-10}, respectively.

Long-range potentials are important in kinetics. For example, the Coulombic potential, equation (7-2), is appropriate for a molecule dissociating to ions (e.g., $KCl \rightarrow K^+ + Cl^-$) and the charge-induced dipole potential, equation (7-8), is the basis of the Langevin theory for ion-molecule reactions (chapter 8, section 8.3.4).

7.2   EMPIRICAL INTERMOLECULAR POTENTIALS

To describe an intermolecular interaction which does not involve the formation of a covalent bond, as for Ar_2, $(H_2O)_2$, or $Li^+(H_2O)$, the r^{-n} long-range attractive potentials described above have been combined with empirical models for the intermediate and short-range regions of the potential. The degree of precision required of the model is determined by the quantity to be represented. For example, a model with more detail is required to fit molecular beam scattering experiments (chapter 9, section 9.1) than is required for transport properties such as diffusion, heat conductivity, thermal diffusion, or viscosity.

The simplest interparticle potential is that for the hard-sphere model, which attempts to account for the short-range repulsive forces but not for the long-range attraction. This potential energy function is given by

$$V(r) = 0 \qquad r > \sigma$$
$$V(r) = \infty \qquad r < \sigma$$

and is depicted in Figure 7-2(a). The two hard-sphere particles come into contact at the internuclear distance σ and the potential energy rises to infinity at this point.

Interparticle potentials with more detail are constructed by combining either of the repulsive terms $V(r) = ae^{-br}$ or $V(r) = a/r^m$ with an attractive r^{-n} term to form

$$V(r) = ae^{-br} - c/r^n \tag{7-13}$$

or

$$V(r) = a/r^m - c/r^n \tag{7-14}$$

One of the most celebrated forms of equation (7-14) is the Lennard-Jones "6–12" potential

$$V(r) - 4\varepsilon\left[\left(\frac{\sigma}{r}\right)^{12} - \left(\frac{\sigma}{r}\right)^6\right] \tag{7-15}$$

where the r^{-6} attractive term represents the dispersion forces, as well as dipole-dipole and dipole-induced dipole interactions. The repulsive part, though purely empirical and chosen for mathematical convenience, does fit many potentials. The potential has a minimum at $r_m = 2^{1/6}\sigma$, with $V(r^m) = -\varepsilon$. A depiction of the Lennard-Jones potential is given in Figure 7-2(b). Values of r_m and ε for noble-gas atoms are given in Table 7-4.

7.3   MOLECULAR BONDING POTENTIALS

The above potential functions describe nonbonding interparticle and intermolecular interactions between atoms, ions, and molecules. Different potential functions are required for bonding interactions between atoms in molecules. For a diatomic molecule the only bonding interaction is a bond stretch between the two atoms. The simplest representation of the stretching potential is that of the harmonic oscillator model, i.e.

$$V(r) = V(r_e) + \frac{1}{2}f_r(r - r_e)^2 \tag{7-16}$$

The vibrational frequency for the harmonic stretch is $\nu_e = (f_r/\mu)^{1/2}/2\pi$, where μ is the diatomic reduced mass. The vibrational energy levels are $E(n) = (n + 1/2)h\nu_e$.

The harmonic oscillator model is an accurate representation of the stretching potential only for small displacements of r from the equilibrium bond length r_e. Differences between the true potential and the harmonic are called "anharmonic effects." These differences are magnified at large r, where, as shown in Figure 7-1, the anharmonic potential approaches D_e while the harmonic potential increases to large positive values.

A way to include anharmonicity is to represent the potential in a Taylor series expansion,

$$V(r) = V(r_e) + \left(\frac{\partial V}{\partial r}\right)_{r=r_e}(r - r_e) + \frac{1}{2}\left(\frac{\partial^2 V}{\partial r^2}\right)_{r=r_e}(r - r_e)^2 + \frac{1}{6}\left(\frac{\partial^3 V}{\partial r^3}\right)_{r=r_e}(r - r_e)^3$$

$$+ \frac{1}{24}\left(\frac{\partial^4 V}{\partial r^4}\right)_{r=r_e}(r - r_e)^4 + \cdots \tag{7-17}$$

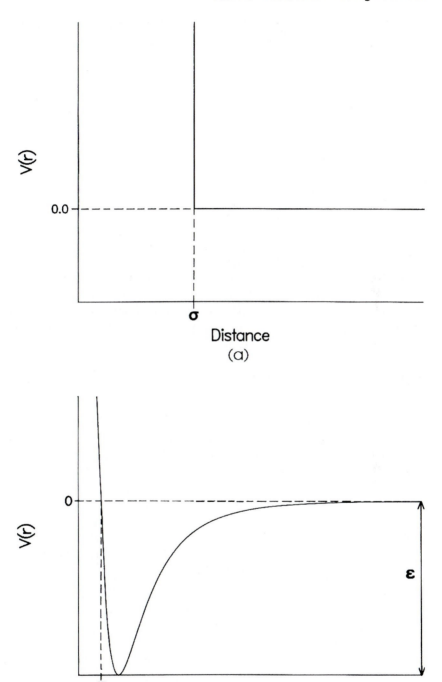

FIGURE 7-2   (a) Hard-sphere potential energy curve. (b) Lennard-Jones 6–12 potential energy curve.

TABLE 7-4 Potential Parameters for Noble Gas Mixtures.

	He	Ne	Ar	Kr	Xe
He	$\left(\begin{array}{c}10.8\\2.97\end{array}\right)$	22	29.4	30.2	28
Ne	3.00	$\left(\begin{array}{c}42.3\\3.08\end{array}\right)$	70	73	75
Ar	3.47	3.43	$\left(\begin{array}{c}143.2\\3.76\end{array}\right)$	167.5	187.4
Kr	3.67	3.58	3.88	$\left(\begin{array}{c}199.9\\4.01\end{array}\right)$	231.1
Xe	3.95	3.75	4.06	4.18	$\left(\begin{array}{c}282.3\\4.36\end{array}\right)$

Source: G. Scoles, *Ann. Rev. Phys. Chem. 31,* 81 (1980). Upper triangle: well depth(K).
Lower triangle: position of potential minimum r_m (Å).

The first derivative term equals zero at the minimum $r = r_e$. The second, third, and fourth derivative terms are called quadratic (harmonic), cubic, and quartic force constants, respectively. Many terms must be included in equation (7-17) to accurately represent the potential energy for large displacements in r from r_e.

A useful approximation to a bond stretch potential for all values of r is the Morse function:

$$V(r) = D_e[1 - \exp\{-\beta_e(r - r_e)\}]^2. \tag{7-18}$$

The three parameters in the Morse function D_e, β_e and r_e are positive and are usually chosen to fit the bond dissociation energy, the harmonic vibrational frequency ν_e, and the equilibrium bond length. The shape of the Morse function is the same as $V(r)$ in Figure 7-1. At $r = r_e$ the Morse function $V(r)$ equals zero. As $r \to \infty$ $V(r)$ approaches D_e so that the Morse function dissociation energy is D_e. For $r \ll r_e$, $V(r)$ is large and positive, corresponding to short-range repulsion. By taking the second derivative of $V(r)$ in equation (7-18) with respect to r at $r = r_e$, it is found the parameters D_e and β_e are related to the harmonic force constant f_r through the relation $f_r = 2D_e\beta_e^2$. Vibrational energy levels for the Morse oscillator are

$$E(n) = (n + 1/2)h\nu_e - (n + 1/2)^2 h\nu_e x_e, \tag{7-19}$$

where the anharmonicity term x_e is given by

$$x_e = h\nu_e/4D_e \tag{7-20}$$

The difference between the classical and quantum dissociation energies D_e and D_0 (See Figure 7-1) is the zero point energy for the diatomic molecule. The harmonic oscillator zero point energy is $h\nu_e/2$, while that for the anharmonic Morse oscillator is $h\nu_e(1 - x_e/2)/2$.

Two shortcomings of the Morse function are that in representing actual diatomic potentials it is not sufficiently repulsive at small r and too attractive at large r. Several different functions have been suggested as improvements to the Morse function.[3-5] A widely used procedure is to represent the β_e term in the exponential as a polynomial in r;[6] e.g.,

$$\beta_e = \beta_0 + a(r - r_e) + b(r - r_e)^2 + c(r - r_e)^3 \qquad (7\text{-}21)$$

7.4   INTERNAL COORDINATES AND NORMAL MODES OF VIBRATION

For molecules with more than two atoms, accurate representation of the potential energy becomes more difficult, since more coordinates and potential parameters are required. To depict the potential either conceptually or mathematically, it is natural and convenient to use internal coordinates such as bond stretches and angle bends. A nonlinear molecule consisting of N atoms has 3N-6 independent internal coordinates. Additional internal coordinates are called "redundant coordinates." The analytic representation of a potential energy surface is often facilitated by including redundant internal coordinates. In general, no difficulty arises in including redundant internal coordinates in a potential energy function as long as the function correctly represents the molecular symmetry.

A nonlinear triatomic molecule such as water, shown in Figure 7-3(a), has three independent internal coordinates which may be chosen as two bond stretches and a valence angle bend. A comparison with the diagrams for formaldehyde in Figures 7-3(b) and 7-3(c) shows that increasing the size of a molecule by only one atom can lead to a significant increase in the complexity of the potential energy. A potential energy function for formaldehyde could include three bond stretching coordinates, three valence angle bends, and an out of plane bend. Figure 7-3(b) illustrates the three valence angle bends and Figure 7-3(c) shows the out of plane bend. Note that a redundant valence angle bend angle is included in this representation of the formaldehyde potential energy function. An important internal coordinate for many molecules is the torsion (dihedral) angle, which is depicted for hydrogen peroxide in Figure 7-3(d).

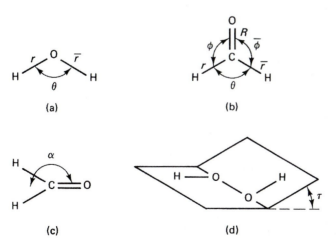

FIGURE 7-3   Different types of internal coordinates.

In chapter 11, section 11.7, internal coordinates are used to define analytic potential energy functions for chemical reactions. They can also be used to represent the potential energy for a polyatomic molecule near its potential energy minimum by expanding the potential in a Taylor series as described by equation (7-17) for a diatomic. Expressing the potential energy for the water molecule, Figure 7-3(a), in this manner gives

$$V = V_0 + \frac{1}{2} f_{rr}(r - r_e)^2 + \frac{1}{2} f_{\bar{r}\bar{r}}(\bar{r} - \bar{r}_e)^2 + \frac{1}{2} f_{r\bar{r}}(r - r_e)(\bar{r} - \bar{r}_e) + \frac{1}{2} f_{\theta\theta}(\theta - \theta_e)^2$$

$$+ \frac{1}{2} f_{r\theta}(r - r_e)(\theta - \theta_e) + \frac{1}{2} f_{\bar{r}\theta}(\bar{r} - \bar{r}_e)(\theta - \theta_e) + \cdots \tag{7-22}$$

where the anharmonic cubic, quartic, and higher order terms have been neglected. The $f_{r\bar{r}}, f_{r\theta}$ and $f_{\bar{r}\theta}$ nondiagonal terms in equation 7-22 are often called "interaction terms," since they describe how the displacement of a particular internal coordinate affects the potential of the remaining coordinates. Force constants may be determined from *ab initio* calculations[7] and by fitting experimental vibrational frequencies.[8] Representative internal coordinate quadratic force constants are listed in Table 7-5.

Normal mode coordinates of vibration, which assume a harmonic (i.e., quadratic) potential energy function as in equation (7-22), are often used to characterize the potential energy minimum for a polyatomic molecule. For a molecule with N atoms the most straightforward way to determine the normal mode coordinates and their frequencies is to diagonalize the 3N × 3N mass-weighted Cartesian force constant matrix $\mathbf{F}$,[9] i.e.

$$\tilde{\mathbf{L}} \mathbf{F} \mathbf{L} = \Lambda \tag{7-23}$$

TABLE 7-5 Diagonal Internal Coordinate Force Constants.[a]

Diagonal force constant	Value[b]
H—C stretch	4.55, 4.64, 4.70
C—C stretch	4.39
C=O stretch	12.99
C—O stretch	5.70
HCH bend	0.540, 0.550
HCC bend	0.645, 0.656
CCC bend	1.130, 1.084, 1.086
HCO or CCO bend	1.166
COC or COH bend	0.651

[a]The force constants were derived from pyruvic acid, acetone, acetic acid, methyl acetate, formic acid, methyl formate, and straight chain aliphatic hydrocarbons; J. H. Schachtschneider and R. G. Snyder, *Spectrochim. Acta 19*, 117 (1963); H. Hollenstein and Hs. H. Günthard, *J. Mol. Spectrosc. 84*, 457 (1980).
[b]The stretching force constants are in units of mdyn/Å, and the bending force constants in units of mdyn-Å/rad².

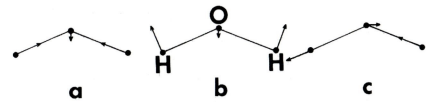

FIGURE 7-4 H_2O normal modes of vibration: (a) symmetric stretch; (b) symmetric bend; (c) asymmetric stretch.

where $\tilde{L}$ is the transpose of the matrix L with $\tilde{L}\,L = 1$. Solving this equation is a standard problem in linear algebra. The solution gives Λ which is a diagonal matrix of the $3N$ eigenvalues λ_i, and L the eigenvector matrix which defines the transformation between normal mode coordinates Q and displacements of the mass-weighted Cartesian coordinates q from the equilibrium geometry q_0; i.e., $[q - q_0] = L\,Q$. Included in Λ are the eigenvalues for translation and external rotation, which are equal to zero. The remaining nonzero eigenvalues are for vibration and equal $4\pi^2\nu_i^2$, when ν_i is the normal mode frequency. The normal modes of vibration for water are illustrated in Figure 7-4.

The mass-weighted Cartesian coordinates $q_{\alpha i}$ are related to Cartesian coordinates $s_{\alpha i}$ according to

$$q_{\alpha i} = \sqrt{m_i}\, s_{\alpha i} \qquad i = 1, \ldots, N$$
$$\alpha = x, y, z \tag{7-24}$$

The elements of the force constant matrix F are denoted by $f_{\alpha i, \alpha' i'}$ and are equal to the second derivatives of the potential energy with respect to $q_{\alpha i}$ and $q_{\alpha' i'}$ i.e. $\partial^2 V/\partial q_{\alpha i}\partial q_{\alpha' i'}$. The second derivatives are determined from the potential energy function, e.g., equation (7-22), by writing the internal coordinates as functions of the Cartesian coordinates. The force constants $f_{\alpha i, \alpha' i'}$ are symmetric with respect to interchanging the indices αi and $\alpha' i'$.

To illustrate a normal mode analysis consider the stretching vibration of a diatomic molecule with the harmonic potential $V = f(r - r_e)^2/2$. (A normal mode analysis of the stretching vibrations for a linear triatomic molecule is given in Problem 7.8). By inserting $x_2 - x_1$ into the potential energy function for the internuclear distance r, the second derivatives of the potential with respect to the Cartesian coordinates x_1 and x_2 are found to be $\partial^2 V/\partial x_1^2 = f$, $\partial^2 V/\partial x_2^2 = f$, and $\partial^2 V/\partial x_1\partial x_2 = \partial^2 V/\partial x_2 x_1 = -f$. Thus, the mass-weighted Cartesian force constant matrix is

$$F = \begin{vmatrix} f/m_1 & -f/(m_1 m_2)^{1/2} \\ -f/(m_1 m_2)^{1/2} & f/m_2 \end{vmatrix} \tag{7-25}$$

If equation (7-23) is written in the form

$$(F - \Lambda)L = 0 \tag{7-26}$$

it is apparent that the two eigenvalues λ_1 and λ_2, are found by solving the secular determinant

$$\begin{vmatrix} f/m_1 - \lambda & -f/(m_1 m_2)^{1/2} \\ -f/(m_1 m_2)^{1/2} & f/m_2 - \lambda \end{vmatrix} = 0 \tag{7-27}$$

The resulting two eigenvalues are $\lambda_1 = 0$ for translation and $\lambda_2 = f/\mu$ for vibration, where the reduced mass μ equals $m_1 m_2/(m_1 + m_2)$. Since $\lambda = 4\pi^2\nu^2$, the harmonic vibrational frequency is $\nu = (f/\mu)^{1/2}/2\pi$.

Each column of $\mathbf{L}$ is the eigenvector for an eigenvalue λ_i. To determine an eigenvector insert the value for λ_i into equation (7-26) to obtain 3N simultaneous homogeneous linear equations, which for the diatomic molecule considered here are

$$(f/m_1 - \lambda)\ell_1 - [f/(m_1 m_2)^{1/2}]\ell_2 = 0$$
$$-[f/(m_1 m_2)^{1/2}]\ell_1 + (f/(m_2 - \lambda)\ell_2 = 0 \tag{7-28}$$

Using the normalization condition $(\ell_1^2 + \ell_2^2) = 1$, equation (7-28) can be solved to give $\ell_{11} = [m_1/(m_1 + m_2)]^{1/2}$ and $\ell_{12} = [m_2/(m_1 + m_2)]^{1/2}$ for $\lambda_1 = 0$, and $\ell_{21} = [m_2/(m_1 + m_2)]^{1/2}$ and $\ell_{22} = -[m_1/(m_1 + m_2)]^{1/2}$ for $\lambda_2 = f/\mu$. Since the normal coordinates are given by $\mathbf{Q} = \tilde{\mathbf{L}}[\mathbf{q} - \mathbf{q_0}]$, ℓ_{11} and ℓ_{12} give the mass-weighted Cartesian coordinate displacements for normal mode 1 (a translation), while ℓ_{21} and ℓ_{22} give the mass-weighted Cartesian coordinate displacements for normal mode 2 (a vibration).

It becomes exceedingly difficult to determine the matrices $\mathbf{\Lambda}$ and $\mathbf{L}$ analytically as the dimensionality of the problem is increased. For polyatomic molecules, these matrices may be obtained numerically with standard computer programs which are readily available.[10] Values for quadratic force constants (Table 7-5) can be found by varying them in a normal mode analysis to fit experimental harmonic frequencies.

The energy levels for a normal mode are given by $E_i = (n_i + 1/2)h\nu_i$. A summation of the individual normal mode energies gives the normal mode approximation to the total vibrational energy of the molecule, i.e.

$$E = \sum_i E_i = \sum_i (n_i + 1/2)h\nu_i \tag{7-29}$$

7.5   POTENTIAL ENERGY SURFACES

As shown in Figure 7-1, the potential energy versus a diatomic's internuclear distance defines a curve for a diatomic molecule. In a chemical reaction, three or more atoms must interact; thus the relevant coordinates define a multidimensional surface. Since it is difficult, if not impossible, for most people to visualize surfaces with more than three dimensions, methods must be used to reduce the dimensionality of the problem. For even the simplest system, the three-atom reaction A + BC, the potential surface is a four-dimensional hypersurface (three independent coordinates plus the energy). This can be reduced to a three-dimensional surface in real space by constraining one of the coordinates, e.g., fixing the angle between the two bonds to a specified value. This value is often taken to be 180°, so that the surface represents collinear A—B—C. This gives the potential surface as a function of the two independently variable bond lengths. Even such a three-dimensional surface is inconvenient to represent on a planar blackboard or the pages of a book; thus the surface is generally presented in the form of a contour diagram. This approach of reducing the potential surface for a triatomic (or any polyatomic) to a three-dimensional surface in real space does not place any restrictions on the surface itself but is merely used for graphical representation.

There are two general types of potential energy surfaces for a collinear $A + BC \rightarrow AB + C$ reaction. One has a potential energy barrier at intermediate A-B and B-C distances, while the other has a potential energy well (minimum). Contour maps for these two types of reactions are depicted in Figure 7-5. Some special $A + BC$ reactions may have both minima and potential energy barriers. The dashed line in Figure 7-5(a) and 7-5(b) represents the reaction path $r^\dagger$ (the term *reaction coordinate* is a number representing a particular position along the reaction path). In the reactant configuration, $A + BC$, the reaction path is primarily the A-B distance, while it is the B-C distance for the products, $AB + C$. In the intermediate region of the surface, the reaction path contains components of both of these bond distances. At the maximum in the potential energy along the reaction path in Figure 7-5(a) there is a pass or saddlepoint. There is a well at the potential energy minimum in Figure 7-5(b). A potential energy profile results from plotting the potential versus reaction path. The reaction path potential energy profiles for the contour maps in Figure 7-5 are plotted in Figure 7-6.

The saddlepoint in Figure 7-5(a) is a stationary point on the surface. Thus, at the saddlepoint, as for a potential energy minimum, the derivative of the potential energy is zero with respect to each Cartesian coordinate; i.e., $\partial V / \partial q_i = 0$. The harmonic vibrational frequencies and eigenvectors at the saddlepoint are found in the same manner as for a potential energy minimum, *viz.*, by use of equations like those in (7-23) through (7-28). At the saddlepoint, the potential energy surface curves downward along the reaction path, so the corresponding force constant is negative. Since the frequency is proportional to the square root of the force constant, this implies an imaginary frequency. Thus, in contrast to the minimum where all 3N-6 frequencies are real, one of the 3N-6 frequencies is imaginary at the saddlepoint. This frequency corresponds to infinitesimal motion along the reaction path. The remaining 3N-7 frequencies and their eigenvectors define harmonic vibrational motion orthogonal to the reaction path. From the 3N-7 real frequencies the zero-point energy at the saddlepoint can be calculated.

A potential energy surface may also be represented by a perspective. Such a perspective and its relationship to a contour map are illustrated in Figure 7-7. Potential energy perspectives and contour maps are described by topographical terms such as pass, saddlepoint, valley, well, depression, ridge, etc., which are evident in Figure 7-7. Hikers and mountain climbers should be particularly adept at analyzing potential energy surfaces. Many different perspectives and contour maps may be constructed for a reaction with more than three atoms, since different sets of coordinates may be constrained. Several types of perspectives and maps for polyatomic systems are depicted in the next section.

7.6   *AB INITIO* CALCULATION OF POTENTIAL ENERGY SURFACES

Potential energy surfaces may be determined by *ab initio* electronic structure calculations. If one makes the Born-Oppenheimer approximation, discussed in the introduction to this chapter, the molecular wave function is written as

$$\Psi = \Psi_e(\mathbf{r}, \mathbf{R})\Psi_n(\mathbf{R}) \qquad (7\text{-}30)$$

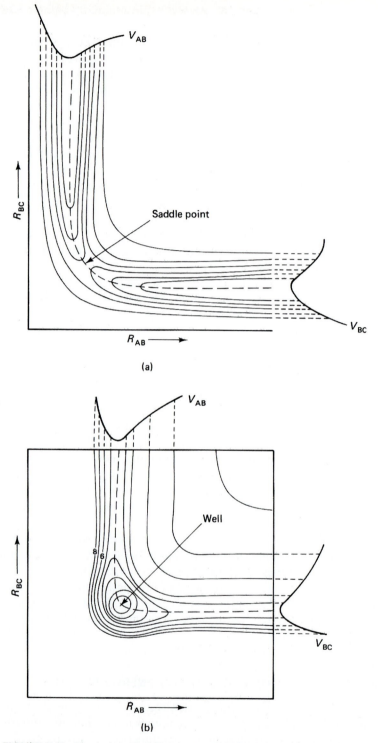

FIGURE 7-5 Depiction of collinear $A + BC \rightarrow AB + C$ contour maps. The dashed lines are reaction paths. (a) Contour map with a potential energy barrier; (b) contour map with a potential energy minimum.

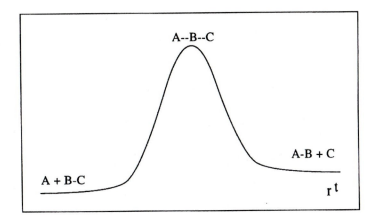

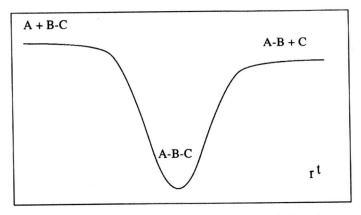

FIGURE 7-6   Potential energy profile (i.e., potential along the reaction path) for the contour maps in Figure 7-5(a) and 7-5(b).

where Ψ_e is the electronic wave function, which depends on the electron coordinates $\mathbf{r}$ and nuclear coordinates $\mathbf{R}$, and Ψ_n is the nuclear wavefunction. Solving the electronic Schrodinger Equation for the molecule, i.e.

$$(T_e + V_{ee} + V_{en})\Psi_e = E_e(\mathbf{R})\Psi_e \qquad (7\text{-}31)$$

at a particular nuclear configuration $\mathbf{R}$ gives the electronic energy E_e for the configuration. In the above equation, T_e, V_{ee}, and V_{en} are the operators for the electron kinetic energy, electron-electron repulsion energy, and electron-nuclear attraction energy. The sum of E_e and the nuclear-nuclear repulsion energy V_{nn} is the total potential energy in which the nuclei move with kinetic energy T_n. By performing the calculation at many different nuclear configurations a complete potential energy surface may be determined. The nuclear wave function may be found by solution of the nuclear Schrodinger equation[1]

$$(T_n + V_{nn} + E_e)\Psi_n = E\Psi_n$$

where T_n is the kinetic energy operator for the nuclei.

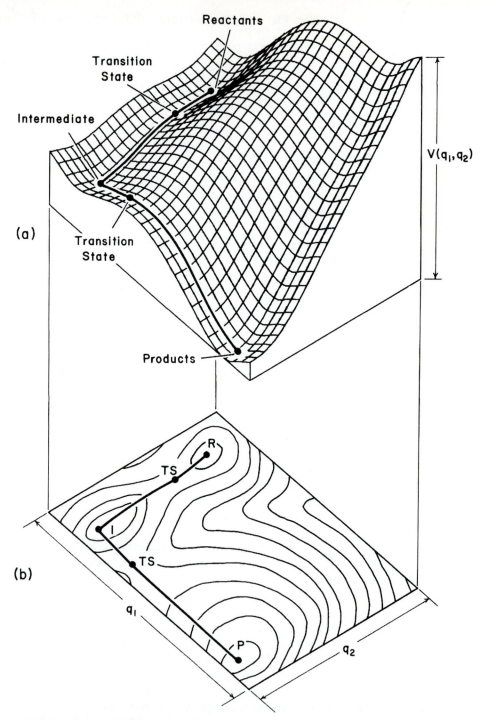

FIGURE 7-7 A three-dimensional perspective (a) and contour map (b) for a model chemical reaction. The solid line is the reaction path. [Adapted from G. M. Maggiora and R. E. Christofferson, in *Transition States of Biochemical Processes,* ed. R. D. Gandour and R. L. Schowen (New York: Plenum, 1978).]

Sophisticated *ab initio* methods, which are based on the variational theorem,[1] have been developed for calculating potential energy surfaces. The Self-Consistent-Field (SCF) method often gives good structural and force constant information for a molecule near its equilibrium geometry, and is widely used for evaluating these properties.[11] However, the SCF method usually gives an inaccurate potential energy surface for a chemical reaction in which bonds are broken and formed so that there are extensive geometry changes. This is because the SCF approximation neglects instantaneous correlations between electronic motions, and, thus, it often can not describe changes in the wavefunction Ψ_e as bonds are broken and formed.[12] This difficulty is well known for reactions as simple as $H_2 \rightarrow 2H$ and $H + H_2 \rightarrow H_2 + H$.[12] To treat electron correlation correctly and obtain an accurate potential energy surface requires a Generalized-Valence-Bond (GVB),[13] Multiconfiguration-SCF (MCSCF),[14] Configuration-Interaction (CI),[15] density-functional,[16] or perturbation[17] treatments. Computer programs are available for calculating *ab initio* potential energy surfaces using these theoretical methods.

Ab initio calculations have provided much of the information known about potential energy surfaces. The accuracy of the calculated result is determined by the number of basis functions and the treatment of electron correlation, both of which depend on the number of electrons in the molecular system. For many reactions the calculated structures for potential energy minima are as accurate as those found experimentally.[18] Also, *ab initio* and experimental harmonic vibrational frequencies often agree to within 5%.[19] Of course, the agreement usually improves as the total number of electrons decreases. For reactions which are comprised of three or fewer second row atoms, activation energies and heats of reaction are usually accurate to within several kJ/mole.[20] A very accurate potential energy surface has been calculated for the $H + H_2 \rightarrow H_2 + H$. reaction.[21] Extensive calculations have also been performed for the $F + H_2 \rightarrow FH + H$. surface,[22] illustrated in Figures 7-8 and 7-9. Other reactions which have received considerable study are listed in Table 7-6 and two additional examples are discussed below.

Ab initio calculations have been widely used to calculate interaction energies between alkali ions and ligands.[23,24] The potential surface for this interaction is often depicted by rotating the ion in a plane about the ligand which is fixed in its equilibrium geometry. Such a surface for the $Li^+ + H_2O$ interaction is given in Figure 7-10.

A somewhat different approach is used to represent *ab initio* calculations of the $H + CH_3$ potential energy surface.[25] As shown in Figure 7-11, the two internal coordinates which are used in representing the 3-dimensional surface are the bond length r and the angle χ. This angle is a displacement from the C_{3v} symmetry axis in a plane that includes C and two H atoms, and bisects the HCH angle of the remaining two H-atoms; i.e.

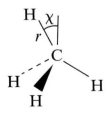

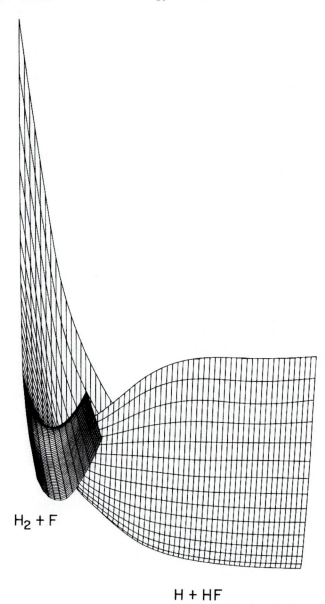

H₂ + F

H + HF

FIGURE 7-8A Plot of the electronic energy of linear FH_2 as a function of F-H and H-H distances. Each small section bounded by four sides represents a square region in space 0.075 bohrs on a side. Thus the F-H distance ranges from 1.4 to 8.0 bohrs and the H-H distance from 1.0 to 7.6 bohrs. [Adapted from C. F. Bender, et al., *Science 176*, 1412 (1972). Copyright 1972 by the AAAS.]

The structure of the CH_3 group was optimized at each value of r with $\chi = 0$, and that structure was retained at other values of χ for the same value of r.

7.7 ANALYTIC POTENTIAL ENERGY FUNCTIONS

Dynamical calculations, whether classical or quantum mechanical, are facilitated by analytic potential energy functions. A variety of analytic functions have been used in such calculations. A function may be based on an empirical model for the chemical

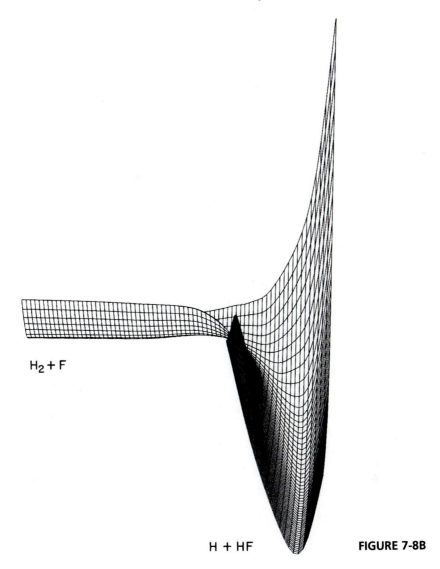

H$_2$ + F

H + HF   **FIGURE 7-8B**

reaction. Parameters for the empirical functions may be chosen to fit experimental equilibrium geometries, vibrational frequencies, bond energies, and activation energies, and the results of *ab initio* calculations. *Ab initio* results are often used to connect smoothly regions of the surface which are known experimentally such as the reactant and product asymptotic limits, and potential energy minima. The parameters for empirical surfaces are often of physical significance.

A dynamical calculation can also be performed by using a table of potential energy points. Splines or other functions[26] may be used to connect analytically the points in the table. One of the drawbacks of using such a table is that it can not be easily cast into an empirical potential energy function, which is often useful in interpreting

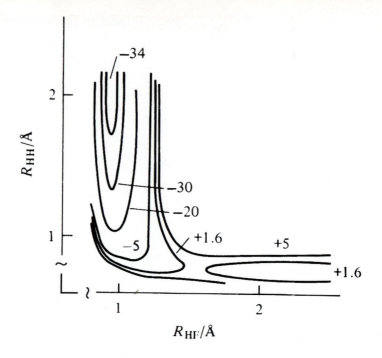

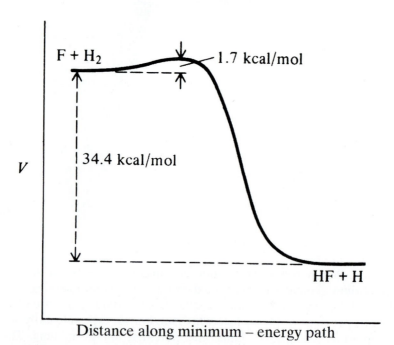

FIGURE 7-9 (a) Contour map of the linear FH_2 surface. The F-H distance varies from 1.4 to 8.0 bohrs and the H-H distance from 1.0 to 7.6 bohrs. [Adapted from C. F. Bender, S. V. O'Neil, P. K. Pearson, and H. F. Schaefer, *Science 176*, 1412 (1972)]; (b) Potential energy profile for the collinear FH_2 system. [Adapted from I. N. Levine, *Physical Chemistry* (McGraw-Hill, New York, 1988). Reproduced with permission of McGraw-Hill.]

TABLE 7-6 Potential Energy Surfaces from *Ab Initio* Calculations.

Reaction	Reference
$H + H_2 \rightarrow H_2 + H$	See Reference 21, p. 212.
$F + H_2 \rightarrow FH + H$	See Reference 22, p. 212.
$H + CO \rightarrow HCO$	J. M. Bowman, J. S. Bittman, and L. B. Harding, *J. Chem. Phys. 85*, 911 (1986); H. -J. Werner et al. *J. Chem. Phys. 102*, 3593 (1995).
$O(^1D) + H_2 \rightarrow OH + H$	T. -S. Ho, T. Hollebeek, H. Rabitz, L. B. Harding, and G. C. Schatz, *J. Chem. Phys. 105*, 10472 (1996)
$H_2CO \rightarrow H + HCO$ $H_2 + CO$	G. E. Scuseria and H. F. Schaefer, III, *J. Chem. Phys. 90*, 3629 (1989); W. H. Green, Jr., A. Willetts, D. Jayatilaka, and N. C. Handy, *Chem. Phys. Lett. 169*, 127 (1990); W. Chen, W. L. Hase, and H. B. Schlegel, *Chem. Phys. Lett. 228*, 436 (1994).
$Li + FH \rightarrow LiF + H$	G. A. Parker, A. Laganà, S. Crocchianti, and R. T. Pack, *J Chem. Phys. 102*, 1238 (1995).
$Cl + H_2 \rightarrow HCl + H$	T. C. Allison, G. C. Lynch, D. G. Truhlar, and M. S. Gordon, *J. Phys. Chem. 100*, 13575 (1996).

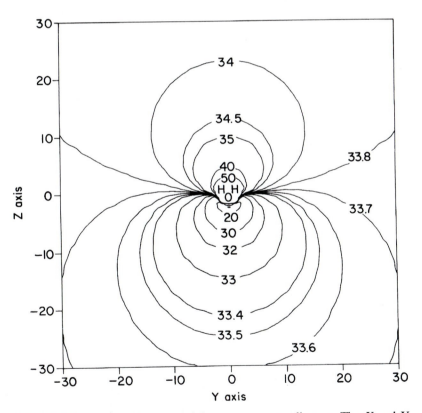

FIGURE 7-10 $Li^+ + H_2O$ potential energy contour diagram. The X and Y axes define the distance of Li^+ from O. The Li^+ ion is in the plane of H_2O molecule. Energy is in units of kcal/mol. [Adapted from W. L. Hase and D. -F. Feng, *J. Chem. Phys. 75*, 738 (1981)].

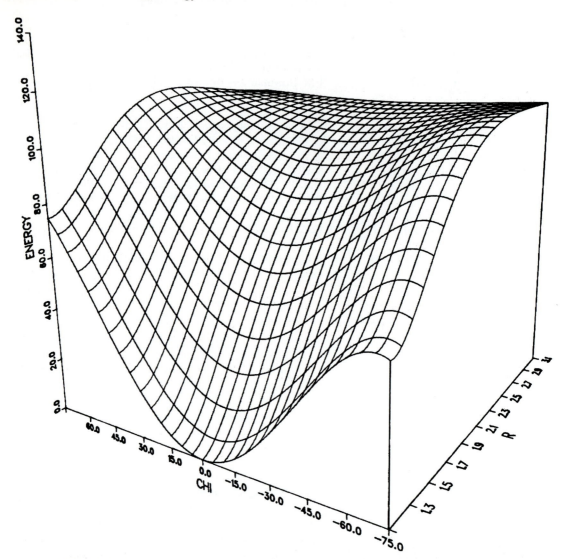

FIGURE 7-11   Plot of the $H + CH_3 \rightarrow CH_4$ potential energy surface, where r is the H—C distance (Å) and χ is the H—CH_3 angle (degrees). The geometry of the CH_3 group is optimized at each value of r. Energy is in kcal/mol. [Adapted from W. L. Hase, S. L. Mondro, R. J. Duchovic, and D. M. Hirst, *J. Am. Chem. Soc. 109*, 2916 (1987)].

the calculated reaction dynamics. Also, in contrast to a table of points, an empirical function can be easily modified by adjusting several parameters.

Rather simple analytic potential energy functions are those based on the r^{-n} intermolecular potential terms discussed in section 7.1. Such potentials have been used to describe ion-molecule (e.g., $Li^+ + H_2O$) and molecule-molecule intermolecular interactions (e.g., $NH_3 + NH_3$), for cases where the intermolecular interaction causes only minor changes in the molecular equilibrium geometry and vibrational frequencies. The complete intermolecular potential is represented by a sum of atom-atom two-body potentials similar to those in equations (7-13) and (7-14). Such potential functions have been used in studies of the structure and thermodynamics of solutions,[27] and of ion-molecule reaction kinetics in the gas phase[28] and in solution.[29]

As an example, consider the intermolecular potential between a metal ion M and a ligand L. The potential is given by

$$V_{M,L} = \sum_i V_{im},$$
(7-32)

where i is a sum over the atoms on the ligand and V_{im} is a two-body potential between M and one of the ligand atoms. A widely used form for V_{im} is

$$V_{im} = \frac{a_{im}}{r^{12}} - \frac{b_{im}}{r^6} + \frac{c_{im}}{r}$$
(7-33)

The first two terms comprise the Lennard-Jones "6–12" potential. The remaining term represents an electrostatic interaction, either repulsive or attractive, between the metal atom and ligand atom. *Ab initio* quantum mechanical calculations are used to derive the parameters a_{im}, b_{im}, and c_{im}.[23,24] The *ab initio* potential energy surface for $Li^+ + H_2O$ in Figure 7-10 is accurately described by equations (7-32) and (7-33).

A potential energy function, similar to that given above, may be used to describe the interaction between two molecules A and B. Here a sum of two-body interactions between the a atoms contained in A and the b atoms contained in B is required: i.e.

$$V_{A,B} = \sum_a \sum_b V_{a,b}$$
(7-34)

An expression similar to equation (7-33) may be used for $V_{a,b}$ and *ab initio* calculations may be used to derive the potential energy parameters.[30,31]

The most widely used empirical analytic potential function for A + BC atom-diatom reactive scattering is the London-Eyring-Polanyi-Sato (LEPS) surface.[32] This function has its genesis in the semiempirical valence bond approach of Heitler and London,[33] which gives $(Q + J)/(1 + S^2)$ and $(Q - J)/(1 - S^2)$ for the bonding and anti-bonding (i.e., repulsive) potential energy curves of two atoms. Q, J, and S are referred to as the Coulomb, exchange, and overlap integrals. Over the years the original function, based on the Heitler-London model, has been modified by including empirical functions and parameters, and has acquired an empirical character.

One widely used form of the LEPS function is given by

$$V(r_{AB}, r_{AC}, r_{BC}) = \frac{Q_{AB}}{1 + S_{AB}} + \frac{Q_{BC}}{1 + S_{BC}} + \frac{Q_{AC}}{1 + S_{AC}} - \left\{ \frac{1}{2} \left[\left(\frac{J_{AB}}{1 + S_{AB}} - \frac{J_{BC}}{1 + S_{AB}} \right)^2 \right. \right.$$

$$\left. \left. + \left(\frac{J_{BC}}{1 + S_{BC}} - \frac{J_{AC}}{1 + S_{AC}} \right)^2 + \left(\frac{J_{AC}}{1 + S_{AC}} - \frac{J_{AB}}{1 + S_{AB}} \right)^2 \right] \right\}^{1/2}$$
(7-35)

The Coulomb and exchange integrals, Q and J, are determined from the Morse and anti-Morse functions via

$$Q_{AB} + J_{AB} = D_{AB}\{\exp[-2\beta_{AB}(r_{AB} - r_{AB}^0)] - 2\exp[-\beta_{AB}(r_{AB} - r_{AB}^0)]\} \qquad (7\text{-}36)$$

$$Q_{AB} - J_{AB} = \frac{1}{2}D_{AB}\{\exp[-2\beta_{AB}(r_{AB} - r_{AB}^0)] + 2\exp[-\beta_{AB}(r_{AB} - r_{AB}^0)]\} \qquad (7\text{-}37)$$

Equation (7-37) was first proposed by Sato[34] to describe the repulsive potential between A and B. He found that changing the sign between the two exponential terms of the Morse potential, equation (7-36), from minus to plus, and dividing the resulting expression by 2 gives a function which reasonably fits the repulsive potential curve of the first excited state $^3\Sigma_u^+$ of H_2. The three parameters S_{AB}, S_{BC}, and S_{AC} in equation (7-35) give the LEPS function flexibility. They may be chosen to fit the location of the potential energy barrier with respect to the B-C and A-B distances, and the height of the barrier. The original LEPS form of Sato[34] has $S_{AB} = S_{BC} = S_{AC}$. In the extended LEPS form[35] these S parameters are allowed to differ. The LEPS potential is 3-dimensional, since it is a function of the three distances r_{AB}, r_{BC}, and r_{AC}.

Though the LEPS function has received wide use, it has severe limitations. Since it has only three adjustable parameters, it has limited flexibility and does not correctly represent surfaces for a number of AB + C reactions. Often, when trying to fit a barrier height and saddlepoint location, spurious wells appear in the surface.[32] This is particularly evident for the D + HCl system.[36] Clearly, the LEPS function can not be used to represent highly accurate *ab initio* calculations.

Another widely used empirical potential energy function for A + BC → AB + C reactions is the BEBO ("bond energy–bond order") model of Johnston and Parr.[37] However, like the LEPS model it also lacks flexibility. The BEBO model is developed from a number of empirical relationships. One is the equation proposed by Pauling[38] to relate bond length r and bond order n:

$$r = r_s - 0.26 \ln n \qquad (7\text{-}38)$$

where r_s is the single-bond length and 0.26 has units of Å. A log-log correlation between bond energy and bond order is assumed, so that the bond energy D is given by

$$D = D_s n^p \qquad (7\text{-}39)$$

where the single-bond energy is D_s and p is a constant. This relationship works quite well for CC, CO, and OO bonds. Furthermore, it is postulated that the bond order is conserved so that

$$n_{AB} + n_{BC} = 1. \qquad (7\text{-}40)$$

The potential energy is written as the dissociation energy of the BC bond minus the sum of the energies of the partially formed AB bond and partially broken BC bond plus the A-C repulsion energy:

$$V = D_{BC} - D_{BC}(n_{BC})^{p_{BC}} - D_{AB}(n_{AB})^{p_{AB}} + V(r_{AC}) \qquad (7\text{-}41)$$

An anti-Morse function, equation (7-37), may be used for $V(r_{AC})$. Equations (7-38) through (7-41) specify the potential energy along the minimum energy path.

Additional terms must be added to give the variation in the potential energy for displacements from the reaction path.

Another procedure for writing analytic potential energy functions uses the concept of a switching function. An example is the A + BC → AB + C analytic function developed by Blais and Bunker,[39] and based on the Morse function and a switching function. The analytic function is

$$V = D_{AB}\{1 - \exp[-\beta_{AB}(r_{AB} - r^0_{AB})]\}^2 + D_{BC}\{1 - \exp[\beta_{BC}(r_{BC} - r^0_{BC})]\}^2$$

$$+ D_{BC}[1 - \tanh(ar_{AB} + c)]\exp[-\beta_{BC}(r_{BC} - r^0_{BC})] \tag{7-42}$$

$$+ D_{AC}\exp[-\beta_{AC}(r_{AC} - r^0_{AC})].$$

The first two terms are Morse potentials for the A-B and B-C bonds. The third term attenuates the attraction between B and C as A approaches by means of the switching function $[1 - \tanh(ar_{AB} + c)]$, which ranges from zero for large r_{AB} to unity for smaller r_{AB}. The final term supplies a repulsive interaction between A and C.

Murrell and co-workers[40] have developed a function for triatomic systems based on a many-body expansion of the potential energy with parameters for the function chosen to match experimental energies and vibrational frequencies, and *ab initio* data. The potential is a sum of three two-body terms and a three-body term, and is written as

$$V = V_{AB}(r_{AB}) + V_{BC}(r_{BC}) + V_{AC}(r_{AC}) + V_{ABC}(r_{AB}, r_{BC}, r_{AC}), \tag{7-43}$$

The role of the three-body term is to fill the difference between the exact potential and the sum of the diatomic parts. It goes to zero if any atom of the triatomic is removed to infinity. The three-body term is expressed as a polynomial (quartic if necessary) multiplied by a product of switching functions:

$$V_{ABC}(r_{AB}, r_{BC}, r_{AC}) = AP(s_1, s_2, s_3) \prod_{i=1}^{3} \left(1 - \tanh \frac{1}{2} \gamma_i s_i \right) \tag{7-44}$$

The s_i are displacements in the bond lengths from a reference configuration; e.g., $s_1 = r_{AB} - r^0_{AB}$. The switching function $(1 - \tanh \gamma_i s_i/2)$ goes to zero for large s_i. Parameters in the function are A, the γ_i, and the coefficients in the polynomial P. In principle, this method can be generalized to tetra-atomic molecules by writing the potential as a sum of two-body and three-body terms, plus a four-body term.

A number of analytic potential functions have been derived for polyatomic reactive systems and some are listed in Table 7-7. In one approach, (e.g., H + C_2H_4 and H + CH_3) polyatomic analytic potential functions have been constructed from internal coordinates, switching functions, and extensions of BEBO concepts. To illustrate, consider the H + $CH_3 \rightleftarrows CH_4$ system depicted in Figure 7-11. The H---C—H diagonal quadratic bending force constant f_χ and equilibrium angle χ_0 along the minimum energy path can be written as

$$f_\chi = S_f(r) f^{CH_4}_\chi \tag{7-45}$$

and

$$\chi_0 = \chi^{CH_4}_0 + [\chi^{CH_4}_0 - 90.00][S_\chi(r) - 1.0]. \tag{7-46}$$

TABLE 7-7   Polyatomic Analytic Potential Energy Surfaces.

Reaction	Reference
$NO + O_3 \rightarrow O_2 + NO_2$	R. Viswanathan and L. M. Raff, *J. Phys. 87*, 3251 (1983); S. Chapman, *J. Chem. Phys. 74*, 1001 (1980).
$SiH_4 \rightarrow SiH_2 + H_2$ $\rightarrow SiH_3 + H$	R. Viswanathan, D. L. Thompson, and L. M. Raff, *J. Chem. Phys. 80*, 4230 (1984).
$Cl^- + CH_3Cl \rightarrow ClCH_3 + Cl^-$	S. R. Vande Linde and W. L. Hase, *J. Phys. Chem. 94*, 2778 (1990); S. C. Tucker and W. L. Hase, *J. Am. Chem. Soc. 112*, 3338 (1990).
$H + CH_4 \rightarrow H_2 + CH_3$	T. Joseph, R. Steckler, and D. G. Truhlar, *J. Chem. Phys. 87*, 7036 (1987).
$H + C_2H_4 \rightarrow C_2H_5$	W. L. Hase, G. Mrowka, R. J. Brudzynski, and C. S. Sloane, *J. Chem. Phys. 69*, 3548 (1978).
$H + CH_3 \rightarrow CH_4$	R. J. Duchovic, W. L. Hase, and H. B. Schlegel, *J. Phys. Chem. 88*, 1339 (1984).
$CN + H_2 \rightarrow H + HCN$	M. ter Horst, G. C. Schatz, and L. B. Harding, *J. Chem. Phys. 105*, 558 (1996).
$C + H_2 \rightarrow CH + H$	L. B. Harding, R. Guadagnini, and G. C. Schatz, *J. Phys. Chem. 97*, 5472 (1993).
$OH + H_2 \rightarrow H_2O + H$	D. C. Clary, *J. Chem. Phys. 96*, 3656 (1992).
$H + CO_2 \rightarrow OH + CO$	K. S. Bradley and G. C. Schatz, *J. Chem. Phys. 106*, 8464 (1997).

In these equations, $S_f(r)$ and $S_\chi(r)$ are switching functions which depend on the rupturing CH bond length r. They are equal to unity at the CH equilibrium distance r_e and go to zero for large r. Parameters for the switching functions may be fit to *ab initio* data. By including other functions similar to those in equations (7-45) and (7-46), a complete analytic function results for the $H + CH_3 \rightleftarrows CH_4$ system.

7.8   EXPERIMENTAL DETERMINATION OF POTENTIAL ENERGY SURFACE PROPERTIES

The determination of interatomic potentials for weakly bound systems (e.g., Ar + Ar) from molecular beam elastic scattering experiments (see chapter 9, section 9.1) is an example of how experiments may be used to derive information about potentials. Here the potential is a curve. For a polyatomic chemical reactive system, where the potential is not a curve but a multidimensional surface, the derivation of potential information from experiments is often less direct. Measurements of kinetic rate parameters, vibrational/rotational transition frequencies, and chemical reaction dynamics are used to deduce potential energy surface properties. In chapter 10, it will be shown how transition state theory may be used to estimate the tightness or looseness of the saddlepoint molecular configuration as compared to that of the reactants. Also, a value for the difference between the zero-point energy levels of the saddlepoint and reactants may be determined from the thermal activation energy using transition state theory. From recent experiments[41] it is possible to determine vibrational frequencies in the transition state region of the potential energy surface.

Potential energy surface properties may be determined from experimental measurements of vibrational and rotational energy levels. For small polyatomic molecules it is possible to extract harmonic normal mode vibrational frequencies from the experimental anharmonic $n = 1 \leftarrow 0$ normal mode transition energies (the harmonic frequencies are usually approximately 5 percent larger than the anharmonic $1 \leftarrow 0$ transition frequencies). Using a normal mode analysis as described by equations (7-23) through (7-28), internal coordinate quadratic force constants (e.g., Table 7-5) may be determined for the molecule by fitting the harmonic frequencies.

For a diatomic molecule the Rydberg-Klein-Rees (RKR) method may be used to determine the potential energy curve $V(r)$ from the experimental vibrational energy levels. This method is based on the Einstein-Brillouin-Keller (EBK) semiclassical quantization of the action integral of the old quantum theory.[42] It may also be derived from the Wentzel-Kramer-Brillouin (WKB) semiclassical limit of quantum mechanics.[43] The classical vibrational energy of a diatomic molecule is $E = p_r^2/2\mu + V(r)$, for which the EBK quantization condition is

$$(n + 1/2)h = 2 \int_{r_{min}}^{r_{max}} p(r)dr = 2(2\mu)^{1/2} \int_{r_{min}}^{r_{max}} [E - V(r)]^{1/2}\, dr \qquad (7\text{-}47)$$

Values of the inner and outer potential energy turning points, r_{min} and r_{max}, are evaluated for each level n, from which the complete $V(r)$ curve is determined.

In contrast to the above RKR method for diatomics, a general practical method has not been developed for determining potential energy surfaces from experimental anharmonic vibrational energy levels of polyatomic molecules. Methods which have been used are based on an analytic representation of the potential energy surface. At low levels of excitation the surface may be represented as a sum of quadratic, cubic, and quartic terms; viz., equations (7-17) and (7-22). A vibrational basis set is chosen for the molecule, and the quantum mechanical variational method is used to calculate vibrational energy levels from the analytic potential. Parameters in the potential are varied until the experimental vibrational energy levels are fit. This variational method is only practical at low levels of excitation. To fit potential surface parameters to highly excited levels, a vibrational self-consistent-field (SCF) method has been proposed.[44] In one version of this approach, the SCF theory and EBK semiclassical quantization theory have been combined.[45]

Dynamical properties of chemical reactions are also used to determine potential energy surfaces. The relationship between reaction dynamics and potential energy surfaces is discussed in considerable detail in chapters 8, 9, and 11. Here this relationship is briefly described. For a bimolecular reaction the dependence of the observed cross section on translational, vibrational and/or rotational energy, velocity and angular momentum scattering angles, and product energy partitioning characterize the reaction's potential energy surface.[46] Dynamical calculations, such as classical trajectories, performed on a potential energy surface can be used to reproduce the experimental dynamical results. By varying adjustable parameters in an analytic function depicting the surface, it is often possible to determine what surface properties (shapes) are required to fit the experimental results. In unimolecular dissociation reactions the partitioning of available energy to vibration, rotation, and translation in the products may be used to deduce qualitative shapes of potential surfaces.[47]

7.9   DETAILS OF THE REACTION PATH

Reaction paths for chemical reactions have been illustrated in Figure 7-5. A reaction path is defined by a single trajectory, which may be found by starting at the saddlepoint and moving towards reactants and products in infinitesimal steps, with the kinetic energy removed after each step. In mass-weighted Cartesian coordinates, (see chapter 8, section 8.3.1), this motion traces out the steepest descent path and can be found by following the negative gradient vector from the saddlepoint.[48] If the surface does not have a saddlepoint, as for many association reactions, the reaction path can be found by choosing a sufficiently large separation between the two reactants and initializing the motion at that point. By following the negative gradient vector one finds the variation in the Cartesian coordinates as the chemical system transverses the reaction path. Internal coordinates such as bond lengths, valence angle bends, and dihedral angles can be determined at any point on the reaction path from the Cartesian coordinates.

There are 3N degrees of freedom for the N-atom chemical system moving along the reaction path. Six of the degrees of freedom are for overall translation and rotation, and one is for the reaction coordinate. The remaining 3N-7 degrees of freedom are vibrations orthogonal to the reaction coordinate. Harmonic normal mode frequencies for those vibrations may be determined at any point along the reaction path.[49] However, only at saddlepoints and potential minima are they determined by diagonalizing the mass-weighted Cartesian force constant matrix $\mathbf{F}$, equation (7-23). At all other points along the reaction path the derivatives of the potential with respect to the Cartesian coordinates $(\partial V/\partial q_{\alpha i})$ are not zero, and instead a projected force constant matrix $\mathbf{F^P}$ must be diagonalized.[49] $\mathbf{F^P}$ is formed from $\mathbf{F}$ by projecting out the direction along the reaction path and the directions corresponding to infinitesimal rotations and translations.

The eigenvalues and eigenvectors of $\mathbf{F^P}$ are determined in the same manner as those for $\mathbf{F}$; i.e., equations (7-23) through (7-28), except $\mathbf{F^P}$ replaces $\mathbf{F}$. $\mathbf{F^P}$ has seven zero eigenvalues which correspond to infinitesimal translations and rotations, and motion along the reaction path. The remaining 3N-7 nonzero eigenvalues give the harmonic frequencies $\nu_i(r^\dagger)$ for the vibrations orthogonal to the reaction path. By evaluating $\mathbf{F^P}$ as a function of the reaction path, the $\nu_i(r^\dagger)$ may be determined along the reaction path. The harmonic frequencies orthogonal to the $H_2CO \rightarrow H_2 + CO$ reaction path are plotted in Figure 7-12. The potential energy along the reaction path is called the vibrationally adiabatic ground state potential, which at the harmonic level is given by

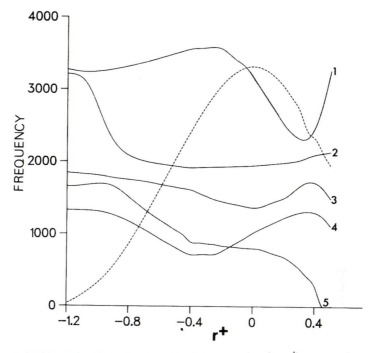

FIGURE 7-12   Plots of vibrational frequencies (cm^{-1}) vs. reaction coordinate $(amu^{1/2} \cdot \text{Å})$ for the reaction $H_2CO \rightarrow H_2 + CO$. Saddlepoint is at $r^\dagger = 0$, and positive and negative $r^\dagger$ lead to products and reactants, respectively. At the H_2CO minimum, $\nu_1 - \nu_5$ are the asymmetric CH stretch, symmetric CH stretch, CO stretch, out-of-plane bend, and CH_2 symmetric deformation. The dashed line depicts the vibrationally adiabatic ground-state potential. [Adapted from W. L. Hase, *Acc. Chem. Res. 16,* 258 (1983).]

$$V_a^G(r^\dagger) = V(r^\dagger) + \sum_{i=1}^{3N-7} h\nu_i(r)/2 \qquad (7\text{-}48)$$

In this expression, $V(r^\dagger)$ is the classical potential energy.

7.10   POTENTIAL ENERGY SURFACES OF ELECTRONICALLY EXCITED MOLECULES

Up to this point, we have considered potential energy surfaces for only the ground electronic state of molecular systems. However, within the framework of the Born-Oppenheimer approximation, every molecular system has a multitude of electronically excited states, each with its own potential energy surface. Each state is identified by its spin angular momentum and orbital symmetry. The electronic states for diatomic systems are further classified by the component of electronic orbital angular momentum along the internuclear axis. Since the electronic density between nuclei is different for each electronic state, each state has its own equilibrium geometry, sets of vibrational frequencies, and bond dissociation energies.

One of the ramifications of the Born-Oppenheimer approximation for diatomic molecules is that potential energy curves for states with the same orbital symmetry and spin do not cross; i.e., the noncrossing rule.[50,51] Energy curves determined from simple (approximate) molecular orbital or valence bond wave functions χ_j do cross. However, to obtain accurate Born-Oppenheimer wave functions ϕ_i and potential energy curves one has to account for electron correlation. This is done by writing the ϕ_i as a mixture of the χ_j which are of the same spin and symmetry, a procedure called "configuration interaction";[15] i.e.,

$$\phi_i = \sum_j c_{ij}\chi_j \qquad (7\text{-}49)$$

Perturbation theory tells us that in general this mixing is most pronounced at nuclear configurations where the simple curves have similar energies. Thus, the mixing between the approximate wave functions is a function of the nuclear geometry.

An illustration of the noncrossing rule is shown in Figure 7-13 where the four lowest $^2A'$ states* of LiH_2 are treated. Since the LiH_2 geometry is kept linear and the H_2 bond length fixed, there is only one adjustable geometric parameter, i.e., the $Li\text{-}H_2$ separation r and the preceding noncrossing rule for diatomics applies. Potential energy curves for the accurate ϕ_i wave functions are shown in Figure 7-13(a), while those for the approximate χ_j wave functions are shown in Figure 7-13(b). In the region of crossing between the approximate energy curves, the approximate electronic wave functions strongly mix. The result of such an interaction is always to push the approximate potential energies apart, and this interaction is strongest at the *avoided crossing* point at $r = r_0$.

The accurate potential energy curves with mixing are called *adiabatic* potentials $V(r)$. The term *diabatic* is used to describe the approximate potential curves $U(r)$. At an avoided crossing the electronic character of a wave function for an adiabatic potential can change significantly. Consider the highest energy adiabatic potential in Figure 7-13(a). At large $Li\text{-}H_2$ separations the wave function has ionic character. At shorter separations it becomes covalent. The wave functions which are ionic at short $Li\text{-}H_2$ separations are those for low energy adiabatic potentials.

The crossing (or non-crossing) of potential energy curves for diatomic molecules depends on only one coordinate, the internuclear separation, which gives rise to the above non-crossing rule. However, the polyatomic molecules there are multiple coordinates to consider and this rule has to be modified.[52–54] This may be illustrated by considering the radical-ion NH_2^+, which is of C_{2v} symmetry and has the triplet electronic states of $^3A_2$ and $^3B_1$ symmetry.[55] The three normal modes of vibration for NH_2^+ are the symmetric stretch Q_1, the symmetric bend Q_2, and the asymmetric stretch Q_3. If only the symmetric Q_1 and Q_2 modes are displaced, so the C_{2v} geometry is maintained, the

---

*The group-theoretical notation used to denote the symmetry of electronic states is discussed in J. I. Steinfeld, *Molecules and Radiation* (Cambridge, MA: MIT Press, 1985), chapter 6.

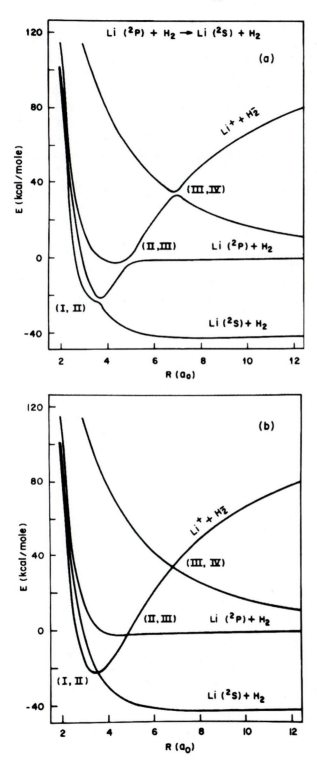

FIGURE 7-13   Four lowest $^2A'$ potential energy curves for LiH_2 as a function of Li-H_2 separation. The orientation of the H_2 internuclear axis is 60° relative to the Li-H_2 direction, and the H_2 internuclear distance is $1.4a_o$ (a_o = bohr radius). (a) Adiabatic representation. (b) Diabatic representation. [Adapted from J. C. Tully, *Dynamics of Molecular Collisions,* Part B, ed. W. H. Miller (New York: Plenum, 1976).]

$^3A_2$ and $^3B_1$ surfaces are allowed to cross since they are of different symmetries. As a result, there is a line of intersection (i.e., seam) between the $^3A_2$ and $^3B_1$ surfaces which is a function of Q_1 and Q_2. The dynamics are much different if the asymmetric stretch mode Q_3 is displaced, so that the molecular symmetry is lowered to C_s and the symmetries of the two surfaces become identical; i.e., $^3A''$. The surfaces will have an avoided crossing for this nonsymmetric displacement. If one of the symmetric coordinates is fixed and two-dimensional potential energy surfaces constructed as a function of one symmetric coordinate (Q_1 or Q_2) and the asymmetric coordinate Q_3, the two surfaces have a double cone geometrical shape known as a *conical intersection* in the vicinity of the crossing point for the two surfaces. This is illustrated in Figure 7-14. For the general polyatomic case the potential energy surfaces depend on n variables (3 for NH_2^+) and the intersection between the two surfaces is a surface of n-2 dimensions.[52-54]

Spin-orbit couplings and a breakdown in the Born-Oppenheimer approximation can lead to a transition between adiabatic potentials. A treatment of such a transition is given in chapter 8, section 8.3.5. These two perturbations can also cause transitions between potential curves of different electronic symmetries and/or spin which do cross. In the *adiabatic approximation* in chemical kinetics, it is assumed that transitions between adiabatic curves are negligible, so that nuclear motion evolves on a single adiabatic potential energy surface.

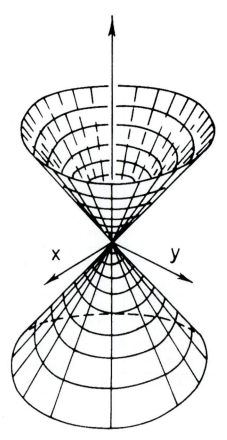

x   y

FIGURE 7-14   Schematic drawing of a conical intersection between two potential energy surfaces [Adapted from D. F. Hirst, *Potential Energy Surfaces* (London: Taylor and Francis, 1985)].

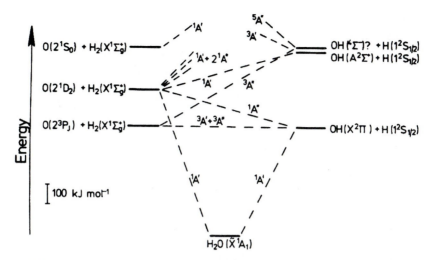

FIGURE 7-15 Correlation diagram connecting the states of O + H$_2$ and OH + H. [Adapted from I. W. M. Smith, *Kinetics and Dynamics of Elementary Gas Reactions* (London: Butterworths, 1980)].

In the context of the Born-Oppenheimer approximation, adiabatic correlation rules may be used to predict reaction pathways.[51] A correlation diagram for the states of O + H$_2$ and OH + H is shown in Figure 7-15. The ground state reactants $[O^3(P) + H_2(^1\Sigma_g^+)]$ correlate adiabatically with both the ground state products OH + H and excited OH plus the ground state H-atom. Within the adiabatic approximation, H$_2$O can not be formed by an $[O^3(P) + H_2(^1\Sigma_g^+)]$ collision. Water can be formed by a collision with electronically excited $O(^1D_2)$ and ground state H$_2$. However, the resulting vibrationally and rotationally excited H$_2$O has sufficient energy to dissociate to ground state OH + H.

One of the limitations of such adiabatic correlations is that they give no indication of the potential energy barriers for chemical reactivity. Experiments indicate there are large barriers on the $^3A'$ and $^3A''$ surfaces connecting $[O^3(P) + H_2(^1\Sigma_g^+)]$ and the ground state products. Formation of the ground state products from $[O(^1D_2) + H_2(^1\Sigma_g^+)]$ occurs without a barrier. The reaction pathway also appears to involve formation of vibrationally/rotationally excited H$_2$O as a reaction intermediate.

REFERENCES

[1] D. M. Hirst, *Potential Energy Surfaces* (London: Taylor and Francis, 1985), p. 1.

[2] R. S. Berry, S. A. Rice, and J. Ross, *Physical Chemistry* (New York: Wiley, 1980), p. 413.

[3] D. Steele, E. R. Lippincott, and J. T. Vanderslice, *Rev. Mod. Phys. 34,* 239 (1962).

[4] P. Huxley and J. N. Murrell, *J. Chem. Soc. Farad. 279,* 323 (1983).

[5] J. N. Murrell, S. Carter, P. Huxley, S. C. Farantos, and A. J. C. Varandas, *Molecular Potential Energy Functions* (Chichester: Wiley, 1984).

[6] P. J. Kuntz and W. N. Whitton, *J. Chem. Phys. 64,* 3624 (1976).

[7] Peter Pulay, in *Applications of Electronic Structure Theory,* edited by H. F. Schaefer III (New York: Plenum, 1977), p. 153.

[8] J. L. Duncan, D. C. McKean, and P. D. Mallinson, *J. Mol. Spectrosc. 45,* 221 (1973).

[9] S. Califano, *Vibrational States* (New York: Wiley, 1976), p. 21.

[10] H. L. Sellers, L. B. Sims, L. Schafer, and D. E. Lewis, *QCPE 13,* 339 (1980).

[11] J. A. Pople, in *Applications of Electronic Structure Theory,* edited by H. F. Schaefer III (New York: Plenum Press, 1977), p. 1.

[12] Hirst, p. 94.

[13] W. A. Goddard III, T. H. Dunning, Jr., W. J. Hunt, and P. J. Hay, *Acc. Chem. Res. 6,* 368 (1973).

[14] P. E. M. Siegbahn, J. Almlöf, A. Heiberg, and B. O. Roos, *J. Chem. Phys. 74,* 2384 (1981).

[15] S. R. Langhoff and E. R. Davidson, *Int. J. Quantum Chem. 8,* 61 (1974); I. Shavitt, in *Methods of Electronic Structure Theory,* edited by H. F. Schaefer III (New York: Plenum, 1977), p. 89.

[16] R. G. Parr and W. Yang, *Density-Functional Theory of Atoms and Molecules* (Oxford: Oxford Univ. Press, 1989).

[17] J. A. Pople, J. S. Binkley, and R. Seeger, *Int. J. Quantum Chem., Symp. 10,* 1 (1976); H. B. Schlegel, *J. Chem. Phys. 84,* 5340 (1986).

[18] W. J. Hehre, L. Radom, P. v. R. Schleyer, and J. A. Pople, *Ab Initio Molecular Orbital Theory* (New York: Wiley, 1986), pp. 135–226.

[19] Langhoff and Davidson, pp. 226–260.

[20] Langhoff and Davidson, pp. 261–323.

[21] P. Siegbahn and B. Liu, *J. Chem. Phys. 68,* 2957 (1978); D. G. Truhlar and C. J. Horowitz, *J. Chem. Phys. 68,* 2466 (1978); A. J. C. Varandas, F. B. Brown, C. A. Mead, D. G. Truhlar, and N. C. Blais, *J. Chem. Phys. 86,* 6258 (1987); A. I. Boothroyd, W. J. Keogh, P. G. Martin, and M. R. Peterson, *J. Chem. Phys. 95,* 4343 (1991).

[22] S. L. Mielke, G. C. Lynch, D. G. Truhlar, and D. W. Schwenke, *Chem. Phys. Lett. 213,* 10 (1993); K. Stark and H. -J. Werner, *J. Chem. Phys. 104,* 6515 (1996).

[23] E. Clementi and H. Popkie, *J. Chem. Phys. 57,* 1077 (1972).

[24] G. Corongiu, E. Clementi, E. Pretsch, and W. Simon, *J. Chem. Phys. 72,* 3096 (1980).

[25] W. L. Hase, S. L. Mondro, R. J. Duchovic, and D. M. Hirst, *J. Am. Chem. Soc. 109,* 2916 (1987).

[26] L. F. Shampine and R. C. Allen, Jr., *Numerical Computing* (Toronto: Saunders, 1973), p. 54; M. J. T. Jordan and M. A. Collins, *J. Chem. Phys. 104,* 4600 (1996); T. -S. Ho, T. Hollebeck, H. Rabitz, L. B. Harding, and G. C. Schatz, *J. Chem. Phys. 105,* 10472 (1996).

[27] W. L. Jorgensen, J. Chandrasekhar, J. D. Madura, R. W. Impey, and M. L. Klein, *J. Chem. Phys. 79,* 926 (1983).

[28] K. N. Swamy and W. L. Hase, *J. Am. Chem. Soc. 106,* 4071 (1984).

[29] J. P. Bergsma, B. J. Gertner, K. R. Wilson, and J. T. Hynes, *J. Chem. Phys. 86,* 1356 (1987).

[30] O. Matsuoka, E. Clementi, and M. Yoshimine, *J. Chem. Phys. 64,* 1351 (1976).

[31] W. L. Jorgensen and M. Ibrahim, *J. Am. Chem. Soc. 102,* 3309 (1980).

[32] C. A. Parr and D. G. Truhlar, *J. Phys. Chem. 75,* 1844 (1971).

[33] F. London, "Probleme der modernen Physik" *Sommerfeld Festschrift* (1928), p. 104.

[34] S. Sato, *J. Chem. Phys. 23,* 592, 2465 (1955).

[35] P. J. Kuntz, E. M. Nemeth, J. C. Polanyi, S. D. Rosner, and C. E. Young, *J. Chem. Phys. 44,* 1168 (1966).

[36] A. Persky and M. Baer, *J. Chem. Phys. 60,* 133 (1974).

[37] H. S. Johnston and C. Parr, *J. Am. Chem. Soc. 85,* 2544 (1963); H. S. Johnston, *Gas Phase Reaction Rate Theory* (New York: Ronald, 1966), pp. 177–183.

[38] L. Pauling, *J. Am. Chem. Soc. 69,* 542 (1947).

[39] N. C. Blais and D. L. Bunker *J. Chem. Phys. 37,* 2713 (1962).

[40] K. S. Sorbie and J. N. Murrell, *Molec. Phys. 29,* 1387 (1975); J. N. Murrell, K. S. Sorbie, and A. J. C. Varandas, *Molec. Phys. 32,* 1359 (1976).

[41] D. E. Manolopoulos, K. Stark, H. J. Werner, D. W. Arnold, S. E. Bradforth, and D. M. Neumark, *Science, 262,* 1852 (1993).

[42] I. C. Percival, *Adv. Chem. Phys. 36,* 1 (1977); D. W. Noid, M. L. Koszykowski, and R. A. Marcus, *Ann. Rev. Phys. Chem. 32,* 267 (1981).

[43] L. I. Schiff, *Quantum Mechanics* (New York: McGraw-Hill, 1968), p. 268.

[44] J. M. Bowman, *J. Chem. Phys. 68,* 608 (1978); K. M. Christoffel and J. M. Bowman, *Chem. Phys. Letters 85,* 220 (1982).

[45] M. A. Ratner and R. B. Gerber, *J. Phys. Chem. 90,* 20 (1986).

[46] R. D. Levine and R. B. Bernstein, *Molecular Reaction Dynamics* (New York: Oxford University Press, 1974), chapters 4 and 6.

[47] W. L. Hase and R. J. Wolf, in *Potential Energy Surfaces and Dynamics Calculations,* edited by D. G. Truhlar (New York: Plenum, 1981), p. 37.

[48] K. Fukui, *J. Phys. Chem. 74,* 4161 (1970); K. Fukui, S. Kato, and H. Fujimoto, *J. Am. Chem. Soc. 97,* 1 (1975); N. Kato, H. Kato, and K. Fukui, *J. Am. Chem. Soc. 99,* 684 (1977); H. F. Schaefer, *Chem. Brit. 11,* 227 (1975); K. Ishida, K. Morokuma, and A. Komornicki, *J. Chem. Phys. 66,* 2153 (1977).

[49] W. H. Miller, N. C. Handy, and J. E. Adams, *J. Chem. Phys. 72,* 99 (1980); S. Kato and K. Morokuma, *J. Chem. Phys. 73,* 3900 (1980).

[50] W. Kauzmann, *Quantum Chemistry* (New York: Academic Press, 1957), pp. 142, 401.

[51] C. A. Mead, *J. Chem. Phys. 70,* 2276 (1979).

[52] E. Teller, *J. Phys. Chem. 41,* 109 (1937).

[53] G. Herzberg and H. C. Longuet-Higgins, *Disc. Faraday Soc. 35,* 77 (1963).

[54] H. C. Longuet-Higgins, *Proc. Roy. Soc. London A344,* 147 (1975).

[55] Ref. 1, pp. 51–55.

BIBLIOGRAPHY

Intermolecular Potentials

HIRSCHFELDER, J. O., CURTISS, C. F., and BIRD, R. R. *Molecular Theory of Gases and Liquids.* New York: Wiley, 1954. Chapters 12–14.

KAUZMANN, W. *Quantum Chemistry.* New York: Academic Press, 1957. Chapter 13.

MARGENAU, H., and KESTNER, N. R. *Theory of Intermolecular Forces.* 2d ed. Oxford: Pergamon, 1969.

SMITH, I. W. M. *Kinetics and Dynamics of Elementary Gas Reactions.* Boston: Butterworths, 1980. Chapter 2.

WESTON, R. E., Jr., and SCHWARZ, H. A. *Chemical Kinetics.* Englewood Cliffs, NJ: Prentice-Hall, 1972. Chapter 2.

Vibrational Force Constants and Frequencies

CALIFANO, S. *Vibrational States.* New York: Wiley, 1976.

STEINFELD, J. I. *Molecules and Radiation.* Cambridge, MA: MIT Press, 1985. Chapters 4 and 8.

WILSON, E. B., Jr., DECIUS, J. C., and CROSS, P. C. *Molecular Vibrations.* New York: McGraw-Hill, 1955.

Properties of Potential Energy Surfaces

LAIDLER, K. J. *Theories of Chemical Reaction Rates.* New York: McGraw-Hill, 1969. Chapter 2.

MAGGIORA, G. M., and CHRISTOFFERSEN, R. E. in *Transition States of Biochemical Processes,* edited by R. G. Gandour and R. L. Schowen, p. 119. New York: Plenum. 1978.

SALEM, L. *Electrons in Chemical Reactions: First Principles* (New York: Wiley, 1982).

Ab Initio and Analytical Potential Energy Surfaces

HEHRE, W. J., RADOM, L., SCHLEYER, P. v. R., and POPLE, J. A. *Ab Initio Molecular Orbital Theory.* New York: Wiley, 1986.

HIRST, D. M. *Potential Energy Surfaces.* London: Taylor and Francis, 1985.

KUNTZ, P. J. in *Dynamics of Molecular Collisions, Part B.,* edited by W. H. Miller, p. 53. New York: Plenum, 1976.

MURRELL, J. N. *Gas Kinetics and Energy Transfer, Vol. 3.* London: The Chemical Society, 1978. Chapter 5.

SCHAEFER, H. F. *Electronic Structure of Atoms and Molecules.* New York: Addison-Wesley, 1972.

TRUHLAR, D. G., STECKLER, R., and GORDON, M. S. *Chem. Rev. 87,* 217 (1987).

Potential Energy Surface Properties from Experiment

BUNKER, D. L. *Acc. Chem. Res. 7,* 195 (1974).

CARNEY, G. D., SPRANDEL, L. L., and KERN, C. W. *Adv. Chem. Phys. 37,* 305 (1978).

HERZBERG, G. *Molecular Spectra and Molecular Structure II* New York: Van Nostrand and Reinhold, 1945.

POLANYI, J. C. *Acc. Chem. Res. 5,* 16 (1972).

Reaction Path

FUKUI, K. *Acc. Chem. Res. 14,* 363 (1981).

MILLER, W. H. *J. Phys. Chem. 87,* 3811 (1983).

Potential Energy Surfaces for Excited Electronic States

DONOVAN, R. J., and HUSAIN, D. *Chem. Rev. 70,* 489 (1970).

HERZBERG, G. *Molecular Spectra and Molecular Structure III. Electronic Spectra and Electronic Structure of Polyatomic Molecules,* p. 442. New York: Van Nostrand Reinhold (1966).

MAHAN, B. H. *Acc. Chem. Research 8,* 55 (1975).

SHULER, K. E. *J. Chem. Phys. 21,* 624 (1952).

TULLY, J. C. in *Dynamics of Molecular Collisions, Part B,* edited by W. H. Miller, p. 217. New York: Plenum, 1976.

YARKONY, D. R., in *Modern Electronic Structure Theory. Part I*, ed. D.R. Yarkony, p. 642. Singapore: World Scientific Publishing (1995).

PROBLEMS

7.1 Compute the factor for converting a quadratic force constant in units of millidyne/angstrom (mdyn/Å) to kJ/mol/Bohr^2.

7.2 Using equations (7-4) through (7-7) derive equation (7-8).

7.3 Derive an expression for the quadratic force constant of a Morse oscillator in terms of D_e and β.

7.4 For a Lennard-Jones "6–12" potential determine the internuclear distance at the potential energy minimum. Using this value, derive an expression for the quadratic force constant.

7.5 From the force constants in Table 7-5, calculate harmonic frequencies for the diatoms C—H, C—C, and C=O.

7.6 Show that for a harmonic oscillator, the EBK quantization condition equation (7-47) leads to the expression $E = (n + 1/2)h\nu$.

7.7 Give the equations which define the relationships between internal bond stretch and angle bend coordinates and Cartesian coordinates of the atomic nuclei.

7.8 A linear triatomic molecule such as HCN has two bond-stretching internal coordinates, r_{HC} and r_{CN}. Assume the potential energy for the linear internal motion is given by

$$V = f_{HC}(r_{HC} - r_{HC}^0)^2/2 + f_{CN}(r_{CN} - r_{CN}^0)^2/2$$

with $f_{HC} = 5.0$ mdyn/Å, $f_{CN} = 14.0$ mdyn/Å, $r_{HC}^0 = 1.08$ Å, and $r_{CN}^0 = 1.15$ Å. Using equations like those in (7-23) through (7-28) calculate the normal mode frequencies and eigenvectors for linear motion. One of the eigenvalues, that for translation, will be zero.

7.9 **(a)** Sketch potential-energy contour maps for the following:

(i) Isomerization of H—C≡C → H—C≡N. Your sketch should be in the r_{HC}, θ plane where θ is given by

(ii) Dissociation of H—C≡N → H + C≡N. Your sketch should be in the r_{HC}, r_{CN} plane for a linear configuration.

(b) Discuss significant differences between the two potential energy surfaces.

7.10 What are the major differences between the $F + H_2 \rightarrow FH + H$ and $F + C_2H_4 \rightarrow C_2H_3F + H$ potential energy surfaces?

7.11 Describe the difference between electronically adiabatic and electronically nonadiabatic processes. What is the noncrossing rule?

7.12 What is the mathematical property common to both minima and saddlepoints on potential energy surfaces?

7.13 **(a)** How is the harmonic zero-point energy of a molecule or saddlepoint (transition state) determined? Give the mathematical expression for calculating the zero-point energy.

(b) Describe with both an equation and diagram the relationship between the harmonic quantum mechanical threshold energy and the classical mechanical threshold energy.

7.14 A widely used LEPS function is that for the "Muckermann V" $F + H_2 \rightarrow FH + H$ potential energy surface: P. A. Whitlock and J. T. Muckermann, *J. Chem. Phys. 61,* 4624 (1974). The LEPS parameters are

$$D_{FH} = 591.1 \text{ kJ/mol} \qquad D_{HH} = 458.2 \text{ kJ/mol}$$

$$\beta_{FH} = 2.2189 \text{ Å}^{-1} \qquad \beta_{HH} = 1.9420 \text{ Å}^{-1}$$

$$r^\circ_{FH} = 0.917\text{Å} \qquad r^\circ_{HH} = 0.7419\text{Å}$$

$$S_{FH} = 0.167 \qquad S_{HH} = 0.106$$

Using the above parameters and equations (7-35) through (7-37) make a contour diagram for the $F + H_2 \rightarrow FH + H$ reaction like that shown in Figure 7-8(a).

CHAPTER 8

Dynamics of Bimolecular Collisions

Once an intermolecular potential energy function for a reactive system is available, either from experiment or theory, scattering amplitudes and cross sections for that system may be found by carrying out appropriate calculations. Classical scattering theory requires the solution of Hamilton's equations of motion on the potential surface (see section 8.3.1); alternatively, a quantum mechanical solution for the wave functions may be effected by solving Schrödinger's equation using the said potential function (see section 8.3.2). Before considering these methods, however, let us first examine a few very simple collision models. Even though these models are only rough approximations to actual reactive scattering, they have nevertheless proven to be very useful in the development of chemical-kinetic theories because they embody the essential physical content of the actual situation, thereby leading to analytic expressions which provide an intuitive grasp of the more complex problem.

8.1 SIMPLE COLLISION MODELS

8.1.1 Constant Cross Section

The simplest possible picture of an encounter between two molecules is that shown in Figure 8-1, in which it is assumed that the molecules are structureless hard spheres and that the collision may be described by a cross section with an energy-independent magnitude

$$\sigma_R(E) = \text{constant} = \pi d^2 \tag{8-1}$$

where $d = 1/2 \, (d_A + d_B)$ is the arithmetic average of the "hard-sphere" diameters of molecules A and B. The collision model employed is that the probability of a "reaction" between A and B occurring is $P_{\text{react}} = 1$ if the intermolecular distance $r_{AB} \leq d$ and $P_{\text{react}} = 0$ otherwise. The effective activation energy for such a reaction is $E_{\text{act}} = 0$. The rate coefficient may be found by substituting equation (8-1) for the cross section into equation (6-7a). Since σ is independent of v, it may be taken outside the integral to give

$$k(T) = \left(\frac{\mu}{2\pi k_B T} \right)^{3/2} \sigma \int_0^\infty v e^{-\mu v^2 / 2k_B T} \, 4\pi v^2 dv$$

$$= \bar{v}\sigma$$

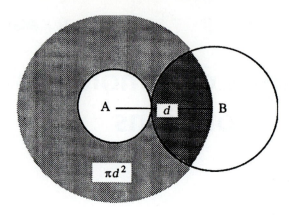

FIGURE 8-1   Collision of two hard-sphere molecules.

or

$$k(T) = \left(\frac{8k_B T}{\pi \mu} \right)^{1/2} \cdot \pi d^2 = \left(\frac{8\pi k_B T}{\mu} \right)^{1/2} d^2 \qquad (8\text{-}2)$$

For a generic molecular cross section ($d \approx 4 \times 10^{-8}$ cm) and a typical molecular velocity, ($\bar{v} \approx 5 \times 10^4$ cm sec^{-1}), equation (8-2) gives a "hard-sphere" or "gas-kinetic" collision rate coefficient $k_{coll} \approx 5 \times 10^{-10}$ cm^3 molecule^{-1} sec^{-1}. In other commonly used units,

$$k_{coll} = \left(\frac{6.022 \times 10^{23} \text{ molecule mole}^{-1}}{10^3 \text{ cm}^3 \text{ liter}^{-1}} \right)(2.5 \times 10^{-10} \text{ cm}^3 \text{ molecule}^{-1} \text{ sec}^{-1})$$

$$= 1.5 \times 10^{11} \text{ liter mole}^{-1} \text{ sec}^{-1}$$

If we multiply instead by the ratio $(6.022 \times 10^{23}$ molecule mole$^{-1})/(22{,}421$ cm^3 mole^{-1} @STP), we obtain a relaxation rate $(p\tau)^{-1} = 6.7 \times 10^9$ atm^{-1} sec^{-1} at 273 K, or approximately 1×10^7 torr^{-1} sec^{-1} at that temperature. This means that τ_{coll}, the mean free time between gas-kinetic collisions, is about 0.1 μsec at a pressure of 1 torr. The corresponding collision frequency would be $\omega = \tau_{coll}^{-1} = 10^7$ sec^{-1}.

One frequently encounters in the literature values of the "effective cross section" for a reaction defined as $\sigma_{eff} = k_{meas}/\bar{v}$. This relationship between rate coefficient and cross section is of course valid only for the hard-sphere model. An even cruder approximation is to define the "reaction probability" as

$$P_{react} = \frac{k_{meas}}{k_{coll}} = \frac{\sigma_{eff}}{\sigma_{gas\text{-}kinetic}} \approx \frac{\sigma_{eff}}{50\text{\AA}^2} \qquad (8\text{-}3)$$

A glance back at Figure 6-2, however, shows that equation (8-3) glosses over three stages of averaging between the reaction probability and the rate coefficient!

One should also be careful about other concepts employed in simple collision models, e.g., "mean free time between collisions." Recall that molecules interact with an ensemble of velocities and instantaneous intermolecular distances and therefore possess a *distribution* of time intervals between collisions. This distribution can be expressed as a Poisson function of n, the number of collisions per time interval T, given by

$$P(n) = e^{-\gamma}\gamma^n/n! \tag{8-4}$$

with mean $\langle n \rangle = \gamma = T/\tau_{\text{coll}}$. Thus, in any gas phase system, some molecules may survive many multiples of the mean free time before undergoing a collision. And this is often important—for example, in measurements of quenching of luminescence or photochemical reaction following excitation by a short laser pulse. In a similar vein, the notion of molecules as structureless hard spheres is an oversimplification: molecules are, after all, *not* simply hard spheres, and they possess long-range attractive forces. This will be discussed in greater detail later in the chapter.

8.1.2   Old Collision Theory (Reactive Hard Spheres)

The hard-sphere collision model just presented is clearly inadequate for describing chemical reactions. For one thing, it predicts a dependence of $T^{1/2}$ for the rate coefficient, while experimentally measured rate coefficients typically follow an Arrhenius behavior, i.e., $\exp\left(-E_{\text{act}}/RT\right)$. Furthermore, the magnitudes of hard-sphere collision rates are the gas-kinetic limit, while most rate coefficients are only a small fraction of that. Clearly, a more elaborate model must be introduced to account for these observations.

Such a model is that of reactive hard spheres. Consider a collision between hard-sphere molecules A and B, with relative velocity $\mathbf{v}$ and kinetic energy (in the center-of-mass coordinate system) $E = 1/2\,\mu v^2$. The hard-sphere minimum approach distance is $d = 1/2\,(d_A + d_B)$, as before. We define an *impact parameter b*, given by the perpendicular distance between parallel lines drawn in the direction of $\mathbf{v}$ and through the centers of molecules A and B, as shown in Figure 8-2. Since $\mathbf{v}$ is a vector quantity, we can resolve it into $v_\perp$ along the line of centers and $v_\parallel$ perpendicular to $v_\perp$. Even though the kinetic energy E is a scalar quantity, it can also be expressed in terms of these velocity components, viz., $E = 1/2\,\mu(v_\parallel^2 + v_\perp^2) = 1/2\mu v_\parallel^2 + 1/2\,\mu v_\perp^2$. The first term is that part

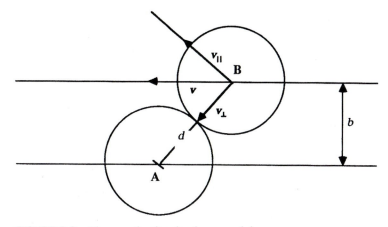

FIGURE 8-2   The reactive hard-sphere model.

of the energy which is tangential to the surfaces of the interacting spheres; the second term, denoted by E_c, represents energy "directed" along the line of centers which is presumed to be effective in bringing about reaction. This "energy along the line of centers" is given from a simple geometrical construction as

$$\frac{E_c}{E} = 1 - \frac{b^2}{d^2} \qquad (b \leq d)$$

$$\frac{E_c}{E} = 0 \qquad\qquad (b > d)$$

(8-5)

Assuming an energy-dependent reaction probability

$$P_R(E_c) = 0 \qquad (E_c < E^*)$$

$$P_R(E_c) = 1 \qquad (E_c \geq E^*)$$

(8-6)

where E^* is the minimum-energy threshold for the reaction to occur, and using a relationship between reaction probability and cross section [equation (8-10)], to be proven in section 8.2, we find the following expression for the cross section:

$$\sigma_R(E) = \int_0^\infty P_R(E(b)) \cdot 2\pi b\, db = \begin{cases} 0 & \text{if } E < E^* \\ \pi d^2 \left(1 - \dfrac{E^*}{E}\right) & \text{if } E \geq E^* \end{cases}$$

(8-7)

The energy dependence of $\sigma_R(E)$ is shown in Figure 8-3.

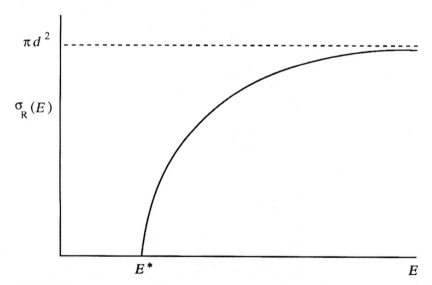

FIGURE 8-3 Energy dependence of the reactive cross section in the reactive hard-sphere model.

Using equation (8-7) for $\sigma_R(E)$ in equation (6-5b) to calculate the thermal rate coefficients yields

$$k_{\text{coll}} = \frac{1}{k_B T} \left(\frac{8}{\pi \mu k_B T} \right)^{1/2} \int_0^\infty E \sigma_R(E) e^{-E/K_B T} \, dE$$

$$k_{\text{coll}} = \frac{1}{k_B T} \left(\frac{8}{\pi \mu k_B T} \right)^{1/2} \int_0^\infty \pi d^2 (E - E^*) e^{-E/k_B T} \, dE \qquad (8\text{-}8)$$

$$k_{\text{coll}} = \pi d^2 \left(\frac{8 k_B T}{\pi \mu} \right)^{1/2} e^{-E^*/k_B T}$$

Note that k_{coll} consists of three factors, viz.,

(hard-sphere cross section) $\times$ (mean $\bar{v}\,(\text{T})$) $\times$ (Arrhenius factor)

Differentiating the last of equations (8-8) with respect to $1/T$ affords a useful comparison with the phenomenological rate coefficient. We have

$$\frac{d(\ln k_{\text{coll}})}{d(1/T)} = -(1/2)T - \frac{E^*}{k_B}$$

while

$$\frac{d(\ln k_{\text{obs}})}{d(1/T)} = -\frac{\tilde{E}_{\text{act}}}{R}$$

Thus, the activation energy is seen to be equal to the threshold energy of the reactive hard-sphere model, give or take a few $k_B T$ per molecule. (The latter will be stated more precisely in chapter 10.) Equation (8-8) also predicts an explicit form for the pre-exponential factor equal to $\pi d^2 \bar{v} \approx 2.5 \times 10^{-10}$ cm^3 molecule^{-1} sec^{-1}, i.e., the gas-kinetic collision rate.

The major problem with the reactive hard-sphere model is that the observed pre-exponential factor is sometimes much less than the gas-kinetic collision rate! This can be taken care of in an ad hoc way by using $P_R(E_c) = p$ in equation (8-6), rather than $P_R(E_c) = 1$, for $E_c = E^*$. The quantity p, known as the *steric factor*, carries right through the integration in equations (8-8) to give a modified rate coefficient

$$k_{\text{coll}} = \pi d^2 \left(\frac{8 k_B T}{\pi \mu} \right)^{1/2} p e^{-E^*/k_B T}$$

For the moment we can consider p a purely empirical correction which in some way is meant to take into account the idea that molecules must equally be in a fairly specific relative orientation in order to react. This orientation requirement is related to the concept of an *entropy* of activation, which is discussed in chapter 10, section 10.5.

While the simple models introduced in this section are only qualitative, they do display the essential elements of a bimolecular rate constant: a collision must occur, some critical energy must be exceeded, and the relative orientations of the colliding particles must be appropriate. Additional useful comments on these models are given by Nordman and Blinder.[1]

8.2 TWO-BODY CLASSICAL SCATTERING

To begin a more rigorous discussion of reactive scattering, let us consider collisions of two structureless particles interacting through a central potential $U(r)$, where r is the distance between the centers of the two particles. Although such an interaction describes only collisions between atoms, such as helium or argon, it is a useful starting point for introducing a number of concepts and quantities which are important for understanding the description of more complex scattering processes. Furthermore, only this "two-body" problem can be solved analytically: collisions involving three or more particles, which are in fact required for a chemical reaction, can be treated only numerically or through the use of approximations. Additional discussion of two-body scattering may be found in a number of introductory texts on chemical kinetics listed in the bibliography, as well as in textbooks on classical mechanics.[2]

The scattering process in laboratory fixed coordinates is shown in Figure 8-4. As previously noted, the interparticle potential energy function $U(r)$ depends only on the distance $r = |\mathbf{r}_A - \mathbf{r}_B|$. The total energy (kinetic plus potential) in the system is

$$E = \frac{1}{2} m_A |\mathbf{v_A}|^2 + \frac{1}{2} m_B |\mathbf{v_B}|^2 + U(r) + E_{A,\text{internal}} + E_{B,\text{internal}}$$

Several generic types of collisions can occur. In an *elastic* collision, the A and B particles retain their identities and no energy is exchanged with internal degrees of freedom of the particles. Since the total energy E is conserved, we must have

$$\frac{1}{2} m_A |\mathbf{v_A}|^2 + \frac{1}{2} m_B |\mathbf{v_B}|^2 + U(r) = E = \frac{1}{2} m_A |\mathbf{v_A'}|^2 + \frac{1}{2} m_B |\mathbf{v_B'}|^2 + U(r)$$

where the primes refer to quantities (velocities, in this case) that exist after the collision.

Inelastic collisions are those in which some of the internal energy in the A and B particles is exchanged between the particles or is coupled into translation. In this case, all the terms in the expression for the total energy must be retained. In monatomic atoms, this energy can only take the form of electronic excitation; molecular species possess vibrational and rotational degrees of freedom as well. We have already noted that *reactive* collisions, i.e., those in which the colliding particles change their identities, cannot strictly take place in a two-body collision. Nevertheless, we shall use this simple dynamical model to derive relationships between some important scattering parameters, such as reaction probability and reactive cross section. In this regard, a *charge-exchange* collision

$$A^+ + B \rightarrow A + B^+$$

is a reactive process which can be well described in terms of two-body scattering.

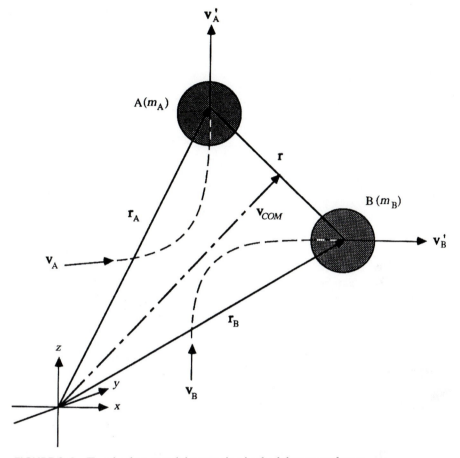

FIGURE 8-4   Two-body potential scattering in the laboratory frame.

A simpler analytic representation of the two-body collision is obtained in a transformed coordinate system. The simplest such transformation is to the *center-of-mass* coordinate system, in which the origin is located at $(m_A \mathbf{r}_A + m_B \mathbf{r}_B)/(m_A + m_B)$ and is moving at constant velocity $\mathbf{v}_{COM}$. In the center-of-mass system, the collision appears to follow a simple linear trajectory as the particles approach each other, contact, and then separate. For a central potential the particle motion is confined to a plane, and a more suitable coordinate representation for potential scattering is the *fixed-center-of-force* coordinate representation. In this representation, a composite particle AB is presumed to have a mass equal to the collision-reduced mass $\mu = m_A m_B/(m_A + m_B)$ and to scatter off a point center of infinite mass through the interaction potential $U = U(r = |\mathbf{r}_A - \mathbf{r}_B|)$. The fixed-center-of-force coordinate system is shown in Figure 8-5.

The relevant variables in this representation, in addition to the interparticle distance r (shown here as the distance from the composite particle AB to the center of

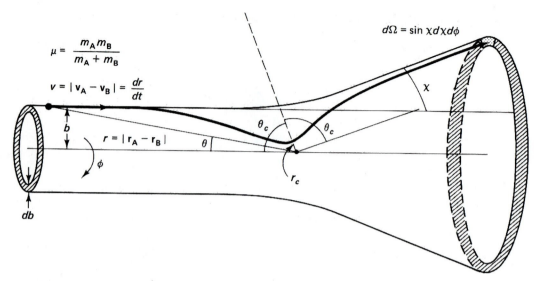

FIGURE 8-5   Two-body potential scattering in the fixed-center-of-force coordinate representation.

force) and the relative velocity $v = dr/dt$, are the impact parameter b (equivalent to the impact parameter b in the reactive hard-sphere model shown in Figure 8-2), the scattering angle θ, and the deflection angle χ. The azimuthal angle ϕ is also shown, but since the coordinate system is cylindrically symmetric, ϕ will be eliminated by a simple integration.

We now proceed to develop the relationships among these variables. The principal relationship derives from the conservation of particle flux. This requires that every system entering from the left in Figure 8-5, with velocity v_0 and impact parameter between b and $b + db$, must leave the scattering center to the right with velocity v' and a deflection angle between χ and $\chi + d\chi$. The explicit relationship between χ and b is determined by the potential function $U(r)$, to be shown shortly. For a hypothetical reactive process (e.g., A + B → C + D), the differential reactive scattering cross section is

$$I_R(\chi,\phi;v\Gamma) = P_R(v,b;\Gamma)d(\pi b^2)/d\Omega \qquad (8\text{-}9)$$

where the *reaction probability* P_R depends on velocity v, internal state Γ, and impact parameter b, and the differential ratio $d(\pi b^2)/d\Omega$ maps the area of the annular region on the left-hand side of Figure 8-5 onto that on the right-hand side. Using equation (6-2), we find the corresponding expression for the total reactive cross section σ_R by integrating over the solid angle $d\Omega$:

$$\sigma_R(v,\Gamma) = \int\int I_R(\chi,\phi;v\Gamma)d\Omega = \int_0^\infty P_R(v,b;\Gamma)\,2\pi b\,db \qquad (8\text{-}10)$$

We next proceed to find the *deflection function* $\chi(b)$. To do so, we first write the expression for the total energy in center-of-force coordinates, i.e.,

$$E = \frac{1}{2}\mu v^2 + U(r) + \frac{L^2}{2\mu r^2} \tag{8-11}$$

where $L = \mu v_0 b = \mu v' b'$ is the classical *angular momentum* associated with the collision trajectory. Both E and L are conserved quantities. The value of E can be readily determined from the boundary conditions of the problem: as the distance $r \to \infty$ (corresponding to the time $t \to -\infty$), any well-behaved potential function $U(r)$ must asymptotically approach zero, because at infinite separation the interaction between two particles must cease to exist. Also, the term $L/2\mu r^2$ obviously goes to zero. Therefore, $E = 1/2\,\mu v_0^2$. L is also equal to the product of the moment of inertia $I = \mu r^2$ and the angular velocity $\omega = d\theta/dt$; this fact will be used in the next step of the calculation.

For a simple central force field, the trajectory $\theta(t)$ can be found by solving the set of equations

$$L = \mu r^2 \frac{d\theta}{dt} \tag{8-12}$$

and

$$E = \frac{1}{2}\mu \left(\frac{dr}{dt}\right)^2 + \frac{L^2}{2\mu r^2} + U(r) \tag{8-13}$$

for the conserved quantities E and L. Doing so, we have

$$d\theta = \frac{L}{\mu r^2}\, dt$$

and

$$dr = -\left[\frac{2}{\mu}\left(E - U(r) - \frac{L^2}{2\mu r^2}\right)\right]^{1/2} dt$$

so that

$$dt = -\left[\frac{2}{\mu}\left(E - U(r) - \frac{L^2}{2\mu r^2}\right)\right]^{-1/2} dr$$

We find $\theta(t)$ by integrating

$$\theta(\tau) = \int_0^\theta d\theta'$$

$$= \int_{-\infty}^r \left(\frac{L}{\mu r^2}\right) dt$$

$$= -\int_{\infty}^{r} \frac{L/\mu r^2 \, dr}{\left[\frac{2}{\mu}\left(E - U(r) - \frac{L^2}{2\mu r^2}\right)\right]^{1/2}}$$

Using $L = \mu v_0 b = b(2\mu E)^{1/2}$, we have

$$\theta(t) = -b \int_{\infty}^{r} \frac{dr}{r^2 \left[1 - \frac{b^2}{r^2} - \frac{U(r)}{E}\right]^{1/2}} \tag{8-14}$$

For any trajectory, there will be a *closest approach distance* r_c at which $r(t)$ has its minimum value. Typically, this occurs when the two particles are interacting via the repulsive part of the potential and all of the kinetic energy has been converted into potential energy. Since $r(t)$ has a minimum at r_c, the velocity must be zero at that point in the trajectory. It is convenient to define $t = 0$ for $r(t) = r_c$ and $\theta(0) = \theta_c$. From Figure 8-5, the *deflection angle* χ is just $\pi - 2\theta_c$. We therefore have an expression for the deflection angle as a function of the impact parameter b and collision energy E, *viz.*,

$$\chi(E, b) = \pi - 2b \int_{r_c}^{\infty} \frac{dr}{r^2 \left[1 - \frac{b^2}{r^2} - \frac{U(r)}{E}\right]^{1/2}} \tag{8-15}$$

Let us now proceed to find $\chi(b)$ for some simple model potentials.

   (a) *Hard-sphere potential.* This is the potential considered earlier in Chapter 7, section 7.2. It can be written explicitly as

$$U(r) = 0 \qquad (r > d)$$
$$U(r) = \infty \qquad (r \leq d)$$

Clearly, $r_c = d$ for collisions at any energy. Using equation (8-15), we have

$$\chi(E, b) = \pi - 2b \int_{r_c}^{\infty} \frac{dr}{r^2[1 - (b^2/r^2) - 0]^{1/2}}$$

$$= \pi - 2b \int_{r_c}^{\infty} \frac{dr}{r(r^2 - b^2)^{1/2}}$$

$$= \pi - 2b \left[\frac{1}{2} \cos^{-1}(b/r)\right]_{r=d}^{r=\infty} \qquad \text{(see reference 3)}$$

$$= \pi - 2[\cos^{-1}(b/\infty) - \cos^{-1}(b/d)]$$

$$= \pi - \pi + 2\cos^{-1}(b/d)$$

so that

$$\chi(E, b) = 2\cos^{-1}(b/d) \tag{8-16}$$

For impact parameters greater than the hard-sphere diameter d, equation (8-16) requires taking the inverse cosine of a quantity greater than one, which does not exist.

In this situation, $\chi = 0$. For impact parameters less than d, $\chi = 2 \cos^{-1}(b/d)$ and varies over the range 0 to π. Note that χ is independent of E for any b and d. The scattering in center-of-force coordinates is shown in Figure 8-6(a), and the corresponding deflection function in Figure 8-6(b). The cross section for this potential can be calculated most conveniently by referring back to Figure 8-5. Recall that all particles scattered from the

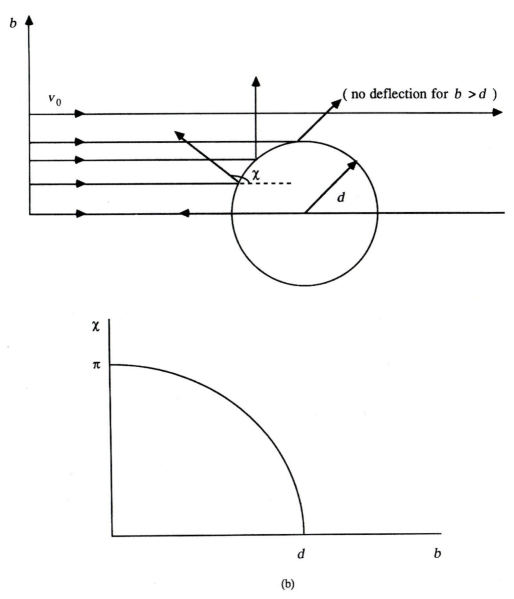

(b)

FIGURE 8-6 (a) Scattering from a hard-sphere potential in center-of-force coordinates. (b) Deflection function $\chi(b)$ for a hard-sphere potential.

range between b and $b + db$ must end up with scattering angles between χ and $\chi + d\chi$. This maps the annular area $2\pi b$ into $I(E, \chi)2\pi \sin \chi d\chi$. (The factor 2π in the range comes from integrating the azimuthal angle ϕ over 2π radians.) Equating the two areas gives

$$I(E, \chi)|2\pi \sin \chi d\chi| = 2\pi b db$$

or

$$I(E, \chi) = \frac{b}{|\sin \chi (d\chi/db)|} \tag{8-17}$$

For the hard-sphere potential with $\chi(b) = 2 \cos^{-1}(b/d)$, we obtain the simple results $I(\chi) = 1/4\; d^2$, which is independent of E and χ (in the center of mass), and $\sigma(E) = \pi d^2$, just the geometrical cross section, which is also independent of E. The latter is just the result [equation (8-1)] we found earlier on a simple intuitive basis. Deriving it in this manner confirms our intuition and, more importantly, provides an example of the use of scattering theory in which a simple analytic form can be obtained.

(b) *Lennard-Jones potential.* A more realistic intermolecular potential energy function is the Lennard-Jones potential, discussed in chapter 7 and given by

$$U(r) = 4\varepsilon \left[\left(\frac{\sigma}{r} \right)^{12} - \left(\frac{\sigma}{r} \right)^6 \right]$$

The deflection function $\chi(b)$ for this potential can be calculated with the use of equation (8-15), but the arithmetic becomes too complicated to reproduce here in detail.[4,5] The results are shown in Figure 8-7 in terms of the reduced impact parameter $b^* = b/\sigma$ and for a range of reduced energies $E^* = E/\varepsilon$ between 5 and 0.2. A set of trajectories is displayed in Figure 8-8. There are three distinct regions of scattering behavior:

1. For nearly head-on collisions ($b \rightarrow 0$), the deflection function is very similar to that for a hard-sphere collision shown in Figure 8-6(a), because the impinging particle passes through the attractive region of the potential and interacts with the repulsive r^{-12} wall. Almost all potential scattering has such a region of *specular* reflection.

2. For *large* impact parameters, the scattering is also similar to that for a hard sphere, namely, the impinging particle is hardly scattered at all, because the attractive $(-r^{-6})$ part of the potential is rapidly falling to zero in this region. This limit is reached more rapidly at higher values of E^*.

3. In the intermediate region of b^* and E^*, we encounter behavior which is not found for hard-sphere scattering. Note the large negative values of χ in Figure 8-8 for some values of b^* between 1 and 2. This results from the impinging particle interacting strongly with the *attractive* part of the Lennard-Jones potential. Included in this behavior is the phenomenon of *rainbow* scattering, which will be discussed shortly.

The cross section may be found from the deflection function with the use of equation (8-17). The angular dependence of the cross section is shown in Figure 8-9 for a value of E^* on the order of unity. Each of the previously described three regions of the deflection function maps onto a portion of the cross section. Region 1 corresponds to

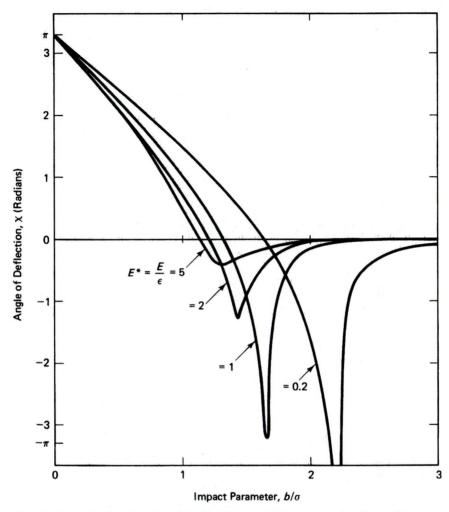

FIGURE 8-7   Deflection function for Lennard-Jones potential. The minima in $\chi(b/\sigma)$ correspond to the "rainbow" angle discussed in the text.

$b \to 0$ and $\chi \to \pi$; since both b and $\sin \chi$ approach zero in a uniform manner, the cross section remains well defined and approaches the limiting hard-sphere value $d^2/4$. In region 2, b increases without limit while χ (and therefore $\sin \chi$) goes to zero; also, $d\chi/db \to 0$. This leads to a strong divergence of I at small χ. In region 3, there is a minimum in the deflection function at some value of χ (not necessarily $-\pi$), at which the derivative $d\chi/db \to 0$. The cross section also diverges at this value of χ, which is called the *rainbow angle* χ_r. The term arises from the mathematical similarity between this phenomenon and the scattering of light from raindrops in the atmosphere which produces the rainbow. Both this divergence and that in region 2 are an artifact of classical mechanics; for example, the classical attractive potential always has some finite (but very small) value at arbitrarily large r, so there is always some nonzero scattering

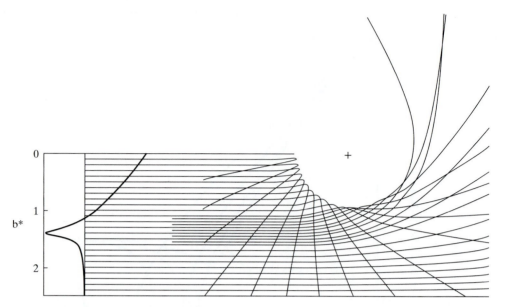

FIGURE 8-8 Classical trajectories for various reduced impact parameters in Lennard-Jones scattering. The corresponding deflection function $\chi(b^*)$ is shown at the left-hand side. [From H. Pauly, "Collision Processes, Theory of Elastic Scattering," in *Physical Chemistry: An Advanced Treatise,* Vol. VI-B, edited by H. Eyring, D. Henderson, and W. Jost (New York: Academic Press, 1975). Reproduced by permission.]

angle. Quantum mechanics damps out these divergences, however; for example, the uncertainty principle places a lower limit on the deflection angle that can be measured with a given apparatus, and the cross section rounds off at that value of χ.

Before leaving the topic of the Lennard-Jones potential, let us note the correction to the gas-kinetic collision rate due to attractive intermolecular forces, alluded to in section 8.1.1. The importance of this correction was noted by Kohlmaier and Rabinovitch[6] and Lin et al.,[7] and has been evaluated rigorously by Hirschfelder et al.[8] Let us consider the collision rate coefficient between a and b molecules in a gas,

$$z_{\text{coll}} = \Omega_{ab}^{(2,2)*}\, \bar{v}_{ab} \pi \sigma_{ab}^2 \qquad (8\text{-}18)$$

where

$$\sigma_{ab} = \frac{1}{2}\left(\sigma_a^{\text{LJ}} + \sigma_b^{\text{LJ}}\right)$$

is the arithmetic average of the Lennard-Jones σ parameters, and the "omega integral" is defined as

$$\Omega_{ab}^{(2,2)*} = \frac{2\pi k_B T}{\mu_{ab}} \int \int_0^{\infty} e^{-\gamma_{ab}^2}\, \gamma_{ab}^7 \left(1 - \cos^2 \chi\right) b\, db\, d\gamma_{ab} \qquad (8\text{-}19)$$

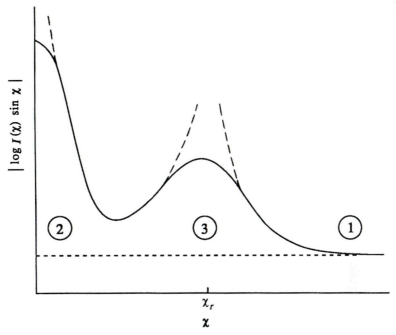

FIGURE 8-9   Differential cross section for Lennard-Jones potential function. The dashed parts of the curve show the classical divergences.

The variables in equation (8-19) are

$$\gamma_{ab} = \left(\frac{\mu_{ab}}{2k_BT} \right)^{1/2} \nu_{ab}$$

and $\chi(b)$ is the deflection function, defined by equation (8-15). Tables of $\Omega_{ab}^{(2,2)*}$ for the Lennard-Jones potential function are given in Hirschfelder et al.[8] At small values of T^* (low temperatures, or sticky molecules with large attractive terms), Z_{coll} is *larger* than the calculated hard-sphere collision rate. At large values of T^*, on the other hand, corresponding to high temperatures or for collisions with He atoms, Z_{coll} may actually be less than the hard-sphere estimate.

8.3   COMPLEX SCATTERING PROCESSES

While the two-body scattering dynamics described in the preceding section was useful for introducing a number of important concepts and methods of scattering theory, there is obviously no way to describe a chemical reaction in terms of only two interacting particles; a minimum of three is required. Unfortunately, the three-body scattering problem is one for which no analytic solution is known. Accordingly, we turn to either approximations or numerical analysis to solve this problem. In this section, we shall consider two methods which are of particular importance in theoretical chemical kinetics: classical trajectory calculations with Monte Carlo sampling, and solution of the time-dependent Schrödinger equation on the potential energy surface.

8.3.1 Classical Trajectory Calculations

A potential surface for a chemical reaction is expressed as a function of the relative coordinates of the constituent atoms (see chapter 7). With such a function available, it is possible to solve the classical equations of motion on the surface for the collision trajectory, which is just the set of time-dependent coordinates of the individual atoms. In general, the solution requires numerical integration of the Hamilton equations of motion for each coordinate Q_j and its conjugate momentum $P_j = \mu_j dQ_j/dt$:

$$\frac{dQ_j}{dt} = \frac{\partial \mathcal{H}}{\partial P_j}, \quad \frac{dP_j}{dt} = -\frac{\partial \mathcal{H}}{\partial Q_j} = -\frac{\partial U}{\partial Q_j} \tag{8-20}$$

These equations are equivalent to Newton's second law of motion, $\mathbf{F} = m\mathbf{a}$, where the force $\mathbf{F}$ is a gradient of the potential energy and the acceleration $\mathbf{a}$ is the second time derivative of the corresponding coordinate.

The essential content of the problem is contained in the Hamiltonian function $\mathcal{H}$, which is the sum of kinetic and potential energies:

$$\mathcal{H} = T + V$$

The kinetic energy T is a function of relative velocities, while the potential energy V is the potential $U(\{\mathbf{r}\})$ expressed in terms of relative coordinates. This separation into position- and velocity-dependent terms is what allows us to simplify the derivative $\partial \mathcal{H}/\partial Q_j$ in equation (8-20) to $\partial U/\partial Q_j$.

The simplest example of a classical trajectory is that for a collinear AB + C collision complex, for which there are only two coordinates, the A-B distance r_1 and the B-C distance r_2. The trajectories for such a system can be represented on a contour diagram such as that shown in Figure 7-5, or even simulated by rolling a steel ball on a machined facsimile of the surface. Even for such a simple system, however, a complication is introduced by transforming from internal molecular coordinates to Cartesian space. The origin of this problem may be seen by writing out the Hamiltonian in detail:

$$\mathcal{H} = T + V$$

$$\mathcal{H} = \frac{1}{2}\frac{m_A m_B}{m_A + m_B}\dot{r}_1^2 + \frac{1}{2}\frac{(m_A + m_B)m_C}{m_A + m_B + m_C}\left(\dot{r}_2 + \frac{m_A}{m_A + m_B}\dot{r}_1\right)^2 + U(r_1, r_2) \tag{8-21}$$

$$\mathcal{H} = \frac{1}{2(m_A + m_B + m_C)}$$

$$\times [m_A(m_B + m_C)\dot{r}_1^2 + 2m_A m_B \dot{r}_1 \dot{r}_2 + m_C(m_A + m_B)\dot{r}_2^2] + U(r_1, r_2) \tag{8-22}$$

The first kinetic energy term in equation (8-21) represents the relative motion of A with respect to B (with reduced mass μ_{AB}), while the second represents motion of AB with respect to C (with reduced mass μ_{AB+C}). We see that when equation (8-21) is expanded to give equation (8-22), a cross term in $\dot{r}_1 \dot{r}_2$ remains. This cross term must be eliminated if the motion of a representative point across the surface is to be that of the

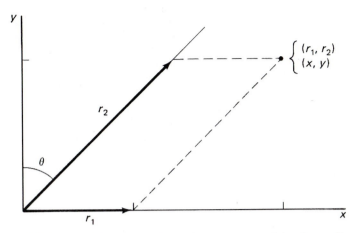

FIGURE 8-10   Affine transformation to mass-weighted coordinate system.

"sliding mass" alluded to previously. This can be done by a further coordinate transformation to a *mass-weighted coordinate system,* known as the *affine transformation,*[9,10] which diagonalizes the kinetic energy of the system. Such a transformation is shown in Figure 8-10. We let $r_1 = x - y \tan \theta$ and $r_2 = \alpha y \sec \theta$. Then if

$$\alpha = \left[\frac{m_A(m_B + m_C)}{m_C(m_A + m_B)} \right]^{1/2}$$

and

$$\sin \theta = \frac{\alpha m_C}{m_B + m_C} = \left[\frac{m_A m_C}{(m_A + m_B)(m_B + m_C)} \right]^{1/2}$$

$T(\dot{r}_1, \dot{r}_2)$ becomes $1/2\, m\dot{x}^2 + 1/2\, m\dot{y}^2$, with $m = m_A(m_B + m_C)/(m_A + m_B + m_C)$. (The reader is encouraged to verify this by direct substitution.) This transformation gives rise to the *skewed coordinate system* often seen in representations of potential surface contour diagrams. For F + H–H, $\theta + 43.6°$ and $\alpha = 0.7255$. The skewed surface, derived from that depicted in Figure 7-8(a), is shown in Figure 8-11.[11]

The power of the classical trajectory method lies in its ability to predict the details of reactive scattering for complex systems on almost any potential surface, even if analytic solutions cannot be obtained. Its major drawback is that it is very computation-intensive, even for simple systems. For example, for a three-body system, such as H + H_2 or F + H_2, there are $3 \times 3 = 9$ degrees of freedom in the problem. Three of these can be immediately eliminated by going to center-of-mass coordinates. But this still leaves six independent coordinates which, with their six conjugate momenta, give rise to 12 coupled linear first-order differential equations having the form given by equation (8-20). The full Hamiltonian for the three-body problem is

$$\mathcal{H}(\mathbf{P},\mathbf{Q}) = \frac{1}{2\mu_{BC}} (P_1^2 + P_2^2 + P_3^2) + \frac{1}{2\mu_{A,BC}} (P_4^2 + P_5^2 + P_6^2) + U(Q_1, \ldots, Q_6) \quad (8\text{-}23)$$

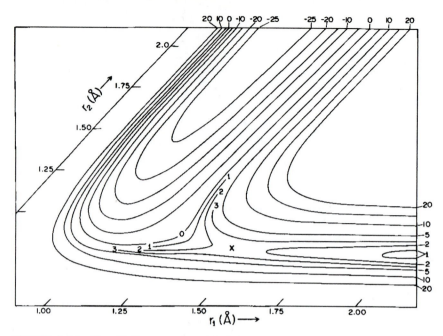

FIGURE 8-11   F + H–H potential surface in mass-weighted coordinate system. [From J. C. Polanyi and J. L. Schreiber, *Faraday Disc. Chem. Soc. 62,* 269 (1977)].

These equations can be readily integrated by one of the numerical techniques discussed in chapter 2. Because each trajectory may involve a large number of successive integration steps, cumulative errors ("global error") may become a problem. A good way of verifying the accuracy of the integration is to run the trajectory *backwards* from the final computed state: by virtue of microscopic reversibility, the initial conditions should be accurately reproduced. The major problem in implementing classical trajectories is that the typical experiment samples a wide distribution of initial conditions. Since each trajectory is run for a specified set of initial coordinates and momenta, a large number of trajectories must be run over a range of initial conditions in order to simulate a particular experimental situation. An efficient choice of initial conditions is essential to getting useful information from trajectory calculations.

The conditions which must be specified can be identified by reference to Figure 8-12. Three of the coordinates refer to the diatomic species BC. These can be written, along with their conjugate momenta, in the form

$$Q_1^0 = R\sin\theta\cos\phi \qquad P_1^0 = -P(\sin\phi\sin\eta - \cos\phi\cos\theta\cos\eta)$$

$$Q_2^0 = R\sin\theta\sin\phi \qquad P_2^0 = P(\cos\phi\sin\eta - \sin\phi\cos\theta\cos\eta)$$

$$Q_3^0 = R\cos\phi \qquad P_3^0 = P(\sin\theta\cos\eta)$$

In these equations, R is the initial B–C distance, $P^2 = J(J + 1)\hbar^2/R^2$ is proportional to the square of the rotational angular momentum of BC, θ and ϕ are the polar and

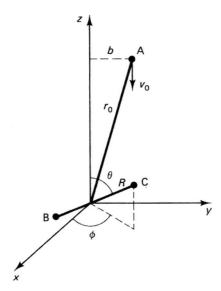

FIGURE 8-12   Classical coordinates for an A + BC collision.

azimuthal angles specified in Figure 6-1 and η is the angle between the vector ($\mathbf{R}_{BC} \times$ $\mathbf{z}$) and the rotational angular momentum of BC. For specifying the coordinates and momenta of the impinging atom A, it is convenient to take the initial relative velocity v_r^0 in the yz-plane. This gives the following simple forms:

$$Q_4^0 = 0 \qquad\qquad P_4^0 = 0$$

$$Q_5^0 = b \qquad\qquad P_5^0 = 0$$

$$Q_6^0 = -(r^2 - b^2)^{1/2} \qquad P_6^0 = \mu_{A,BC}\, v_r^0$$

In most trajectory calculations, the reactant molecule BC is specified to be in an initial vibrational and rotational state, which determines P and allows R to be set to the maximum bond extension compatible with the total vibrational energy. The initial relative velocity v_r^0 may be varied systematically, or it may be chosen at random from a Boltzmann distribution $f(v) = 4\pi v^2 e^{-\mu v^2/2k_B T}$. The orientation angles, which specify rotational phase, are selected at random, as is the impact parameter b. Finally, the vibrational phase is accounted for by selecting r_0 at random from the interval between r_0 and $r_0 + v_r^0 \tau$, where τ is the vibrational period of BC. In any case, r_0 must be chosen so that the intermolecular interaction is initially zero, and the trajectory is integrated forward until $r(t)$ is sufficiently large that the interaction is once again zero. At that point the final state of the system is determined by inspecting which atom is bound to which other atom, and final energies and angular momenta are determined by calculating classical expectation values over a few cycles of the motion. The use of random sampling for many of the initial conditions has given rise to the term "Monte Carlo method" in connection with these calculations.

The basic quantity derived from such calculations is the reaction probability for a specified initial relative velocity (i.e., initial relative kinetic energy), rotational state, and impact parameter. This is operationally defined as the ratio of the number of reactive trajectories to the total number of trajectories after a sufficiently large number of trajectories has been run:

$$P_R\left(v_r^0, J, b\right) = \lim_{\substack{N \to \infty \\ (\$ \to 0)}} \frac{N_R(v_r^0, J, b)}{N_{\text{total}}\left(v_r^0, J, b\right)} \tag{8-24}$$

Although a large number of trajectories is required to obtain good statistics, the practical limitation is usually the availability and cost of computer time. The reactive cross section is found by averaging the reaction probability over the impact parameter and rotational state:

$$\sigma_R(v_r^0, J) = 2\pi \int_0^{b_{\max}} P_R\left(v_r^0, J, b\right) b \, db$$

$$\sigma_R^0(v_r^0) = \sum_{J=0}^{J_{\max}} \sigma_R\left(v_r^0, J\right) f_B\left(J, T\right)$$

Finally, the thermally averaged rate coefficient is the velocity-averaged, velocity-weighted cross section, as discussed in chapter 6:

$$k(T) = \int_0^\infty dv_A \int_0^\infty dv_{BC}\, f_A\left(v_A\right) f_{BC}\left(v_{BC}\right) v_r^0\, \sigma_R(v_r^0) \tag{8-25}$$

In the averaging equations shown above, we have assumed that trajectories have been run for a random sampling of initial relative velocities, rotational states, and impact parameters, and have included the appropriate distribution functions explicitly. The reason for doing this is practical: it turns out that most reactive collisions tend to occur at small b and large v_r^0, so that sampling at random from the distributions $f(b) = 2\pi b$ and $f(v) = 4\pi v^2 e^{-\mu v^2/2k_B T}$ would tend to give a preponderance of nonreactive trajectories. It is more efficient to sample from a random distribution of b's and v_r^0's and then account for ensemble averages by weighting with the proper distribution function. This type of biasing, known as *importance sampling*,[12-14] is an important aid in performing efficient trajectory calculations.

The first fully realized classical trajectory calculation of a rate coefficient was by Karplus and co-workers[15] on $H + H_2$. Representative reactive and nonreactive trajectories from that work are shown in Figure 8-13. One microscopic quantity which can be determined by inspection of such plots is the "collision duration," which can be estimated as that period of time during which the approaching or departing atom deviates significantly from straight-line trajectories. From the plots in the figure, this appears to be on the order of 10^{-14} sec. Most "impulsive" collisions have durations on this order; longer lived collision complexes are also possible and are discussed briefly in section 8.3.3.

The trajectory results for the $H + H_2$ reaction probability and reactive cross section are shown in Figure 8-14(a) and (b), respectively. Note that the reaction probability does not resemble the old-collision-theory, reactive-hard-sphere model; the b dependence is much more like $P_R^0 \cos\left(\pi b/2b_{\max}\right)$, with $P_R^0 \approx 0.4$ and $b \approx 0.98$Å. Nevertheless, the reactive cross section shown in Figure 8-14(b) approximates very well to the reactive-hard-sphere form. The computed thermally averaged rate coefficient for the $H + H_2$

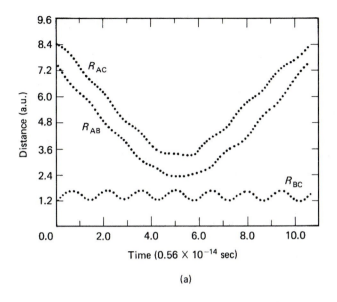

(a)

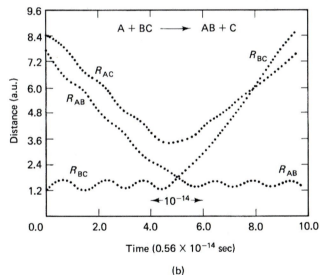

(b)

FIGURE 8-13 $H + H_2$ scattering trajectories, represented as plots of interatomic distance vs. time. (a) Nonreactive trajectory with $J_i = 0, v_i = 0, v_r^0 = 1.32 \times 10^6$ cm sec^{-1}. (b) Reactive trajectory with same initial conditions. The difference between trajectories (a) and (b) appears to be a minuscule difference in the vibrational phase of the original (BC) hydrogen molecule. The "crossing" of the plots in (b) indicates rotational excitation in the products, with $J' \approx 5$. [From Karplus et al., *J. Chem. Phys. 43*, 3259 (1965).]

reaction, over the range 300 K to 1,000 K, agrees with the experimentally determined values to within the accuracy of the measurements.* On the one hand, such an *ab initio* calculation of a rate coefficient is a remarkable *tour de force;* on the other hand, it illustrates vividly how rate constants, like all thermally averaged properties of a reaction, are very insensitive to the details of interaction potentials and molecular dynamics.

———————

*For a discussion of experimental techniques for determining $H + H_2$ reaction rates, the reader is referred to Johnston.[16]

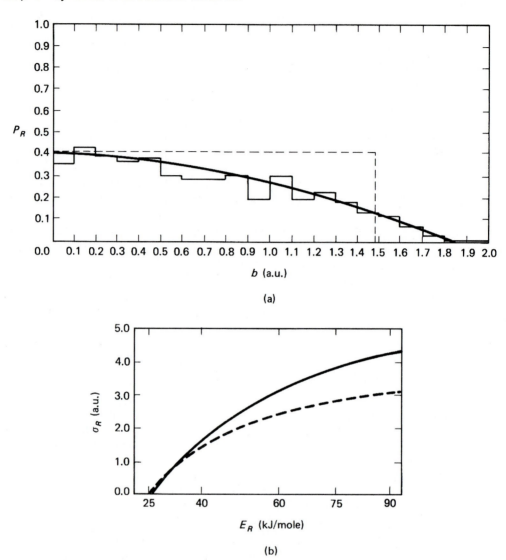

FIGURE 8-14   Results of classical trajectory calculations for $H + H_2$ reaction. (a) Reactive probability P_R vs. impact parameter. The solid curve is a fit to the empirical function $P_R^0/\cos(\pi b/2b_{max})$, while the dashed curve is the reactive-hard-sphere form [see equation (8-6)] with $b^* = 1.5$ a.u. and steric factor $p = 0.4$. (b) Reaction cross section vs. relative kinetic energy. The solid line is the (smoothed) result of the classical trajectory calculations, while the dashed curve is the reactive-hard-sphere (old-collision-theory) result given in equation (8-7). [From Karplus et al., *J. Chem. Phys. 43*, 3259 (965).]

In fact, the real usefulness of trajectory calculations is their ability to provide, in great detail, information on product final states, scattering angles, and other dynamical quantities, which may then be compared with experimental results obtained by using the techniques to be discussed in the next chapter. Trajectory calculations at this level have been carried out for a number of systems, including $F + H_2$.[11] Tested and documented trajectory programs are available from the Quantum Chemistry Program Exchange at the University of Indiana.

8.3.2   Quantum Scattering Calculations

In the preceding section, we have described how solution of the classical equations of motion may be used to find reactive probabilities and cross sections. Such calculations are more properly termed *quasiclassical* because the quantization of vibrational and rotational energy levels is accounted for by choosing initial conditions for the reactants in accord with their vibrational/rotational energy level and, after the calculation, by "binning" product states possessing suitable ranges of internal energies into groups according to corresponding v and J quantum numbers. We know, however, that atoms and molecules obey the laws of quantum mechanics— specifically, that their motion is governed by a wave equation (the Schrödinger equation) rather than by the deterministic laws of classical motion. The questions we now address are, first, how do we set up the scattering problem quantum mechanically and, second, does the quantum solution display significant differences from the classical treatment? Some necessary background for this section may be found in References 9 and 17-20.

We begin with the time-dependent Schrödinger equation,

$$\mathcal{H}\Psi = i\hbar\,\frac{\partial\Psi}{\partial t} \tag{8-26}$$

The wave function Ψ depends on both the coordinates $\mathbf{r}$ and the time t. The Hamiltonian function is

$$\mathcal{H} = \frac{p^2}{2\mu} + V(\mathbf{r}) \tag{8-27}$$

The potential function $V(\mathbf{r})$ is the same as in the classical treatment, but since the momentum is a differential operator, viz.,

$$\hat{p}_j = \frac{\hbar}{i}\,\frac{\partial}{\partial q_j} \tag{8-28}$$

the entire Hamiltonian is an operator rather than an algebraic function such as equation 8-21. That is to say, $\mathcal{H}$ is written as

$$\mathcal{H} = \frac{-\hbar^2}{2\mu}\nabla^2 + V(\mathbf{r}) \tag{8-29}$$

The second major difference from classical mechanics is that not all coordinates and momenta can be specified simultaneously, as was the case in setting up initial conditions for the classical trajectory calculations. This is a consequence of the Heisenberg uncertainty relationship

$$\Delta p_j\,\Delta q_j > \hbar \tag{8-30}$$

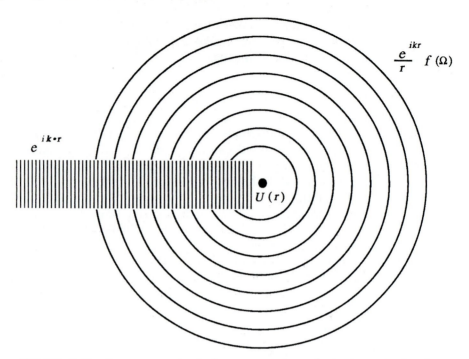

FIGURE 8-15 Quantum-mechanical representation of elastic scattering in a center-of-force coordinate system, with a spherically symmetric potential.

which follows from the operator nature of the momentum given by equation (8-28). Thus, in Figure 8-5, for example, we cannot specify both the impact parameter b and the relative velocity v for an individual collision.

If we wish, we can retain the center-of-force coordinate system used in Figure 8-5, but the interacting-particle representation must then be replaced by a wave picture, as shown in Figure 8-15. We can specify only the wave vector $k = p/\hbar$; the interaction potential $V(\mathbf{r})$ remains the same as in the classical calculation. In the case of elastic scattering, such as the interaction of two rare-gas atoms, the time-dependent wave function can be written as a simple product, $viz.$,

$$\Psi(\mathbf{r}, t) = \Psi(\mathbf{r})e^{-iEt/\hbar}$$

where $E = \hbar^2 k^2/2\mu$. For a central interaction potential, the time-independent part of the wave function may be represented as a superposition of an incoming plane wave and an outgoing spherically scattered wave, i.e.,

$$\Psi(\mathbf{r}, \Omega) = e^{ikz} + f(\Omega)\,\frac{e^{ikr}}{r} \tag{8-31}$$

where the propagation of the incident wave is along the z axis and Ω denotes the angular variables θ and ϕ for the scattered wave.

The quantity $f(\Omega)$, the *scattering amplitude,* is the probability that the incident particle will emerge in a direction defined by θ and ϕ as a result of the collision.

Experimental observables in quantum mechanics are expressed as expectation values of the wave function. The pertinent expectation value for the scattering problem is called the *current* or *probability flux*. This quantity represents the probability per unit time that a particle crosses a unit area normal to its direction of motion. It is defined by

$$J = \frac{\hbar}{2i\mu} \left(\Psi^* \nabla \Psi - \Psi \nabla \Psi^* \right) \tag{8-32}$$

In evaluating the incident and product fluxes we may treat the two terms of equation (8-31) separately, since there are no interferences between the incident and scattered waves in the regions where the fluxes are measured. This is the correct procedure because the incident flux is measured at large negative values of z where the scattered wave is negligible due to its $1/r$ dependence, and because in a real experiment the scattered flux is measured outside the finite transverse limits of the incident beam. Thus, from equation (8-32), the probability flux for the incident plane wave e^{ikz} is

$$J_{\text{incident}} = \hbar k / \mu$$

Similarly, for the outgoing scattered wave $f(\Omega)e^{ikr}/r$ one finds

$$J_{\text{scattered}} = \frac{\hbar k}{\mu} \frac{|f(\Omega)|^2}{r^2}$$

since, in computing the gradient of the scattered wave, only the radial component is important at the very large r's for which equation (8-31) is valid. At this point it is worthwhile noting that for this situation J_{incident} and $J_{\text{scattered}}$ are each simply the probability density $\Psi^*\Psi$ times the velocity, and have units of probability per unit area and unit time.

The probability per unit time that the scattered particle crosses a surface element of area dA normal to its direction of motion is

$$J_{\text{scattered}} \, dA = \frac{\hbar k}{\mu} |f(\Omega)|^2 \, d\Omega$$

since, by definition, dA/r^2 is equal to the element of solid angle $d\Omega$ (i.e., the total surface area A is $\int dA = \int r^2 d\Omega = 4\pi r^2$). Thus, the relative probability that the scattered particle will emerge into the solid angle $d\Omega$ is

$$\frac{J_{\text{scattered}} \, dA}{J_{\text{incident}}} = |f(\Omega)|^2 \, d\Omega$$

This relative probability has units of area and $d\sigma$ is the differential of the collision cross section. Therefore, the differential cross section defined in chapter 6 and in section 8-2, is

$$I(\theta, \phi) = \frac{d\sigma}{\delta\Omega} = |f(\Omega)|^2 \tag{8-33}$$

The calculation of $I(\theta, \phi)$ is seen to be a matter of calculating $f(\Omega)$.

To find $f(\Omega)$, we must solve the time-independent Schrödinger equation

$$\mathcal{H}\Psi(r, \Omega) = \frac{-\hbar^2}{2\mu} \nabla^2\Psi + V\Psi = E\Psi \tag{8-34}$$

This is done by matching the incident wave with the scattered wave in such a way as to obtain a solution to this Schrödinger equation that behaves like equation (8-31) for $r \to \infty$. Since equation (8-31) has no ϕ dependence for a spherically symmetric potential, it suffices here to consider only solutions which are independent of ϕ. There are many techniques for effecting such a solution, but the one most appropriate for this problem is a *partial wave analysis*. We expand the wave function Ψ in partial waves corresponding to the orbital angular momentum quantum number l. (For large l, $\hbar l = \mu v b$ so $l = |k| b$). This expansion has the form

$$\Psi(r, \Omega) = \sum_{l=0}^{\infty} c_l P_l(\cos\theta) \frac{\Psi_l(r)}{r} \tag{8-35}$$

where the c_l are expansion coefficients, and the P_l, which are Legendre polynomials, and the Y_l solve the angular and radial parts of the Schrödinger equation, respectively. Defining a reduced radial wave function $Y_l(r) = \Psi_l(r)/r$ and factoring out the angular part, equation (8-34) becomes

$$\frac{-\hbar^2}{2\mu} \frac{d^2}{dr^2} Y_l + \left(\frac{\hbar^2 l(l+1)}{2\mu r^2} + V(r) \right) Y_l = EY_l \tag{8-36}$$

Note that the term in $l(l+1)$ takes the place of the centrifugal potential term in equation (8-13). If $V(r)$ were zero, the radial wave function $Y_l(r)$ would become the well-known form for a free particle, which, for large r, takes the form

$$\sin\left(kr = \frac{\pi l}{2} \right)$$

Since $V(r)$ becomes negligible at large r, the presence of the potential cannot change this functional form but can change the phase of the radial wave function. Thus, one can write as $r \to \infty$

$$Y_l(r) = \sin\left(kr - \frac{\pi l}{2} + \eta_l \right) \tag{8-37}$$

where η_l is the collision-induced phase shift of the wave function for the lth partial wave.

Our final step is to express the scattering amplitude and cross section in terms of the phase shift. This is done by choosing the expansion coefficients c_l in such a way that equation (8-35), with equation (8-37) substituted for $Y_l(r)$, has precisely the form of equation (8-31) as r becomes very large. The scattering amplitude is then found to be

$$f(\Omega) = \frac{1}{2k} \sum_{\ell=0}^{\infty} (2l+1)(1 - \exp(2i\eta_l)) P_l(\cos\theta) \tag{8-38}$$

which can be rewritten as

$$f(\Omega) = \frac{1}{k} \sum_{\ell=0}^{\infty} (2l + 1)\, e^{i\eta_l} \sin \eta_l \, P_l(\cos \theta) \tag{8-39}$$

Using equation (8-33) and the orthonormality property of the Legendre polynomials leads to an expression for the cross section:

$$\sigma = \int \int |f(\Omega)|^2 \, d\Omega = \int |f(\Omega)|^2 \sin \theta \, d\theta \, d\phi$$

$$\sigma = \frac{\pi}{k^2} \sum_{l=0}^{\infty} (2l + 1)\, |1 - \exp(2i\eta_l)|^2 = \frac{4\pi}{k^2} \sum_{l=0}^{\infty} (2l + 1)\, \sin^2 \eta_l \tag{8-40}$$

The scattering amplitude depends only upon the angle between the direction of incidence and the direction of scattering, and is completely determined once the η_l are known. The relationship between the strength of the interaction potential and the magnitudes of the phase shifts is rather complicated. However, if there is no interaction, so that the particles move freely, then by definition the phase shifts are all zero and the scattering amplitude and cross section are seen to vanish, as they must.

An important way to treat quantum scattering is to use the *scattering matrix,* i.e. the **S** matrix. For each value of the initial angular momentum quantum number l, we have a different **S** matrix. An element of the **S** matrix for initial angular momentum quantum number l is denoted as S_{ij}^l. The magnitude of S_{ij}^l squared, $|S_{ij}^l|^2$, represents the probability of a transition from the reactant's quantum state i to the product's quantum state j. The S_{ij}^l element determines the amplitude of the outgoing (i.e., scattered) wave in the jth channel for an incoming (incident) wave of unit amplitude in the ith channel. Microscopic reversibility means that **S** is a symmetric matrix; i.e.

$$S_{ij}^l = S_{ji}^l \tag{8-41}$$

Conservation of flux requires that

$$\sum_j |S_{ij}^l|^2 = 1, \text{ for all } i \tag{8-42}$$

For the elastic scattering problem described above $i = j$, and for each l there is only one matrix element $S^l(k)$, where k is the wave vector. By considering the relationship between $Y_l(r)$ in equations (8-35) to (8-37) and the scattering matrix element $S^l(k)$, it can be shown that

$$S^l(k) = \exp(2i\eta_l) \tag{8-43}$$

Thus, $f(\Omega)$ and σ in equations (8-38) and (8-39) can be expressed in terms of $S^l(k)$, and we find

$$\sigma = \frac{\pi}{k^2} \sum_{\ell=0}^{\infty} (2l + 1)\, |1 - S^l(k)|^2 \tag{8-44}$$

The above procedure has been used to analyze elastic scattering, in which neither energy nor constituent particles are exchanged between the collision partners. The results of this treatment are useful chiefly for inverting molecular beam-scattering data to obtain intermolecular potential functions, e.g., in rare gas - rare gas collisions. The energy dependence of such scattering cross sections is implicit, appearing in the phase shifts η_l. The principal limitation of this calculation is that one may need to include 100 or more partial waves (values of l) for molecule-molecule collisions. In the above equations we have assumed a symmetric intermolecular potential, which gives a constant orbital angular momentum. Typically, the intermolecular potential is not symmetric, so that changes in l during the elastic collision must be considered. Additional details of the theory may be found in texts on quantum mechanics.

In the case of inelastic or reactive scattering, the situation is even more complex, because the molecular basis states change between the initial and final states of the collision. The potential $V(r, R)$ includes terms coupling internal coordinates R with relative translational coordinates r. The total energy

$$E = \frac{\mathcal{H}_0 \Phi}{\Phi} + \frac{\hbar^2 k^2}{2\mu}$$

is, of course, conserved in the collision ($\mathcal{H}_0$ is the internal Hamiltonian of the molecule, e.g., BC in an A + BC reaction, and Φ is the asymptotic form of the wave function as $t \to \infty$ and $r \to \infty$). The case where the orbital angular momentum is conserved proceeds just as in the above elastic case and the cross section for a transition state from state i to state j is found to be

$$\sigma_{i \to j} = \frac{\pi}{k_i^2} \sum_{l=0}^{\infty} (2l + 1) \left| S_{ij}^l - \delta_{ij} \right|^2 \tag{8-45}$$

where the Kronecker δ function in equation (8-45) is equal to zero for $i \neq j$, i.e., if the collision is reactive or inelastic. For $i = j$, the collision is elastic and we regain the factor $1 - S_{ij}^l$ which appears in equation (8-44).

Generalizing equation (8-45) to the more complex situation where l changes during the inelastic or reactive collision leads to a "coupled-channel" calculation, which is formidable because of the large number of basis states which must be included. The calculations on the H + H$_2$ reaction by Schatz and Kupperman[21] are a good example. As a result of advances in computer technology and computer algorithms, quantum scattering calculations have become possible for more complex reactions such as OH + H$_2$ → H + H$_2$O,[22-24] OH + CO → H + CO$_2$,[25] and H + H$_2$O → H$_2$ + OH.[26,27] However, these calculations remain extremely difficult for reactions with more than four atoms if all degrees of freedom are treated explicitly.

Results of quantum scattering calculations carried out to date have indicated several important differences from purely classical calculations. The possibility of tunneling through a potential barrier (see chapter 10) is one such effect. The consequence of tunneling is the appearance of a threshold cross section slightly below the barrier height energy, which in turn affects the activation energy for the reaction and can lead to non-Arrhenius behavior of the rate coefficient. Interference between scattered waves can lead to amplitude fluctuations in the differential cross section $I(\Omega)$ and also rounds off the classical divergences indicated in Figure 8-9. Classical trajectories also experience difficulties at threshold; when the total energy is slightly above the minimum energy required to cross the potential barrier, the vibrational zero-point energy tends to be converted into translational kinetic energy, which is not allowed quantum mechanically. With these exceptions, however, it is found that classical trajectories usually provide a good approximation to the motion of the corresponding quantum mechanical wave packet across the potential energy surface. A comparison of semiclassical and quantum mechanical calculations for the H + H$_2$ reaction will be discussed in chapter 9, Section 9.3.

8.3.3   Long-Lived Collision Complexes

From Figure 8-15, we estimated the "collision duration" (a concept which is not well defined quantum mechanically) to be about 10^{-14} sec. If the potential surface possesses a strong attractive well, a long-lived collision complex may be formed. Classically, this would result from an "orbiting" collision, in which the impinging particle appears to be captured by the scattering center. The quantum analogue is a "resonance" which can occur when the collision energy matches that of a bound state in the potential well. When such a complex collision takes place, products tend to be isotropically scattered in the center-of-mass system, and there may also be near-statistical energy release in the reaction products. An example is discussed in chapter 11.

8.3.4   Ion-Molecule Scattering

Thus far, the discussion has focused almost exclusively on reactions between neutral, i.e., uncharged, atoms and molecules. An important class of reactions involves charged species (ions) interacting with neutral species. Such reactions occur in ionized gas plasmas, radiation chemistry, electric discharges (as in electron-beam pumped lasers), flames, and the upper atmosphere. They can also occur within a mass spectrometer, giving rise to additional peaks which complicate the interpretation of mass spectra. For dynamical studies (see chapter 9), the charge provides a convenient "handle" on the ion by means of which the kinetic energy in the collision can be varied over a wide range. The theoretical treatment of such collisions, first developed by Langevin[28] in 1905, is quite straightforward,[29] and we proceed to develop it here.

An ion (M^+ or M^-) and a neutral molecule interact through the ion-induced-dipole potential

$$U(R) = -\frac{1}{2}\,\frac{\alpha e^2}{r^4} \tag{8-46}$$

where α is the polarizability of the neutral molecule and e is the elementary electronic charge (4.8×10^{-10} esu). The collision is best analyzed in the center-of-force coordinate system (see section 8.2, and cf. especially Figure 8-5). The effective potential, including the centrifugal term, is

$$V_{\text{eff}}(r) = U(r) + \frac{L^2}{2\mu r^2} = -\frac{1}{2}\,\frac{\sigma e^2}{r^4} + \frac{L^2}{2\mu r^2} \tag{8-47}$$

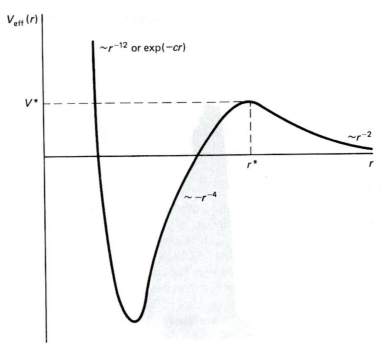

FIGURE 8-16 Effective potential for the interaction between an ion and a neutral, nonpolar molecule.

where $E = 1/2\,\mu v_0^2$ and $L = \mu v_0 b$. We can thus write

$$V_{\text{eff}}(r) = E\left(\frac{b}{r}\right)^2 - \frac{1}{2}\frac{\alpha e^2}{r^4}$$

A plot of $V_{\text{eff}}(r)$ given by equation (8-47) is shown in Figure 8-16.

Note that the potential has a maximum at $r^* = (\alpha e^2/E)^{1/2}/b$, with $V^*(r^*) = 1/2\,E^2 b^4/\alpha e^2$. Whenever $E < V^*$, the centrifugal barrier cannot be penetrated (if tunneling is neglected), and so r_c always remains greater than r^*. The collision is presumed to be elastic, in that the internal states of the ion and the target molecule remain unchanged. For $E \geq V^*$, the particle can move inside the centrifugal barrier and indeed can penetrate to the inner repulsive wall of the potential. In this case, we assume that reaction occurs with a probability of unity; if E is exactly equal to V^*, then "orbital capture" can occur.

The critical impact parameter at which reaction can occur is found from the expression for V^* to be

$$b^* = (2\alpha e^2/E)^{1/4}$$

Since the reaction probability is stipulated to be unity for $b < b^*$, the cross section for reaction is just

$$\sigma_L(E) = \pi(b^*)^2 = \pi\left(\frac{2\alpha e^2}{E}\right)^{1/2} \tag{8-48}$$

The quantity σ_L is often referred to as the *Langevin cross section*. We can estimate its magnitude by using a typical polarizability for small polyatomic molecules. (From Table 7-2, $\alpha = 3 \times 10^{-24}$ cm^3 and $E = k_B T = 4 \times 10^{-14}$ erg/molecule.) With these values, equation (8-44) gives

$$\sigma_L = \pi(3.5 \times 10^{-29})^{1/2} = 2 \times 10^{-14} \text{ cm}^2$$

The Langevin cross section is thus on the order of a few hundred square Ångstroms. It depends only on the collision energy E and the polarizability of the collision partner; no other chemical or structural information is incorporated into the expression. As in the reactive-hard-sphere model (see section 8.1.2), this represents an upper limit for the ion-molecule reaction cross section; for a given system, the accurate cross section will depend on the permanent dipole moment, activation barriers along the reaction coordinate, and other chemical properties of the ion and the molecule.

The rate constant for the Langevin model may be found by inserting the equation (8-48) for the cross section into equation (6-6) and substituting $v = (2E/\mu)^{1/2}$. This gives

$$k = \int v\sigma(v)\, f(v) dv$$

$$= \int \left(\frac{2E}{\mu}\right)^{1/2} \left(\frac{2\alpha e^2}{E}\right)^{1/2} \pi f(v)\, dv$$

Noting that the $E^{1/2}$ factors cancel and that the velocity distribution function $f(v)$ is normalized gives the simple result

$$k = \left(\frac{4\pi^2 \alpha e^2}{\mu}\right)^{1/2} \tag{8-49}$$

which is velocity (relative kinetic energy)-independent.

Detailed discussions of ion-molecule reactions may be found in several review articles.[30–33]

8.3.5 Electronically Nonadiabatic Processes: Surface Crossing

The scattering processes which we have been considering thus far have all been adiabatic, in the sense that they take place on a single electronic potential energy surface. Another important class of reactions is *nonadiabatic* reactions, which involve two or more interacting potential surfaces and the possibility of moving between these surfaces during a collision. Examples of such processes include reactions involving electronically excited species, such as

$$\text{N*}(^2\text{P}) + \text{O}_2(^3\Sigma) \rightarrow \text{NO}(^2\Pi, v) + \text{O}(^3\text{P}, {}^1\text{D})$$

and near-resonant electronic-energy-transfer reactions, such as

$$\text{I}(^2\text{P}_{3/2}) + \text{O}_2^*(a^1\Delta) \rightarrow \text{I}(^2\text{P}_{1/2}) + \text{O}_2(^3\Sigma)$$

The latter reaction is important, for example, in the chemically pumped oxygen-iodine laser system.[34] Potential energy surfaces for electronically excited molecules have been described previously in chapter 7.

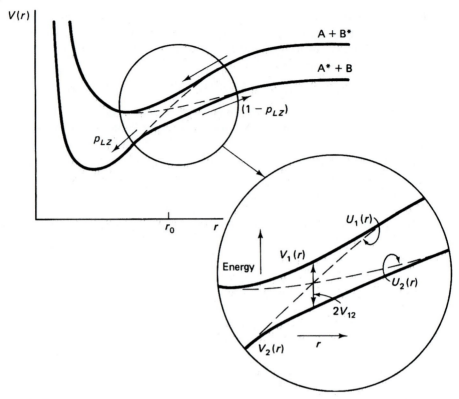

FIGURE 8-17   Interacting potential energy curves, showing detail of avoided-crossing region. An electronically nonadiabatic trajectory is shown which crosses with probability P_{LZ} from the upper to the lower surface on the way in and remains on the lower surface with probability $1 - P_{LZ}$ on the way out.

The simplest description of electronically nonadiabatic processes was developed by Landau[35] and Zener[36] in the early 1930's. The Landau-Zener model estimates the probability of surface crossing or surface hopping at the avoided crossing between two interacting one-dimensional potential energy curves (see Figure 8-17). This probability has the form[37]

$$P_{LZ} = \exp\left(-\frac{4\pi V_{12}^2}{hv|s_1 - s_2|}\right) \tag{8-50}$$

where $v = dr/dt$, the velocity at the crossing point r_0; s_1 and s_2 are the slopes dU/dr of the *unperturbed* potential curves at r_0, which must be calculated from the potential model; and V_{12} is one-half the splitting between the actual potential curves at r_0.

The net nonadiabatic process involves two traversals of the avoided-crossing region, one as the particles approach each other and a second as they separate. In order for the collision products to end up on a potential surface different from the one on which they started, one of these traversals must be nonadiabatic with probability

P_{LZ}, and the other must be adiabatic with probability $1 - P_{LZ}$. The net transition probability in the Landau-Zener model is thus

$$p_{12}(v) = 2P_{LZ}(1 - P_{LZ}) \tag{8-51}$$

and the thermally averaged nonadiabatic rate coefficient is

$$k(T) = \pi r_0^2 \int p_{12}(v)f(v)v\,dv \tag{8-52}$$

The velocity dependence in equation (8-46) gives rise to a very non-Arrhenius form for $k(T)$. This model has been applied to a number of electronically nonadiabatic systems, such as the $O_2^* - I$ transfer process mentioned earlier.[38]

More exact treatments of these processes are based on the classical trajectory method described in section 8.3.1. The interacting potential energy curves are approximated as sections of the actual surfaces normal to the collision trajectory, and the Landau-Zener model is used to estimate the surface-hopping probability at the crossing seams. Further details may be found in the article by Tully.[39]

REFERENCES

[1] C. E. Nordman and S. M. Blinder, *J. Chem. Ed. 51,* 790 (1974).

[2] H. Goldstein, *Classical Mechanics* (Reading, Mass.: Addison-Wesley, 1950).

[3] C. D. Hodgman, ed., *CRC Standard Mathematical Tables,* 12th Ed., p. 284, Integral No. 21 (Cleveland, Ohio: Chemical Rubber Publishing Co., 1960).

[4] H. Pauly, "Collision Processes, Theory of Elastic Scattering", in *Physical Chemistry: An Advanced Treatise,* Vol. VI-B, edited by H. Eyring, D. Henderson, and W. Jost (New York: Academic Press, 1975), pp. 553–628.

[5] H. Pauly, "Elastic Scattering Cross Sections I: Spherical Potentials", in *Atom-Molecule Collision Theory,* edited by R. B. Bernstein (New York: Plenum Press, 1979), pp. 111–199.

[6] G. H. Kohlmaier and B. S. Rabinovitch, *J. Chem. Phys. 38,* 1692 (1963).

[7] H.-M. Lin, M. Seaver, K. Y. Tang, A. E. W. Knight, and C. S. Parmenter, *J. Chem. Phys. 70,* 5442 (1979).

[8] J. O. Hirschfelder, C. M. Curtiss, and R. B. Bird, *Molecular Theory of Gases and Liquids* (New York: Wiley, 1954).

[9] R. D. Levine and R. B. Bernstein, *Molecular Reaction Dynamics* (Oxford University Press, 1974), pp. 98–100.

[10] B. H. Mahan, *J. Chem. Ed. 51.* 308 (1974).

[11] J. C. Polanyi and J. L. Schreiber, *Faraday Disc. Chem. Soc. 62,* 26 (1977).

[12] I. W. M. Smith, "Quasiclassical Trajectory Methods", in *Modern Gas Kinetics: Theory, Experiment, and Application,* edited by M. J. Pilling and I. W. M. Smith (Oxford: Blackwell, 1987), pp. 55–72.

[13] M. B. Faist, J. T. Muckerman, and F. E. Schubert, *J. Chem. Phys. 69,* 4087 (1978).

[14] J. T. Muckerman and M. B. Faist, *J. Phys. Chem. 83,* 79 (1979).

[15] M. Karplus, R. N. Porter, and R. D. Sharma, *J. Chem. Phys. 43,* 3259 (1965).

[16] H. S. Johnston, *Gas Phase Reaction Rate Theory* (New York: Ronald Press, 1966), pp. 158–169.

[17] R. M. Eisberg, *Fundamentals of Modern Physics* (New York: John Wiley and Sons, 1967).

[18] D. S. Saxon, *Elementary Quantum Mechanics* (San Francisco: Holden-Day, 1968).

[19] R. H. Dicke and J. P. Wittke, *Introduction to Quantum Mechanics* (Menlo Park, CA: Addison-Wesley, 1960).

[20] G. C. Schatz and M. A. Ratner, *Quantum Mechanics in Chemistry* (Englewood Cliffs, N.J.: Prentice Hall, 1993).

[21] G. C. Schatz and A. Kupperman, *J. Chem. Phys. 65,* 4668 (1976).

[22] U. Manthe, T. Seidemann, and W. H. Miller, *J. Chem. Phys. 101*, 4759 (1994).

[23] W. H. Thompson and W. H. Miller, *J. Chem. Phys. 101*, 8620 (1994).

[24] D. H. Zhang and J. Z. H. Zhang, *J. Chem. Phys. 101*, 1146 (1994).

[25] E. M. Goldfield, S. K. Gray, and G. C. Schatz, *J. Chem. Phys. 102*, 8807 (1995).

[26] D. H. Zhang and J. Z. H. Zhang, *J. Chem. Phys. 103*, 6512 (1995).

[27] D. H. Zhang and J. C. Light, *J. Chem. Phys. 104*, 4544 (1996).

[28] P. Langevin, *Ann. Chem. Phys. 5,* 245 (1905).

[29] A. Henglein, "Elastic and Reactive Scattering of Ions on Molecules", in *Physical Chemistry: An Advanced Treatise,* Vol. VI-B, edited by H. Eyring, D. Henderson, and W. Jost (New York: Academic Press, 1975), pp. 509–551.

[30] P. Ausloos, "Ion-molecule reactions in radiolysis and photoionization of hydrocarbons", in *Progr. Reaction Kinetics,* edited by G. Porter (Oxford: Pergamon Press, 1970), pp. 113–179.

[31] M. T. Bowers, ed., *Gas Phase Ion Chemistry* (New York: Academic Press, 1979).

[32] A. G. Brenton, R. P. Morgan, and J. H. Beynon, *Ann. Rev. Phys. Chem. 30,* 51 (1979).

[33] D. C. Clary, *Mol. Phys. 53,* 3 (1984).

[34] W. E. McDermott, N. R. Pchelkin, D. J. Benard, and R. R. Bousek, *Appl. Phys. Letts. 32,* 469 (1978).

[35] L. Landau, *Physik Z. Sowjetunion 2,* 46 (1932).

[36] C. Zener, *Proc. Roy. Soc. London A137,* 696 (1933); ibid. *A140,* 660 (1933).

[37] H. Eyring, J. Walter, and E. W. Kimball, *Quantum Chemistry* (New York: Wiley, 1944), pp. 326–330.

[38] J. I. Steinfeld and D. G. Sutton, *Chem. Phys. Letts. 64,* 550 (1979).

[39] J. C. Tully, in *Dynamics of Molecular Collisions,* Part B, edited by W. H. Miller (New York: Plenum Press, 1976), pp. 217-267.

PROBLEMS

*8.1 Consider the following "single beam" experiment, which enables the total cross section for elastic scattering to be measured directly. A beam of molecules of species B at a high translational energy E is passed through a gas of species G which is at very low pressures; E is sufficiently large that the atoms of G are essentially stationary with respect to B. Let the velocity of B be v.

   (a) From the definition of the cross section, show that the total number of deflections (collisions) of B per unit time per G atom is $\sigma n_B v$, where n_B is the density of B and σ is the total elastic cross section for B–G collisions. (Assume that only elastic collisions are possible.)

   (b) Show that the total number of B–G collisions per unit time in a volume V is $\sigma V v n_B n_G$, where n_G is the density of G.

(c) Suppose that the volume in question has (constant) area A and length dx. Suppose also that the intensity of B is $J =$ (density of B) $\cdot v$ and is initially $J = J_0 = n_B v$. Show that the change in intensity over dx is $dJ = -J\sigma n_G dx$.

(d) Show that the overall change in intensity over a finite distance x is given by

$$\ln(j_0/J) = n_G \sigma X$$

It is plain that the measurement of this intensity change enables the total cross section to be determined from one experiment (see also chapter 9, problem 1.)

8.2 Gas viscosity is given by

$$\eta = \frac{5\pi}{32\sqrt{2}} \frac{\mu\bar{v}}{Q}$$

[R. D. Present, *Kinetic Theory of Gases,* New York: McGraw-Hill, 1958], where the average velocity $\bar{v} = (8k_BT/\pi\mu)^{1/2}$, μ, is the collision-reduced mass, and

$$Q = \frac{1}{32} \left(\frac{\mu}{2k_BT}\right)^4 \int_0^\infty dv \int_0^{2\pi} d\phi \int_0^\pi d\theta \sin^3\theta I(\theta) v^7 \exp\left(-\frac{\mu v^2}{4k_BT}\right) dv$$

Given that the differential cross section of a hard sphere of diameter d is $I(\theta) = d^2/4$, find the viscosity of a gas of hard spheres.

*8.3 Show that Hamilton's equations of motion

$$-\frac{\partial\mathcal{H}}{\partial q_i} = \frac{dp_i}{dt} \qquad \frac{\partial\mathcal{H}}{\partial p_i} = \frac{dq_i}{dt}$$

with $\mathcal{H}(q_i p_i) = \Sigma_i(p_i^2/2\mu_i) + V(q_1, \ldots, q_N)$ are equivalent to Newton's law

$$\frac{d^2q_i}{dt^2} = -\frac{\partial V}{\partial q_i}$$

*8.4 Consider motion in two dimensions, with positions x, y and momenta p_x, p_y. Suppose that the potential V is a function only of $r = (x^2 + y^2)^{1/2}$. Find the expression for the Hamiltonian in terms of the new positions and momenta

$$r = (x^2 + y^2)^{1/2}$$
$$\theta = \tan^{-1}(y/x)$$
$$p_r = p_x\cos\theta + p_y\sin\theta$$

and

$$p_\theta = -p_x r\sin\theta + p_y r\cos\theta$$

(i.e., the radial and angular momenta). Hence, show from Hamilton's equations that angular momentum is conserved.

8.5 Suppose molecules A and B are characterized by Boltzmann velocity distributions with temperatures T_A and T_B, respectively ($T_A \neq T_B$). If d is the A–B collision diameter, show that the average rate of bimolecular collisions per unit volume is

$$\pi d^2 \sqrt{\frac{8k_BT_A}{\pi m_A} + \frac{8k_BT_B}{\pi m_B}} \cdot C_A \cdot C_B$$

8.6 Let d and D denote the collision diameters of A–B and AB–C, respectively. The function

$$h(d - r)h(D - R)$$

where

$$r = |\mathbf{R}_A - \mathbf{r}_B|$$

and

$$R = \left| \mathbf{r}_C - \frac{m_A \mathbf{r}_A + m_B \mathbf{r}_B}{m_A + m_B} \right|$$

is 1 if A, B, and C are experiencing a three-body collision, and is zero otherwise. Average the time derivative of the function over Boltzmann distributions of molecules A, B, and C, and show therefrom that the average three-body collision rate per unit volume is

$$\bar{v}_{AB} \pi d^2 \frac{4}{3} \pi D^3 + v_{AB-C} \pi D^2 \frac{4}{3} \pi d^3$$

where

$$\bar{v}_{AB} = \sqrt{\frac{8k_B T}{\pi} \frac{(m_A + m_B)}{m_A m_B}}$$

and

$$\bar{v}_{AB-C} = \sqrt{\frac{8kT}{\pi} \frac{(m_A + m_B + m_C)}{(m_A m_B) m_C}}$$

*8.7 For the case of classical scattering of two particles with a repulsive coulomb potential given by $V = +B/r$,

$$\chi(E, b) = 2 \operatorname{cosec}^{-1} \left[1 + \left(\frac{2bE}{B} \right)^2 \right]^{1/2}$$

Show, then, that

$$I(E, \chi) = \left(\frac{B}{4E} \right)^2 \operatorname{cosec}^4 \left(\frac{\chi}{2} \right)$$

This result is that used (and derived) by Rutherford in his original α-scattering experiments. Note that the classical total cross section is infinite and that the same result is obtained in an exact quantum treatment.

*8.8 Consider a reaction between two dilute species in an inert gas, and assume that only one initial and one final state are involved. The rate coefficient is then related to the total cross section by

$$k = 8\pi \mu^{-1/2} (2\pi k_B T)^{-3/2} \int_0^\infty E\sigma(E) \exp(-E/k_B T) dE$$

The "old-collision-theory" cross section is

$$\sigma(E) = 0 \qquad\qquad \text{if } E < E_0$$

$$\sigma(E) = \pi d^2 \left(1 - \frac{E_0}{E} \right) \qquad \text{if } \geq E_0$$

(a) Evaluate k for T $= 1000$, $1,500$ and $2,000$ K for a system where $\mu = 10$ daltons, $d = 0.1$ nm, and $E_0 = 100$ kJ mol^{-1}. N.B. units!

(b) Plot your three values from (a) on an Arrhenius plot and comment.

*8.9 The classical deflection function is given by

$$\chi(b) = \pi - 2b \int_{r_0}^{\infty} dr\, r^{-2} \left[1 - \frac{V(r)}{E} - \frac{b^2}{r^2} \right]^{-1/2}$$

where E is the kinetic energy and r_0 is the classical distance of closest approach. Derive $\chi(b)$ for a hard-sphere potential given by $V(r) = 0$ if $r \geq d$ and $V(r) = \infty$ if $r < d$.

*8.10 Ion-molecule reactions such as $N^+ + O_2 \rightarrow ND^+O$ are important in the upper atmosphere and in interstellar space. In this problem, we use a simple model, called the *Langevin model* (Section 8.3.4), for *reactive* scattering in such reactions. The following diagram shows results for the reaction $D_2^+ + N \rightarrow ND^+ + D$; the points are from experiment, while the line is the Langevin model prediction [from McClure et al., *J. Chem. Phys.* **66**, 2079 (1977)].

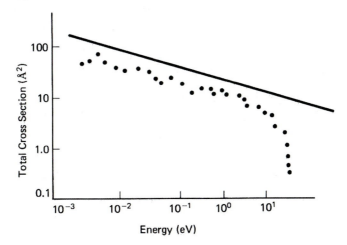

(a) Plainly, the Langevin model breaks down at high energies. Suggest a reason for this.

(b) Evaluate b^* for the $N^+ + D_2$ reaction, assuming that $\alpha + 10^{-24}$ cm^3 and the translational energy $E = k_B T$ for $T = 300$ K. Comment on your value of b^* compared to the value of a typical molecular van der Waals radius, i.e., the value of d in the Lennard-Jones potential $V(r) = 4\varepsilon[(d/r)^{12} - (d/r)^6]$.

*8.11 For a particular collinear reaction $A + BC \rightarrow AB + C$, it is found that the classical Hamiltonian can be represented by

$$\mathcal{H} = (p_s^2 + p_u^2)/2m + \frac{1}{2}ku^2 + A\exp(-Bs^2)$$

where (s, u) are the curvilinear reaction coordinates.

(a) Sketch a contour diagram of the potential in (i) (s, u) space and (ii) (R_{AB}, R_{BC}) space.

(b) Consider the classical probability of reaction $\chi(p_s, p_u, u, s = \infty)$. This probability is always 0 or 1; find the conditions for it to be 0 and those when it is 1.

(c) Would the rate coefficient show an Arrhenius form? Justify your answer.

(d) Give two reasons why the preceding Hamiltonian cannot be accurate for a real system.

*8.12 Consider two particles colliding under influence of a *purely repulsive* potential (e.g., $V = C/r^{12}$). Sketch the deflection function $\chi(b)$ and classical cross section $I(\chi)$, explaining your answers briefly. No mathematical derivation is required. Note that $\chi(b)$ and $I(\chi)$ are significantly different from their counterparts for, say, a Lennard-Jones potential.

*8.13 **(a)** For the attractive part of a Lennard-Jones potential, one finds that $\chi(b) \approx -C/b^6$, where $\chi(b)$ is the deflection function and C is a constant. Given this result, derive an expression for the classical differential cross section $I(\chi)$.

(b) Will the expression derived in (a) be valid for large or small angles and for large or small energies? Explain your answers briefly.

*8.14 Consider the following cross sections, the first two for the reactive system K + HI and the third for the isoelectronic K + Xe:

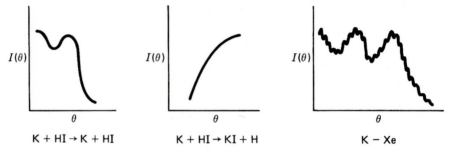

| $K + HI \rightarrow K + HI$ | $K + HI \rightarrow KI + H$ | $K - Xe$ |

(a) Why are no "wiggles" present for the K–HI elastic cross section shown in the first diagram?

(b) Why does the elastic cross section for K–HI drop off much more rapidly at large θ than it does for K–Xe?

(c) Does reaction occur mainly at small or large impact parameters?

---

*These problems have been provided by Dr. R. G. Gilbert, Department of Theoretical Chemistry, University of Sydney, N.S.W., Australia (known as the "Surfing Professor").

Experimental Chemical Dynamics

In chapters 7 and 8, we learned about theoretical methods for generating detailed reaction dynamics, i.e., cross sections and reaction probabilities, from potential energy surfaces at the state-to-state level. In this chapter, we shall consider some of the important experimental techniques for measuring such quantities,[1] including molecular beam scattering and state-resolved spectroscopic techniques. We then examine the prototypical $H + H_2$ and $F + H_2$ reactions using these techniques. Some general principles relating potential energy surfaces to reaction properties are discussed, and we consider how application of the principle of detailed balancing can be used to relate specific energy release in exothermic reactions to selective energy consumption in endothermic reactions.

9.1   MOLECULAR BEAM SCATTERING

One of the first approaches to moving the experimental study of chemical reactions from the level of macroscopic kinetics to that of detailed dynamics was the crossed molecular beam method. This is essentially a realization of the scattering experiment discussed in section 6.1. The principles of the method can be discussed with the aid of Figure 9-1. The experiment consists of intersecting two beams of molecules (A and B) and catching the product molecules C in a suitable detector. While this may seem simple, there are a number of constraints which must be met in order for the experiment to be successful.

The first of these constraints reflects the fact that the critical pieces of information obtained from the experiment are the angular distribution of product molecules and the velocity with which these product molecules leave the scattering center. In order for this information to be preserved, the product molecules must not undergo collisions with background gas molecules before they reach the detector. This means that the mean free path in the gas must be greater than the distance from the scattering center to the detector, which may be on the order of 10 to 30 cm. That is to say,

$$\text{Mean free path} = (2^{1/2} \pi n d^2)^{-1} > 30 \text{ cm}$$

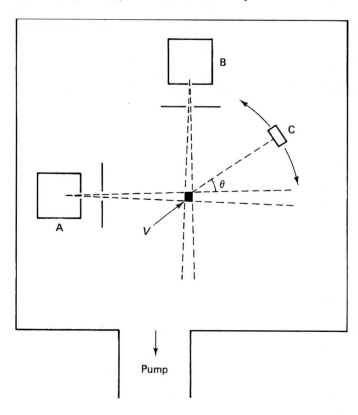

FIGURE 9-1   Schematic of a molecular beam-scattering apparatus. The reactant molecules, A and B, intersect in a small scattering volume V. Product molecules C are collected in the detector, which can be rotated around the scattering center. Various devices may be inserted in the beam paths A $\rightarrow$ V, B $\rightarrow$ V, and V $\rightarrow$ C to select reactant and/or product species according to velocity or other properties. [From Steinfeld and Kinsey, *Progr. Reaction Kinetics 5*, 1 (1970).]

Therefore, the gas density must be

$$n < 1/(1.4 \times 3 \times 30 \times 30 \times 10^{-16})$$
$$< 3 \times 10^{12} \, cm^{-3} \approx 10^{-4} \, \text{torr}$$

This is not a severe constraint; molecular beam-scattering chambers ("cans", in the trade) are typically run at pressures below 10^{-6} torr. In fact, *differential pumping* is frequently employed to reduce pressures in parts of the chamber to 10^{-10} torr or less, for reasons that will be described shortly.

The second constraint is more serious and has to do with the net flux of product molecules. This can be estimated[1] from equation (6-1), which is reproduced as follows for convenience:

$$\frac{dN_C}{dt} = n_A f_A(v_A) n_B f_B(v_B) I_R(v, \theta, \phi) v V \tag{9-1}$$

Let us take typical beam densities of 10^{11} molecules cm^{-3} and make use of the entire velocity distribution $f(v)$. Also, let the reactive cross section $I_R = 10^{-15} \, cm^{-2}$, the average relative velocity $v = 10^5$ cm sec^{-1}, and the beam intersection volume $V = 10^{-2} \, cm^3$. This gives a total production rate of 10^{10} C molecules per second. Only a small fraction of these are intercepted by the detector, however, which may subtend a solid angle element $d\Omega = 10^{-4}$, so that a flux of 10^6 molecules sec^{-1} *at the detector* may be anticipated. This may still seem like a lot, but it must be considered in light of available detection techniques.

The simplest collection technique that can be imagined is actual collection of the product. However, 10^6 molecules sec^{-1} amounts to only a few nanograms of material per year, or perhaps a few monolayers per year coated onto a plate. Nevertheless, this method has been used on occasion. A far more efficient procedure is to convert the molecular current into an electrical current which can then be measured. There are two principal ways of doing this. Historically, the first method to be employed was a thermal ionization technique: when alkali metal atoms impinge on a hot metal filament, they lose an electron which can be collected as a negative current. Furthermore, by judiciously damaging the filament with hydrocarbon vapors, it can be rendered selective, so that only atomic alkalis, and not their salts, will be detected. This surface ionization technique is very efficient; unfortunately, it is applicable only to the alkali metals (K, Rb, Cs), which explains why virtually all the early molecular beam studies were carried out using these species. A more general method, the so-called universal detector, uses mass spectrometry: electron-impact ionization followed by a quadrupole field to select individual ions by their masses. The problem with this technique, however, is that the ionization efficiency is only about 10^{-4}, so that the original 10^6 product molecules per second have now become a few hundred ion counts per second. In order to measure these accurately, the detection region of the beam chamber must be pumped down to 10^{-10} torr or less to eliminate background ions, as mentioned earlier.

An alternative detection technique makes use of laser-excited fluorescence or multiphoton ionization of the product molecules[2] and will be discussed further in section 9.2. While this may be a much more efficient detection method than electron impact ionization, it, too, is restricted to a small set of product molecules.

The first beam experiments were done using *effusive* beams, that is, molecular flow from low-pressure gas reservoirs or ovens. It was found that much higher beam fluxes could be obtained using *supersonic* sources,[3] which result from releasing a high-pressure gas through a small nozzle. Under such conditions, the flow is hydrodynamic rather than molecular. One consequence of this kind of flow is that the molecules are accelerated to many times their thermal velocity in the forward direction; hence the name supersonic. Another consequence is that the energy required for this acceleration is drawn from the transverse velocity components and the internal (vibrational, rotational) energy of the molecules. The net effect is a fast but very cold molecular beam. This cooling has been put to advantage in the spectroscopy of complicated molecules;[4] one disadvantage is that, at these low internal temperatures, the molecules tend to stick together to form clusters. This can be inconvenient if one is trying to perform scattering experiments on individual, uncomplexed molecules. Molecular beamists are nothing if not resourceful, however, and an industry has developed to study the spectroscopy and scattering of molecular clusters.

Other than supersonic acceleration, few methods are available for controlling the relative velocities of the beam particles. Amdur and co-workers[5] used an ionization-acceleration-neutralization scheme to produce fast beams of inert gas atoms, and thus to measure repulsive interaction potentials. This method works well only for high particle velocities, however. Portions of the velocity distributions can be selected by inserting rotating slotted disks in the beam paths which pass only molecules moving at certain speeds; doing so, however, throws away a substantial fraction of the total beam flux and reduces even further the signal at the detector.

To see what sort of information can be obtained from a molecular beam-scattering experiment, let us consider one of the very first such reactions studied,[6] viz.,

$$K + CH_3I \rightarrow KI + CH_3$$

First, we need to account for the total energy available in the reaction system. The major contribution is made by the reaction exothermicity ΔH^0, which can be estimated as the difference in energy between the newly formed K–I bond and the broken H_3C–I bond, approximately 88 kJ/mol. To this is added the relative translational energy of the particles in the two beams and the mean thermal excitation of the reactants, in this case vibration and rotation of the CH_3I. This total available energy Q must be partitioned between the translational energy of the products and their internal, i.e., vibrational and rotational, energy. If thermal effusive beams are employed, the reactant translational energy is about 10 kJ/mol, and CH_3I vibration and rotation add another 2 kJ/mol, to give

$$Q \approx 88 + 10 + 2 = 100 \text{ kJ/mol}$$

Interpretation of molecular beam-scattering experiments is most easily done with the aid of the Newton diagram introduced by Herschbach.[6] Such a diagram is shown for the K + CH_3I reaction in Figure 9-2. This is essentially a velocity space construction in the laboratory rest frame. The two initial velocity vectors are at right angles to each other, since this is the orientation of the beam sources in the experimental apparatus; the K atom vector is much longer than that of CH_3I, because it is produced in a heated oven and in fact contributes most of the relative translational kinetic energy in the reaction. The velocity of the center of mass, a constant of the motion, is given by

$$\mathbf{v}_{com} = \frac{m_1\mathbf{v}_1 + m_2\mathbf{v}_2}{m_1 + m_2} = \frac{m_3\mathbf{v}_3' + m_4\mathbf{v}_4'}{m_3 + m_4} \tag{9-2}$$

where the subscripts 1 and 2 refer to the reactants (K and CH_3I) and the subscripts 3 and 4 and the primed velocities refer to the products (KI and CH_3). The relative velocities in the center-of-mass coordinate system are

$$\mathbf{w}_R = \mathbf{v}_1 - \mathbf{v}_2$$

and

$$\mathbf{w}_R' = \mathbf{v}_3' - \mathbf{v}_4' \tag{9-3}$$

The relative velocity components are found by weighting these quantities by the inverse mass ratio, i.e.,

$$\mathbf{w}_1 = \mathbf{w}_R \frac{m_2}{M} \qquad \mathbf{w}_3' = \mathbf{w}_R' \frac{m_4}{M}$$

$$\mathbf{w}_2 = -\mathbf{w}_R \frac{m_1}{M} \qquad \mathbf{w}_4' = -\mathbf{w}_R' \frac{m_3}{M} \tag{9-4}$$

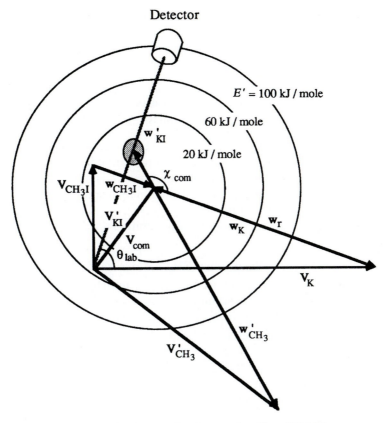

Detector

FIGURE 9-2   Newton diagram for the reaction K + CH₃I. The concentric circles are labeled by the amount of translational energy in the products.

where the total mass $M = m_1 + m_2 = m_3 + m_4$.

We now use the two basic conservation laws of classical mechanics to find the range of final velocity vectors and scattering angles (θ_{lab}, χ_{COM}) compatible with the available reaction energy. The first law, conservation of linear momentum, is automatically satisfied by the definitions of the relative velocity components given in equation (9-4):

$$m_1 w_1 = -m_2 w_2$$
$$m_3 w_3' = -m_4 w_4'$$
(9-5)

The second law, conservation of total energy, provides an important additional constraint:

$$m_1 w_1^2 + m_2 w_2^2 = 2E_{\text{transl}}$$
$$m_3 w_3'^2 + m_4 w_4'^2 = 2E'_{\text{transl}} = Q - E'_{\text{vib,rot}}$$
(9-6)

The concentric circles in Figure 9-2 show the allowable ranges of $\mathbf{w}'_{KI}$ for various specified values of E'_{transl}.

What is observed experimentally is that the KI product peaks at angles θ of $70°-90°$ in the laboratory frame. This is consistent with at least two possible scenarios:

1. a very low value of E'_{transl}, say, 10–20 kJ/mol, with final velocity vectors peaked sharply *backwards* in the center-of-mass frame.

2. a value of E'_{transl} near the maximum of 100 kJ/mol, with χ_{COM} near 90°.

The molecular beam method, in its simplest form, cannot distinguish between these possibilities, or, for that matter, any intermediate possibilities. On the other hand, the kinematic constraints expressed by equations (9-5) and (9-6) preclude the possibility of the KI product being forward-scattered in the center-of-mass frame, and, indeed, no such forward scattering is observed.

In order to remove this ambiguity, additional measurements must be performed. These include:

1. Measurement of the product CH_3, but this is very difficult.

2. Velocity analysis of the product KI, by inserting rotating slotted disks between the beam intersection volume and the detector.

3. Measurement of out-of-plane scattering. The experiment described uses coplanar beam sources and detectors. Because of the cylindrical symmetry of the center of mass (see Figure 8-5), there will be significant out-of-plane scattering. In the first scenario presented previously, the KI product will be confined to small out-of-plane angles, while in the second scenario, the out-of-plane scattering will be a much larger fraction of the total.

4. As an alternative to these kinematic techniques, the product *internal* energy distributions can be measured by a technique such as laser-induced fluorescence.

Results of the investigations of a number of alkali halide-forming reactions, primarily using methods (2) and (4), reveal that most of the product is in fact backward-scattered, with a small value of E'_{transl}. Most of the reaction exothermicity appears in vibration of the newly formed metal-halogen bond. One way of presenting the data is by transforming to center-of-mass coordinates using the Jacobian transformation[7]

$$I_R(\chi) = I_R(\theta_{lab}) \left| \frac{d\Omega_{lab}}{d\Omega_{COM}} \right| \tag{9-7}$$

A presentation of this type is shown in Figure 9-3.

In addition to the kinematics of the reaction, we also want to find the magnitude of the reactive cross section. There are several methods available for doing this.[8] The most direct method would appear to be simply to measure the total product flux; combined with a knowledge of particle densities in the beams, use of equation (9-1) should then provide values for the differential cross section I_R. The equation for I_R can then be integrated with respect to the scattering angle to provide the total reactive cross section σ_R. There are a number of problems with this procedure, however. The most important is the difficulty of attaining absolute calibration of the detectors, so that absolute particle fluxes cannot be established with precision. Also, it is not generally

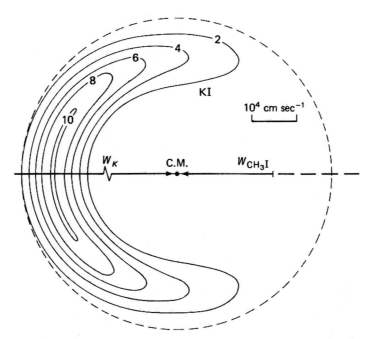

FIGURE 9-3 Contour representation of reactive scattering in center-of-mass coordinates for $K + CH_3I$ system at a translational energy $E_r = 11.6$ kJ/mol. [From A. M. Rulis and R. B. Bernstein, *J. Chem. Phys.* 57, 5497 (1972)].

possible to measure I_R over a complete range of scattering angles, and the beam intersection volume V is difficult to define precisely.

Perhaps the most accurate way of determining total reactive scattering cross sections is by carrying out *elastic* scattering measurements on the same system. The principle is as follows. Most molecular interactions can be approximated by a Lennard-Jones or similar potential function (see section 7.2.). The deflection function, and therefore the elastic scattering cross section, from a Lennard-Jones potential is well known (see Figures 8-7 and 8-9). Since scattering at small deflection angles, corresponding to large impact parameters, is essentially completely elastic, a potential function can be fit to small-angle scattering in a reactive system. Alternatively, a simulant can be used to determine the elastic scattering, such as Xe atoms for CH_3I. When either of these methods is employed, it is observed that elastic scattering at large angles falls below the predicted curve in reactive systems, the difference being due to reactive scattering at the smaller impact parameters. Figure 9-4 shows typical data[9] for $K + CH_3I$. The procedure is then to calculate the reaction probability at a specified relative kinetic energy $\overline{E}$ over a range of scattering angles χ from the calculated elastic scattering cross section $I_{el}(\overline{E}, \chi)$ and the observed cross section $I_{obs}(\overline{E}, \chi)$ using the formula

$$P_R(\overline{E}, \chi) = \frac{I_{el}(\overline{E}, \chi) - I_{obs}(\overline{E}, \chi)}{I_{el}(\overline{E}, \chi)} \tag{9-8}$$

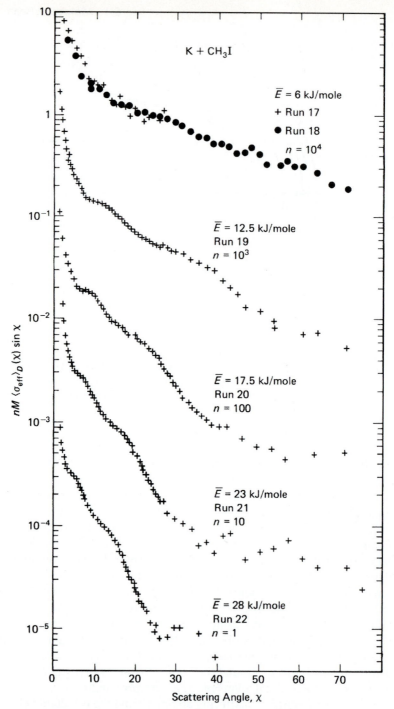

FIGURE 9-4 Experimental differential elastic scattering cross sections for the system K + CH₃I. The factor of n is introduced to displace successive plots vertically. [From M. Ackerman et al., Ninth Symp. (Int'l.) on Combustion (New York: Academic Press, 1963).]

Subsequent application of equations (8-9) and (8-10) allows determination of the total reactive cross section σ_R.

For the reaction $K + CH_3I$, the reaction probability P_R is found to be nearly equal to unity at small impact parameters and drops off a little ways beyond the Lennard-Jones interaction parameter σ_{LJ}. The drop-off is not as sharp as in the reactive-hard-sphere collision model: the reaction threshold energy is 5–10 kJ/mol, and the cross section attains a maximum value of approximately 40 $Å^2$ at relative kinetic energies above 50 kJ/mol. Application of molecular-beam techniques to the $H + H_2$ and $F + H_2$ reactions is discussed in sections 9.3 and 9.4, respectively.

9.2 STATE-RESOLVED SPECTROSCOPIC TECHNIQUES

A second class of techniques for determining microscopic properties of chemical reactions is spectroscopic measurement. These techniques complement the molecular beam method discussed in the preceding section in that they can provide rotational and vibrational state distributions of the product molecules. In principle, the combination of spectroscopic detection with molecular beam techniques can furnish both energy and angular resolution. Although experiments using these techniques are difficult, they are now being carried out in a number of laboratories. The principles of the methods discussed in this section are essentially the same as those described in chapter 3 on the techniques of kinetic measurements, which should be reviewed at this point. A good pedagogical survey of these techniques may be found in the review by Kovalenko and Leone.[10]

9.2.1 Infrared Chemiluminescence

Observation of radiation from product molecules is a clear indication that these molecules are produced in excited states. If the excitation is primarily vibrational, then the radiation will appear in the infrared region of the spectrum, i.e., between 3 μm and 15 μm (see chapter 3). This technique has been applied extensively to hydrogen-halogen reactions, producing HX molecules in excited (ν, J) states. In order to be able to identify populations in individual states, two experimental criteria must be met. First, the spectral bandwidth at the detector must be sufficiently narrow to resolve the emission bands; to accomplish this, the infrared filter indicated in Figure 3-8 must be replaced by a monochromator or a tunable filter. The second requirement is dictated by the rapid deactivation of excited vibration-rotation states due to collisions. In order to observe the *nascent* distribution of reaction products, that is, the distribution unrelaxed by collisions, the molecules must be able to radiate prior to suffering a deactivating collision. Since radiative lifetimes in the infrared are on the order of milliseconds, this means in turn that the experiments must be carried out at extremely low pressures, which of course reduces the signal strength. One way of ensuring that both of these conditions are met is through the use of the *arrested relaxation* method devised by Polanyi.[11] In this technique, the reaction is carried out in a chamber whose walls are maintained at a very low temperature, e.g., 20 K (by means of liquid H_2). This has the effect of *cryo-pumping* the chamber so that all products and unreacted starting materials are condensed on the walls; only those product molecules that radiate before being pumped away are observed. The low temperature also has the effect of reducing thermal noise and black-body background in the detector and filters, making it less difficult to observe weak infrared chemiluminescence signals. Results on product distributions in the reaction $F + H_2$, obtained using this technique, will be discussed in section 9.4. Information can also be obtained from observing partially relaxed distributions by means of a Master Equation analysis, discussed in the Appendix to this chapter.

9.2.2   Laser-induced Fluorescence

The LIF technique described in chapter 3 is very well suited to the determination of product state distributions, since the spectroscopic resolution is afforded by the narrow-band dye laser. The principal limitation, of course, is that the technique is restricted to those product species which display visible- or ultraviolet-excited fluorescence. Also, in order to observe the nascent distribution, the molecular density must be kept sufficiently low so that collisional relaxation does not occur. Improvements in laser technology are steadily expanding the range of species that can be studied by LIF.

In addition to providing information on internal-state distributions, spectroscopic techniques such as LIF can also provide information on the velocity distribution of the products. In order to carry out such measurements, a collimated, velocity-aligned beam of reactant molecules is made to interact with a second beam or, to study photodissociation dynamics, with a collimated beam of laser light. The product molecules are interrogated by an ultra-narrow-bandwidth laser which is scanned across the Doppler lineshape[4] of the molecular transition. The velocity distribution can be obtained from the fluorescence excitation profile (or its Fourier Transform).[12] Some representative examples of Doppler spectroscopy are photodissociation of ethylene[13] to acetylene and H_2 and electronic-to-translation energy transfer[14] from excited Ba atoms to H_2, O_2, N_2, and NO.

9.2.3   Raman Techniques

An alternative to LIF which has proven useful in a number of cases is the use of Raman spectroscopy for product detection. While spontaneous Raman scattering is much too weak to be useful for observing the low densities of product molecules required to suppress collisional relaxation, laser-based Raman techniques, such as stimulated Raman gain and coherent anti-Stokes Raman spectroscopy (CARS),[15] can provide the necessary sensitivity. As with LIF, the state resolution is provided by the narrow-bandwidth lasers used to excite the Raman spectrum.

9.2.4   Resonant Multiphoton Ionization

The resonant multiphoton ionization (REMPI) technique described in chapter 3 is also a very good probe for product state distributions, since ionization currents can be detected with high sensitivity. While single-photon ionization spectra are typically broad and featureless, offering little or no state specificity, the presence of a resonant intermediate state in the MPI process permits isolation of individual features with a resolution comparable to that of LIF. An example is shown in Figure 9-5, which illustrates detection of nitric oxide in excited vibrational states produced by the reaction[16]

$$N(^4S) + O_2 \rightarrow NO(v) + O$$

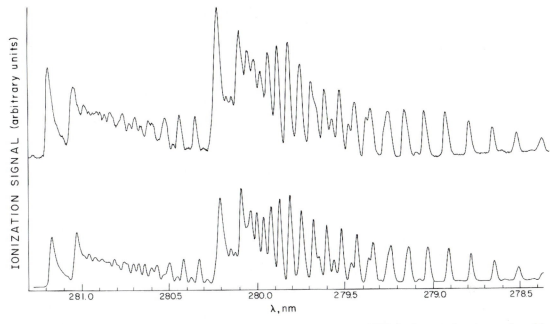

FIGURE 9.5   Resonant multiphoton ionization spectrum of NO in the $v = 1$ state produced in the flowing afterglow reaction of nitrogen atoms with O_2. The upper trace is the experimental spectrum, and the lower trace is the synthesized spectrum calculated for saturated two-photon resonant ionization [From I. C. Winkler, et at., *J. Chem. Phys. 85, 890* (1986)].

In addition to the vibrational states, rotational and spin-orbit states are resolved in the MPI spectrum.

9.2.5   Reaction Product Imaging

In a further development of product detection by MPI, a two-dimensional image of the product angular distribution may be obtained by capturing the photoions produced at the scattering center on a position-sensitive detector. Such a detector typically consists of an array of microchannel plates coupled to a phosphor screen which lights up when struck by an ion; the luminescence is then captured by a charge-coupled-device (CCD) camera, which in turn transmits the data to image analysis software. This technique was originally developed to image fragment angular distributions following photodissociation;[17,18] the apparatus shown in Fig. 9-6 has been used to study the dynamics of the $H + D_2$ reaction, discussed in greater detail in Section 9.3.

9.2.6   Femtochemistry

While the preceding discussion has emphasized use of narrow-band, high-resolution lasers to elucidate vibrational, rotational, translational, and angular distributions of reaction products, a rather different approach is that represented by the use of very-short-pulse (and therefore wide-bandwidth) lasers to probe real-time reaction dynamics. This technique, called "femtochemistry,"[19] will be discussed in chapter 10, in the context of attempts to observe the reaction path of a chemical reaction.

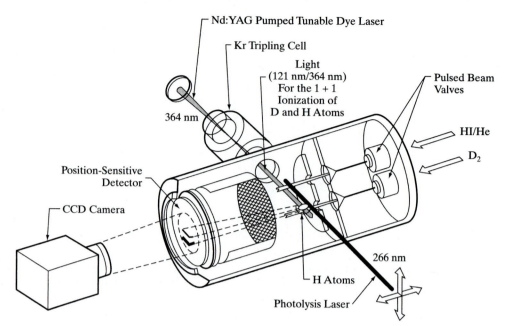

FIGURE 9-6   Schematic of a reaction product imaging apparatus. A photolytically produced beam of H atoms is crossed with a supersonic molecular beam of D_2 molecules. Product D atoms are ionized and accelerated toward a position-sensitive detector consisting of microchannel plates coupled to a phosphor screen. Ion positions appearing on the phosphor screen are recorded with a CCD camera. [From Kitsopoulos *et al., Science 260,* 1605 (1993)]; reproduced with permission.]

9.3   MOLECULAR DYNAMICS OF THE H + H$_2$ REACTION

In Section 8.3.1, we saw that quasi-classical trajectory (QCT) calculations on a reasonably accurate potential energy surface (PES) were capable of reproducing thermal rate coefficients for the H + H$_2$ exchange reaction to within experimental error. Since that work was carried out, virtually every experimental and theoretical method described in this and the preceding chapter has been applied to this, the simplest of all reaction systems. The H + H$_2$ reaction is the most amenable to theory of all reactions, since the quantum-chemistry PES calculations involve only three interacting electrons; similarly, molecular dynamics calculations are readily carried out for the three light hydrogen nuclei moving on such a surface. An important question for the QCT calculations, however, is whether quantum-mechanical effects are important for such light atoms. As is often the case, theoretical convenience goes hand-in-hand with experimental difficulty: molecular beam scattering experiments[20] place enormous technical demands on detectors, vacuum systems, and data acquisition, while LIF or REMPI measurements had to await development of non-linear optical methods to upconvert available laser sources to the required vacuum ultraviolet wavelengths.

In order to obtain observable results, isotopic variants such as

$$D + H_2 \rightarrow HD + H$$

or

$$H + D_2 \rightarrow HD + D$$

were studied, but these are treated as easily as $H + H_2$ in either semi-classical or quantum-mechanical theories.

Work on the $H + H_2$ system through 1976 has been reviewed by Truhlar and Wyatt.[21] Much of the experimental data available at the time were in the form of thermal rate coefficients. In the mid-1980's, molecular-beam scattering[20] and "hot-atom" techniques (see Ref. 22 and Problem 9.3) were used to obtain total reaction cross sections and energy threshholds. Gerlach-Meyer et al.[23] detected H and D atoms with vacuum-ultraviolet radiation at the Lyman-α wavelength, generated by frequency-tripling the output of an excimer-pumped dye laser. They found $\sigma_R = 1.1 \pm 0.2$ Å² at a collision energy of 1.5 eV (145 kJ/mole), in good agreement with the molecular beam result of 1.7 ± 0.8 Å² at the same energy and with the QCT calculations.

Several years after these results appeared, a burst of activity in both experimental and theoretical work on the "H/D + H₂ → H₂/HD + H" reaction provided definitive results for this system.[24,25] Earlier, Gerrity and Valentini[26] had measured product distributions in the $H + D_2 \rightarrow HD$ (v, J) + H reaction using CARS (see Section 9.2.3), but these experiments began with a thermal distribution of reactant energies. Zare and coworkers[27,28] used supersonic molecular beams, stimulated Raman pumping of H₂ into a specific rovibrational level, and photolysis of DI as a source of monoenergetic D atoms, to carry out measurements on single quantum levels with a specified center-of-mass collision energy. The experimental results were then compared with QCT and quantum-mechanical (QM) calculations.[29,30] As shown in Fig. 9-7, nearly perfect agreement was

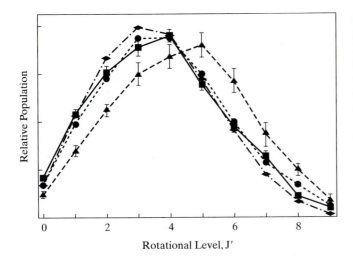

Relative Population

Rotational Level, J′

FIGURE 9-7   HD ($v' = 1, J'$) rotational distributions from the $D + H_2$ reaction at a center-of-mass collision energy of about 1.05 eV: experiment (solid curve), where the error bars represent one standard deviation; QM calculations (dotted and dash-dotted curves); and QCT calculations (dashed curve), where the error bars represent one standard deviation in the sampling statistics. All four distributions have been normalized to the sum of the common populations [From D. A. V. Kliner, K.-D. Rinnen, and R. N. Zare, *Chem. Phys. Letts.* 166, 107 (1990). Reproduced with permission.]

found with the QM calculations, but the QCT calculation predicted a rotational distribution of product HD molecules which is too high by one or two rotational quantum units. Differential cross sections for this reaction have also been measured using reactant product imaging[31] and an elegant parallel molecular beam technique.[32] As before, the QM calculations provided slightly better agreement with these experiments than did the QCT results. The conclusion derived from these studies is "that although quasiclassical methods may be good enough to predict the broad outlines of [a] reaction, quantum mechanics is needed to provide the details. 'The world really is quantum mechanical on the molecular level'."[25] But since the observed quantum effects are small and subtle for a system as light as H + H$_2$ (or D$_2$), where such effects are expected to be the most noticeable, an argument may also be made that for virtually all other reactions, which involve at least one atom which is heavier than hydrogen, a quasi-classical description may be perfectly adequate. This has, in fact, been found[33] for the Cl + H$_2$ → HCl + H reaction, in which QM and QCT calculations give equally good agreement with experiment.

9.4   STATE-TO-STATE KINETICS OF THE F + H$_2$ REACTION

While the H + H$_2$ (or D$_2$) reaction is of great fundamental interest, it might be argued that a "real" chemical reaction, in which the products actually differ from the reactants, would be a more significant test for both experiment and theory. One such reaction, which has been scrutinized in nearly as much detail,[34] is the exothermic reaction F + H$_2$ → HF + H. In this section, we shall consider several aspects of this reaction. In section 9.4.1, we consider the reaction dynamics, specifically product state distributions. We then review some general conclusions about dynamics of exothermic reactions gleaned from studying this and similar reactions. The relation between exothermic and endothermic processes is considered in section 9.4.3, and the exploitation of the F + H$_2$ reaction in chemical lasers is discussed briefly in section 9.4.4.

9.4.1   Dynamics of the F + H$_2$ → HF + H Reaction

The reaction of F with H$_2$ proceeds along the potential energy surface depicted in Figures 7-8 and 7-9. Not only is the surface exothermic, but it is also highly asymmetric, in that the local maximum along the reaction coordinate (the transition state, as defined in chapter 10) occurs while the reaction complex is very close to the F + H$_2$ configuration. The consequence of this is that most of the reaction exothermicity is channeled into vibration of the product HF molecule. This may be understood by examining the trajectory shown in Figure 9-8; energy is released while the H-F distance is decreasing, and this energy appears in vibration of the H-F bond.

The reaction exothermicity may be calculated from the difference between the H-F and H-H bond dissociation energies, viz., 5.869 eV − 4.478 eV = 1.391 eV = 130 kJ/mol. In addition, there is a small activation barrier along the reaction coordinate, amounting to an additional 5 kJ/mol. If this entire amount of energy were channeled into product vibration, it would be sufficient to produce HF in the $\nu = 3$ state, as indicated in Figure 9-9. In addition to product vibration, some of the energy can appear as rotation of the HF molecule.

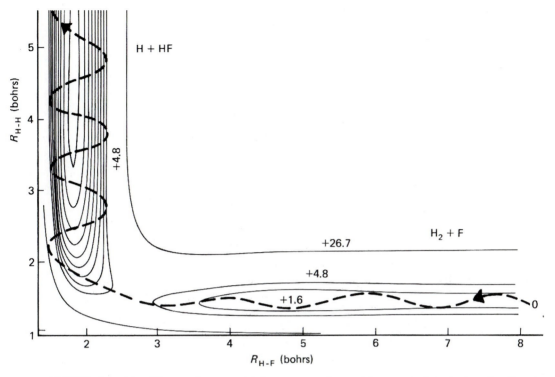

FIGURE 9-8   F + H₂ reactive potential energy surface with a typical trajectory leading to vibrational excitation of HF.

The HF product energy distributions have been measured[35] using the infrared chemiluminescence technique described earlier. The partition of energy among the various degrees of freedom of the reaction product can be conveniently expressed by the fractions $f_v, f_R,$ and f_T, where, for F + H₂, the average values are

$$\langle f_v \rangle = \frac{\text{product vibrational energy}}{Q} = 0.66$$

$$\langle f_R \rangle = \frac{\text{product rotational energy}}{Q} = 0.08$$

and

$$\langle f_T \rangle = \frac{\text{product translational energy}}{Q} = 0.26$$

Q is the total amount of energy available in the reaction and is equal to the sum of the reaction exothermicity, the activation energy, and any initial reactant excitation; clearly,

$$f_v + f_R + f_T = 1 \tag{9-9}$$

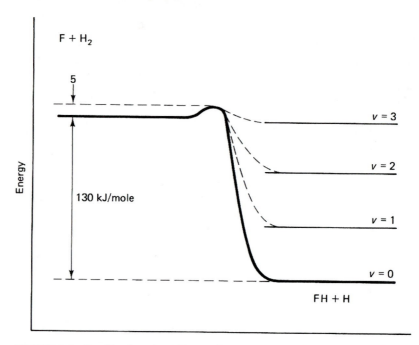

FIGURE 9-9   Profile along F + H_2 reaction coordinate, showing accessible HF vibrational energy levels.

A convenient way of displaying the preceding quantities is the *triangle plot,* shown in Figure 9-10 for the F + H_2 reaction. In such a plot, f_v and f_R are plotted on orthogonal axes; then, by virtue of equation (9-9), the f_T scale must lie along the bisector of the triangle. The contours on the plot indicate the region of f-space in which most of the product is found. Of course, it must be remembered that the accessible energy states are not continuously distributed on the f_v-f_R plane, because vibrational and rotational energies are quantized. The value of the contour line at any discrete v, J-state gives the probability of a HF molecule being produced in that state. Plainly, for the F + H_2 reaction, the HF molecules are produced with appreciable vibrational excitation, but relatively little rotation. By contrast, a statistical distribution of reaction products would favor rotational and translational excitation of the products because of the higher statistical weights associated with those degrees of freedom. F + H_2 is thus an excellent example of *specific energy release* in a chemical reaction, in which energy is channeled into degrees of freedom having relatively low statistical weights. The departure from the statistical, or "prior," distribution is considered quantitatively in chapter 13.

Other measurements of the rates and product distributions of the F + H_2 reaction have been made using flow-tube (see chapter 3) and molecular beam techniques. The results of the latter deserve comment. Lee and co-workers studied F + $D_2 \rightarrow DF(v)$ + D in a crossed molecular beam,[36] and the DF product translation energy was deduced

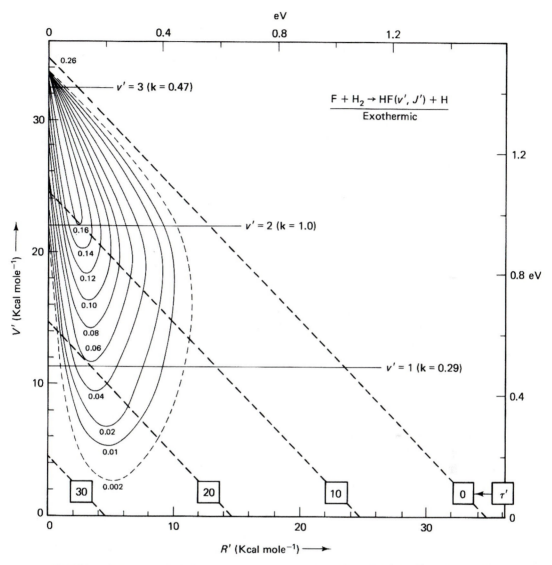

FIGURE 9-10   Triangle plot representation of $F + H_2 \rightarrow HF(v', J') + H$ product energy distribution. The $v' = 1, 2$, and 3 levels are drawn in at their quantized energies. [Adapted from Polanyi and Schreiber, *Faraday Disc. Chem. Soc. 62,* 267 (1977).]

from the scattering angle by means of a Newton diagram. In this reaction, vibrational levels up to $v = 4$ can be populated because the DF vibration frequency is 27% smaller than that of HF. Assuming that there is little rotational excitation in the product, the vibrational energy distribution can be obtained using equation (9-9). The results are as follows:

	Branching ratios, $k(v)/k_{total}$, for $F + D_2 \rightarrow DF(v) + D$					
$v =$	0	1	2	3	4	Reference
Infrared chemiluminescence	—	0.10	0.24	0.38	0.27	35
Molecular beam	0.01	0.02	0.04	0.21	0.73	36

The molecular beam results clearly differ from those obtained by infrared chemilumi-nescence. The latter, of course, cannot detect molecules in $v = 0$, but all measurements and calculations indicate that few molecules are produced in that state. The difference is that there is a peak at $k(3)$, while the molecular beam results show a monotonic increase with most molecules produced in $v = 4$. The reason for the difference is that this set of beam measurements was carried out using fast F and D_2 beams, so that the center-of-mass collision energy was about 11 kJ/mol. By contrast, the infrared chemi-luminescence measurements were carried out at or below room temperature, so that the average relative kinetic energy was only about 2.3 kJ/mol. This illustrates the importance of taking account of experimental conditions when attempting to compare measurements using different techniques.

9.4.2   Some General Conclusions Concerning Energy Disposition in Chemical Reactions

In order to attain a greater understanding of the dynamics of simple reactions of the kind A + BC, such as the $F + H_2$ reaction discussed in the preceding section, Polanyi has carried out an extensive series of calculations on model potential surfaces.[37–39] In particular, classical trajectory calculations are carried out on a simple LEPS potential surface. The height and location of the energy barrier in the reaction coordinate are systematically varied, and the effects on the results of the trajectory calculations are noted. The purpose of such calculations is not to simulate an actual chemical system, but rather to identify those features of the potential energy surface that affect energy consumption and disposal in bimolecular reactions.

The feature of the energy surface that appears to be most important in determin-ing energy distribution in reaction products is the location of the barrier along the reaction coordinate. An "early" or "attractive" barrier, which occurs in the entrance channel while the reactants are approaching each other, generally tends to favor vibra-tional excitation of the product. Conversely, a "late" barrier, which occurs in the exit channel where the products are separating, tends to lead to low product vibrational excitation. The basis for this is relatively easy to see. In the first case, the energy of the exothermic reaction is released while the A-B distance, which is the new chemical bond, is changing. In the second case, the energy is released only after the A-B bond has formed, and the B-C distance is changing; this motion corresponds to separation of products, so the energy is channeled into relative translation. In order to make such visualizations work reliably, however, we must use properly mass-weighted coordi-nates (see chapter 8). The affine transformation which accounts for the various mass combinations tends to compress the angle between the entrance and exit channels, coupling the vibrational and translational degrees of freedom. Thus, the product

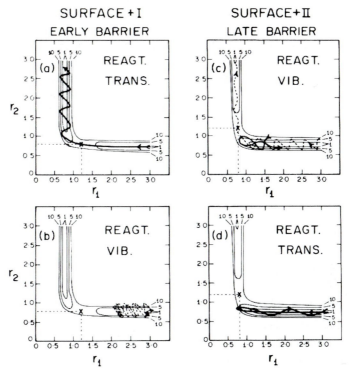

SURFACE + I
EARLY BARRIER

SURFACE + II
LATE BARRIER

FIGURE 9-11   Specimen trajectories on a model potential surface, illustrating selective energy consumption (barrier height = 29.3 kJ mol^{-1} in all cases). In (a) and (b), the potential energy barrier is located in the reactant (entrance) channel side of the reaction coordinate. In (a), molecules with 37 kJ mol^{-1} of translational energy easily surmount the barrier, and 79% of that energy is converted into product vibration. In (b), molecules with 60 kJ mol^{-1} of vibrational energy are unable to cross the barrier. In (c) and (d), the barrier is located in the product (exit) channel. Now molecules with 67 kJ mol^{-1} of translational energy (d) cannot cross the barrier, while molecules with 31 kJ mol^{-1} of vibrational energy (c) can do so, if the vibrational phase is favorable. Note that reactant vibrational energy is converted into product translational energy in this case. [From J. C. Polanyi and W. H. Wong, *J. Chem. Phys. 51,* 1439 (1969).]

energy distribution in a reaction in which a light atom attacks a heavy molecule (such as H + F$_2$) differs from that in which the masses are reversed (F + H$_2$).

The location of the energy barrier also plays an important role in selecting the distribution of reactant energy most likely to lead to reaction. Translational energy is most effective for passage across an "early" barrier in the entrance channel, whereas reactant vibrational energy far in excess of the barrier height may be ineffective for reaction. Conversely, a "late" barrier is best surmounted by vibrational rather than translational energy in the reactants. The dynamical basis for these tendencies can be seen with the aid of Figure 9-11. In the case of the early barrier, a vibrationally excited molecule (b) is just too busy rattling from side to side and does not have enough energy left to reach the top of the barrier; the molecule in (a), however, has all its energy in motion along the reaction coordinate and easily surmounts the barrier.* In the late barrier case, vibrationally excited molecules (c) with the correct phase can find their way to the top of the barrier, while molecules with rapid r_{AB} motion (d) simply slam into the repulsive inner wall of the potential surface and bounce back into the entrance channel.

Polanyi[39] has pointed out that these general principles were anticipated in 1955 by Hammond,[40] who postulated that for endothermic reactions the transition state** resembles the reaction products, while for exothermic reactions the transition state resembles the reactants. In the endothermic case this implies that the maximum of the potential barrier occurs in the product channel (late barrier), so that vibrational energy in the reactants will be more effective than translational energy in crossing the barrier. In exothermic reactions, where the barrier occurs in the reactant channel (early barrier), translational energy is sufficient to reach the product side. Bauer[41] has reviewed the effects of energy accumulation and disposal on chemical reaction, and how laser excitation of reactants may influence the rates.

9.4.3 Detailed Balance Revisited

We have encountered the principle of detailed balance several times in this text. In chapter 2, it was used to relate the forward and reverse rate constants of a reversible or opposing set of reactions [equation (2-17)]. In chapter 6, we saw how detailed balancing could be derived from the more general principle of microscopic reversibility, which embodies the invariance of classical physics under time reversal [equation (6-10)]. Here, we shall use an explicit form of the latter equation to find the relationship between detailed state-to-state rate constants for an exothermic reaction and those for the reverse endothermic reaction at the same total energy. This relationship was derived by Polanyi and co-workers[42] in 1969. It may be written as

$$k_R(v', J') = \frac{2\bar{J} + 1}{2J' + 1} \left(\frac{\mu}{\mu'} \right)^{2/3} \left(\frac{\bar{E}_T}{E_T'} \right)^{1/2} k_f(v', J') \qquad (9\text{-}10)$$

where relative translational energy $\bar{E}_T = E_{act} + 3/2\,RT$, $E_T' = \bar{E}_T - \Delta H°$, $\bar{J}$ is the most populated rotational level at the initial temperature T, and μ' and J' refer to the products of the exothermic, or forward, reaction. Essentially, equation (9-10) says that, if the exothermic reaction favors products with high internal (v' or J') and low translational excitation (as in the $F + H_2$ reaction), then the reverse or endothermic reaction at the same total energy will be favored by the same distribution of energy in the reactants (HF + H in this case). That is, the *specific energy disposal* of the exothermic process is mapped into *selective energy consumption* in the reverse, endothermic process.

---

*There is probably a moral here, if one looks hard enough.

**The concept of a transition state is thoroughly discussed in chapter 10; for the present purpose, we can simply regard it as the configuration of the reacting species at the top of the potential energy barrier in the reaction coordinate.

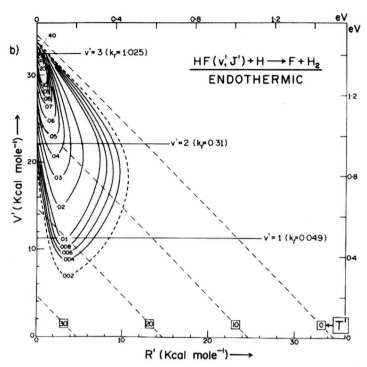

FIGURE 9-12 Energy consumption map for the endothermic reaction HF(v', J') + H → F + H₂ at a total energy of 140 kJ/mol, obtained by applying equation (9-11) to the data in Figure 9-7. [From Polanyi and Tardy, *J. Chem. Phys. 51,* 5717 (1969).]

Figure 9-12 shows the energy consumption map for the reaction HF + H → F + H₂ at a total energy of 140 kJ/mol. Clearly, reaction is most probable from the $v = 3$ state of HF, with little rotational or translational excitation. Such states would be very inefficiently populated by simple thermal heating of a mixture of HF and H atoms; there would instead tend to be equipartition of energy, with approximately equal amounts of excitation in translational, rotational, and vibrational degrees of freedom. Observations such as these have led to suggestions for "laser-selective chemistry,"[43] in which reactions could be promoted by excitation of specific molecular states with a narrow-band laser. Attempts to find such reactions have met with limited success, however, primarily because of the rapid inter- and intramolecular relaxation processes which redistribute energy within the molecule. This will be discussed further in chapter 11.

9.4.4   Chemical Lasers

The product HF vibrational population distribution described in section 9.4.1 is remarkable in that more molecules are produced in excited than in ground vibrational states. That is, $n(v = 2) > n(v = 1) > n(v = 0)$ for the nascent distribution. Such a distribution is said to be *inverted* with respect to the Boltzmann equilibrium distribution, in which the low-lying energy states are preferentially populated. A population inversion can be used to create a laser[44] if an allowed radiative transition connects the inverted levels, as is certainly the case for HF. Polanyi[45] first suggested the use of hydrogen-halogen reactions to create a laser pumped by an exothermic chemical reaction, which he termed a "chemical laser."[46] Following this suggestion, a number of such systems were developed, with the hydrogen-fluorine system being one of the most efficient. The output of this laser occurs on HF vibration-rotation bands in the 2.8–3.0-μm region of the infrared.

The behavior of the HF chemical laser depends on a complex interplay of chemical kinetics, gas flow, and optics. Pimentel and co-workers have made use of this interaction to learn about kinetic processes from the behavior of the HF laser under various operating conditions. The HF laser is also notable in that it can be scaled to very high output powers by simply increasing the flow of gases through the laser cavity. This observation has led to proposals that "directed energy weapons" based on chemical lasers, which could destroy missiles and satellites over vast distances in space, should be moved from the realm of science fiction[47] to that of military reality. While such schemes have been repeatedly dismissed on technical, political, and strategic grounds,[48,49] the concept has nevertheless shown a stubborn persistence. This is one example of how fundamental scientific questions can be translated into the realm of global policy; another, more cogent example will be described in Chapter 15.

9.5 WARNING: INFORMATION OVERLOAD!

In this chapter we have seen that the techniques of modern experimental and computational physical chemistry can produce an enormous amount of detail on energy-, angle-, and state-resolved chemical reaction rates and cross sections.[50] Indeed, it may now be possible to generate too much detail: the quantitative information produced by such measurements may be greater than the capacity of human minds to assimilate, or of traditional journals to publish. Attempts have thus been made to systematize the data by finding some general rules and principles that govern the behavior of chemical reactions. In chapter 13 we shall consider the information-theoretic approach to compacting such information by reducing large arrays of state-to-state kinetic quantities to one or a few parameters. Before doing so, however, we shall take up two more traditional approaches to chemical kinetics. These approaches treat reacting systems as being close to thermal equilibrium, rather than far removed from it. For bimolecular reactions, this treatment results in transition state theory, discussed in chapter 10; for unimolecular reactions, the corresponding treatment is the Rice-Ramsperger-Kassel-Marcus, or RRKM, theory, developed in chapter 11.

REFERENCES

[1] Y. T. Lee, *Science 236,* 793 (1987).

[2] J. L. Kinsey, *Ann. Rev. Phys. Chem. 28,* 349 (1977).

[3] J. B. Anderson, R. P. Andres, and J. B. Fenn, in *Molecular Beams,* edited by J. Ross, *Adv. Chem. Phys. 10,* 275–317 (1966).

[4] J. I. Steinfeld, *Molecules and Radiation* (Cambridge, MA: M.I.T. Press, 1985) pp. 307–309.

[5] I. Amdur and J. E. Jordan in *Molecular Beams,* edited by J. Ross, *Adv. Chem. Phys. 10,* 29–73 (1966).

[6] D. R. Herschbach, *Disc. Faraday Soc. 33,* 149 (1962).

[7] R. D. Levine and R. B. Bernstein, *Molecular Reaction Dynamics* (Oxford: University Press, 1974), pp. 60–63.

[8] J. I. Steinfeld and J. L. Kinsey, *Progr. Reaction Kinetics 5,* 1 (1970).

[9] M. Ackerman, E. F. Greene, A. L. Moursund, and J. Ross, *Ninth Symp. (Int'l.) on Combustion,* (New York: Academic Press, 1963), p. 669.

[10] L. J. Kovalenko and S. R. Leone, *J. Chem. Education 65,* 681 (1988).

[11] K. G. Anlauf, P. J. Kuntz, D. H. Maylotte, P. D. Pacey, and J. C. Polanyi, *Disc. Faraday Soc. 44,* 183 (1967); J. C. Polanyi et al., *J. Chem. Phys. 57,* 1547, 1561, 1574, 4988 (1972).

[12] J. L. Kinsey, *J. Chem. Phys. 66, 2560* (1977).

[13] E. F. Cromwell, A. Stolow, M. J. J. Vrakking, and Y. T. Lee, *J. Chem. Phys. 97,* 4094 (1992).

[14] A. G. Suits, P. de Pujo, O. Sublemontier, J. -P. Visticot, J. Berlande, J. Cuvellier, T. Gustavsson, J. -M. Mestdagh, P. Meynadier, and Y. T. Lee, *J. Chem. Phys. 97,* 4094 (1992).

[15] Steinfeld, *op. cit.,* pp. 429–438.

[16] I. C. Winkler, R. A. Stachnik, J. I. Steinfeld, and S. M. Miller, *J. Chem. Phys. 85,* 890 (1986).

[17] D. W. Chandler and P. L. Houston, *J. Chem. Phys. 87,* 1445 (1987).

[18] A. J. R. Heck and D. W. Chandler, *Ann. Rev. Phys. Chem. 46,* 335 (1995).

[19] A. H. Zewail, *Science 242,* 1645 (1988); L. R. Khundkar and A. H. Zewail, *Ann. Rev. Phys. Chem. 41,* 15 (1990).

[20] R. Gotting, H. R. Mayne, and J. P. Toennies, *J. Chem. Phys. 85,* 6396 (1986).

[21] D. G. Truhlar and R. E. Wyatt, *Ann. Rev. Phys. Chem. 27,* 1 (1976).

[22] R. Wolfgang, *Progr. Reaction Kinetics 3,* 97 (1965).

[23] U. Gerlach-Meyer, K. Kleinermanns, E. Linnebach, and J. Wolfrum, *J. Chem. Phys. 86,* 3047 (1987).

[24] W. H. Miller, *Ann. Rev. Phys. Chem. 41,* 245 (1990).

[25] R. Pool, *Science 247,* 413 (1990).

[26] D. Gerrity and J. J. Valentini, *J. Chem. Phys. 81,* 1298 (1984); D. Gerrity and J. J. Valentini, *ibid. 82,* 1323 (1985).

[27] D. A. V. Kliner and R. N. Zare, *J. Chem. Phys. 92,* 2107 (1990).

[28] D. E. Adelman, N. E. Shafer, D. A. V. Kliner, and R. N. Zare, *J. Chem. Phys. 97,* 7323 (1992).

[29] D. A. V. Kliner, K. -D. Rinnen, and R. N. Zare, *Chem. Phys. Letts. 166,* 107 (1990).

[30] N. C. Blais, M. Zhao, D. G. Truhlar, D. W. Schwenke, and D. J. Kouri, *Chem. Phys. Letts. 166,* 11 (1990).

[31] T. N. Kitsopoulos, M. A. Buntine, D. P. Baldwin, R. N. Zare, and D.W. Chandler, *Science 260,* 1605 (1993).

[32] L. Schnieder, K. Sukamp-Rahn, J. Borkowski, E. Wrede, K. H. Welge, F. J. Aoiz, L. Banares, M. J. D'Mello, V. J. Herrero, V. Saez Rabanos, and R. E. Wyatt, *Science 269,* 207 (1995).

[33] M. Alagia, N. Balucani, L. Contechini, P. Casavecchia, E. H. van Kleef, G. G. Volpi, F. J. Aoiz, L. Banares, D. W. Schwenke, T. C. Allison, S. L. Mielke, and D. G. Truhlar, *Science 273,* 1519 (1996).

[34] J. C. Polanyi and J. L. Schreiber, *Faraday Disc. Chem. Soc. 62,* 267 (1977).

[35] J. C. Polanyi and K. Woodall, *J. Chem. Phys. 57,* 1574 (1972).

[36] T. P. Schafer, P. E. Siska, J. M. Parson, F. P. Tully, Y. C. Wang, and Y. T. Lee, *J. Chem. Phys. 53,* 3385 (1970).

[37] J. C. Polanyi, *Accts. Chem. Research 5,* 161 (1972).

[38] J. C. Polanyi and J. L. Schreiber, "The Dynamics of Bimolecular Reactions" in *Kinetics of Gas Reactions: Physical Chemistry—An Advanced Treatise,* Vol. 6A, edited by H. Eyring, W. Jost, and D. Henderson (New York: Academic Press, 1974) p. 383.

[39] J. C. Polanyi, *Science 236,* 680 (1987).

[40] G. S. Hammond, *J. Am. Chem. Soc. 77,* 334 (1955).

[41] S. H. Bauer, *Chem. Reviews 78,* 147 (1978).

[42] K. G. Anlauf, D. H. Maylotte, J. C. Polanyi, and R. B. Bernstein, *J. Chem. Phys. 51,* 5716 (1969); J. C. Polanyi and D. C. Tardy, *ibid.,* p. 5717.

[43] A. M. Ronn, *Scientific American* v. 290, No. 5, 114 (May, 1979).

[44] Steinfeld, *op. cit.,* pp. 308-320.

[45] J. C. Polanyi, *Appl. Opt. Suppl. 2,* 109 (1965).

[46] A good summary of the practical aspects of chemical lasers may be found in R. W. F. Gross and J. Bott, eds., *Handbook of Chemical Lasers* (New York: Wiley-Interscience, 1976).

[47] "Many think that in some way they [the Martians] are able to generate an intense heat in a chamber of practically absolute non-conductivity. This intense heat they project by means of a polished parabolic mirror of unknown composition, in a parallel beam against any object they choose . . . However it is done, it is certain that a beam of heat is the essence of the matter . . . Whatever is combustible flashes into flame at its touch, lead runs like water, it softens iron, cracks and melts glass, and when it falls on water incontinently that explodes into steam." [H. G. Wells, "The War of the Worlds," *Pearson's Magazine,* London (1897)]

[48] American Physical Society Study Group (N. Bloembergen and C. K. Patel, co-chairmen), "Report to the American Physical Society of the Study Group on Science and Technology of Directed Energy Weapons," *Rev. Mod. Phys.* 59(3), Part II (July 1987).

[49] P. M. Boffey, W. J. Broad, L. H. Gelb, C. Mohr, and H. B. Noble, *Claiming the Heavens* (New York: Time Books, 1988).

[50] Many additional examples are given in a special "Reaction Dynamics" issue of *Science* magazine (20 March 1998). Specific references include: R. F. Service, *Science 279,* 1847 (1998); R. N. Zare, *ibid.,* p. 1875; D. C. Clary, *ibid.,* p. 1879; M. L. Chabinyc, S. L. Craig, C. K. Regan, and J. I. Brauman, *ibid.,* p. 1862; H. Tributsch and L. Pohlmann, *ibid.,* p. 1891.

PROBLEMS

*9.1 The total cross section for elastic scattering can be obtained from a single-beam experiment, in which a beam of molecules of species A at high velocity is passed through a rarefied gas of species B and the attenuation of the beam is measured. Although the molecules of the rarefied gas have random thermal velocities, if the velocity of the beam is sufficiently high, then the *relative* velocity, v_{AB}, is well approximated by that of the beam A. Let the densities of the beams be n_A and n_B, respectively, and let the beam of A have area α. Consider the attenuation in the beam intensity over a small increment of length dx.

    B. Brunetti, G. Liuti, E. Luzzatti, and F. Vecchiocattivi [*J. Chem. Phys. 74,* 6734 (1981)] carried out single-beam studies for a beam of O_2 passed through an O_2 bath gas. Their data show that at 0.01 Pa pressure and a temperature of 85 K, with a path length of 10 cm, for a beam velocity of 1.0×10^3 m s^{-1}, the beam has an attenuation of 92% (i.e., $I/I_0 = 0.08$). For a velocity of 1.5×10^3 m s^{-1}, they found an attenuation of 86%.

(a) Evaluate σ for these two velocities.

(b) Do you think that the values obtained in (a) correspond to an ordinary molecular "size"?

(c) You will notice that σ decreases with increasing velocity in these experiments. Explain this in terms of collision dynamics.

(d) Brunetti et al. used their data on $\sigma(E)$ to deduce the O_2/O_2 intermolecular potential function. List two physical properties and one chemical property of oxygen that could be predicted from this potential function.

9.2 When a beam of Rb atoms from a source at 600 K intersects a beam of CH_3I at 90° from a source at 300 K (with no velocity selection), reaction occurs according to the formula

$$Rb + CH_3I \rightarrow CH_3 + RbI$$

and the final kinetic energy E' of relative motion of the products peaks at about 8.3 kJ/mole. Assuming that the angle between the final relative velocity v'_r and the velocity v_c of the center of mass is 90°,

(a) Construct a vector diagram (Newton diagram) and indicate the in-plane laboratory angle where the peak intensity of RbI would be expected to appear.

(b) Account for any major differences between the figure obtained in (a) and that for $K + CH_3I$ (Figures 9-2 and 9-3).

(c) Assume that the in-plane peak intensity predicted in (a) has been experimentally observed. What additional experimental observations would be expected if a substantial fraction of the products had values of E' equal to about 21 kJ/mole?

9.3 A direct measurement of the threshold energy for the $D + H_2$ reaction may be obtained by photolyzing DI in the presence of H_2 and measuring the appearance of DH [Kupperman and White, *J. Chem. Phys. 44*, 4353 (1966)].

(a) If DI is photolyzed by ultraviolet light having a wavelength of 3,000 Å, what is the velocity of the D atom in the laboratory coordinate system? (The bond dissociation energy of DI is 3.098 eV.)

(b) What would be the relative kinetic energy in the collision center-of-mass system if the D atom produced in (a) struck an H_2 molecule? Assume the H_2 molecule is at rest in the laboratory frame.

(c) If the *longest* photolysis wavelength at which HD is observed is 3,340 Å, what is the energy threshold for the $D + H_2$ reaction? How does this value compare with the experimental activation energy and the theoretical potential barrier height?

*9.4 An experimental colleague in the Department of Quintessential Chemistry makes the following proposal to you for a research project. Correct his errors and misconceptions, suggest an explanation for the observed effect, and suggest how your explanation could be checked by a proper calculation. "We pass protons from a radioactive source into ethylene at low pressure and monitor the formation of C_2H_5. From the observed rate coefficient and from the hard-sphere formula $k = \pi d^2 (8k_B t/m\pi)^{1/2}$, we deduce the value of the total cross section $\sigma = \pi d^2$. This assumes that the critical energy for the reaction is zero. The resulting cross section is for the reaction $H^+ + C_2H_4 \rightarrow C_2H_5^+$; it is found to be considerably greater than the hard-sphere cross section for typical systems. We suspect that this is due to tunneling. We would like to carry out classical trajectory calculations to confirm this. For these trajectory calculations, we can use a potential function obtained from infrared spectral studies of $C_2H_5^+$ in polar solvents."

9.5 Reactions involving N_2, O_2, N, and O are important processes in the upper atmosphere. Answer each of the following questions concerning these reactions, using the potential-energy diagrams at the end of this problem where appropriate. Show briefly how you arrived at each numerical result requested in parts (a) through (i); the expected precision is that to which you can read off the diagrams. Note that 1 eV = 8,065.73 cm^{-1} = 23,061 cal/mole = 11,605 degrees K $\cdot$ $k_B = 1.6 \times 10^{-12}$ ergs/molecule.

(a) Estimate the energy released in the reaction

$$O_2(X^3\Sigma_g^-) + N(^4S°) \rightarrow NO(X^2\Pi) + O(^3P) \tag{1}$$

(b) Estimate the energy released in the reaction

$$O_2(X^3\Sigma_g^-) + N(^2D°) \rightarrow NO(X^2\Pi) + O(^3P) \tag{2}$$

(c) What is the highest vibrational level, v''_{max}, of $NO(X^2\Pi)$ which could be populated in reaction (1)? In reaction (2)?

(d) Is the reaction

$$N_2(X^1\Sigma_g^-) + O(^3P) \rightarrow NO(X^2\Pi) + N(^4S°) \tag{3}$$

likely, on energetic grounds?

(e) Is the reaction

$$N_2^*(A^3\Sigma_u^+) + O(^3P) \rightarrow NO(X^2\Pi) + N(^4S°) \tag{4}$$

likely, on energetic grounds?

(f) Is NO stable with respect to disproportionation to N_2 and O_2?

(g) Estimate the longest wavelength of light that can be used to excite fluorescence from an electronically excited state of NO. Assume that the NO is originally in $v'' = 0$ of the X state.

(h) Estimate the longest wavelength of light that would be effective for ionization of NO in a one-photon transition.

(i) Estimate the longest wavelength of light that would be effective for resonant multi-photon ionization of NO.

9.6 V-V relaxation in a gas of diatomic molecules involves the reaction

$$AB(v_1) + AB(v_2) \rightarrow AB(v_3) + AB(v_4), k(v_1v_2|v_3v_4)$$

in which the number of vibrational quanta is not changed, i.e., $v_1 + v_2 = v_3 + v_4$. The rates are related by detailed balancing as follows:

$$\frac{k(v_1v_2|v_3v_4)}{k(v_1v_2|v_3v_4)} = \exp[-(E_{v3} + E_{v4} - E_{v1} - E_{v2})/k_BT]$$

The quasi-steady distribution $n_v^{(0)}$ can be found by the Chapman-Enskog expansion procedure, which states that the exchange collision terms, when equated to zero, determine the quasi-steady distribution.

(a) Write the master equation for the evolution with time of the vibrational populations. [See *Appendix: The Master Equation* which follows.]

(b) For harmonic oscillators, $E_v^{H.O.} = v\hbar\omega$. Show that $n_v^{(0)} = n_0^{(0)}e^{-E_v/k_BT}$ is a solution to the master equation.

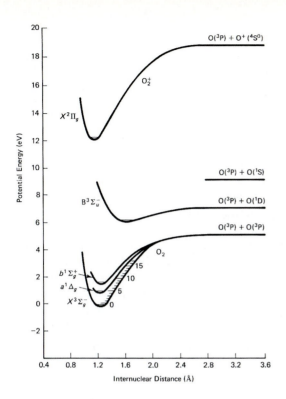

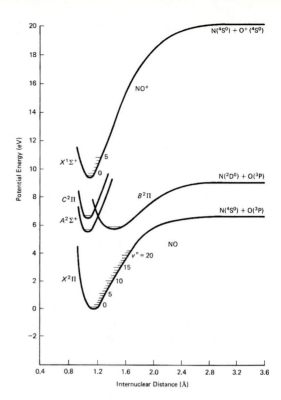

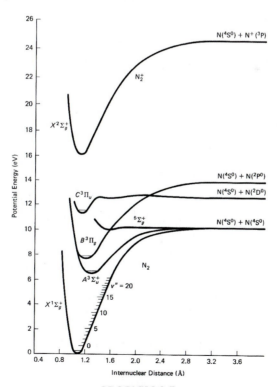

PROBLEM 9.5

(c) For anharmonic oscillators, $E_v = v\hbar\omega(1-x) - v^2\hbar\omega x < E_v^{\text{H.O.}}(x \ll 1)$. Show that $n_v^{(0)} = n_0^{(0)}e^{-v\gamma}\,e^{-E_v/K_BT}$ is the solution in this case.

(d) Find γ in terms of ω, x, and T.

APPENDIX: THE MASTER EQUATION

In this chapter, we have considered in some detail the vast increase in the array of kinetic information available when microscopic quantum states of the reactants and products are considered. While many exceedingly clever (and often correspondingly difficult) experiments have been devised to garner this information, in many cases it is not feasible to carry out the measurements on isolated quantum states or nascent product state distributions. An example is the partially relaxed distributions observed in infrared chemiluminescence experiments at moderate pressures. For such situations, we need methods for bridging the gap between complete state resolution and the Boltzmann distributions characteristic of thermal equilibrium ensembles. The master equation is such a method, and we review it briefly here. For greater detail and a discussion of the mathematical properties of the master equation, the reader is referred to references A1 and A2.

The master equation is actually a set of complex integro-differential equations that describe the evolution with time of a physical system. This set of equations can be written in a simple form for a system consisting of a set of N discrete levels; the canonical example is that of harmonic-oscillator vibrational levels. The state-to-state kinetic equations for such a system are

$$\text{AB}(v) + \text{M} \rightleftharpoons \text{AB}(v') + \text{M}$$

with a rate equal to

$$R(v \to v') = k(v \to v';T)\rho_M$$

where ρ_M is the total density of collision partners inducing a change in oscillator state with rate coefficient $k(v \to v'; T)$.

The master equation for the level populations $n_v(t)$ $[v = 0, 1, \ldots, N]$ can be written as

$$\frac{dn_v(t)}{dt} = \rho_M \sum_{v' \neq v} [k(v' \to v; T)n_{v'}(t) - k(v \to v'; T)n_v(t)] \qquad (9\text{A-}1)$$

The first set of terms in the summand of equation (9A-1) represents the population of level $[v]$ from all other levels; the second term represents the depopulation of $[v]$ to all other levels. Let us define

$$-\sum_{v' \neq v} k(v \to v') \equiv k(v \to v) = k(v) \qquad (9\text{A-}2)$$

---

*These problems have been provided by Prof. R. G. Gilbert, School of Chemistry, University of Sydney, Australia.

in which we no longer write the temperature T explicitly. Then, with this definition, we have

$$\sum_{\text{all } v} k(v' \rightarrow v) = 0$$

which ensures that $\sum_{\text{all } v} n_v(t) = $ constant and that the occupation probabilities are normalized and their sum is conserved, i.e., that

$$\sum_{\text{all } v} P_v(t) = \sum_v \frac{n_v(t)}{\sum_v n_v(t)} = \sum_v \frac{n_v(t)}{N} = 1$$

at all times. We can then write equation (9A-1) formally as

$$\frac{dP_v(t)}{dt} = \rho_{\text{M}} \sum_{v' \neq 0}^{N} k(v' \rightarrow v) \, P_{v'}(t) \tag{9A-3}$$

The principle of detailed balancing is an important help in reducing the master equation to soluble form. This principle states that, as the system approaches equilibrium (time $\rightarrow \infty$), the equilibrium occupation probabilities no longer change in time; that is,

$$P_v(\infty) = P_v^{\text{eq}} = p(v|T)$$

and

$$\frac{dP_v(\infty)}{dt} = 0 \tag{9A-4}$$

Thus, from equation (9A-1),

$$\sum_{v' \neq v} k(v' \rightarrow v; T) \, P_{v'}(\infty) = \sum_{v' \neq v} k(v \rightarrow v'; T) \, P_v(\infty)$$

Since this must be true for every v, the terms in the sum must be pairwise equal, and

$$\frac{k(v \rightarrow v')}{k(v' \rightarrow v)} = \frac{p(v'|T)}{p(v|T)} = \frac{g_{v'}}{g_v} e^{-(E_{v'} - E_v)/k_B T} \tag{9A-5}$$

We use the Boltzmann distribution to calculate the equilibrium occupation probabilities $p(v|T)$.

A closed-form analytic solution can be found for the master equation (9A-3) in only a few cases, the most important being the relaxation of harmonic oscillators with $E_v = (v + 1/2)\hbar\omega_0$ and $g_v = 1$. For the $k(v \rightarrow v')$, let us assume a Landau-Teller collision model,[A3-A5] i.e., $\Delta v = v' - v = \pm 1$ (so that $\Delta E = \pm\hbar\omega_0$) and $k(v \rightarrow v - 1; T) = vk(1 \rightarrow 0; T)$. Then the detailed balancing equation (9A-5) can be written as

$$\frac{k(v - 1 \rightarrow v; T)}{k(v \rightarrow v - 1; T)} = \frac{p(v|T)}{p(v - 1|T)} = e^{-\hbar\omega_0/k_B T} \equiv \alpha$$

and the equilibrium population distribution is

$$p(v|T) = \alpha^v(1 - \alpha)$$

The master equation can be then written as

$$\frac{dP_v(t)}{dt} = \frac{1}{\tau(1 - \alpha)} = \{\alpha v P_{v-1}(t) - [v + \alpha(a + 1)]P_v(t) + (v + 1)P_{v+1}(t)\} \qquad (9A\text{-}6)$$

where the vibrational relaxation time

$$\tau = [(1 - \alpha)\rho_M k(1 \rightarrow 0; T)]^{-1}$$
$$= [\rho_M \{k_{10}(T) - k_{01}(T)\}]^{-1}$$

For $\hbar\omega_0 T \gg 1$, $\alpha \rightarrow 0$ and $\tau \approx [\rho_M k_{10}(T)]^{-1}$.

The detailed solution of the set of N coupled linear first-order differential equations (9A-6) depends on the initial conditions $P_v(t = 0) = P_v(0)$. If we take this to be the so-called canonical distribution

$$P_v(0) = \frac{\exp[-E_v/k_B T(0)]}{Q[T(0)]}$$

where

$$Q[T(0)] = \sum_v e^{-E_v/k_B T(0)}$$

then the solution is

$$P_v(t) = \frac{\exp[-E_v/k_B T(t)]}{Q[T(t)]}$$

That is, if the set of harmonic oscillators can be initially described by a temperature (in other words, if the set is in a Boltzmann distribution), then it can be described by a time-varying temperature $T(t)$ throughout its relaxation. In order to find the evolution with time of T, we calculate the average energy per oscillator, viz.,

$$\langle E_v(t)\rangle = \sum_v E_v P_v(t)$$

$$= \hbar\omega_0 \sum_v v P_v(t)$$

$$= \hbar\omega_0 \langle v(t)\rangle$$

so that

$$\langle E_v(\infty)\rangle = \sum_v E_v P_v(v|T)$$

$$= \hbar\omega_0(1 - \alpha) \sum_v v\alpha$$

$$= \hbar\omega_0 \frac{\alpha}{1 - \alpha} = \hbar\omega_0 \langle v(\infty) \rangle^{\dagger}$$

Thus,

$$\frac{d}{dt} \langle E_v(t) \rangle = \frac{d}{dt} \sum_v E_v P_v(t)$$

$$= \sum_v E_v \frac{d}{dt} P_v(t)$$

$$= \sum_v \left\{ \frac{\hbar\omega_0 v}{\tau(1 - \alpha)} [\alpha v P_{v-1} - \{v + \alpha(v + 1)\} P_v(t) + (v + 1) P_{v+1}(t)] \right\}$$

or

$$\frac{d}{dt} \langle E_v(t) \rangle = \frac{1}{\tau} [\langle E_v(\infty) \rangle - \langle E_v(t) \rangle] \tag{9A-7}$$

Integrating equation (9A-7) gives

$$\langle E_v(t) \rangle = \langle E_v(\infty) \rangle + [\langle E_v(0) \rangle - \langle E_v(\infty) \rangle] e^{-t/\tau} \tag{9A-8}$$

That is, for harmonic oscillators, the energy content relaxes exponentially for any initial set of $P_v(0)$. If the initial distribution is canonical (i.e., if it is a Boltzmann distribution), then there is a temperature proportional to $\langle E_v \rangle$ that relaxes exponentially as well. That is,

$$T(t) = T(\infty) + [T(0) - T(\infty)] e^{-t/\tau} \tag{9A-9}$$

Closed-form solutions can also be found for the case of relaxation of anharmonic oscillators [see Treanor et al.[A6] and problem (9-6)].

For systems that cannot be simply described as involving the relaxation of harmonic (or anharmonic) oscillators, the situation is much more complicated. In general, the equation for such a system must be solved by numerical integration of the coupled differential equations, starting from some initial distribution $\{n_v(0)\}$. The numerical solution can be greatly facilitated by an eigenvalue technique, which we now proceed to develop.

Let us write equation (9A-1) as

$$\frac{dn_v(t)}{dt} = \sum_{v'} A(v, v') n_{v'}(t) \tag{9A-10}$$

where

$$A(v, v') = \rho_M[k(v' \to v; T) - \delta_{vv'} k(v \to v'; T)] \tag{9A-11}$$

---

†Note that $\displaystyle\sum_v v\alpha^v = \alpha \frac{\partial}{\partial\alpha} \left(\sum_v \alpha^v \right) = \alpha \frac{\partial}{\partial\alpha} \left(\frac{1}{1 - \alpha} \right) = -\frac{\alpha}{(1 - \alpha)^2} \frac{\partial}{\partial\alpha} (1 - \alpha) = \frac{\alpha}{(1 - \alpha)^2}$

when the sum over v goes from 0 to ∞.

Similarly, equation (9A-6) can be written in matrix form as

$$\dot{\mathbf{n}}(t) = \mathbf{A} \cdot \mathbf{n}(t) \qquad (9A\text{-}12)$$

where $\mathbf{n}(t)$ is the N-dimensional time-dependent population vector and $\mathbf{A}$ is the $N \times N$ array defined by equation (9A-11). Equation (9A-12) has the formal solution

$$\mathbf{n}(t) = \exp[\mathbf{A}t] \cdot \mathbf{n}(0) \qquad (9A\text{-}13)$$

in which the matrix exponential may be taken to be its Taylor-series representation, i.e.,

$$\exp[\mathbf{A}t] = 1 + \mathbf{A} + \frac{1}{2}(\mathbf{A} \cdot \mathbf{A})t^2 + \cdots$$

Equation 9A-13 can be written out by components as

$$n_v(t) = n_v(\infty) + \sum_{j=1}^{N} C_j(v)\exp(\lambda_j t) \qquad (9A\text{-}14)$$

in which the λ_j are the N eigenvalues of the matrix A, and the $C_j(v)$ are the corresponding eigenvectors, which depend on the set of initial conditions $\{n_v(0)\}$. Note that $\{n_v(\infty)\}$ is the unique eigenvector corresponding to $\lambda_0 = 0$; to see this, merely write

$$\frac{dn_v(\infty)}{dt} = \sum_{v'} A(v', v')n_{v'}(\infty) = \lambda_o n_v(\infty) = 0$$

Since for $j > 0$ all the λ_j are negative, all the terms in equation (9A-14) decay to zero and $n_v(\infty) = n(v|T) = Np(v|T)$. The series in this equation usually converges rapidly, so that the lowest one or two eigenvalues dominate the kinetics. Also, matrix diagonalization routines are readily available for most computers, so this procedure is a practical alternative to direct numerical solution of the coupled differential equations.

REFERENCES

A1 I. Oppenheim, K. E. Shuler, and G. H. Weiss, *Adv. Mol. Relaxation Processes, 1,* 13 (1967–68).

A2 I. Oppenheim, K. E. Shuler, and G. H. Weiss, *Stochastic Processes in Chemical Physics: The Master Equation* (Cambridge, MA: M.I.T. Press, 1977).

A3 G. G. Hammes, *Principles of Chemical Kinetics* (New York: Academic Press, 1978), chapter 6.

A4 J. T. Yardley, *Introduction to Molecular Energy Transfer* (New York: Academic Press, 1980).

A5 L. Landau and E. Teller, *Physik. Z. Sowjetunion 10,* 34 (1936).

A6 C. E. Treanor, J. W. Rich, and R. G. Rehm, *J. Chem. Phys. 48,* 1798 (1968).

Statistical Approach to Reaction Dynamics Transition State Theory

In the preceding chapters, we have seen how reaction rates can be calculated from molecular dynamics, given a potential energy surface for the reaction. These calculations, besides being complex and frequently difficult, are often inexact because (1) classical dynamics is only an approximation to the quantum behavior of atomic systems, and (2) sampling of initial conditions is limited. Exact quantum calculations are possible, but they are limited to reactive systems and energies for which the total number of available quantum states is small. Also, the results of such classical and quantum calculations are frequently far more detailed than the available experimental data, which may consist only of a thermal rate constant. Information on state-to-state reactive cross sections, scattering angles, and product energy partitioning may be generated in the calculations, but those properties are measured in only the most painstaking experiments.

The level of detail in a theoretical model should be consistent with that of the experimental observations to which the model will be applied. One approach to matching the information content of theory and experiment will be discussed in chapter 13; in this chapter, an approach is described which is based on the concept of a *transition state* intermediate between reactants and products. The reactants are assumed to be in thermal equilibrium with the transition state. The resulting theory, called *transition-state theory*, takes into account only the statistical properties of reactive systems, not the microscopic details of molecular collisions. Transition-state theory rate constants are determined using the principles of statistical mechanics.

10.1 MOTION ON THE POTENTIAL SURFACE

For a chemical reaction involving N atoms there are $3N - 6$ vibrational degrees of freedom at the saddle point. As discussed in section 7.5, one of these degrees of freedom corresponds to infinitesimal motion along the reaction path. The remaining $3N - 7$ degrees of freedom define vibrational motion orthogonal to the reaction path.

To study properties of the saddle point in more detail, consider a potential energy contour diagram for a collinear chemical reaction such as the one shown in

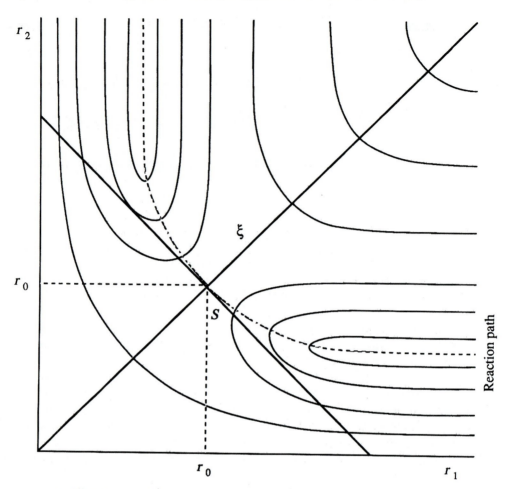

FIGURE 10-1   Saddle-point region of a symmetrical potential energy surface.

Figure 7-5(a). For simplicity, consider three identical atoms (such as $H + H - H$) in a linear configuration (Figure 10-1). This collinear system has two orthogonal internal degrees of freedom, for which the coordinates are

$$\xi = r_2 + r_1$$

$$s = r_2 - r_1$$

The saddle point has a symmetric configuration at which $r_2 = r_1 = r_0$, and $\xi = 2r$ and $s = 0$. As shown in Figure 10-1, at the saddle point s defines the reaction path and ξ is perpendicular to the reaction path. Curves of the potential energy versus s, $U(s)$, and versus ξ, $U(\xi)$, are shown in Figures 10-2(a) and 10-2(b), respectively, for these s and ξ. For the reaction path potential $U(s)$, there is a maximum at the saddle point. The potential for the coordinate orthogonal to the reaction path, $U(\xi)$, has a minimum at the saddle point.

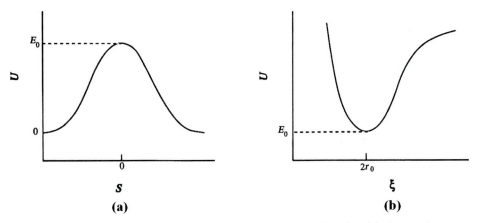

FIGURE 10-2   Sections of the potential energy surface showing (a) the maximum along the reaction path and (b) the minimum perpendicular to the reaction path.

The first and second derivatives of the potential at the saddle point are

$$\frac{\partial U}{\partial s} = 0, \qquad \frac{\partial^2 U}{\partial s^2} < 0$$

and

$$\frac{\partial U}{\partial \xi} = 0, \qquad \frac{\partial^2 U}{\partial \xi^2} > 0$$

since $\partial^2 U/\partial s^2 < 0$, the vibrational frequency for motion along the reaction path will be imaginary (see chapter 7, section 7.5). The vibrational frequency for the ξ coordinate will be positive. In the vicinity of the saddle point, a Taylor's series expansion of the potential is

$$U(\xi, s) \approx \frac{1}{2} C_\xi (\xi - 2r_0)^2 - \frac{1}{2} C_s s^2 + E_0 \qquad (10\text{-}1)$$

The transition state theory (TST) is based on the postulate that the rate of transformation of systems from reactants to products is given solely by passage in the forward direction from coordinates $r_1 > r_2 (s < 0)$ to $r_1 < r_2$ $(s > 0)$. We now proceed to develop that theory.

10.2   BASIC POSTULATES AND DERIVATION OF TRANSITION STATE THEORY

Transition state theory, introduced by Eyring[1] and by Evans and Polanyi[2] in 1935, provided the first theoretical attempt to determine absolute reaction rates. In this theory, a transition state separating reactants and products is used to formulate an expression for the thermal rate constant. The relationship between transition state theory and dynamical theories was first discussed by Wigner[3] who emphasized that the theory was a model essentially based on classical mechanics. A number of assumptions are made in deriving the TST rate expression. The two most basic are the separation of electronic and nuclear motions, equivalent to the Born-Oppenheimer approximation in quantum mechanics, and the assumption that the reactant molecules are distributed among their states in accordance with the Maxwell-Boltzmann distribution. However, the following additional assumptions, which are unique to the theory, are also required:

1. Molecular systems that have crossed the transition state in the direction of products cannot turn around and reform reactants.
2. In the transition state, motion along the reaction coordinate may be separated from the other motions and treated classically as a translation.
3. Even in the absence of an equilibrium between reactant and product molecules, the transition states that are becoming products are distributed among their states according to the Maxwell-Boltzmann laws.

In a rigorous formulation of transition state theory, the third assumption is not necessary, since it follows from the first.

It is customary to derive transition state theory by postulating a quasi-equilibrium between transition state species and reactants. This approach focuses on the third assumption listed. Another approach follows the dynamical formulation of Wigner and uses the first assumption, which is considered the "fundamental assumption" of transition state theory. Both derivations are given here, and they use the conventional transition state theory model in which the transition state is located at the saddle point on the potential energy surface. Later in the chapter, variational transition state theory, a more accurate method for choosing the transition state, is discussed.

In many presentations the term *activated complex* is used synonymously with "transition state." It is avoided here, however, because first, the word "complex" implies an entity which has a chemically significant lifetime, which the transition state does not have, and second, the collision complex which may be formed when two molecules collide (as in a molecular beam experiment) is often called an activated complex.

The third assumption, the "quasi-equilibrium hypothesis," is based on a simple physical interpretation. Suppose we have an elementary reaction

$$A + B \rightarrow X^{\ddagger} \rightarrow C \qquad (10\text{-}2)$$

where $X^{\ddagger}$ is the transition state, A and B are reactants, and C is the product. Suppose also that it is possible to define a small region at the top of the potential energy barrier such that all systems entering from the left pass through to products without turning back, and similarly all products entering from the right must pass through to reactants. This small region is often referred to as the *dividing surface* and is orthogonal to the reaction pathway. We consider two parallel dividing surfaces separated by a very small distance δ (see Figure 10-3). Systems within this small distance are by definition transition states; to the right are products, to the left reactants. Consequently, for a system in which reactants are in equilibrium with the products, there are two types of transition states: those moving from reactant to products, and those moving in the opposite direction from products to reactant. We designate the concentration of these by $N_f^{\ddagger}$ and $N_b^{\ddagger}$, where f and b stand for forward and backward, respectively. At equilibrium the rate of

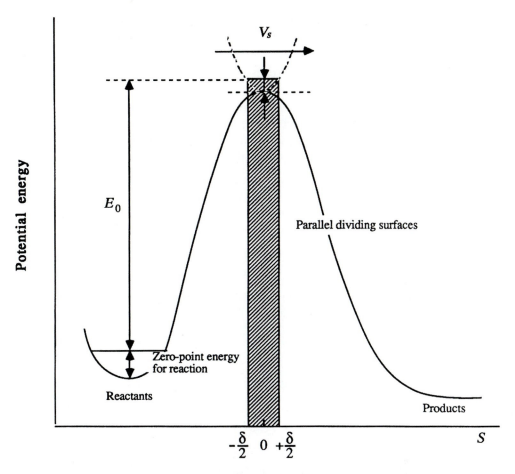

Reaction coordinate

FIGURE 10-3   Reaction coordinate profile.

the forward reaction must be the same as the backward reaction; consequently, the concentrations of the transition states moving toward reactants and toward products must be equal. Hence,

$$N_f^\ddagger + N_b^\ddagger = N^\ddagger = K^\ddagger[A][B] \tag{10-3}$$

Now, if all the products are suddenly removed from equilibrium, that is, if

$$N_b^\ddagger = 0 \tag{10-4}$$

then, since the forward rate of reaction must be the same regardless of the back reaction rate, it follows that the concentration of the forward-moving transition states is the same as it would be for full equilibrium; hence,

$$N_f^\ddagger = K^\ddagger[A][B]/2 \tag{10-5}$$

This is the "quasi-equilibrium hypothesis" of transition state theory.

To calculate the net rate of reaction, the rate (average velocity) at which transition states pass over the barrier to products is needed; i.e., we derive

$$\frac{dN}{dt}(\text{reactants} \rightarrow \text{products}) = \frac{\delta N^\ddagger}{\delta t} \tag{10-6}$$

The number of transition states per unit volume having velocity between v and $v + dv$ in *one* direction is represented by $\delta N^\ddagger$. The average time δt for the $\delta N^\ddagger$ transition states to cross the barrier is equal to the thickness δ of the dividing surface divided by the average velocity $\bar{v}_s$ at which the transition states traverse the dividing surface; that is,

$$\delta t = \frac{\delta}{\bar{v}_s} \tag{10-7}$$

Hence,

$$\frac{dN}{dt} = \delta N^\ddagger \frac{\bar{v}_s}{\delta} \tag{10-8}$$

From equation (10-5), the number of transition states crossing the dividing surface in the direction of products is one-half the total number $N^\ddagger$ of transition states, i.e.,

$$\delta N^\ddagger = \frac{N^\ddagger}{2} \tag{10-9}$$

Therefore, the number of transition states crossing the barrier per unit volume in unit time is

$$\frac{dN}{dt} = \frac{N^\ddagger}{2} \frac{\bar{v}_s}{\delta} \tag{10-10}$$

On the assumption that there is an equilibrium distribution of velocities, the average velocity of the transition state moving in one direction, e.g., the forward direction, is

$$\bar{v}_s = \frac{\int_0^\infty v_s e^{(-\mu_s v_s^2/2k_BT)}\, dv_s}{\int_0^\infty e^{(-\mu_s v_s^2/2k_BT)}\, dv_s} = \left(\frac{2K_BT}{\pi\mu_s}\right)^{1/2} \tag{10-11}$$

where μ_s is the reduced mass for motion through the dividing surface. Inserting this value for $\bar{v}_s$ into equation (10-10) gives

$$\frac{dN}{dt} = \frac{N^\ddagger}{2}\left(\frac{2k_BT}{\pi\mu_s}\right)^{1/2}\frac{1}{\delta} \tag{10-12}$$

Since the transition state is in equilibrium with the reactants, it is possible to obtain $N^\ddagger$ from equilibrium statistical mechanics, namely, as

$$K^\ddagger = \frac{N^\ddagger}{[A][B]} \tag{10-13}$$

This statement of equilibrium does *not* imply that the transition state is long-lived. This equilibrium constant can be expressed in terms of partition functions as

$$K^{\ddagger} = \frac{N^{\ddagger}}{[A][B]} = \frac{Q^{\ddagger}_{tot}}{Q_A Q_B} e^{-E_0/k_B T} \tag{10-14}$$

where $Q^{\ddagger}_{tot}$ is the equilibrium partition function per unit volume for the transition state, Q_A and Q_B are the same functions for the reactants, and E_0 is the energy of the lowest level of the transition state relative to that of the lowest level of the reactants (see Figure 10-3). Combining equations (10-12) and (10-14), we have, for the reaction rate,

$$\frac{dN}{dt} = \left(\frac{2k_B T}{\pi \mu_s} \right)^{1/2} \frac{1}{2\delta} \frac{Q^{\ddagger}_{tot}}{Q_A Q_B} e^{-E_0/k_B T}[A][B] \tag{10-15}$$

In the approaches of Eyring[1] and Evans and Polanyi,[2] the partition function for the reaction coordinate in the transition state is considered to be a translational function. Since this motion is assumed to be separable (as a consequence of the second assumption), the translational partition function is therefore separable as

$$Q^{\ddagger}_{tot} = Q_s Q^{\ddagger} \tag{10-16}$$

where Q_s is the partition function for the reaction coordinate motion and $Q^{\ddagger}$ is the partition function for all other $3N - 1$ degrees of freedom in the transition state. The translational partition function for motion in *one* dimension in a system of length δ is

$$Q_s = (2\pi \mu_s k_B T)^{1/2} \delta/h \tag{10-17}$$

This Q_s is not divided by 2, since it is the partition function for all the transition states [see equation (10-14)]. Substituting this expression into equation (10-16) and then the result into equation (10-15) gives

$$\frac{dN}{dt} = \frac{k_B T}{h} \frac{Q^{\ddagger}}{Q_A Q_B} e^{-E_0/k_B T}[A][B] \tag{10-18}$$

From chapter 1, we know that the experimental rate equation for a bimolecular reaction is

$$\frac{dN}{dt} = k[A][B] \tag{10-19}$$

Comparison of equation (10-18) and equation (10-19) gives the absolute rate coefficient for A + B → products,

$$k = k_{abs} = \frac{k_B T}{h} \frac{Q^{\ddagger}}{Q_A Q_B} e^{-E_0/k_B T} \tag{10-20}$$

Note that the artificial constructs δ and μ_s, as well as the factor 1/2, have cancelled out of the absolute rate theory expression. These may be avoided by treating the motion along the reaction coordinate as a low-frequency vibration; this leads to a somewhat simpler derivation of equation (10-20), as shown in Problem 10.2. The ratio $k_B T/h$, having the units of frequency and the magnitude 6.25×10^{12} sec^{-1} (for $T = 300$ K), is frequently termed the *frequency factor*. It is of the magnitude of the encounter frequency between the molecules in liquids, or of slow molecular vibrations (300 K is equivalent to 200 cm^{-1}), but the period $(k_B T/h)^{-1}$ is a little longer than gas-kinetic collision durations.

In order to be able to evaluate equation (10-20), we must be able to calculate the transition state partition function $Q^{\ddagger}$, using the statistical mechanical techniques discussed subsequently in section 10.6. To do that, we need to know the structural parameters of the transition state (specifically, the moment of inertia $I^{\ddagger}$), as well as its $3N - 7$ vibrational frequencies $\{\nu^{\ddagger}\}$. In practice, such parameters can only be estimated approximately. With detailed information about potential surfaces now becoming available from *ab initio* theory, however, we may expect to see more precise calculations of rate coefficients from transition state theory.

The energy E_0 is the difference in zero point energy between the transition state and the reactants. It is useful to point out the relationship between E_0 and the experimental activation energy E_a. As shown by Tolman (see Fowler and Guggenheim),[4]

$$E_a = \langle E_r(T) \rangle - \langle E(T) \rangle \tag{10-21}$$

where $\langle E_r(T) \rangle$ is the average energy of molecules undergoing reaction and $\langle E(T) \rangle$ is the average energy of all reactant molecules. In transition state theory, $\langle E_r(T) \rangle$ is given by the average energy of the transition state plus E_0.

10.3 DYNAMICAL DERIVATION OF TRANSITION STATE THEORY

In the preceding derivation of transition state theory, we have postulated a dividing surface perpendicular to the reaction coordinate at the saddle point on the potential surface. The reactants exist on one side of the dividing surface, and products exist on the other. An important step in the derivation of the rate is the calculation of the number of transition states that lie at the critical surface, as well as the calculation of their velocity as they move across the saddle point. The dynamical derivation of transition state theory[5,6] follows the same line of argument. However, in this derivation the equilibrium flux through the transition state is evaluated. By using the "fundamental assumption" of transition state theory, this flux is equated to the reactive flux. Since the dynamical derivation uses classical mechanical language, the classical origin of transition state theory becomes evident. Appendix 2 gives a general discussion of classical statistical mechanics.

With the assumption that the reactants are in a state of thermodynamic equilibrium, one can determine the number of activated species that lie in a volume element of phase space on the dividing surface. For a reactant system defined by n atoms, there are six dimensions for each atom in the system (three coordinates and three conjugate momenta), and the volume is specified as $d\tau = dq_1^{\ddagger} \cdots dq_{3n}^{\ddagger} dp_1^{\ddagger} \cdots dp_{3n}^{\ddagger}$, where $q_1^{\ddagger}$ is the reaction coordinate and $p_1^{\ddagger}$ is its conjugate momentum. As shown by equations (A2-1) through (A2-4) in Appendix 2, the number of states in the volume element $d\tau$ is $d\tau/h^{3n}$. The fraction of reactant species lying in this volume element on the dividing surface is given by the following standard expression from classical statistical mechanics [equation (A2-26) in Appendix 2]:

$$\frac{dN^*}{N} = \frac{e^{-H/k_BT}dp_1^\ddagger \cdots dp_{3n}^\ddagger dq_1^\ddagger \cdots dq_{3n}^\ddagger/h^{3n}}{\int^R e^{-H/k_BT}dp_1 \cdots dp_{3n}dq_1 \cdots dq_{3n}/h^{3n}} \tag{10-22}$$

In this expression, H is the classical Hamiltonian of the system which defines the total energy expressed in coordinates and conjugate momenta. The asterisk on dN^* denotes that only one volume element on the dividing surface is under consideration. The differential for all volume elements on the dividing surface will be denoted by dN.

The integral in the denominator of equation (10-22) is over all values of coordinates and momenta which are associated with the reactant (R) molecules. As shown by equation (A2-25) in Appendix 2, this integral is the partition function of the reactants. For reactants A and B, it can be written as $Q_A VQ_B V$, where the Q's are partition functions per unit volume and V is the volume of the container. The quantity N is the total number of reactant pairs and equals $N_A N_B$, the product of the numbers of individual A and B molecules. Thus, for an A + B bimolecular reaction, equation (10-22) becomes

$$dN^* = \frac{[A][B]e^{-H/k_BT}dp_1^\ddagger \cdots dp_{3n}^\ddagger dq_1^\ddagger \cdots dq_{3n}^\ddagger/h^{3n}}{Q_A Q_B} \tag{10-23}$$

where [A] and [B] have replaced the two N/V terms.

If the fundamental assumption is made that molecular systems do not recross the transition state, the reactant-to-product flux across the transition state becomes the reactive flux. The rate of passage of a representative point through the phase-space volume on the dividing surface associated with dN^* is then

$$\frac{dN^*}{dt} = [A][B]\frac{dq_1^\ddagger}{dt}\frac{e^{-H/k_BT}dq_2^\ddagger \cdots dq_{3n}^\ddagger dp_1^\ddagger \cdots dp_{3n}^\ddagger/h^{3n}}{Q_A Q_B} \tag{10-24}$$

To obtain the total rate of reaction, integration must be carried out over all the phase space on the dividing surface. Also, this integration must be over all *positive* values of $p_1^\ddagger$, since it is only this motion which takes the system from the reactant to the product side of the dividing surface. To carry out the integration, the total Hamiltonian at the dividing surface is separated into two parts and is given by

$$H = H^\ddagger + H' \tag{10-25}$$

where $H^\ddagger$ defines the energy of the reactant system along the reaction coordinate $q_1^\ddagger$ and H' gives the energy along all coordinates orthogonal to the reaction coordinate.

The Hamiltonian for translational motion along the reaction coordinate is

$$H^\ddagger = \frac{(p_1^\ddagger)^2}{2\mu_1} \tag{10-26}$$

where μ_1 is the reduced mass and

$$p_1^\ddagger = \mu_1\frac{dq_1^\ddagger}{dt} \tag{10-27}$$

Inserting equations (10-25) through (10-27) into equation (10-24) and integrating over all the phase-space volume on the dividing surface gives

$$\frac{dN}{dt} = \frac{[A][B]\frac{1}{\mu_1 h} \int_0^\infty p_1^\ddagger e^{-(p_1^{\ddagger 2}/2\mu_1 k_B T)}\, dp_1^\ddagger \cdots \int e^{-H'/k_B T}\, dq_2^\ddagger \cdots dq_{3n}^\ddagger dp_2^\ddagger \cdots dp_{3n}^\ddagger/h^{3n-1}}{Q_A Q_B}$$

(10-28)

which is the total reactant-to-product flux through the dividing surface. The result of integrating over $p_1^\ddagger$ gives

$$\frac{dN}{dt}\frac{1}{[A][B]} = \frac{k_B T}{h}\frac{\int E^{-H'/k_B T} dq_2^\ddagger \cdots dq_{3n}^\ddagger dp_2^\ddagger \cdots dp_{3n}^\ddagger/h^{3n-1}}{Q_A Q_B}$$

(10-29)

If the potential energy at the transition state is E_0, one can write

$$H' = E_0 + H''$$

(10-30)

and make this substitution in the integral in equation (10-29). By comparing the resulting integral with equation (A2-25) in Appendix 2, the integral is clearly recognized as the partition function for the transition state's $3n - 1$ degrees of freedom multiplied by $\exp(-E_0/k_B T)$. Thus, one gets

$$\frac{dN}{dt}\frac{1}{[A][B]} = \frac{k_B T}{h}\frac{Q^\ddagger V}{Q_A Q_B}\exp(-E_0/k_B T)$$

(10-31)

where $Q^\ddagger$ is the transition state's partition function per unit volume. Bringing V to the left side, we obtain the rate of disappearance of reactant pairs expressed in concentration units. The resulting expression for the rate constant is

$$\frac{Rate}{[A][B]} = \frac{1}{V}\frac{dN}{dt}\frac{1}{[A][B]} = \frac{k_B T}{h}\frac{Q^\ddagger}{Q_A Q_B}\exp(-E_0/k_B T)$$

(10-32)

which is the same as equation (10-20) found before, but derived with fewer assumptions.

In the dynamical derivation of transition state theory, the quasi-equilibrium assumption (the third assumption) is not required. In fact, this assumption should be viewed as a corollary of the fundamental assumption that trajectories do not recross the transition state. To illustrate this, consider the situation where there is thermal equilibrium between the reactants and the products. Obviously, there will be a thermal equilibrium at the transition state for trajectories crossing toward products as well as for those crossing toward reactants. Now consider the situation at the transition state when all trajectories emanating from the reactants are removed. If trajectories do not recross the transition state, none of the trajectories crossing the transition state toward reactants will have been removed. As a result, thermal equilibrium will be maintained between the reactants and trajectories proceeding through the transition state toward products in the absence of thermal equilibrium between the reactants and the products. Thus, the quasi-equilibrium assumption follows from the fundamental assumption of TST that trajectories do not recross the dividing surface.

10.4   QUANTUM MECHANICAL EFFECTS IN TRANSITION STATE THEORY

The foregoing dynamical derivation illustrates that transition state theory is a classical mechanical model. The theory is exact if classical trajectories passing through the transition state do not turn back and form reactant molecules again. Is there an equivalent model for a quantum mechanical transition state theory? Regretfully, the answer is no. Advances have been made toward a truly quantum transition state theory,[6-7] but much additional work needs to be done.

A basic difficulty in the generalization of classical transition state theory to quantum mechanics is that the reaction criterion cannot be formulated as the condition that a trajectory pass through a critical surface. This is connected with the fact that, in a quantum mechanical observation, the coordinates and momenta of a system cannot be assigned simultaneously. Consider the Heisenberg uncertainty relations $\Delta p \Delta q \gtrsim \hbar$ and $\Delta E \Delta t \gtrsim \hbar$ for the reaction coordinate. If Δp is replaced by h/λ, where λ is the de Broglie wavelength, we have the relation $\Delta q > \lambda/2\pi$. Thus, the uncertainty in the value of the reaction coordinate at the transition state must be at least the size of the wavelength associated with the motion along the reaction coordinate. In other words, quantum mechanically, the transition state is not localized. The uncertainty relation $\Delta E \Delta t \gtrsim \hbar$ may be analyzed in a similar way. For the thermal averaging in transition state theory to have a meaning, it is necessary that the translational energy E in the reaction coordinate be much less than $k_B T$. As a result, the lifetime Δt of the transition state must be larger than $\hbar/k_B T$. Therefore, in the quantum case, the transition state cannot be considered a definite configuration of nuclei during an infinitesimal interval of time. Thus, for the calculation of the reactive flux in the quantum case, it is necessary to consider explicitly the dynamics of the trajectories' motion in the region Δq or to follow the evolution with time of the system for time Δt.

In classical transition state theory, the potential energy is constant and the reaction coordinate motion is separated from the remaining internal motions at the localized position along the reaction coordinate which defines the transition state. Quantum mechanical delocalization of the transition state along the reaction coordinate can lead to two problems. First, if the potential is not flat in the region Δq, so that the system is not freely moving, it is incorrect to treat the reaction coordinate as being a classical translational motion. Rather, the potential will usually have a concave-down shape, and quantum mechanical tunneling (described in more detail shortly) will occur through the potential.

The second problem is more critical. If there is curvature along the reaction coordinate in the region Δq, the reaction coordinate is not separable from the remaining internal degrees of freedom.[6,8-10] Thus, the rate constant expression cannot be factored into a frequency $k_B T/h$ for the reaction coordinate and a partition function for the remaining degrees of freedom, as is done in equation (10-16). Also, since the curvature couples the reaction coordinate with the remaining modes, tunneling cannot be treated as a one-dimensional reaction coordinate barrier-penetration problem. Instead, there will be a multitude of tunneling paths which involve all the coordinates.[11-13]

Clearly, then, there are many difficulties in deriving a quantum mechanical transition state theory expression which does not include the separable reaction coordinate approximation. To obtain a working transition state theory expression, the separable approximation is assumed and corrections are made to the classical rate constant of equation (10-20) to account for quantum effects. It is well known that quantum partition functions are required for most vibrational motions. By contrast, for translational and most rotational motions, classical mechanical partition functions are accurate. The exception for rotation is at low temperatures and for molecules like H_2 and D_2 with small moments of inertia. With the vibrations treated quantum mechanically, the barrier for reaction is not the classical barrier, but the difference in zero-point energies between the transition state and the reactants. Finally, a correction for reaction coordinate motion must be included to account for tunneling effects.

Despite these reservations, transition state theory has been found to do a remarkably good job of accounting for the magnitudes and temperature dependences of a wide range of reactions. The most significant quantum deviations appear when light atoms (i.e., hydrogen) are moving through the transition state region, in which case a straightforward correction for quantum-mechanical tunneling may be applied. Classically, the barrier crossing probability is zero for all $E < E_0$; as a result of tunneling, there is a finite nonzero probability for $E < E_0$ which connects smoothly to the classical result for E much larger than E_0.

The probabilities can be calculated for any barrier from the Schrödinger equation

$$\frac{\partial^2 \Psi}{\partial s^2} + \frac{8\pi^2 m}{h^2}[E - V(s)]\Psi = 0 \tag{10-33}$$

We have assumed a one-dimensional barrier, which is allowable since, with the separable approximation, there is reaction only along one coordinate, that is, the reaction pathway. A solution of equation (10-33) is obtained for a suitable choice of potential $V(s)$ to describe the barrier. The resulting expression for the probability of tunneling through the barrier is given by[14]

$$G(E) = \exp\left(-\frac{4\pi\sqrt{m}}{h}\int_{s_1}^{s_2}(V(s) - E)^{1/2}\, ds\right) \tag{10-34}$$

where $G(E)$, called the *permeability function*, depends on the exact shape of the barrier along the reaction coordinate, and s_1 and s_2 are turning points, i.e., the coordinates of the reaction path for which $V(s) = E$. The tunneling correction, often denoted by $\Gamma(T)$, is obtained by integrating over a Maxwell-Boltzmann distribution,[15,16] i.e.,

$$\Gamma(T) = e^{E_0/k_BT}\int_0^\infty \frac{1}{k_BT} e^{-E/k_BT}G(E)\, dE \tag{10-35}$$

Even for simple barrier shapes, a closed solution for the permeability function is difficult to obtain; and, in any case, the integration over all energies must be carried out numerically. However, one barrier shape for which an exact expression for $G(E)$ is known and is physically meaningful is the symmetrical Eckart barrier.[17] The potential for this barrier has the form

$$V(s) = E_0 \, \text{sech}^2(\pi s/\ell) \tag{10-36}$$

The permeability function for this barrier is

$$G(E) = \frac{\cosh(4\pi\alpha) - 1}{\cosh(4\pi\alpha) + \cosh(2\pi\delta)} \tag{10-37}$$

where

$$\alpha = \frac{1}{h}(2mE)^{1/2}$$

and

$$\delta = \frac{1}{h}\left(8mE_0 - \frac{h^2}{4\ell^2}\right)^{1/2}$$

and where ℓ is the width of the barrier. For small barrier heights and widths ($1.2 - 1.8$Å), the tunneling correction approaches an asymptotic form[18] given by

$$Q_{\text{tunnel}} = 1 - \frac{1}{24}\left(\frac{h\nu_s}{k_B T}\right)^2\left(1 + \frac{k_B T}{E_0}\right) \tag{10-38}$$

where ν_s is the imaginary frequency of the transition state at the top of the barrier (see Section 10.1). For reactions involving $H + H_2$ and $D + D_2$, some tunneling corrections are given in Table 10-1. As is evident from the trends in the tunneling correction factors, quantum mechanical tunneling is important, and the lighter isotope reacts more rapidly.

The experimental evidence for tunneling comes from Arrhenius plots of $\ln k$ versus $1/T$ (see Figure 10-4). If the Arrhenius plots show significant nonlinearity, then tunneling may be the cause. The curvature in the Arrhenius plots of $\ln k$ vs. $1/T$ tends to be concave upward when tunneling is important, because the calculated reaction rate deviation is positive and increases for large values of $1/T$.

For some reactions, additional tunneling paths besides the one along the reaction path must be included. Several approximate ways to include these have been considered.[11–13]

TABLE 10-1   Tunnel Effect Correction Factors for H and D Atom Reactions.[18]

$T(°K)$	Q_{tunnel}	
	$H + H_2$	$D + D_2$
200	5.180	3.090
250	3.706	2.353
300	2.900	1.950
400	2.093	1.546
600	1.507	1.253
800	1.297	1.148
1000	1.198	1.094
1200	1.142	1.071

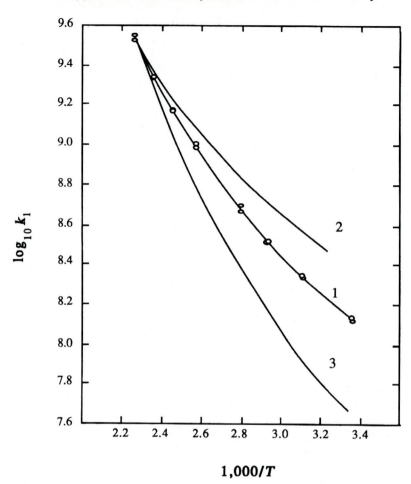

FIGURE 10-4 Arrhenius plots of ln k vs. $1/T$. Curve 1: experimental results for the reaction H + H exhibiting non-Arrhenius behavior. Curve 2: theoretical curve for an Eckart barrier with $E_0 = 15.05$, $\nu_s = 2,573i$. Curve 3: theoretical curve for an Eckart barrier with $E_0 = 14.35$, $\nu_s = 1,964i$. [Adapted from W. R. Schulz and D. J. LeRoy, *J. Chem. Phys.* **42**, 3869 (1965)].

10.5 THERMODYNAMIC FORMULATION OF TRANSITION STATE THEORY

The transition state derived rate constant can be reformulated in thermodynamic terms, since it is sometimes more useful to work with the rate constant in this form than with partition functions. The rate constant expression given by equation (10-20), viz.,

$$k = \frac{k_B T}{h} \frac{Q^\ddagger}{Q_A Q_B} e^{-E_0/k_B T}$$

can be written as

$$k = \frac{k_B T}{h} K_c^\ddagger \qquad (10\text{-}39)$$

where

$$K_c^{\ddagger} = \frac{Q^{\ddagger}}{Q_A Q_B} e^{-E_0/K_B T} \tag{10-40}$$

is the equilibrium constant for formation of the transition state. The subscript c indicates that molar concentration units are being used. If the equilibrium constant is expressed in terms of the molar Gibbs standard free energy using the van't Hoff relation

$$\Delta G^{\ddagger o} = -RT \ln K_c^{\ddagger} \tag{10-41}$$

then equation (10-20) can be written as

$$k = \frac{k_B T}{h} e^{-\Delta G^{\ddagger o}/RT} \tag{10-42}$$

where the superscript o refers to the standard states of the reactants and products. $\Delta G^{\ddagger o}$ may be expressed in terms of standard enthalpy and entropy changes by

$$\Delta G^{\ddagger o} = \Delta H^{\ddagger o} - T\Delta S^{\ddagger o} \tag{10-43}$$

allowing equation (10-42) to be divided into two terms:

$$k = \frac{k_B T}{h} e^{\Delta S^{\ddagger o}/R} e^{-\Delta H^{\ddagger o}/RT} \tag{10-44}$$

For rate constants expressed in units of mol/dm^3 (concentration) and seconds (time), the appropriate standard state for $K^{\ddagger o}$, $\Delta G^{\ddagger o}$, $\Delta H^{\ddagger o}$, and $\Delta S^{\ddagger o}$ is 1 mol/dm^3. [1 m^3 = 1,000 dm^3.]

Equation (10-44) is similar to the Arrhenius equation

$$k = A e^{-Ea/RT} \tag{10-45}$$

and the thermodynamic parameters can be related to the Arrhenius parameters. The Arrhenius activation energy is defined by

$$\frac{d(\ln k)}{dT} \equiv \frac{E_a}{RT^2} \tag{10-46}$$

Taking the logarithm of equation (10-39) and differentiating with respect to T gives

$$\frac{d(\ln k)}{dT} = \frac{1}{T} + \frac{d(\ln K_c^{\ddagger})}{dT} \tag{10-47}$$

Since $K_c^{\ddagger}$ is a concentration equilibrium constant, its variation with temperature is given by the Gibbs-Helmholtz equation

$$\frac{d(\ln K_c^{\ddagger})}{dT} = \frac{\Delta E^{\ddagger o}}{RT^2} \tag{10-48}$$

Inserting this equation into equation (10-47) and comparing the result with equation (10-46) gives

$$E_a = RT + \Delta E^{\ddagger o} \tag{10-49}$$

The relationship between the standard thermodynamic energy and enthalpy for a constant-pressure process is

$$\Delta H^{\ddagger o} = \Delta E^{\ddagger o} + P(\Delta V^{\ddagger o}) \tag{10-50}$$

since $H = E + PV$. The quantity $\Delta V^{\ddagger o}$ is known as the *standard volume of activation*. With equation (10-50), equation (10-49) becomes

$$E_a = \Delta H^{\ddagger o} + RT - P(\Delta V^{\ddagger o}) \tag{10-51}$$

For a unimolecular reaction, there is no change in the number of molecules in going from the reactants to the transition state, so $\Delta V^{\ddagger o}$ is zero. Also, to a very good approximation, $\Delta V^{\ddagger o}$ is zero for reactions in solution or in a condensed phase. Therefore, in these cases,

$$E_a = \Delta H^{\ddagger o} + RT \tag{10-52}$$

Insertion of this equation into equation (10-44) leads to

$$k = \frac{k_B T}{h} e^{(1 + \Delta S^{\ddagger o}/R)} e^{-E_a/RT} \tag{10-53}$$

Hence,

$$A = \frac{k_B T}{h} e^{(1 + \Delta S^{\ddagger o}/R)} \tag{10-54}$$

For gas phase reactions other than unimolecular, the relationships between the Arrhenius parameters and thermodynamic terms are different from the relationships just given. If the ideal gas relation is assumed, i.e.,

$$P\Delta V^{\ddagger o} = \Delta n^{\ddagger} RT \tag{10-55}$$

then, from equation (10-51), one obtains

$$E_a = \Delta H^{\ddagger o} + RT - \Delta n^{\ddagger} RT = \Delta H^{\ddagger o} + (1 - \Delta n^{\ddagger})RT \tag{10-56}$$

and the Arrhenius pre-exponential factor is

$$A = \exp(-(\Delta n^{\ddagger} - 1)) \frac{k_B T}{h} e^{\Delta S^{\ddagger o}/R} \tag{10-57}$$

10.6 APPLICATIONS OF TRANSITION STATE THEORY

10.6.1 Evaluating Partition Functions by Statistical Mechanics

Determining rate constants from the canonical level of transition state theory is an invaluable aid in both elucidating mechanisms and evaluating rate constants. Calculating the rate constant for a reaction requires that we know how to evaluate the partition functions that arise in the transition state theory derivation. Real molecules may have translational, rotational, vibrational, and electronic energies. Therefore, the total partition function associated with the internal motion for each molecule is given by the product of partition functions associated with independent rotational, vibrational, electronic, and translational motions:

$$Q = Q_{rot}Q_{vib}Q_{elec}Q_{trans}$$

To calculate the individual partition functions, one needs to know moments of inertia, vibrational frequencies, and electronic states. Such information for reactants A and B can be obtained from spectroscopic data, but, for the transition state, we must make use of semiempirical or *ab initio* calculations.

10.6.1.1 Electronic partition function.

The electronic partition function is given by

$$Q_{elec} = \sum_i g_i e^{-E_i/k_BT} \tag{10-58}$$

where g_i is the degeneracy and E_i is the energy above the lowest state of the system. In most reactions, few electronic energy levels other than the ground state need to be considered. For reactions involving doublet and triplet systems, the degeneracy factor is the corresponding spin degeneracy which would contribute a factor of two or more to the calculated rate constant.

10.6.1.2 Translational partition function.

For the translational motion for a particle of mass m moving in a one-dimensional box of length l, the energy derived from the Schrödinger equation is

$$E_n = \frac{n^2h^2}{8ml^2} \tag{10-59}$$

where n is the quantum number for nondegenerate energy levels and h is Planck's constant. The partition function for one dimension is

$$Q_{trans} = \sum_{n=1}^{\infty} \exp\left(\frac{-n^2h^2}{8ml^2k_BT}\right) \tag{10-60}$$

When the energy levels are closely spaced, that is, when their separation $E_{n+1} - E_n = (h^2/8ml^2)(2n + 1)$ is $\ll k_BT$, the sum can be replaced by an integral:

$$Q_{trans} = \int_0^{\infty} \exp\left(\frac{-n^2h^2}{8ml^2k_BT}\right)dn$$

$$Q_{trans} = \frac{1}{2}\left(\frac{8m\pi l^2 k_BT}{h^2}\right)^{1/2} = \frac{(2\pi mk_BT)^{1/2}l}{h} \tag{10-61}$$

The total translation energy in three dimensions is given by

$$E_{trans} = E_{trans,x} + E_{trans,y} + E_{trans,z} \tag{10-62}$$

so that the total translational partition function is a product of one-dimensional partition functions for each component, i.e.,

$$Q_{trans} = Q_{trans,x}Q_{trans,y}Q_{trans,z} \tag{10-63}$$

or

$$Q_{trans} = \frac{(2\pi m k_B T)^{3/2} l^3}{h^3} \tag{10-64}$$

Since l^3 is the volume V,

$$\frac{Q_{trans}}{V} = (2\pi m k_B T/h^2)^{3/2} \tag{10-65}$$

A derivation of the classical Q_{trans} is given in Appendix 2.

10.6.1.3 Vibrational partition function. To find partition functions for the vibrational contribution, we evaluate

$$Q_{vib} = \sum_i g_i \exp(-E_i/k_B T) \tag{10-66}$$

For a simple harmonic oscillator, the vibrational energy has nondegenerate levels

$$E_v = \left(v + \frac{1}{2}\right) hc\bar{\nu}_s \tag{10-67}$$

where $\bar{\nu}_s$ (in cm^{-1}) is the harmonic frequency of a single vibrational level and v is the quantum number. Each nondegenerate level has multiplicity 1, so to evaluate the partition function for the vibrational contribution, we must sum over all levels. Taking the zero-point level as the zero of energy, this leads to

$$Q_{vib} = \frac{1}{1 - \exp(-hc\bar{\nu}/k_B T)} \tag{10-68}$$

A polyatomic molecule has $s = 3N - 6$ vibrational modes if it is nonlinear and $3N - 5$ modes if it is linear; therefore, the vibrational partition function for a polyatomic molecule is

$$Q_{vib} = \prod_{i=1}^{s} \frac{1}{1 - \exp(-hc\bar{\nu}_i/k_B T)} \tag{10-69}$$

10.6.1.4 Rotational partition function. For a linear rigid rotor, the energy levels are given by

$$E_J = \frac{J(J + 1)\hbar^2}{2I} \tag{10-70}$$

where $\hbar^2/2I$ is the rotational constant and $J = 0, 1, 2, \ldots$ is a rotational quantum number. The multiplicity of each level is $2J + 1$, so

$$Q_{rot} = \sum_0^\infty (2J + 1) \exp[-J(J + 1)\hbar^2/2I k_B T] \tag{10-71}$$

If the level intervals are small compared to $k_B T$, the sum can replaced by an integral:

$$Q_{\text{rot}} = \int_0^\infty (2J + 1)\exp[-J(J + 1)\hbar^2/2Ik_B T]\,dJ$$

so

$$Q_{\text{rot}} = \frac{8\pi^2 I k_B T}{h^2} \qquad (10\text{-}72)$$

Equation (10-72) is also a reasonably good approximation for real nonrigid molecules. For a polyatomic molecule with moments of inertia I_a, I_b, and I_c about its principal axes, the rotational partition function is

$$Q_{\text{rot}} = \pi^{1/2}\left(\frac{8\pi^2 I_a k_B T}{h^2}\right)^{1/2}\left(\frac{8\pi^2 I_b k_B T}{h^2}\right)^{1/2}\left(\frac{8\pi^2 I_c k_B T}{h^2}\right)^{1/2} \qquad (10\text{-}73)$$

Usually, in a sample at room temperature the lowest vibrational level is populated, so the rotational partition function should be evaluated using parameters for the structure of this state. For reactions involving molecules in excited vibrational states, it is appropriate to use rotational constants for the particular vibrational state. A derivation of the classical Q_{rot} is given in Appendix 2.

With the aid of equations (10-65), (10-69), (10-73), and (10-80) [following section], the effect of isotopic substitution on the rate coefficient can be fairly well estimated. While the shape of the potential surface is essentially unaffected by isotopic substitution (thanks to the Born-Oppenheimer approximation), isotopic substitution will change the zero-point energies, which will in turn be reflected in the effective value of E_a. Isotopic masses will also affect the vibrational frequencies in the transition state, and thus the pre-exponential factor (or entropy of activation $\Delta S_0^\ddagger$). An example of such a kinetic isotope effect has been given in Section 1.5. Additional details, and many examples, may be found in the sources listed in the Bibliography for this chapter.

10.6.2 Symmetry and Statistical Factors

If molecules involved in a reaction have elements of symmetry, then in calculating the rate expression, the symmetry must be accounted for in the partition function. For molecules which have rotational symmetry, it is customary to divide the rotational partition function for each molecule by its appropriate symmetry number σ, defined as the number of equivalent arrangements that can be obtained by rotating the molecule. For example, consider Cl_2. The number of identical atoms in the molecule is 2, and by rotating we get two equivalent arrangements: $Cl^1 - Cl^2$ and $Cl^2 - Cl^1$. Therefore, the symmetry number is 2. For planar NO_3 radical, the symmetry number is 6. In taking into account the symmetry numbers in the rotational partition function, the standard procedure is to divide by the symmetry number. This procedure is correct when calculating rate constants, but it has some limitations. A good example which illustrates this is the reaction

$$H + H_2 \rightarrow [H \cdots H \cdots H]^\ddagger \rightarrow H_2 + H$$

The symmetry number for H_2 is 2, and so is the symmetry number of the complex; consequently, the rate constant is

$$k_1 = \frac{k_B T}{h} \frac{Q^{\ddagger}/2}{(Q_H/1)(Q_{H_2}/2)} e^{-E_0/k_B T} \tag{10-74}$$

The factors cancel out to give

$$k_1 = \frac{k_B T}{h} \frac{Q^{\ddagger}}{Q_H Q_{H_2}} e^{-E_0/k_B T} \tag{10-75}$$

If we compare the rate obtained for the reaction

$$D + H_2 \rightarrow [D \cdots H \cdots H]^{\ddagger} \rightarrow DH + H$$

we see that the symmetry number for the complex is 1; therefore,

$$k_2 = \frac{k_B T}{h} \frac{Q^{\ddagger}/1}{(Q_D/1)(Q_{H_2}/2)} e^{-E_0/k_B T} \tag{10-76}$$

and we have

$$k_2 = \frac{k_B T}{h} \frac{2Q^{\ddagger}}{Q_D Q_{H_2}} e^{-E_0/k_B T} \tag{10-77}$$

The conclusion is that the second reaction is favored over the first by a factor of 2. However, since both reactions are extracting hydrogen atoms from H_2, the rates cannot differ by a factor of 2; this factor should appear in both equations!

The best way to resolve this dilemma is by using a statistical factor, defined as the number of different transition states that can be formed if all identical atoms are labeled to distinguish them from one another. With this definition, we can omit symmetry numbers from the partition functions and multiply the rate expression by the statistical factor $L^{\ddagger}$, i.e.,

$$k = L^{\ddagger} \frac{k_B T}{h} \frac{Q^{\ddagger}}{Q_A Q_B} e^{-E_0/k_B T} \tag{10-78}$$

For the reaction

$$H^1 + H^2 - H^3 \begin{cases} \nearrow [H^1 \cdots H^2 \cdots H^3]^{\ddagger} \rightarrow H^1 - H^2 + H^3 \\ \searrow [H^1 \cdots H^3 \cdots H^2]^{\ddagger} \rightarrow H^1 - H^3 + H^2 \end{cases} \tag{10-79}$$

the statistical factor for forming the transition state is 2. Similarly, the reaction

$$H^1 + D^2 - D^3 \begin{cases} \nearrow [H^1 \cdots D^2 \cdots D^3]^{\ddagger} \rightarrow H^1 - D^2 + D^3 \\ \searrow [H^1 \cdots D^3 \cdots D^2]^{\ddagger} \rightarrow H^1 - D^3 + D^2 \end{cases} \tag{10-80}$$

also has a statistical factor of 2. Therefore, the correct rate constant for *both* reactions is given by equation (10-77). Great care must be used in the choice of a transition state with the correct degree of symmetry. For further reading, see Bishop and Laidler,[19] Pollak and Pechukas,[20] and Coulson.[21]

10.6.3   Collisions between Atoms

The simplest application of transition state theory is to collisions between two atoms A and B, where a rate constant k_c may be calculated for attaining a critical A-B separation r_c. In this simple case, the energy profile along the reaction coordinate is identical to the one-dimensional interatomic potential curve; we shall see that this leads to a cancellation of partition functions and a particularly simple result. The problem is somewhat artificial, because if the intermediate AB retains the energy of bond formation, it will dissociate after one vibrational period. If there is a mechanism for releasing this energy, however, a stable molecule may be formed and k_c can be related to the rate of formation of AB. An example of this is mentioned at the end of the section.

The transition state $AB^{\ddagger}$ for a collision between two atoms has three translational and two rotational degrees of freedom. The A-B relative center-of-mass motion (vibration) is the reaction coordinate. Thus, the transition state partition function includes contributions from translation and rotation and is

$$Q^{\ddagger} = \frac{[2\pi(m_A + m_B)k_B T]^{3/2}}{h^3} \frac{8\pi^2 I^{\ddagger} k_B T}{h^2} \tag{10-81}$$

where

$$I^{\ddagger} = \frac{m_A m_B}{m_A + m_B} r_c^2 \tag{10-82}$$

The atoms A and B have only translational contributions to their partition functions. After inserting these partition functions into equation (10-20), we find the transition state theory value for k_c to be

$$k_c = \pi r_c^2 \left(\frac{8k_B T}{\pi m} \right)^{1/2} e^{-E_0^{\ddagger}/k_B T} \tag{10-83}$$

This equation is identical with equation (8-8), derived for a line-of-centers collision model.

A case in which an atom-atom collision can lead to a real chemical event is radiative recombination. Consider the collision between an electronically excited atom A* and a ground-state atom B. Once inside the critical separation r_c, the electronically excited diatom (AB)* may emit a photon of radiation to form the stable ground-state molecule AB, i.e.,

$$AB^* \rightarrow AB + h\nu \tag{10-84}$$

The rate constant for formation of AB by this process can be written as the rate constant k_c multiplied by a probability of radiative emission from AB*, which in turn depends on the Franck-Condon overlap density and the oscillator strength of the transition. Some examples of this process are discussed by Smith.[22]

10.6.4   Application to the $F + H_2$ Reaction

The reaction $F + H_2 \rightarrow HF + H$ has been of particular interest in chemical kinetics because it is the rate-limiting step in the chain reaction $H_2 + F_2 \rightarrow 2HF$, which plays an important role in the underlying kinetics of the HF chemical laser (see chapter 9). It is also of special theoretical interest, because it is one of the simplest examples of an exothermic chemical reaction.

We can calculate the rate of reaction of the F atom with an H_2 molecule by transition state theory using equation (10-20). Consider a linear activated complex

Properties of this complex are calculated using a LEPS surface[23,24,25] and are summarized in Table 10-2. There should be three real vibrational frequencies and one imaginary frequency. The frequency at 4007.6 cm^{-1} corresponds to the H_2 stretch against the F atom in the complex, and the degenerate bending frequency is 397.9 cm^{-1}. The imaginary frequency at 310.8i corresponds to passage over the barrier. Using this information, together with properties of the reactants, we can calculate a rate constant for the $F + H_2$ reaction. First, we can rewrite the expression for partition functions in equation (10-20) as a product of partition function ratios for the translational, vibrational, rotational and electronic motion:

$$k = L^{\ddagger} \frac{k_B T}{h} \left(\frac{Q^{\ddagger}}{Q_F Q_H} \right)_{vib} \left(\frac{Q^{\ddagger}}{Q_F Q_{H_2}} \right)_{rot} \left(\frac{(Q^{\ddagger}/V)}{(Q_F/V)(Q_{H_2}/V)} \right)_{trans} \left(\frac{Q^{\ddagger}}{Q_F Q_{H_2}} \right)_{elec} e^{-E_0/RT}$$

(10-85)

It is convenient to evaluate each of these ratios separately.

We first note that the electronic degeneracy for the fluorine atom in its ground electronic state ($^2P_{3/2}$) is 4, and that for the linear FH_2 complex ($^2\Pi$) is the same. We neglect contributions from the ($^2P_{1/2}$) spin-orbit state of F at 404 cm^{-1}. The degeneracy for the H_2 molecule in its ground state ($^1\Sigma_g^+$) is 1. Consequently, the electronic parti-

TABLE 10-2   Properties of the Reactants and Transition State for the $F + H_2$ Reaction.

Parameters	$F \cdots H_a - H_b$	F	H_2
$r_2(F-H)$,Å	1.602		
$r_1(H-H)$,Å	0.756		$r(H-H) = 0.7417$Å
$\bar{\nu}_1$, cm^{-1}	4007.6		4395.2
$\bar{\nu}_2$, cm^{-1}	397.9		
$\bar{\nu}_3$, cm^{-1}	397.9		
$\bar{\nu}_4$, cm^{-1}	310.8i		
E_0(kJ/mole)	6.57		
m(amu)	21.014	18.9984	2.016
I(amu Å^2)	7.433		0.277
g_{elec}	4	4	1

tion function contribution is unity. The translational partition function ratio, using equation (10-65), is

$$\left(\frac{(Q^{\ddagger}/V)}{(Q_F/V)(Q_{H_2}/V)}\right)_{trans} = \left(\frac{m_F + m_{H_2}}{m_F m_{H_2}}\right)^{3/2}\left(\frac{h^2}{2\pi k_B T}\right)^{3/2}$$

and inserting the values given in Table 10-2 gives

$$\left(\frac{(Q^{\ddagger}/V)}{(Q_F/V)(Q_{H_2}/V)}\right)_{trans} = (6.01 \times 10^{39}\,kg^{-3/2})(6.926 \times 10^{-71}\,kg^{3/2}\,m^3)$$

so that

$$\left(\frac{(Q^{\ddagger}/V)}{(Q_F/V)(Q_{H_2}/V)}\right)_{trans} = 4.16 \times 10^{-31}\,m^3 \tag{10-86}$$

The rotational and vibrational partition functions are dimensionless, and contain no contribution from the mono-atomic fluorine species. The rotational partition function ratio is

$$\left(\frac{Q^{\ddagger}}{Q_{H_2}}\right)_{rot} = \left(\frac{8\pi^2 k_B T I^{\ddagger}/h^2}{8\pi^2 k_B T I_{H_2}/h^2}\right) = \frac{I^{\ddagger}}{I_{H_2}} \tag{10-87}$$

Substituting numerical values, we obtain

$$\left(\frac{Q^{\ddagger}}{Q_{H_2}}\right)_{rot} = 26.8 \tag{10-88}$$

In evaluating the vibrational partition function ratio we obtain the expression

$$\left(\frac{Q^{\ddagger}}{Q_{H_2}}\right)_{vib} = \frac{1 - \exp(-h\nu_{H_2}/k_B T)}{\prod\limits_{i=1}^{3}[1 - \exp(h\nu_i^{\ddagger}/k_B T)]} \tag{10-89}$$

Since

$$\frac{h\nu_{H_2}}{k_B T} = \frac{(6.626 \times 10^{-34}\,J \cdot s)(4395.2\,cm^{-1})(2.99 \times 10^{10}\,cm/s)}{(1.281 \times 10^{-23})J \cdot K^{-1} \times 300K} = 21.0$$

is much larger than 1, the term $1 - \exp(-h\nu_{H_2}/k_B T)$ is close to unity. For the complex, we have the real vibrational frequencies

i	$\bar{\nu}_1$, cm^{-1}	$h\nu_i/k_B T$	$[1 - \exp(h\nu_i/k_B T)]$
1	4007.6	19.2	≈ 1.0
2, 3	397.9	1.9	0.850

Thus, the vibrational partition function ratio term is

$$\left(\frac{Q}{Q_{H_2}}\right)_{vib} = \frac{1.0}{[1 - \exp(-h\nu_1/k_B T)][1 - \exp(-h\nu_2/k_B T)]^2} = 1.39 \tag{10-90}$$

The statistical factor $L^{\ddagger}$ contributes a factor of 2.

Inserting the corresponding values for the partition function ratios, we obtain for the rate constant

$$k = (1.95 \times 10^{-16} \, \text{m}^3\text{s}^{-1}) \exp(-E_0/k_B T)$$

Since these are molecular units, we multiply by Avogadro's number, $6.022 \times 10^{23} \, \text{mol}^{-1}$, to obtain

$$k = 1.17 \times 10^8 \, \text{mol}^{-1} \text{m}^3 \text{s}^{-1} \text{e}^{-6570/8.31T}$$

$$= 1.17 \times 10^{11} \exp(-790/T) \text{liter mol}^{-1} \sec^{-1}$$

The experimental data[26,27] are best represented by $k(T) = 2 \times 10^{11} \exp(-800/T)$ liter mol^{-1} sec^{-1}. This is very reasonable agreement, for such a simple model.

10.7 MICROCANONICAL TRANSITION STATE THEORY

In the previous sections we were concerned with the development of a statistical theory of reaction rates in which the population of the reactant molecular states is defined by the Boltzmann distribution and a single temperature. A collection of molecules forming such a distribution is referred to as a *canonical* ensemble. The underlying basis of the Boltzmann distribution is that all molecular states at the same energy are equally probable. An assembly of molecules, for which there is an equal population of all states at a particular energy, is called a *microcanonical* ensemble. In the microcanonical approach to transition state theory, a rate constant is determined for the microcanonical ensemble of reactant molecules.[9,28,29] Microcanonical transition state theory is particularly useful for unimolecular reactions, which are discussed in chapter 11.

The microcanonical rate constant $k(E)$ is for reactants whose total energy is E. The canonical transition state theory rate constant $k(T)$ is a Boltzmann average over $k(E)$ and is given by

$$k(T) = \int_0^\infty k(E) P(E) \, dE \tag{10-91}$$

where the reactants have energy E, $P(E)$ is the normalized Boltzmann probability, and $k(E)$ is the specific rate constant for reactants with total energy between E and $E + dE$.

To derive the microcanonical transition state theory rate constant $k(E)$, it is assumed, as in canonical transition state theory, that the reaction coordinate is separable from the remaining degrees of freedom. Thus, the Hamiltonian is written as in equation (10-25). Microcanonical transition states are reactive systems which have a total energy $H = E$ and a value for the reaction coordinate q_1 which lies between $q_1^\ddagger$ and $q_1^\ddagger + dq_1^\ddagger$. The reaction coordinate potential at the transition state is E_0.

For a bimolecular reaction, the combined states of the reactant molecules at total energy E define a supermolecule. In a unimolecular reaction the supermolecule is simply the single reactant. The fraction of supermolecules that lie at the transition state (i.e., at q_1 ranging from $q_1^\ddagger$ to $q_1^\ddagger + dq_1^\ddagger$) with reaction coordinate momenta in the range from $p_1^\ddagger$ to $p_1^\ddagger + dp_1^\ddagger$ may be found from statistical mechanics. From classical statistical mechanics [equation (A2-7) in Appendix 2], this fraction is

$$\frac{dN(q_1^\ddagger, p_1^\ddagger)}{N} = \frac{dq_1^\ddagger dp_1^\ddagger \displaystyle\int_{H=E-E_1^\ddagger-E_0} \cdots \int dq_2^\ddagger \cdots dq_{3n}^\ddagger dp_2^\ddagger \cdots dp_{3n}^\ddagger}{\displaystyle\int_{H=E} \cdots \int dq_1 \cdots dq_{3n} dp_1 \cdots dp_{3n}} \qquad (10\text{-}92)$$

where $E_1^\ddagger$ is the translational energy in the reaction coordinate. In this equation, $dq_1^\ddagger dp_1^\ddagger$ divided by Planck's constant is the number of translational states in the reaction coordinate and the surface integral in the numerator divided by h^{3n-1} is the density of states for the $3n-1$ degrees of freedom orthogonal to the reaction coordinate. Similarly, the surface integral in the denominator is the density of reactant states multiplied by h^{3n}.

All supermolecules that lie within $q_1^\ddagger$ and $q_1^\ddagger + dq_1^\ddagger$ and with positive $p_1^\ddagger$ will cross the transition state toward products in the time interval $dt = \mu_1 dq_1^\ddagger/p_1^\ddagger$. Thus, from equation (10-92), the reactant-to-product flux through the transition state for momentum $p_1^\ddagger$ is

$$\frac{dN(q_1^\ddagger, p_1^\ddagger)}{dt} = \frac{N \dfrac{p_1^\ddagger}{\mu_1} dp_1^\ddagger \displaystyle\int_{H=E-E_1^\ddagger-E_0} \cdots \int dq_2^\ddagger \cdots dq_{3n}^\ddagger dp_2^\ddagger \cdots dp_{3n}^\ddagger}{\displaystyle\int_{H=E} \cdots \int dq_1 \cdots dq_{3n} dp_1 \cdots dp_{3n}} \qquad (10\text{-}93)$$

The translational energy along the reaction coordinate is given by equation (10-26), so that $dE_1^\ddagger = p_1^\ddagger dp_1^\ddagger/\mu_1$. Therefore, equation (10-93) may be written

$$\frac{dN(q_1^\ddagger, p_1^\ddagger)}{dt} = \frac{N dE_1^\ddagger \displaystyle\int_{H=E-E_1^\ddagger-E_0} \cdots \int dq_2^\ddagger \cdots dq_{3n}^\ddagger dp_2^\ddagger \cdots dp_{3n}^\ddagger}{\displaystyle\int_{H=E} \cdots \int dq_1 \cdots dq_{3n} dp_1 \cdots dp_{3n}} \qquad (10\text{-}94)$$

To find the total reaction flux, equation (10-94) must be integrated between the limits $E_1^\ddagger$ equal to 0 and $E - E_0$, so that

$$\frac{dN}{dt} = \frac{N \displaystyle\int_0^{E=E_0} \left[\int_{H=E-E_1^\ddagger-E_0} \cdots \int dq_2^\ddagger \cdots dq_{3n}^\ddagger dp_2^\ddagger \cdots dp_{3n}^\ddagger \right] dE_1^\ddagger}{\displaystyle\int \cdots \int_{H=E} dq_1 \cdots dq_{3n} dp_1 \cdots dp_{3n}} \qquad (10\text{-}95)$$

which gives

$$\frac{dN}{dt} = \frac{N \displaystyle\int_{H=0}^{H=E-E_0} \cdots \int dq_2^\ddagger \cdots dq_{3n}^\ddagger dp_2^\ddagger \cdots dp_{3n}^\ddagger}{\displaystyle\int_{H=E} \cdots \int dq_1 \cdots dq_{3n} dp_1 \cdots dp_{3n}} \qquad (10\text{-}96)$$

Since $dN/dt = k(E)N$, the ratio of integrals in equation (10-96) is the microcanonical transition state theory rate constant $k(E)$. This becomes readily apparent (see Appendix 2) when the numerator is identified as the sum of states at the transition state $G^{\ddagger}(E - E_0)$ multiplied by h^{3n-1}. As discussed, the denominator is the reactant density of states times h^{3n}. Thus, $k(E)$ becomes

$$k(E) = \frac{G^{\ddagger}(E - E_0)}{hN(E)} \tag{10-97}$$

The reactant density $N(E)$ has units of inverse energy for a unimolecular reaction and (energy $\times$ volume)$^{-1}$ for a bimolecular reaction.

The canonical transition state theory rate constant results from inserting equation (10-97) into equation (10-91). Since, from equation (A1-3) in Appendix 1, the Boltzmann probability $P(E)$ equals $N(E) \exp(-E/k_BT)/Q$, we have

$$k(T) = \frac{1}{hQ} \int_0^{\infty} G^{\ddagger}(E - E_0) \exp(-E/k_BT)\, dE \tag{10-98}$$

where Q is the total partition function for the reactants. Writing $E = E^{\ddagger} + E_0$, we obtain, for equation (10-98),

$$k(T) = \frac{e^{-E_0/k_BT}}{hQ} \int_0^{\infty} G^{\ddagger}(E^{\ddagger}) \exp(-E^{\ddagger}/k_BT)\, dE^{\ddagger} \tag{10-99}$$

The integration with respect to $E^{\ddagger}$ may be done by parts or, more simply, by noting that the integral is the Laplace transform* of $G^{\ddagger}(E^{\ddagger})$.[30] The resulting value for the integral is $k_BTQ^{\ddagger}$, where $Q^{\ddagger}$ is the partition function for the $3n - 1$ degrees of freedom orthogonal to the reaction coordinate. A comparison of the resulting $k(T)$ with k_{abs} in equation (10-20) shows that the two are the same.

10.8   VARIATIONAL TRANSITION STATE THEORY

As discussed in section 10.3, classical transition state theory gives the exact rate constant (in a classical universe) if the net rate of reaction equals the rate at which trajectories pass through the transition state. These two rates are equal, however, only if trajectories do not recross the transition state: any recrossing makes the reactive flux smaller than the flux through the transition state. Thus, the classical transition state theory rate constant may be viewed as an upper bound to the correct classical rate constant.

The effect of recrossings on the reaction rate is illustrated by the six schematic trajectories in Figure 10-5. The left side and right side represent the reactants and products, respectively. The solid line shows the transition state dividing surface at the saddle point, while the dashed line is an alternative transition state dividing surface. There are six crossings of the saddle point in the direction from reactant to products, and transition state theory counts all of them as contributing to the reactive flux.

---

*See chapter 2 for a discussion of Laplace transforms.

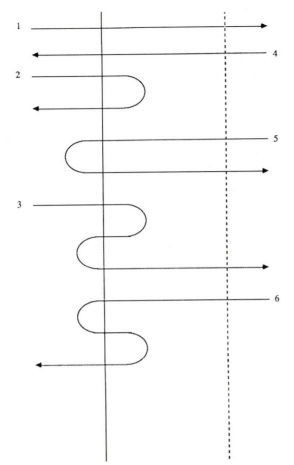

FIGURE 10-5   Schematic trajectories showing different ways of crossing the transition state indicated by the solid vertical line. An alternative transition state is identified by the dashed vertical line.

However, only two trajectories in fact contribute to the reactive flux. Thus, in this example the transition state theory rate constant is too large by a factor of three. This is reflected in the fact that, of the six trajectories in the figure, only numbers 1 and 4 cross the saddle point once.

The alternative transition state denoted by the dashed line suggests how transition state theory may be improved. There are only three crossings for the alternative transition state in the direction from reactants to products, as compared with the two reactive trajectories. Thus, this transition state gives a rate constant that is only 50 percent too large. Clearly, then, the alternative transition state is a much better bottleneck for the reaction. Accordingly, one can consider different positions for the transition state along the reaction path and calculate the rate constant corresponding to each. The minimum rate so obtained is the closest to the truth, assuming that quantum effects related to tunneling and nonseparability are negligible. This procedure is called *variational transition state theory.*[4,10,19,31–38] There are both canonical and microcanonical versions of variational transition state theory. In the canonical approach, the minimum in the canonical transition state theory constant given by equation (10-20) is found along the reaction path

$$\frac{dk(T)}{dq^{\ddagger}} = 0 \tag{10-100}$$

Since the canonical rate constant is related to the free energy according to equation (10-42), the canonical variational transition state will be located at the *maximum* in the free energy along the reaction path.

Canonical variational transition state theory gives an approximate upper bound to the rate constant. To obtain the correct upper bound, microcanonical variational transition state theory must be used. Here, the minimum in the microcanonical rate constant given by equation (10-97) is found along the reaction path

$$\frac{dk(E)}{dq^{\ddagger}} = 0 \tag{10-101}$$

so that a different transition state is found for each energy. The thermal rate constant results from a Boltzmann average of the microcanonical rate constants as given by equation (10-97). The individual microcanonical rate constants are minimized at the point along the reaction path where the sum of the states $G^{\ddagger}(E - E_0)$ is a minimum.

Intuitively, the canonical and microcanonical results are what one expects. The best bottleneck along the reaction path in the canonical sense should be where the free energy is a maximum. In the microcanonical approach, the reaction bottleneck is located where there is the smallest sum of states.

To apply canonical variational transition state theory, the transition state's partition function (or free energy) must be calculated as a function of the reaction path. In the microcanonical approach, the transition state's sum of states must be calculated along the reaction path. Each of these calculations requires values for the classical potential energy, zero-point energy, vibrational frequencies, and moments of inertia as a function of the reaction path. The extraction of this information from a potential energy surface is described in chapter 7, section 7.9.

Variational transition state theory is particularly important for reactions without saddle points, e.g., $H + CH_3 \rightarrow CH_4$. Results of a canonical variational transition state theory calculation[39] for this reaction are listed in Table 10-3. A particularly interesting result is the shorter $H\text{---}CH_3$ bond length and thus "tighter" transition state as the temperature is increased. The vibrational frequency for the degenerate $H\text{---}CH_3$ rocking motions increases as the transition state tightens. A tightening of the transition state with increase in temperature is a common result for reactions without potential energy barriers. The calculated rate constant for $H + CH_3 \rightarrow CH_4$ is in good agreement with the experimental value of $2.8 \times 10^8 \text{ m}^3\text{mol}^{-1}\text{s}^{-1}$ for $T = 300\text{--}600 \ K$.

10.9   EXPERIMENTAL OBSERVATION OF THE TRANSITION STATE REGION

Despite the caveats in sections 10.2 and 10.4, that the transition state does not possess a chemically significant lifetime, and that quantum-mechanical uncertainty limitations preclude localizing the transition state in time or in configuration space, it is nevertheless possible to think about measuring the approximate structure and energy of a reacting system as it passes across the potential barrier separating reactants and prod-

TABLE 10-3   Canonical Variational Transition State Theory Results for the Reaction $H + CH_3 \rightarrow CH_4$.

Temp (K)	$R^{\ddagger a}$	$\nu_{\text{rock}}^{\ddagger b}$	$E_0^{\,c}$	k^d
200	3.54	95	−0.51	1.24
400	3.33	138	−1.05	1.56
600	3.19	173	−1.62	1.70
800	3.09	204	−2.22	1.78
1000	3.00	233	−2.84	1.83

[a]$H\!-\!CH_3$ bond length (Å) at the transition state.
[b]Harmonic vibrational frequency for the degenerate $H\!-\!CH_3$ rocking modes (cm^{-1}).
[c]Difference in classical potential energy between the transition state and reactants.
[d]Rate constant in units of $10^8 m^3 mol^{-1} s^{-1}$.

ucts. To accomplish this, we need a measurement method capable of "capturing" this information on the time scale the system is in the transition state region, which as Figure 8-14(b) shows, can be as short as 10^{-14} sec. Experimental techniques with this time resolution now exist, and the results of such experiments are now beginning to provide important confirmation of the transition state model, as well as structural information for reacting systems.

Early attempts at probing the transition state were carried out by John Polanyi,[40] using the fluorine + sodium reaction studied by his father, Michael Polanyi, in the 1930's. The reaction

$$F + Na_2 \rightarrow [NaNaF]^{\ddagger} \rightarrow Na^* + NaF$$

produces an electronically excited sodium atom which emits at the characteristic Na D-line, 589 nm. A very small fraction of the emission originates from the electronically excited $Na^* \cdots NaF$ configuration, and this can be detected as weak luminescence red-shifted from the D-line. From the intensity of this emission, John Polanyi was able to deduce the form of the potential in the neighborhood of the NaNaF transition state, thereby confirming the model postulated by his father fifty years earlier[2].

More direct probes of the transition state are now possible using ultrashort laser pulses. These experiments are essentially analogues of the laser photolysis experiments discussed in section 3.2.2, but using laser pulses as short as 10^{-14} sec (10 femtoseconds).[41] Using successive delays between initiation (pump) and interrogation (probe) pulses, one can obtain a "stroboscopic" picture of a reaction as it proceeds in real time. This technique, termed "femtochemistry,"[42,43] has now been applied to a number of systems. One system of particular chemical interest is the detection of the diradical intermediate in the decomposition of cyclic ketones.[44,45] A typical overall reaction is

The reaction is initiated by absorption of two photons from a femtosecond-pulse laser operating at a wavelength of 280 ~ 320 nm. This provides enough energy to eliminate CO, leaving the diradical $\cdot CH_2CH_2CH_2CH_2\cdot$, which then either breaks apart into two ethylene molecules or cyclizes to form cyclobutane, C_4H_8. The diradical is ionized by a delayed femtosecond probe pulse and detected in a time-of-flight mass spectrometer (see Fig. 10-6). Buildup and decay of the diradical occurs on a time scale of several hundred femtoseconds, or 0.1 to 1.0 picoseconds. This observation verifies that the diradical is indeed an intermediate in this reaction and provides information about its dynamic behavior. It should be noted, however, that the diradical is in fact a transient intermediate which exists in a shallow potential minimum at the top of the reaction barrier. This accounts for its relatively long lifetime, which in turn makes its detection possible.

Another approach to probing the transition state, which does not require the use of femtosecond lasers, is negative ion photodetachment spectroscopy (NIPS).[46] A stable negative ion is formed having the composition of a bimolecular collision complex; irradiation with light causes ejection of an electron, leaving an uncharged, unstable complex. If the negative ion possesses a structure close to that of the reacting system, information on the structure and dynamics of the transition state may be obtained from the wavelength dependence of the NIP spectrum.

Neumark and co-workers have carried out NIPS experiments on the FH_2^- ion.[47,48] This is an especially productive system to study, since the FH_2^- ion has a structure very close to that of the transition state of the well-studied $F + H_2$ reaction, viz., $r(H\!-\!H) = 0.77$ Å and $r(F\!-\!H) = 1.69$ Å, as compared with 0.76 Å and 1.60 Å for $[FH_2]^{\ddagger}$ (Table 10-2). Agreement between the measured spectrum and that calculated from the best available *ab initio* $F + H_2$ potential energy surface[49] is excellent, indicating that a quantitative description of the $[FH_2]^{\ddagger}$ transition state has now been achieved. The assignment of the FH_2^- NIPS spectrum[47] reveals that it consists of states with 1–5 quanta in a H_2 hindered rotor motion and 0 or 1 quantum in H_2 stretching; states consistent with the loose transition state for the $F + H_2$ reaction.

The NIPS technique should provide us with similar information about other reaction systems including large molecules of which cyclooctatetraene is a recent example.[50,51] However, the states observed in the NIPS spectrum must be sufficiently sparse and long-lived so that there is not extensive broadening and overlapping of the spectral lines, which would make their assignment impossible. A similar effect exists for superposition states prepared in femtochemistry experiments.[42] In these experiments the vibrational dynamics will be resolved if the superposition state resides in the transition state region for a sufficient time that there are recurrences in its time-dependence.[52] Otherwise, the decay time simply gives the lifetime in the transition state region.

10.10   CRITIQUE OF TRANSITION STATE THEORY

In this chapter, we have considered an approach to the theory of rate processes which is both microscopic and statistically based. While this theory has been remarkably successful in systematizing a large amount of data, it is appropriate to look back and ask ourselves how well justified are the basic premises of the theory. The following points need to be considered:

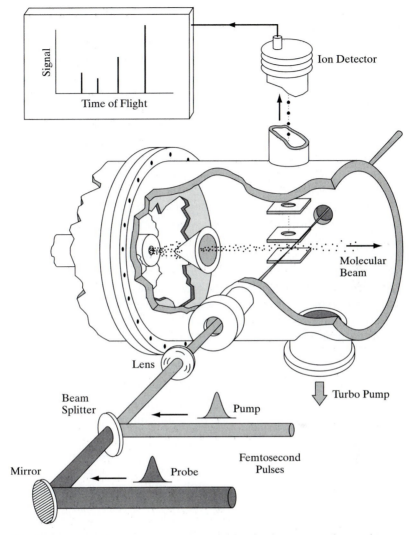

FIGURE 10-6 Schematic apparatus used in the femtosecond experiments on cyclopentanone decomposition. A molecular beam of cyclopentanone emerges from the reservoir on the left. It is dissociated to diradical + CO by an ultraviolet femtosecond "pump" pulse; the diradical is ionized by a time-delayed femtosecond "probe" pulse and the resulting ions are detected in a time-of-flight mass spectrometer. See Refs. 41–45 for further experimental details. [Adapted from Figure 2 of S. Pedersen, J. L. Herek, and A. H. Zewail, *Science 266,* 1359 (1994); reproduced with permission.]

1. Transition state theory makes use of classical mechanics in evaluating partition functions and canonical averages. We have considered, in section 10.4, the difficulties of arriving at a quantum mechanical formulation of transition state theory and the effect of tunneling as a specifically quantum phenomenon. Tunneling appears to be important for reactions involving very light species, such as H or D atoms.

2. Transition state theory simplifies the motion on the potential surface to one-dimensional motion along the reaction coordinate. However, the actual motion may include components normal to the reaction coordinate. To the extent that this is so, the assumption of separability of coordinates (see section 10.2) is invalidated. The reaction path may lie above the minimum energy configuration across the potential surface, with the result that the effective activation energy is somewhat higher than the barrier height E_0.

3. The use of harmonic potentials at the saddle point is reasonably accurate at low temperatures, since large-amplitude motions are not frequently encountered in the transition state configuration. At higher temperatures, however, when the average energies of the molecules and the transition state are higher, anharmonicity corrections could be important.

4. By the first assumption in section 10.2, systems pass through the transition state only once. In that case, exactly one-half of the systems yield products [see equation (10-9)]. By contrast, it is often found experimentally that the effective yield of products is much smaller than this, due to multiple crossings of the transition state region (see section 10.8). This is commonly represented by an empirical "transmission coefficient" $\kappa < 1$, analogous to the "steric factor" $p < 1$ in the old collision theory (see chapter 8). To the extent that recrossing the transition state region (or, equivalently, failure to react in a bimolecular encounter) can be attributed to the need for the molecules to approach each other in a particular orientation or relative configuration, this can also be represented by a negative entropy of activation $\Delta S^{\ddagger}$, which "soaks up all the p."[53]

5. Again, the local equilibrium postulate (the first assumption of section 10.2) is a consequence of the absence of any recrossing of the transition state. But if recrossing occurs, as evidenced by a transmission coefficient less than 1, then local equilibrium is not maintained and the calculation of equilibrium partition functions as in section 10.6.1 may not be valid.

6. A fundamental assumption in transition state theory is that both the reactants and the products are maintained in Maxwell-Boltzmann equilibrium. The examples presented in chapter 9, however, amply demonstrate that in bimolecular gas reactions one can have both selective energy consumption (from reactants) and specific energy release (in products), that is, highly non-Boltzmann distributions over energy states. Just recrossing the dividing surface of the transition state can lead to deviations from Boltzmann distributions in an ensemble initially at thermodynamic equilibrium. The equilibrium condition can be maintained, however, if relaxation of both reactants and products is rapid compared with the reaction rate.[54]

7. It might appear from the foregoing that transition state theory would be more applicable to reactions in liquid solutions than in low-pressure gases, because relaxation times in solution may be on the order of picoseconds. However, while there is a region of low solvent density and viscosity in which rapid relaxation may be significant, in general, solvent effects lead to deviations from transition state theory, as is discussed in chapter 12. These effects are manifested as the *diffusion limit* (see chapter 4) of reactants toward each other, which is on the order of 10^{-10} sec.

Notwithstanding all of these limitations, transition state theory does a remarkably good job describing a wide variety of chemical reactions.

REFERENCES

[1] H. Eyring, *J. Chem. Phys. 3,* 107 (1935).

[2] M. G. Evans and M. Polanyi, *Trans. Faraday Soc. 31,* 875 (1935).

[3] E. Wigner, *Trans. Faraday Soc. 34,* 29 (1938).

[4] R. H. Fowler and E. A. Guggenheim, *Statistical Thermodynamics* (Cambridge: University Press, 1939).

[5] J. Horiuti, *Bull. Chem. Soc. Jpn. 13,* 210 (1938).

[6] B. H. Mahan, *J. Chem. Ed. 51,* 709 (1974); W. H. Miller, *Acc. Chem. Res. 9,* 306 (1976).

[7] W. H. Miller, *J. Chem. Phys. 61,* 1823 (1974).

[8] H. S. Johnston and D. Rapp, *J. Am. Chem. Soc. 83,* 1 (1961).

[9] P. Pechukas and F. J. McLafferty, *J. Chem. Phys. 58,* 1622 (1973).

[10] D. G. Truhlar and B. C. Garrett, *Acc. Chem. Res. 13,* 440 (1980).

[11] R. A. Marcus and M. E. Coltrin, *J. Chem. Phys. 67,* 2609 (1977).

[12] W. H. Miller, N. C. Handy and J. E. Adams, *J. Chem. Phys. 72,* 99 (1980).

[13] R. T. Skodje, D. G. Truhlar and B. C. Garrett, *J. Chem. Phys. 85,* 3019 (1981); B. C. Garrett and D. G. Truhlar, *J. Chem. Phys. 81,* 309 (1984).

[14] E. E. Nikitin, *Theory of Elementary Atomic and Molecular Processes in Gases* (Oxford: Clarendon, 1974), pp. 20–23.

[15] H. S. Johnston, *Gas Phase Reaction Rate Theory* (New York: Ronald, 1966) pp. 38–47; B. C. Garrett and D. G. Truhlar, *J. Phys. Chem. 83,* 2921 (1979).

[16] D. L. Bunker, *Theory of Elementary Gas Reaction Rates* (New York: Pergamon, 1966) pp. 29–32.

[17] H. S. Johnston and J. Heicklen, *J. Phys. Chem. 66,* 532 (1966).

[18] I. Shavitt, *J. Chem. Phys. 31,* 1359 (1959).

[19] D. M. Bishop and K. J. Laidler, *Trans. Faraday Soc. 66,* 1685 (1970).

[20] E. Pollak and P. Pechukas, *J. Am. Chem. Soc. 100,* 2984 (1978).

[21] D. R. Coulson, *J. Am. Chem. Soc. 100,* 2992 (1978).

[22] I. W. M. Smith, *Kinetics and Dynamics of Elementary Gas Reactions* (Boston: Butterworths, 1980) pp. 224–227.

[23] R. L. Jaffe and J. B. Anderson, *J. Chem. Phys. 54,* 2224 (1971).

[24] H. F. Schaefer, *J. Phys. Chem. 89,* 5336 (1985).

[25] M. J. Frisch, B. Liu, J. S. Binkley, H. F. Schaefer and W. H. Miller, *Chem. Phys. Lett. 114,* 1 (1985).

[26] K. H. Homann, W. C. Solomon, J. Warnatz, H. Gg. Wagner, and C. Zetzsch, *Ber. Bunsenges. Phys. Chem. 74,* 585 (1970).

[27] N. Cohen and K. Westberg, *J. Phys. Chem. Ref. Data 12,* 531 (1983).

[28] B. C. Garrett and D. G. Truhlar, *J. Chem. Phys. 70,* 1593 (1979).

[29] R. A. Marcus, *J. Chem. Phys. 45,* 2138 (1966); ibid. 2630 (1966).

[30] W. Forst, *Theory of Unimolecular Reactions* (New York: Academic Press, 1973) p. 154.

[31] J. C. Keck, *J. Chem. Phys. 32,* 1035 (1960).

[32] J. C. Keck, *Adv. Chem. Phys. 13,* 85 (1967).

[33] D. L. Bunker and M. Pattengill, *J. Chem. Phys. 48,* 772 (1968).

[34] J. B. Anderson, *J. Chem. Phys. 58,* 4684 (1973).

[35] J. B. Anderson, *J. Chem. Phys. 62,* 2446 (1975).

[36] M. Quack and J. Troe, *Ber. Bunsenges. Phys. Chem. 78,* 240 (1974); ibid. *81,* 329 (1977).

[37] W. J. Chesnavich, T. Su and M. T. Bowers, *J. Chem. Phys. 72*, 2641 (1980).

[38] W. L. Hase, *Acc. Chem. Res. 16*, 258 (1983).

[39] W. L. Hase, S. L. Mondro, R. J. Duchovic and D. M. Hirst, *J. Am. Chem. Soc. 109*, 2916 (1987).

[40] H. -J. Foth, J. C. Polanyi, and H. H. Telle, *J. Phys. Chem. 86*, 5027 (1982).

[41] G. R. Fleming, *Chemical Applications of Ultrafast Spectroscopy* (Oxford: University Press, 1986).

[42] A. H. Zewail, *Science 242*, 1645 (1988).

[43] L. R. Khundkar and A. H. Zewail, *Ann. Rev. Phys. Chem. 41*, 15 (1990).

[44] J. A. Berson, *Science 266*, 1338 (1994).

[45] S. Pedersen, J. L. Herek, and A. H. Zewail, *Science 266*, 1359 (1994).

[46] R. B. Metz, A. Weaver, S. E. Bradforth, T-N. Kitsopoulos, and D. M. Neumark, *J. Phys. Chem. 94*, 1377 (1990).

[47] G. C. Schatz, *Science 262*, 1828 (1993).

[48] D. E. Manolopoulos, K. Stark, H. -J. Werner, D. W. Arnold, S. E. Bradforth, and D. M. Neumark, *Science 262*, 1852 (1993).

[49] P. J. Knowles, K. Stark, and H. -J. Werner, *Chem. Phys. Letts. 185*, 555 (1991).

[50] D. M. Neumark, *Science 272*, 1446 (1996).

[51] P. G. Wenthold, D. A. Hrovat, W. T. Borden, and W. C. Lineberger, *Science 272*, 1456 (1996).

[52] J. L. McHale, *Molecular Spectroscopy* (Upper Saddle River, NJ: Prentice-Hall, 1999).

[53] I. Amdur, private communication.

[54] B. Widom, *Adv. Chem. Phys. 5*, 353 (1963).

BIBLIOGRAPHY

Basic Postulates and Derivation of Transition State Theory

BAUER, S. H. *J. Chem. Ed. 63*, 377 (1986).

GLASSTONE, S., LAIDLER, K. J., and EYRING, H. *The Theory of Rate Processes.* New York: McGraw-Hill, 1941.

JOHNSTON, H. S. *Gas Phase Reaction Rate Theory.* New York: Ronald, 1966.

LAIDLER, K. J. *Theories of Chemical Reaction Rates.* New York: McGraw-Hill, 1969.

LEVINE, R. D., and BERNSTEIN, R. B. *Molecular Reaction Dynamics.* Oxford: Oxford University Press, 1974.

NIKITIN, E. E. *Theory of Thermally Induced Gas Phase Reactions.* Bloomington, IN: Indiana University Press, 1966.

TRUHLAR, D. G., and Gordon, M. S., *Science 249*, 491 (1990).

Statistical Thermodynamics

DAVIDSON, N. *Statistical Mechanics.* New York: Harper and Row, 1973.

HILL, T. L. *Introduction to Statistical Mechanics.* Reading, MA: Addison-Wesley, 1972.

RICE, O. K. *Elements of Statistical Mechanics.* San Francisco, CA: Freeman, 1967.

Quantum Mechanical Tunneling

BELL, R. P. *The Tunnel Effect in Chemistry.* London: Chapman and Hall, 1980.

Microcanonical and Variational Transition State Theory

HORIUTI, J. *Bull. Chem. Soc. Jpn. 13*, 210 (1938).

KECK, J. C. *Adv. Chem. Phys. 13*, 85 (1967).

PECHUKAS, P. In *Dynamics of Molecular Collision (Part B)*, edited by W. H. Miller. New York: Plenum, 1976, p. 299.

PECHUKAS, P. *Ann. Rev. Phys. Chem. 32*, 159 (1981).

SMITH, I. W. M. *Kinetics and Dynamics of Elementary Gas Reactions.* London: Butterworths, 1980, p. 129.

TRUHLAR, D. G., and GARRETT, B. C. *Ann. Rev. Phys. Chem. 35*, 159 (1984).

TRUHLAR, D. G., HASE, W. L., and HYNES, J. T. *J. Phys. Chem. 87*, 2664 (1983).

WARDLAW, D. W., and MARCUS, R. A., *Adv. Chem. Phys. 70*, 231 (1987).

Other Applications of Transition State Theory

TSAO, J. Y., "TST for quantum and classical particle escape from a square well", *Am. J. Phys. 57*, 269 (1989).

PITT, I. G., GILBERT, R. G., and RYAN, K. R., "Applications of TST to gas-surface reactions in Langmuir systems", *J. Chem. Phys. 102*, 3461 (1995).

PROBLEMS

10.1 Find expressions for s and ξ in a collinear $A + B{-}C$ system for a point (r_1, r_2) where the reaction path on the potential surface is taken to be a circle of radius R centered at $r_1 = a_0, r_2 = b_0$. Over what range of r_1 and r_2 would this be a satisfactory approximation to the true reaction path?

10.2 The derivation of transition state theory given in section 10.2 follows the original treatment by Eyring and M. Polanyi, in which motion along the reaction coordinate is treated as a classical translation. A somewhat simpler derivation, which avoids the constructs δ and μ_s, can be achieved by treating this motion as a vibration with frequency $\nu^{\ddagger}$. Rederive equation (10-20) using this approach.

*10.3 Consider two collinear reactions of the form $A + BC \rightarrow AB + C$ which have similar potential surfaces except that the transition state frequencies are 3000 cm^{-1} for the first reaction and 200 cm^{-1} for the second.

(a) Qualitatively sketch the potential contours for each.

(b) Which has the larger frequency factor? (A qualitative argument only is required here.)

*10.4 Consider the collinear reaction

$$M + F_2 \rightarrow MF + F$$

where $M = H$ or D. Using transition state theory, find an expression for the *ratio* of H:D rate coefficients at a given temperature. Note that

$$k^{TST} = \frac{k_B T}{h} \frac{Q^{\ddagger}}{Q} e^{-E_0/k_B T}$$

and assume that the reactant and transition states are harmonic oscillators with force constants k_R and k_T, respectively. Also, $\nu = (k/m)^{1/2}/(2\pi c)$, and the transformation from

Cartesian to (q_1, q_2) coordinates in a collinear ABC system yields $H = p_1^2 + p_2^2/2m + V$, where $m = [m_A m_B m_C/(m_A + m_B + m_C)]^{1/2}$.

*10.5 Consider the reaction $A + BC \rightarrow AB + C$ *in three dimensions.* Write down all the degrees of freedom (translational, rotational, and vibrational) of (i) the reactants and (ii) the transition state. Take the transition state to be linear (i.e., it will have two rotational degrees of freedom, not three as would a nonlinear molecule). Use the formulae for partition functions given in section 10.6.1.

(a) What are the units of each of the partition functions?

(b) Show that equation (10-20) gives the correct units for a second-order rate coefficient.

(c) Consider the specific reaction $H + Br_2 \rightarrow HBr + Br$. Specify precisely what data are required to carry out an evaluation of the rate coefficient using equation (10-20).

10.6 On the basis of transition state theory, what temperature dependence of the pre-exponential factor would be expected for each of the following reactions?

(a) $H_2 + F_2 \rightarrow 2HF$

(b) $NO_2 + F_2 \rightarrow NO_2F + F$

(c) $2NOCl \rightarrow 2NO + Cl_2$

(d) $2NO_2 + O_2 \rightarrow 2NO + 3$

(e) $C_2H_4 + HCl \rightarrow C_2H_5Cl$

10.7 Use transition state theory to determine an expression for the pre-exponential term and the rate constant for the following hypothetical reactions:

$$\text{RXN 1: atom} + \text{linear} \rightarrow \text{linear activated complex}$$

$$\text{RXN 2: atom} + \text{linear} \rightarrow \text{nonlinear activated complex}$$

$$\text{RXN 3: atom} + \text{nonlinear} \rightarrow \text{nonlinear activated complex}$$

Express the pre-exponential factor in terms of appropriate partition functions. Also, cite one real chemical reaction of the kind of each of the three hypothetical reactions.

10.8 The following reaction was done experimentally at various temperatures in order to obtain rate constants:

$$O + OH_2 \rightarrow OH + O_2$$

$$T_1 = 298 \text{ K} \qquad k_1 = 6.1 \times 10^{-11} \text{ cm}^3/\text{molecule} \cdot s$$

$$T_3 = 229 \text{ k} \qquad k_2 = 7.57 \times 10^{-11} \text{ cm}^3/\text{molecule} \cdot s$$

(a) Using the experimental data, find the activation energy E_{act}.

(b) Calculate the following thermodynamic properties at $T = 298$ K using transition state theory: $\Delta E^{\ddagger}, \Delta H^{\ddagger}, \Delta S^{\ddagger}, \Delta G^{\ddagger}, A$.

10.9 **(a)** Using transition state theory, formulate the rate constant for the exchange reaction

$$H + H_2 \rightleftharpoons H_3^{\ddagger} \rightarrow H_2 + H$$

in terms of the appropriate functions. Assume that $H_3^{\ddagger}$ has a linear symmetric configuration with a ground state electronic degeneracy of 2 and a symmetry number of 2. The electronic degeneracies for ground state H_2 and H are 1 and 2, respectively.

For the $H_3^{\ddagger}$ transition state, theoretical calculations suggest that the $H-H$ distance is 0.93 Å and the fundamental frequencies are 2193, 978, and 978 cm^{-1}. The barrier height may be taken to be 37.6 kJ/mole. The $H-H$ distance in H_2 is 0.74 Å, and its fundamental vibrational frequency is 4395 cm^{-1}.

Calculate numerical values for the transition state theory rate constants for reaction 1 at 300 and 1,000 K. The experimentally measured rate coefficient is $(2.1 \pm 0.6) \times 10^{-12}$ cm^3 molecule^{-1} sec^{-1} at 1,000 K.

(b) Calculate pre-exponential factors for an "old collision theory" rate constant for the reaction $H + H_2 \rightarrow H_2 + H$, and compare the rates calculated in this way at 300 K and 1,000 K with the values found in part (a). Assume a hard-sphere diameter $d = 2.5$ Å.

(c) Estimate the Wigner tunneling correction for the $H + H_2$ reaction.

(d) Compare your results with any other calculated rate coefficients (classical or quantum) you can readily find in the literature.

10.10 Repeat problem 9 for the reaction $D + D_2 \rightarrow D_2 + D$. Assume that the geometry and potential surface for $D + D_2$ are the same as those for $H + H_2$, and use isotope-effect relationships to find the vibrational frequencies and moments of inertia.

10.11 Consider the reaction

$$R-H + Cl \rightarrow R + HCl$$

where R is an alkyl group. Using the following reactant data and assumptions as well as the BEBO scheme discussed in chapter 7 to estimate the necessary molecular parameters for the transition state, calculate the Arrhenius A-factor for this reaction at 400 K. *Data:* The $H-Cl$ bond length and vibrational frequency are 1.27 Å and 3000 cm^{-1}, respectively. The $R-H$ bond length is 1.09 Å and the stretching and bending frequencies are 3000 cm^{-1} and 1500 cm^{-1}, respectively. The mass and moment of inertia of the R group (assumed to be spherical) are 30 amu and 20 amu Å^2, respectively. Assume that the $R-H$ and $H-Cl$ bond orders are both 1/2 at the transition state, and make any other reasonable assumptions that you find necessary.

10.12 The reaction of CH with N_2 is of importance in the chemistry of planetary atmospheres and hydrocarbon flames. By understanding the basic reaction, other complex systems such as combustion in the environment can be studied. The reaction is

$$CH + N_2 \rightarrow (CHN_2)^{\ddagger} \rightarrow HCN + N$$

Using transition state theory, find the partition functions for each molecule and then the rate constant at 298 K. Also, specify all the degrees of freedom and the temperature dependence of the A factor. From experiment, the second-order rate constant at 298 K has been found to be $k_{exp} = 7.1 \times 10^{-14}$ cm^3/molecule · sec.

Properties of the Transition State + Molecules

	CH	N_2	HCN$_2$ (assume linear adduct)
M (amu = g/mol)	13.019	28.013	41.032
I (10^{-40} g cm^2)	1.935	13.998	73.2
E_0			$\approx$221 kJmol
ν (cm^{-1})	2733	2330	3130; 2102; 1252; 1170; 564; 401
g_{elec}	2	1	2

*These problems have been provided by Dr. R. G. Gilbert, Department of Theoretical Chemistry, University of Sydney, Australia.

CHAPTER 11

Unimolecular Reaction Dynamics

Developments in the field of chemical kinetics often arise from the interplay and exchange of ideas between theoretical and experimental research. This relationship is especially well illustrated in the area of unimolecular reaction dynamics. The development of unimolecular theoretical models has been fostered by new experimental studies which have provided more detail about chemical reactivity and molecular motion than was previously available. Likewise, formerly unknown microscopic properties of unimolecular reactions have been discovered by theoretical computer simulation studies and later observed in experiments designed to probe for such properties.

A unimolecular reaction is denoted by

$$A^* \longrightarrow \text{Products} \tag{11-1}$$

where the asterisk indicates that A contains sufficient internal vibrational energy to decompose. The energy is denoted by the symbol E and must be greater than the unimolecular decomposition threshold energy E_0. There are three general types of potential energy profiles for unimolecular reactions (see Figure 11-1). One type is for an isomerization reaction, such as

$$CH_3NC \longrightarrow CH_3CN \tag{11-2}$$

where there is a substantial potential energy barrier separating the two isomers. The other two types are for a unimolecular dissociation reaction. In one case, as for example,

$$C_2H_5Cl \longrightarrow HCl + C_2H_4 \tag{11-3}$$

or

$$C_2H_5 \longrightarrow H + C_2H_4 \tag{11-4}$$

there is an activation energy for the reverse recombination reaction. In the other, for example,

$$C_2H_6 \longrightarrow 2CH_3 \tag{11-5}$$

there is no activation energy for recombination.

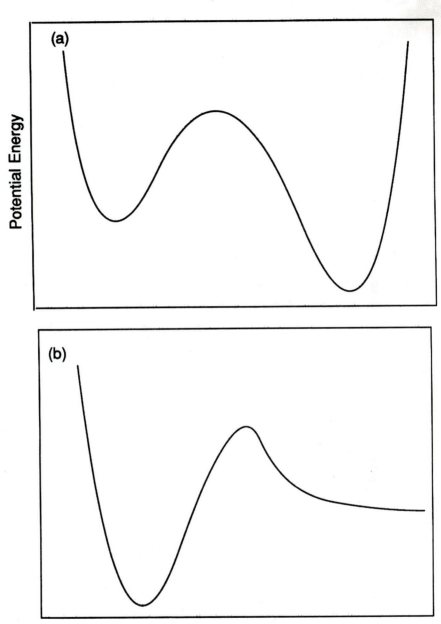

FIGURE 11-1   Schematic potential energy profiles for three types of unimolecular reaction. (a) Isomerization. (b) Dissociation where there is a high energy barrier and a large activation energy for reaction in both the forward and reverse directions. (*Continued*)

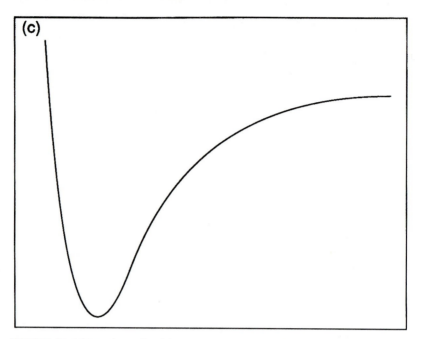

FIGURE 11-1 (*Continued*) (c) Dissociation where the potential energy rises monotonically as for rotational ground state species, so that there is no barrier to the reverse association reaction.

To understand the dynamics and kinetics of unimolecular reactions a number of different topics must be treated and mastered. Some of the most important are (1) the formation of an energized molecule with energy E, (2) the intramolecular vibrational/rotational motion of highly energized molecules; (3) unimolecular decomposition lifetime distributions and rate constants; (4) intermolecular energy transfer to and from highly energized molecules; and (5) energy partitioning between unimolecular dissociation fragments.

There are a number of ways to present these topics. The one partially adopted here is based on a historical perspective of the development of unimolecular reaction dynamics. Its strength is that the special relationship between theory and experiment which has existed in this field is emphasized. In the presentation given, the primary focus will be on unimolecular reactions which occur on the ground-state potential energy surface.

11.1   FORMATION OF ENERGIZED MOLECULES

Before a unimolecular reaction can occur, a molecule must be energized above the reaction threshold energy E_0. A number of different experimental methods have been used to accomplish this energization. For example, energization can take place by transfer of energy in a bimolecular collision, as in the reaction

$$CH_3NC + Ar \longrightarrow CH_3NC^* + Ar \tag{11-6}$$

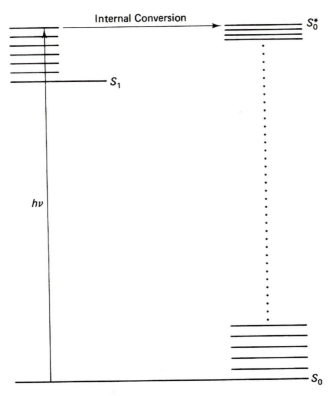

FIGURE 11-2 Preparation of a vibrationally excited ground-state molecule S_0^* by internal conversion from a vibrational level of S_1.

Another method which involves molecular collisions is chemical activation. Here the collision is reactive, and the molecule is excited by potential energy release as in the reaction

$$\mathrm{CH_3 + CF_3 \longrightarrow CH_3CF_3^*} \qquad (11\text{-}7)$$

The reaction coordinate potential energy profile for the $\mathrm{CH_3 + CF_3}$ system is similar to the one shown in Figure 11-1(c). The excited $\mathrm{CH_3CF_3}$ molecule can redissociate to radicals, or go on to form new products such as $\mathrm{CH_2 = CF_2 + HF}$.

Absorption of electromagnetic radiation is another means of preparing vibrationally excited molecules. A widely used method involves initial electronic excitation by absorption of one photon of visible or ultraviolet radiation. After this excitation, many molecules undergo a rapid internal conversion to the ground electronic state S_0, which converts the energy of the absorbed photon into vibrational energy. Such an energization scheme is depicted in Figure 11-2 for formaldehyde, where the complete excitation/decomposition mechanism is

$$\mathrm{H_2CO}(S_0) + h\nu \rightarrow \mathrm{H_2CO}(S_1) \longrightarrow \mathrm{H_2CO^*}(S_0) \rightarrow \mathrm{H_2 + CO} \qquad (11\text{-}8)$$

Molecules also acquire vibrational energy by the absorption of infrared photons. In contrast to visible and ultraviolet radiation, the energy provided by a single infrared photon is usually much less than the unimolecular threshold energy E_0. However, by using a sufficiently intense monochromatic radiation source such as an infrared laser, a molecule may be compelled to absorb multiple infrared photons so that its total vibrational energy is in excess of E_0.[1] This is discussed further in section 11.13.

Two additional excitation methods, both of which employ lasers, have also been developed: overtone excitation[2] and stimulated emission pumping (SEP).[3] In overtone excitation, states are prepared by direct single-photon absorption in which an MH or MD bond contains n quanta of vibrational energy. M is a massive atom in contrast to H or D. For states with large n, the energy in the bond may exceed the molecule's unimolecular threshold. Overtone excitation of CH and OH bonds has been used to decompose molecules. The limitation to the technique is the very weak oscillator strengths of overtone absorptions.

Stimulated emission pumping is an even more selective excitation process than overtone excitation. SEP involves the use of two lasers; the excitation mechanism is depicted in Figure 11-3. The first laser prepares a vibrational level of an excited electronic state. Before this state undergoes a spontaneous radiative or nonradiative tran-

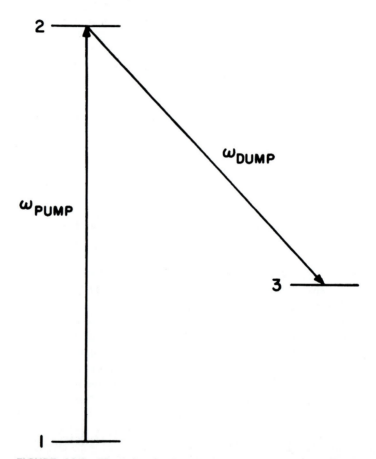

FIGURE 11-3 Three-level stimulated emission pumping (SEP) scheme, showing "PUMP" and "DUMP" transitions. The "PUMP" laser excites a vibrational/rotational level of an excited electronic state. The "DUMP" laser then stimulates emission from this state to an excited vibrational/rotational level of the ground electronic state. [Adapted from C. Kittrell, et al. *J. Chem. Phys. 75*, 2056 (1981).]

sition, a second laser with a lower energy photon stimulates emission of light. The stimulated emission forms a highly excited vibrational level S_0. Because of the high resolution of the lasers, individual rotation-vibration states of S_0 may often be prepared.

The foregoing excitation methods differ in the extent of the vibrational state selectivity inherent in their energization processes. Later in this chapter it will be seen that, as the vibrational selectivity of the experiment is increased, an accurate unimolecular rate theory often requires a more detailed description of molecular properties.

11.2   SUM AND DENSITY OF STATES

At the high energies to which unimolecular reactants are energized, there is often a large number of vibrational states in an energy interval dE. The number of these vibrational states is denoted by $W(E)$, which is related to the sum of the states, $G(E)$, and the density of the states, $N(E)$. $G(E)$ is defined as the number of vibrational states in the energy range 0 to E. The number of states in the energy interval from E to $E + dE$ is $W(E) = N(E)dE$. Thus, the density of the states, $N(E)$, is given by $[G(E + dE) - G(E)]/dE$, or simply $dG(E)/dE$ if $G(E)$ is a continuous function.

The calculation of $G(E)$ and $N(E)$ may be illustrated by using the low-energy harmonic vibrational energy levels of H_2O depicted in Figure 11-4. For $E = 5000 \text{ cm}^{-1}$ $G(E)$ is 6, and it rises to 19 for $E = 10000 \text{ cm}^{-1}$. Clearly $N(E)$ is not a continuous function. For higher energies and for molecules with more vibrational degrees of freedom, $G(E)$ becomes very large and approximates a continuous function. This is illustrated in Figure 11-5, where values of $G(E)$ for CH_4 are plotted against E. Large values of harmonic $G(E)$ and $N(E)$ are easily enumerated on a computer with the very efficient Beyer-Swinehart algorithm.[4] Continuous densities of states are illustrated in Figure 11-6.

If all the frequencies in the molecule are identical, evaluating the harmonic $W(E)$ and $G(E)$ terms becomes an elementary combinatorial problem. For a molecule having s oscillators, each with frequency ν and total molecular energy above the zero-point level $E = jh\nu$, $W(E)$ and $G(E)$ are found by considering the number of ways of distributing the j quanta among the s oscillators; specifically,

$$W(E) = \frac{(j + s - 1)!}{j!(s - 1)!} \tag{11-9}$$

and

$$G(E) = \frac{(j + s)!}{j!s!} \tag{11-10}$$

It is worthwhile to consider the classical equivalents to $G(E)$ and $N(E)$, since they provide a basis for analytic approximations to the quantum mechanical sum and density of states. As shown in Appendix 2, the classical sum of states is

$$G(E) = \frac{E^s}{s!\prod\limits_{i=1}^{s} h\nu_i} \tag{11-11}$$

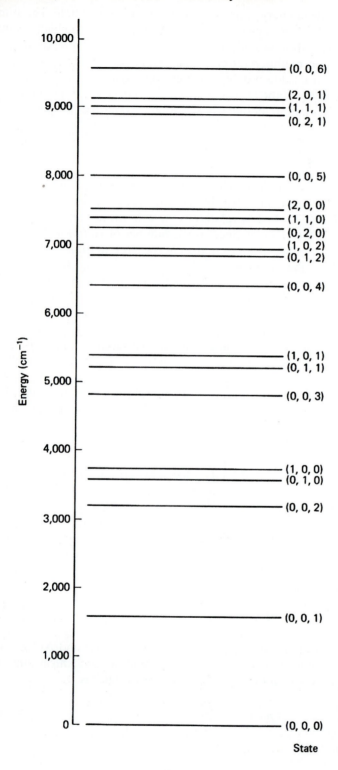

FIGURE 11-4 Harmonic vibrational levels for H_2O. ν_1 (asymmetric stretch) = 3750 cm^{-1}, ν_2 (symmetric stretch) = 3650 cm^{-1}, and ν_3 (bend) = 1600 cm^{-1}.

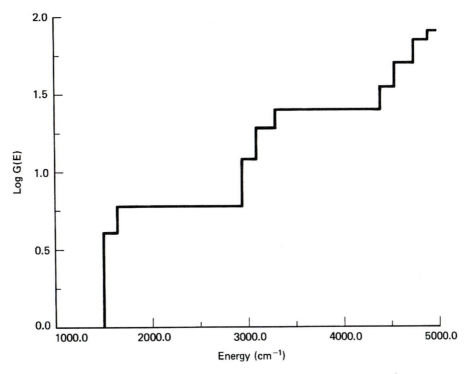

FIGURE 11-5   Direct count of $G(E)$ for CH_4. Energy is in cm^{-1}.

The classical density of states $dG(E)/dE$ is then

$$N(E) = \frac{E^{s-1}}{(s-1)! \prod_{i=1}^{s} h\nu_i} \tag{11-12}$$

As illustrated in Table 11-1, the classical expression for the sum severely underestimates the quantum mechanical harmonic value. Similar errors are observed for the density of states. The error in the classical approximation decreases with energy, and, according to the principle of correspondence, classical and quantum theories should become equivalent in the high-energy limit. However, for energies of interest in unimolecular reactions, the classical sum and density of states are woefully inadequate.

To improve on the classical approximation, Marcus and Rice[5] suggested that the classical energy should be measured from the minimum of the vibrational potential. Thus, the quantum sum or density at energy E should be compared with the classical sum or density at energy $E + E_z$, where E_z is the molecular zero-point energy of $\sum h\nu_i/2$. The resulting "semiclassical" expressions for the sum and density of states are the same as equations (11-11) and (11-12), except E is replaced by $E + E_z$. Semiclassical values for $G(E)$ are given in Table 11-1. Plainly, this approach gives a more accurate approximation than the classical expression given by equation (11-11); however, the number of states is overestimated.

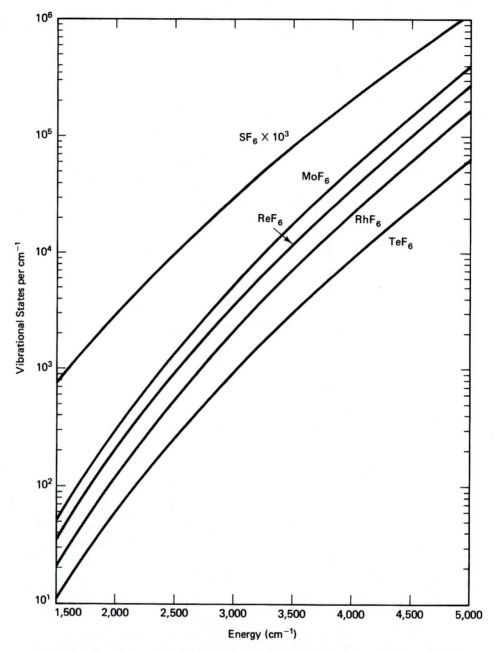

FIGURE 11-6 Density of vibrational states for MoF_6, ReF_6, RhF_6, TeF_6, and SF_6. [Adapted from D. Jackson, *Statistical Thermodynamic Properties of Hexafluoride Molecules,* Report LA-6025-MS, Los Alamos Scientific Laboratory (1975).]

TABLE 11-1   Values of Harmonic $G(E)$ for Cyclopropane by Various Methods[a].

E (kJ/mol)	Classical (11-11)	Marcus-Rice (11-13)	Whitten-Rabinovitch (11-16)	Exact Count
41.8	0.00	5.45×10^2	7.17×10^2	8.02×10^2
125	0.02	20.9×10^6	2.65×10^6	2.69×10^6
209	8.20×10^2	21.5×10^8	6.15×10^8	6.12×10^8
418	1.72×10^9	9.94×10^{12}	5.90×10^{12}	5.84×10^{12}
627	8.58×10^{12}	4.03×10^{15}	3.02×10^{15}	3.00×10^{15}
837	3.61×10^{15}	4.27×10^{17}	3.56×10^{17}	3.54×10^{17}

[a]Taken from Table 5.11 of M. Quack and J. Troe, *Ber. Bunsengesellschaft Phys. Chem.* 78, 240 (1974). The cyclopropane harmonic vibrational frequencies are 3221(6), 1478(3), 1118(7), 878(3), 749(2) cm^{-1}. The values in parentheses are the degeneracies associated with each frequency.

Although the Marcus-Rice expression is not sufficiently accurate, it leads to a greatly improved semiclassical type of approach. Rabinovitch and co-workers[6] suggested that only an appropriate fraction of the zero-point energy should be included in the classical energy. Specifically, they proposed that

$$G(E) = \frac{(E + aE_z)^s}{s! \prod_{i=1}^{s} h\nu_i} \tag{11-13}$$

The empirical factor a, which has the limits of 0 and 1 for $E \to 0$ and $E \to \infty$, respectively, is written as a function of $E' = E/E_z$ and the dispersion in the molecular vibrational frequencies. Explicitly,

$$a = 1 - \beta w(E') \tag{11-14}$$

where

$$\beta = \frac{s-1}{s} \frac{\langle \nu^2 \rangle}{\langle \nu \rangle^2}$$

and

$$w = [5.00E' + 2.73(E')^{0.50} + 3.51]^{-1} \qquad (0.1 < E' < 1.0)$$
$$w = \exp[-2.4191(E')^{0.25}] \qquad (1.0 < E' < 8.0) \tag{11-15}$$

Differentiating equation (11-13) with respect to E gives, for the density of states,

$$N(E) = \frac{(E + aE_z)^{s-1}}{(s-1)! \prod_{i=1}^{s} h\nu_i} \left[1 - \beta \frac{dw(E')}{dE'} \right] \tag{11-16}$$

As shown in Table 11-1, equation (11-16) gives values for the sum of states that are in excellent agreement with the actual quantum mechanical values.

11.3 LINDEMANN-HINSHELWOOD THEORY OF THERMAL UNIMOLECULAR REACTIONS

Perrin[7] proposed in 1919 that reactant molecules acquire energy by absorbing radiation from the walls of the reaction vessel. Many textbooks on kinetics assert that this *radiation hypothesis* was refuted by Langmuir,[8] who showed that the amount of energy radiated to the gas would be insufficient to account for observed unimolecular reaction rates. Actually, a careful reading of Langmuir's paper reveals that the hypothesis is not disproved by his arguments. Langmuir calculated the radiation flux only for $h\nu = hc/\lambda > E_0$, i.e., for $\lambda < 400$ nm, corresponding to $E_0 \approx 300$ kJ/mol. Since the black-body radiation density is essentially zero for wavelengths below 400 nm, for a typical oven temperature of 1,000 K, there would of course be insufficient radiation to energize the molecules.

What Langmuir neglected in this analysis is that most molecules (with the exception of homonuclear diatomics, such as I_2) absorb strongly in the infrared region of the spectrum, and the black-body radiation curve actually peaks in the infrared, near $\lambda = 3$ μm at $T = 1,000$ K. To be fair, infrared absorption by molecules was essentially unknown in the early twentieth century, when this debate was being conducted. If Langmuir's calculation is redone for infrared absorption (see problem 11.4), it becomes clear that there is more than enough energy radiated into and absorbed by the gas to account for observed unimolecular decomposition rates.

Despite the role that could be played by infrared radiation, the radiation hypothesis is in fact untenable for thermal unimolecular reactions, except in the zero-pressure limit where collisions cannot participate in the activation process. (The role of activation by infrared radiation for thermal unimolecular reactions in the zero-pressure limit is discussed later, in section 11.13). There are several reasons for this:

1. Langmuir's argument is applicable to reactions of homonuclear diatomic molecules such as iodine, which dissociates readily at 800–1,000 K but absorbs light only at wavelengths less than 600 nm. Since this case refutes the hypothesis, it cannot be generally applicable.

2. The radiation hypothesis predicts a rate dependence on the surface-to-volume ratio in the reactor, or on the presence of nonreactive absorbers, which is at variance with experimental results.

3. Most importantly, a radiation-driven reaction would have a rate truly independent of gas pressure. However, as discussed subsequently, there is a pronounced falloff of the rate at low gas pressures. It is primarily this observation which forces us to abandon the radiation hypothesis in favor of a collisional-activation mechanism developed by Lindemann and Hinshelwood.

In 1922, Lindemann[9] proposed a general theory for thermal unimolecular reactions, which forms the basis for the current theory of thermal unimolecular rates. He proposed that molecules become energized by bimolecular collisions, as in equation (11-6), with a time lag between the moment of collisional energization and the time the molecule decomposes. Energized molecules could then undergo deactivating collisions before decomposition occurred. A major achievement of Lindemann's theory is its ability to explain the experimental finding that the reaction rate changes from first to second order in going from the high- to low-pressure limit.

According to these concepts, the mechanism for the unimolecular reaction may be written as

$$A + M \xrightarrow{k_1} A^* + M$$

$$A^* + M \xrightarrow{k_{-1}} A + M \qquad (11\text{-}17)$$

$$A^* \xrightarrow{k_2} \text{Products}$$

Here, A* represents a molecule with sufficient energy to react and M is some collision partner, which could be A itself. It is assumed that each A* + M collision is "strong" and thus leads to de-energizing of A; this is known as the *strong collision assumption* for de-energizing collisions. Therefore, k_{-1} is equated with the gas-kinetic collision number Z_1. If the steady-state hypothesis is applied to the concentration of A*, the overall rate of reaction becomes

$$\text{rate} = k_{\text{uni}}[A] = k_2[A^*] = \frac{k_1 k_2[A][M]}{k_{-1}[M] + k_2} \qquad (11\text{-}18)$$

Clearly, the reaction rate is first order at high pressures, where $[M] \to \infty$, and second order at low pressures, where $[M] \to 0$. The high- and low-pressure expressions for k_{uni} are

$$k_\infty = \frac{k_1 k_2}{k_{-1}} \qquad (11\text{-}19)$$

and

$$k_0 = k_1[M] \qquad (11\text{-}20)$$

The rate constant ratio k_1/k_{-1} is the ratio $[A^*]/[A]$ if the activation and deactivation steps are in equilibrium. Therefore, the high-pressure rate constant is simply the rate constant for unimolecular decomposition multiplied by the probability that the molecule is energized. At low pressure, the rate-determining step is bimolecular activation and the rate constant is linearly dependent on the pressure of [M]. The transition from the high-pressure rate constant $k_{\text{uni}} = k_\infty$ to the low-pressure linear decrease in k_{uni} is called the *fall-off region*. The pressure at which $k_{\text{uni}}/k_\infty = 1/2$ is denoted by $p_{1/2}$. From equations (11-18) and (11-19), it follows that

$$p_{1/2} \propto [M]_{1/2} = k_\infty/k_1 \qquad (11\text{-}21)$$

A plot of k_{uni} vs. p is illustrated in Figure 11-7.

To calculate a theoretical Lindemann plot of k_{uni} vs. [M], k_{uni} in equation (11-18) is written as

$$k_{\text{uni}} = \frac{k_\infty}{1 + k_\infty/k_1[M]} \qquad (11\text{-}22)$$

The value of k_∞, the first-order rate constant at high pressure, is found from experiment. A working assumption that can be made in applying the Lindemann theory is to calculate k_1 from the line-of-centers collision theory expression (see chapter 8, section 8.1.2)

$$k_1 = Z_1 \exp\left(-E_0/k_B T\right) \qquad (11\text{-}23)$$

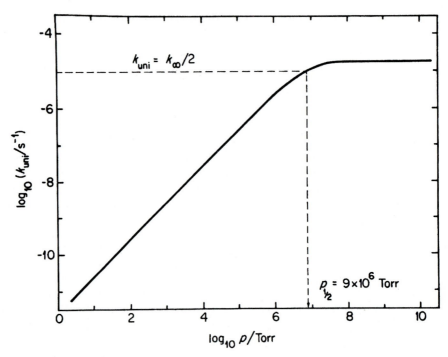

FIGURE 11-7 Theoretical Lindemann plot for *cis-* $\rightarrow$ *trans*-but-2-ene at 469°C. [Adapted from P. J. Robinson and K. A. Holbrook, *Unimolecular Reactions* (New York: Wiley-Interscience, 1972). Reproduced with permission of John Wiley and Sons Limited.]

where E_0 is simply equated to the high-pressure activation energy E_∞. In all cases, such an interpretation of the Lindemann theory leads to a gross discrepancy with experiment. The theory predicts the fall-off in k_{uni} to occur at much higher pressures than those observed experimentally. Since there can be no doubt about k_∞, which is an experimental quantity, the error must be in the estimation of k_1. The Hinshelwood modification to the Lindemann mechanism, described shortly, gives a much larger value for k_1, and theoretical fall-off curves that are in much better agreement with experiment are obtained.

The line-of-centers energization rate constant of equation (11-23) gives the frequency of collisions in which the relative kinetic energy along the line of centers is greater than or equal to E_0. No account is taken of the internal energy of the reactant molecule. In 1926 Hinshelwood[10] proposed that, in addition to the relative motion, the internal degrees of freedom can contribute to the threshold energy E_0. The probability that a molecule contains energy greater than or equal to E_0 clearly increases with the number of internal degrees of freedom. As a result, the energization rate constant k_1 is larger for a complex reactant molecule than for a simple one. For a reactant molecule with s classical degrees of freedom, Hinshelwood found that

$$k_1 = \frac{Z_1}{(s-1)!} \left(\frac{E_0}{k_B T}\right)^{s-1} \exp\left(-E_0/k_B T\right) \tag{11-24}$$

Thus, the energization rate constant is a product of two terms: the line-of-centers hard-sphere collision rate constant Z_1 (see chapter 8) and the probability that the total energy of s classical harmonic oscillators for the reactant molecule exceeds E_0 [see equation (A2-33) in Appendix 2].

To derive the Hinshelwood expression for k_1, one starts with equation (A2-31) in Appendix 2, which gives the classical probability that a molecule has internal energy in the range from E to $E + dE$. This probability also represents the fraction of molecules energized in this range. The differential energizing rate constant for formation of reactant molecules with energy lying between E and $E + dE$ is dk_1. At high concentrations (i.e., where $[M] \to \infty$), where dissociation of energized molecules is negligible, an equilibrium is maintained between the energizing and deenergizing steps in equation (11-17). Under these conditions dk_1/k_{-1} is the statistical fraction of molecules with internal energy between E and $E + dE$, so that

$$\frac{dk_1}{k_{-1}} = \frac{1}{(s-1)!} \left(\frac{E}{k_B T} \right)^{s-1} e^{-E/k_B T} \left(\frac{dE}{k_B T} \right) \tag{11-25}$$

Equation (11-25) is rigorously valid only in the high-pressure limit. To use it for all concentrations, the *strong collision assumption* must be made. The application of this assumption for de-energizing collisions has already been discussed. According to the strong collision assumption, large amounts of energy are transferred in the molecular collisions so that energizing and de-energizing may be viewed as single-step processes, in contrast to ladder-climbing processes. This assumption also requires the collision to be so violent that the distribution of energy resulting from energizing or de-energizing collisions is random. As a result, the distribution of energized molecules with energy in the range from E to $E + dE$ may be determined directly from statistical considerations, yielding equation (11-25).

To find the Hinshelwood expression for $k_1/k_{-1}, dk_1/k_{-1}$ equation (11-25) is integrated between the limits $E = E_0$ and ∞ to obtain

$$\frac{k_1}{k_{-1}} = \int_{E=E_0}^{\infty} \frac{dk_1}{k_{-1}} = \int_{E=E_0}^{\infty} \frac{1}{(s-1)!} \left(\frac{E}{k_B T} \right)^{s-1} e^{-E/k_B T} \left(\frac{dE}{k_B T} \right) \tag{11-26}$$

For the normal situation where E_0 is much greater than $(s-1)k_B T$, the result of the integration is approximately [equation (A2-33) in Appendix 2]

$$\frac{k_1}{k_{-1}} = \frac{1}{(s-1)!} \left(\frac{E_0}{k_B T} \right)^{s-1} e^{-E_0/k_B T} \tag{11-27}$$

For $k_{-1} = Z_{-1} = Z_1$, the Hinshelwood expression for k_1, equation (11-24), results. Inserting equation (11-24) into equation (11-19) gives the high-pressure rate constant according to the Hinshelwood-Lindemann theory, i.e.,

$$k_\infty = \frac{k_2}{(s-1)!} \left(\frac{E_0}{k_B T} \right)^{s-1} e^{-E_0/k_B T} \tag{11-28}$$

Equations (11-24) and (11-28) form the basis of what is usually called the Hinshelwood-Lindemann theory of thermal unimolecular dissociation. Unimolecular falloff curves calculated with this theory are far superior to those calculated with the

Lindemann theory. However, significant differences are still found between theoretical and experimental k_{uni} curves versus [M], particularly at low pressures.

Since k_1 increases with s in the Hinshelwood-Lindemann theory, the dissociation rate constant k_2 decreases with s. This may be illustrated using the typical parameters $A_\infty = 10^{13}$ sec^{-1} and $E_\infty/RT = 40$. The value for k_2 is given by $k_\infty(k_{-1}/k_1)$; for $s = 1, 5$, and 10, $k_2 = 10^{13}, 10^{9.5}$, and $10^{6.5}$ sec^{-1}, respectively. Thus, the lifetime $\tau = 1/k_2$ of the energized molecule increases when the molecule can store energy among a greater number of degrees of freedom. Intuitively, this is what is expected; however, one also expects k_2 to depend on the energy content of the energized molecule A*. According to the Hinshelwood-Lindemann theory, k_2 is a constant that is independent of energy.

In the next section we discuss how statistical theories may be used to calculate k_2 as a function of energy. Making k_2 energy-dependent, expressed as $k(E)$, requires a generalized Hinshelwood-Lindemann mechanism in order to consider energizing, de-energizing, and decomposition for an energy interval from E to $E + dE$:

$$A + M \xrightarrow{dk_1} A*(E, E + dE) + M$$

$$A*(E, E + dE) + M \xrightarrow{k_{-1}} A + M \qquad (11\text{-}29)$$

$$A*(E, E + dE) \xrightarrow{k(E)} \text{products}$$

In this mechanism $k(E)$ has replaced the energy-independent rate constant k_2. Applying the steady-state approximation to the energized intermediate A*$(E, E+ dE)$ leads to the differential unimolecular rate constant

$$dk_{uni}(E, E + dE) = \frac{k(E)(dk_1/k_{-1})}{1 + k(E)/k_{-1}[M]} \qquad (11\text{-}30)$$

The total thermal unimolecular rate constant is obtained by integration thus:

$$k_{uni} = \int_{E_0}^{\infty} \frac{k(E)(dk_1/k_{-1})}{1 + k(E)/k_{-1}[M]} \qquad (11\text{-}31)$$

As discussed previously, it is assumed that for all pressures dk_1/k_{-1} represents the equilibrium probability that the reactant molecule A* has energy in the range from E to $E + dE$. This probability may be denoted $P(E) dE$. Also, $k_{-1}[M]$ is the collision frequency ω between an A* molecule and bath molecules. Making these two notational changes, equation (11-31) becomes the familiar expression for the thermal unimolecular rate constant,

$$k_{uni} = \omega \int_{E_0}^{\infty} \frac{k(E)P(E)dE}{k(E) + \omega} \qquad (11\text{-}32)$$

11.4 STATISTICAL ENERGY-DEPENDENT RATE CONSTANT $k(E)$

Two quite different approaches may be taken to determine an energy-dependent expression for k_2 denoted by $k(E)$. One is to consider the explicit nature of the intramolecular motion of highly energized molecules. This was the approach taken by Slater[11] in 1939, in which he pictured a molecule as an assembly of harmonic oscillators. Decomposition is assumed to occur when a critical coordinate (e.g., a bond length or bond angle) attains a critical displacement. A discussion and critique of the Slater theory, along with a general discussion of the intramolecular dynamics of highly energized molecules, is given in section 11-12.

In contrast to determining $k(E)$ in terms of intramolecular motion, an expression may be formulated totally on the basis of statistical assumptions. This is the approach used in the RRK (Rice-Ramsperger-Kassel) theory and its extension, which is referred to as the RRKM (Rice-Ramsperger-Kassel-Marcus) theory. These two theories picture a molecule as a collection of coupled harmonic oscillators which exchange energy freely under the assumptions that

1. All internal molecular states of A* at energy E are accessible and will ultimately lead to decomposition products, and

2. Vibrational energy redistribution within the energized molecule is much faster than unimolecular reaction.

These two assumptions are embodied in the postulate of the RRK and RRKM theories that complete intramolecular vibrational energy redistribution (IVR) occurs on a time scale much shorter than $1/k(E)$.

Both the RRK and the RRKM theory assume that initially, when $\tau = 0$, there is a statistical population of reactant internal states at a fixed total energy. A collection of molecules forming such a distribution is called a *microcanonical ensemble*. The assumption requires that each state have equal probability of decomposing. Thus, a microcanonical ensemble will be maintained as the reactant molecules decompose, and, as a result, the unimolecular decomposition reaction will be described by only one time-independent rate constant $k(E)$, which is that for the microcanonical ensemble of reactant molecules. Such a unimolecular system obeys the Ergodic Principle of statistical mechanics.[12,13]

The assumption of rapid and complete intramolecular vibrational energy redistribution implies that A* will have a random lifetime distribution given by[14]

$$P(\tau) = k(E)\exp[-k(E)\tau] \tag{11-33}$$

where $P(\tau)$ is the probability per unit time that A*, which contains energy E, has a lifetime τ. Equation (11-33) is similar to that for the probability of radioactive decay, which is what one expects if the decomposition of A* is purely statistical. With random lifetimes there is no particular tendency for all of the A* molecules to decompose soon after their formation, or even to exist for some particular length of time before decomposing.

Slater and Bunker[14,15] have shown that equation (11-32), which embodies the Lindemann-Hinshelwood theory with the strong collision assumption, requires the random lifetime distribution. To demonstrate this, let τ be the time required for a certain A* molecule with energy E to reach dissociation, by virtue of its natural motion in the absence of collisions. For an equilibrium distribution of molecular states at E, there will be a distribution $P(\tau)$ of these natural lifetimes. The probability that A* will *not* undergo a collision in time τ is $\omega\exp(-\omega\tau)$ if the average collision frequency is ω. In that case the rate of reaction at energy E is

$$k_{\text{uni}}(E) = \omega \int_0^\infty \exp(-\omega\tau)P(\tau)d\tau \tag{11-34}$$

and the complete thermal unimolecular rate constant is

$$k_{\text{uni}} = \int_0^\infty k_{\text{uni}}(E)P(E)dE \tag{11-35}$$

If we compare equation (11-35) with equation (11-32), we can see by trial that correspondence is complete if $P(\tau)$ is given by equation (11-33).

11.5   RRK THEORY

The theory which is now known as RRK theory was developed independently and nearly simultaneously by Rice and Ramsperger[16] and by Kassel[17] in 1927 and 1928. Both classical and quantum versions of RRK theory were developed, and in the limit of a large excitation energy E the two versions become identical.

The classical theory is based on the notion that the probability that a molecule of s classical oscillators with total energy E has energy greater than or equal to E_0 in one chosen oscillator, which is the critical mode leading to reaction. This probability is given by the number of ways to attain this particular distribution divided by the total number of ways to distribute E between the s oscillators. The latter number is given in turn by the classical density of states, i.e., $N(E)$ of equation (11-12). The former is the convolution of the density of states in which the critical oscillator contains energy $E' + E_0$ and the density of states for energy $E - E' - E_0$ in the remaining $s - 1$ oscillators. As a result, the probability for the critical oscillator having energy greater than or equal to E_0 is

$$\int_0^{E-E_0}\left[\left(\frac{(E - E' - E_0)^{s-2}}{(s-2)!\prod_{i=1}^{s-1}h\nu_i}\,\frac{dE'}{h\nu_s}\right)\right]\Bigg/ \frac{E^{s-1}}{(s-1)!\prod_{i=1}^{s}h\nu_i} \tag{11-36}$$

which after integration becomes

$$\left(\frac{E - E_0}{E}\right)^{s-1} \tag{11-37}$$

The classical RRK rate constant is simply this probability multiplied by the vibrational frequency for the critical oscillator, i.e.,

$$k(E) = \nu\left(\frac{E - E_0}{E}\right)^{s-1} \tag{11-38}$$

The assumptions used to derive the quantum RRK rate constant are in essence very similar to those for the classical theory. In the quantum theory it is assumed there are s identical oscillators in the molecule, all having frequency ν. The energized molecule contains a total of j quanta, so that its energy is $E = jh\nu$. The critical oscillator must contain m quanta for dissociation to occur ($m = E_0/h\nu$). To find the quantum rate constant $k(E)$, two different statistical weights are required: the total number of ways to distribute the j quanta among the s oscillators, given by equation (11-9); and the total number of ways to attain the particular distribution which leads to reaction;

i.e., a number of quanta in the critical oscillator which ranges from m to j. The latter statistical weight is the sum of states, equation (11-10), for a maximum of $j - m$ quanta in the remaining $s - 1$ oscillators, and is given by

$$\frac{(j - m + s - 1)!}{(j - m)!\,(s - 1)!} \tag{11-39}$$

The probability that a particular oscillator has m quanta and all s oscillators have j quanta is the ratio of these two statistical weights:

$$\frac{(j - m + s - 1)!\,j!}{(j - m)!\,(j + s - 1)!} \tag{11-40}$$

The quantum RRK rate constant is the product of this probability and the vibrational frequency ν and is given by

$$k(E = jh\nu) = \nu\,\frac{(j - m + s - 1)!\,j!}{(j - m)!\,(j + s - 1)!} \tag{11-41}$$

In the classical limit, where $j \gg s$ and $j - m \gg s$, the quantum RRK rate constant becomes

$$k(E = jh\nu) = \nu\left(\frac{j - m}{j}\right)^{s-1} \tag{11-42}$$

If both the numerator and denominator in this equation are multiplied by $(h\nu)^{s-1}$, the equation becomes the classical RRK rate constant in equation (11-38).

The RRK rate constant depends on both the energy in excess of the unimolecular threshold and the number of vibrational degrees of freedom. To illustrate, the classical RRK expression for $k(E)$ divided by ν is plotted for several values of s in Figure 11-8 for a model system for which $E_0 = 40.0$ kcal/mole. Clearly, the rate constant increases with energy approaching its limiting value of ν. For a particular value of E, the rate constant decreases as s is increased. The physical explanation of this is that for larger s there are more ways of distributing the energy and hence less chance for energy to be localized in the critical oscillator.

There are both classical and quantum RRK expressions for the pressure-dependent thermal unimolecular rate constant k_{uni}, which result from using either the classical or quantum expression, respectively, for $dk_1/k_{-1} = P(E)\,dE$ and $k(E)$. The classical formulation of k_{uni} is the one most widely applied. It is found by inserting the classical mechanical expressions for $k(E)$ [equation (11-38)] and $P(E)dE$ [equation (11-25)] into equation (11-32). The resulting equation may be conveniently reduced by making the substitutions

$$x = \frac{E - E_0}{RT}$$

and

$$b = \frac{E_0}{RT}$$

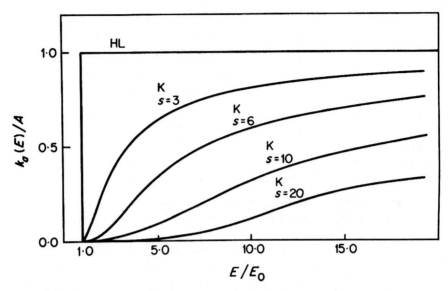

FIGURE 11-8 Comparison of $k_a(E)$ from Hinshelwood-Lindemann and classical Kassel theories. [Adapted from P. J. Robinson and K. A. Holbrook, *Unimolecular Reactions* (New York: Wiley-Interscience, 1972.) Reproduced with permission of John Wiley and Sons, Limited.]

which gives

$$k_{\text{uni}} = \frac{v e^{-E_0/RT}}{(s-1)!} \int_0^\infty \frac{x^{s-1} e^{-x} dx}{1 + \dfrac{v}{\omega}\left(\dfrac{x}{b+x}\right)^{s-1}} \tag{11-43}$$

In the quantum mechanical formulation of k_{uni}, there is a summation rather than the integral of equation (11-43). As shown in Appendix 1, the quantum expression for $P(E)$ is

$$P(E) = \frac{W(E)\exp(-E/k_B T)}{Q_{\text{vib}}} \tag{11-44}$$

where $W(E)$ is given by equation (11-9).

In the high-pressure limit ($\omega \to \infty$), both the classical and quantum RRK expressions for k_{uni} become the Arrhenius equation

$$k_\infty = v \exp(-E_0/k_B T) \tag{11-45}$$

This is readily apparent for the classical case, since in the limit $\omega \to \infty$ the integral in equation (11-43) defines $(s-1)!$. Thus, according to the RRK model, the parameters E_0 and v needed to evaluate $k(E)$ are simply given by the high-pressure Arrhenius equation. Note that, if equation (11-25) is used for dk_1/k_{-1}, equation (11-38) is the only expression that will give rise to equation (11-45).[18]

The parameters required to calculate a curve of k_{uni} vs. ω (i.e., pressure) by either the classical or quantum RRK theories are s, v, and E_0. These theories are a major

TABLE 11-2 High-Pressure Arrhenius Parameters.

Reaction	Temperature, K	$\log A_\infty$	E_∞	Reference
$C_2H_6 \rightarrow 2CH_3$	840–913	16.72 ± 0.17	88.85 ± 0.68	a
Cyclopropane $\rightarrow$ propylene	690–1,038	15.50 ± 0.14	65.97 ± 0.50	b
$(CHO)_2 \rightarrow 2CO + H_2$	1,100–1,540	14.03	55.1	c
$CH_3OH \rightarrow CH_3 + OH$	—	15.97	89.9	d
Cyclobutane $\rightarrow 2C_2H_4$	1,308–1,498	15.6	62.5	e
Oxetan $\rightarrow C_2H_4 + H_2CO$	693–753	15.71 ± 0.31	63.02 ± 0.84	f
$C_3H_8 \rightarrow CH_3 + C_2H_5$	1,400–2,300	17.17	86.4	g
$(CH_3)_4C \rightarrow (CH_3)_3C + CH_3$	1,140–1,300	17.23	84.0	h

(a) A.B. Trenwith, *Far. Trans. I 75*, 614 (1979); (b) H. Furue and P. D. Pacey, *Can. J. Chem. 60*, 916 (1982); (c) K. Saito, T. Kakumoto, and I. Murakami, *J. Phys. Chem. 88*, 1182 (1984); (d) H. G. Wagner, *Ber. Bunsenges. Phys. Chem. 86*, 2 (1982); (e) D. K. Lewis, A. S. Feinstein, and P. M. Jeffers, *J. Phys. Chem. 81*, 1887 (1977); (f) K. A. Holbrook and R. A. Scott, *Far. Trans. I 71*, 1849 (1975); (g) M. Z. Al-Alami and J. H. Kiefer, *J. Phys. Chem. 87*, 499 (1983); (h) D. Bernfeld and G. B. Skinner, *J. Phys. Chem. 87*, 3732 (1988).

improvement over the Lindemann-Hinshelwood treatment. In fact, the RRK theories gave for the first time a treatment which could reproduce experimental results with reasonable accuracy. However, to obtain agreement with experiment with the classical theory, a value for s of about one-fourth to two-thirds the total number of normal modes in the molecule is needed, depending on the temperature. This feature arises from the serious errors that result from using classical statistical mechanics. The quantum RRK theory gives good results by equating s to the total number of normal modes.

The RRK theory assumes that the Arrhenius high-pressure thermal A-factor is given by the frequency for the critical oscillator, which is usually in the range of 10^{13} to 10^{14} sec^{-1}. However, as shown in Table 11-2 for many reactions, Arrhenius high-pressure thermal A-factors are in fact larger than 10^{14} sec^{-1}. The inability to explain these experimental results is one of the shortcomings of the RRK theory that is overcome by the RRKM theory.

11.6 RRKM THEORY

The RRKM theory was developed using the RRK model and extending it to consider explicitly vibrational and rotational energies and to include zero-point energies.[19] Several minor modifications of the theory have been made since its conception, primarily as a result of improved treatments of external degrees of freedom.[20-26] Detailed discussions of the RRKM theory are given in books by Robinson and Holbrook,[27] Forst,[28] Gilbert and Smith,[29] and Baer and Hase.[30]

RRKM theory is a microcanonical transition state theory,[31] and as such, it gives the connection between statistical unimolecular rate theory and the transition state theory of thermal chemical reaction rates discussed in chapter 10. Isomerization or dissociation of an energized molecule A* is assumed in RRKM theory to occur via the mechanism

$$A^* \xrightarrow{k(E)} A^\ddagger \longrightarrow products \qquad (11\text{-}46)$$

where $A^{\ddagger}$ is the transition state. Bunker[21] has argued that *critical configuration* is a more appropriate name for $A^{\ddagger}$, since it represents a unique molecular configuration intermediate between reactant molecule and products. As discussed in chapter 10, trajectories passing from A^* to products, or from products to A^*, are assumed to pass through the transition state only once.

The internal degrees of freedom of A^* and $A^{\ddagger}$ are designated as either active or adiabatic. Active modes are assumed to exchange energy freely, while an adiabatic mode is one that remains in the same quantum state during the reaction. Conservation of angular momentum usually requires the external rotational modes to be treated as adiabatic. By contrast, the vibrational and internal rotational modes are considered to be active.

In applications of RRKM theory A^* and $A^{\ddagger}$ are usually represented as symmetric tops with rotational energy

$$E_r(J, K) = [J(J + 1) - K^2]\hbar^2/2I_a + K^2\hbar^2/2I_c \qquad (11\text{-}47)$$

In a simplified RRKM model the J-dependent term of the rotational energy, $E_r(J) = J(J + 1)\hbar^2/2I_a$, is assumed to be adiabatic while the K-dependent term, $E_r(K) = K^2\hbar^2(I_c^{-1} - I_a^{-1})/2$, is assumed to be active.[23,24,29] The total energy E of the energized molecule may then be written as the sum $E_v + E_r$, where E_v is the vibrational/internal rotational energy and the external rotational energy E_r equals $J(J + 1)\hbar^2/2I$. Molecular motion from A^* to $A^{\ddagger}$ involves a change in potential energy. As a result, the total energy of the critical configuration is

$$E^{\ddagger} + E_r^{\ddagger} = E - E_0 \qquad (11\text{-}48)$$

where E_0 is the potential energy difference between A^* and $A^{\ddagger}$. Part of the energy $E^{\ddagger}$, denoted by $E_v^{\ddagger}$, is associated with vibration, while another part, denoted by $E_t^{\ddagger}$, consists of translational motion in the reaction coordinate. Their sum $E_v^{\ddagger} + E_t^{\ddagger}$ equals $E^{\ddagger}$. These energies are illustrated in Figure 11-9.

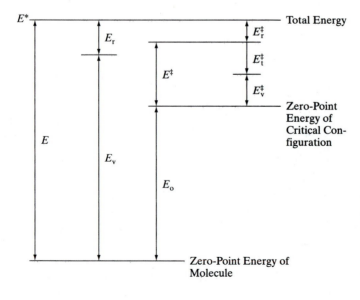

FIGURE 11-9 Diagram of the energies for the RRKM theory. [Adapted from W. L. Hase, in *Modern Theoretical Chemistry, Dynamics of Molecular Collisions, Part B*, edited by W. H. Miller (New York: Plenum, 1976), p. 121.]

The RRKM rate constant $k(E)$ is the microcanonical transition state theory rate constant, Equation (10-97), and is given by

$$k(E) = \frac{1}{h} \frac{G(E^{\ddagger})}{N(E_v)} \tag{11-49}$$

where $G(E^{\ddagger})$ is the sum of states for the active degrees of freedom in the transition state, $N(E_v)$ is the density of states for the active degrees of freedom in the reactant, and the reaction path degeneracy is incorporated into $G(E^{\ddagger})$. If the classical mechanical equations (11-11) and (11-12) are used for $G(E^{\ddagger})$ and $N(E_v)$, it is easily shown that the RRKM $k(E)$ becomes that of classical RRK theory, viz. equation (11-38). Note that the density in equation (11-49) is for all vibrational/internal rotational degrees of freedom in the reactant, while the sum does not include the reaction coordinate degree of freedom.

For many applications the RRKM rate constant is written as an explicit function of the rotational energy, i.e.,

$$k(E, E_r) = \frac{1}{h} \frac{G(E^{\ddagger})}{N(E - E_r)} \tag{11-50}$$

As shown in Figure 11-9, the energy for the sum of states in equation (11-50) may be written as

$$E^{\ddagger} = E_v - E_0 + E_r - E_r^{\ddagger} \tag{11-51}$$

The external rotational modes are assumed to be adiabatic, with their angular momentum $L = \sqrt{J(J+1)}\hbar$ conserved during the unimolecular decomposition. For the above simplified model of external rotation

$$E_r - E_r^{\ddagger} = \frac{L^2}{2I} - \frac{L^2}{2I^{\ddagger}} = E_r\left(1 - \frac{I}{I^{\ddagger}}\right) \tag{11-52}$$

Thus, the RRKM rate constant in equation (11-50) may be written as an explicit function of E_v and E_r:

$$k(E_v, E_r) = \frac{G(E_v + E_r[1 - I/I^{\ddagger}] - E_0)}{hN(E_v)} \tag{11-53}$$

In a more detailed and accurate treatment of external rotational energy, the RRKM rate constant is written as a function of total energy E and angular momentum quantum number J;[25,26,32,33] i.e.,

$$k(E, J) = \frac{G^{\ddagger}(E, J)}{hN(E, J)} \tag{11-54}$$

where $G^{\ddagger}(E, J)$ is the transition state's sum of states and $N(E, J)$ is the energized molecule's density of states. The K-dependent term in equation (11-47) is treated as an active degree of freedom by placing the proper limits on the quantum number K. For a particular set of J and K the active energies of the energized molecule and transition state, are $E - E_r(J, K)$ and $E - E_0 - E_r^\ddagger(J, K)$, respectively, so that the density and sum of states for E, J and K may be written as

$$N(E, J, K) = N[E - E_r(J, K)] \tag{11-55}$$

$$G^\ddagger(E, J, K) = G^\ddagger[E - E_0 - E_r^\ddagger(J, K)] \tag{11-56}$$

The density of states for A^* and sum of states for $A^\ddagger$ are found for E and J by summing over contributions from all possible values of K, i.e.

$$G^\ddagger(E, J) = \sum_{K=-J}^{J} G^\ddagger(E, J, K) \tag{11-57}$$

$$N(E, J) = \sum_{K=-J}^{J} N(E, J, K) \tag{11-58}$$

In this more accurate treatment of rotational energy the total energy cannot be written as the sum $E_v + E_r$, since the rotational energy changes for each value of K.

To determine an RRKM rate constant requires evaluating the sum of states for the transition state and the density of states for the reactant molecule. The internal vibrational degrees of freedom are usually treated as quantum harmonic oscillators, and the sum and density are then evaluated by either state counting[4] or the accurate Whitten-Rabinovitch approximation given by equations (11-13) through (11-16). Anharmonic corrections may be made to the harmonic sum and density.[34,35]

The information needed for calculating the sum and density of states in equation (11-54) includes the reactant total energy E and angular momentum J, the unimolecular threshold E_0, the harmonic vibrational frequencies and moments of inertia for both the reactant and the transition state. Usually E, J, and the reactants' moments of inertia and harmonic vibrational frequencies are known from experiment. However, obtaining the potential energy E_0 at the transition state and the transition state's structure (i.e., moments of inertia) and harmonic vibrational frequencies is usually less direct.

Determining the transition state's properties E_0, $I_i^\ddagger$, and $\nu_i^\ddagger$ depends on the nature of the transition state. There are two general situations: (1) a saddle point exists between the reactants and the products, at which the transition state is located; (2) either no saddle point exists, as in the reaction $CH_4 \rightarrow H + CH_3$, or the saddle-point region is very flat, so that the transition state's position on the potential energy surface is energy dependent and must be found by microcanonical variational transition state theory (see chapter 10). Finding transition state properties for the former situation is more straightforward and will be considered first.

The most widely used method of choosing properties for a transition state located at the saddle point is to adopt a model which reproduces the thermal high-pressure frequency factor A_∞ and activation energy E_∞. These experimental parameters are related to E_0 and the entropy difference $\Delta S^\ddagger$ via the transition state theory equations (10-49) and (10-54). Parameters for such model transition states are chosen using procedures such as BEBO concepts[36] or Benson's "thermochemical rules."[37] Although a unique transition state is not determined, it has been found that the RRKM rate constant is not particularly sensitive to the choice of vibrational frequencies and moments of inertia as long as the E_0 used is in agreement with E_∞ via equation (10-49) and $\Delta S^\ddagger$ agrees with A_∞ via equation (10-54).

To illustrate this approach, consider the decomposition of C_2H_5 to $H + C_2H_4$ and the four-centered elimination of HCl from C_2H_5Cl. For C_2H_5 decomposition, the transition state is depicted as

$$
\begin{array}{c}
\text{H} \\
\vdots \\
\text{H}_2\,\text{C} \;=\!=\!=\!=\; \text{CH}_2
\end{array}
$$

The H—C stretch is taken as the reaction coordinate. The principal vibrational frequency changes in going from C_2H_5 to the transition state occur for the C—C stretch (an increase) and the two $H—C_2H_4$ bends (a decrease). If a transition state structure (i.e., a set of moments of inertia) is chosen, these three vibrational frequencies and E_0 may be adjusted to fit A_∞ and E_∞.

For HCl elimination from C_2H_5Cl, the proposed transition state structure is as follows:

$$
\begin{array}{c}
\text{H} \,\text{-----}\, \text{Cl} \\
\vdots \qquad \vdots \\
\text{H}_2\,\text{C} \;=\!=\!=\!=\; \text{C}\;\text{H}_2
\end{array}
$$

Using BEBO arguments, a transition state with ring bond orders of 1.9, 0.9, 0.1, and 0.1 for the C—C, C—Cl, C—H, and H—Cl bonds, respectively, is found to fit the experimental A_∞ frequency factor.[38] A ring breathing mode is assumed to be the reaction coordinate. (Problem 11.12 makes use of the transition state for this reaction.)

Ab initio calculations may also be used to determine properties of a fixed transition state at the saddle point. However, the calculated unimolecular threshold energy E_0 is usually several kJ/mol in error. For many cases a more accurate value for E_0 is found from the experimental activation energy using equation (10-21). The strength of the *ab initio* approach is that it gives a unique structure and set of vibrational frequencies for the transition state. Overall, thermal high-pressure frequency factors calculated from *ab initio* transition state structures and vibrational frequencies are in good agreement with experimental values. If *ab initio* values are also used for the reactant structure and vibrational frequencies, errors present in the *ab initio* results tend to cancel when the transition state's sum of states is divided by the reactant's density of states in evaluating the RRKM $k(E)$.

To apply the variational version of RRKM theory (i.e., microcanonical variational transition state theory—see chapter 10, section 10-7) potential energy and vibrational frequencies are required along the reaction path in order to find the minimum in the sum of states [equation (10-10)] which identifies the transition state.[39] These reaction path properties may be either derived from a postulated model or determined by an *ab initio* calculation. For reactions without saddle points, such as $C_2H_6 \rightarrow 2CH_3$, the transition state structure is strongly energy-dependent. This is illustrated in Figure 11-10. A major accomplishment of the variational form of RRKM theory is that it explains the "loose" transition states and large thermal A-factors experimentally (see Table 11-2) for reactions without potential energy barriers.

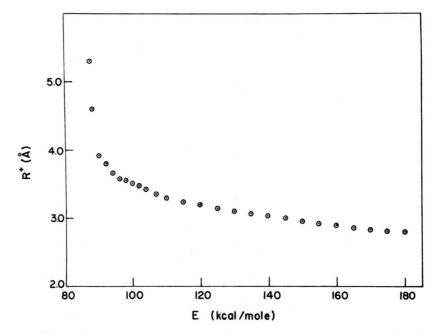

FIGURE 11-10   Relationship between the variational transition state C—C bond length $R^{\ddagger}$ and vibrational energy for a model of $C_2H_6 \rightarrow 2CH_3$ dissociation. [Adapted from W. L. Hase, in *Modern Theoretical Chemistry, Dynamics of Molecular Collisions, Part B*, edited by W. H. Miller (New York: Plenum, 1976).]

To illustrate the details of an RRKM calculation with a fixed transition state structure, Table 11-3 makes use of the unimolecular isomerization reaction $CH_3NC \rightarrow CH_3CN$. The vibrational frequencies and moments of inertia for CH_3NC and the transition state are taken from the *ab initio* calculations of Saxe et al.[40] Calculated rate constants are listed versus E_v for $E_{rx} = E_{ry} = E_{rz} = 1.25$ kJ/mole. The rotational energy difference between the reactant and transition state is found from

$$E_r - E_r^{\ddagger} = \sum_{i=x,y,z} E_{ri}(1 - I_i/I_i^{\ddagger}) \tag{11-59}$$

The transition state energy $E^{\ddagger}$ is given by equation (11-51); the sum and density of states are found from the Whitten-Rabinovitch approximation. The threshold energy E_0 is 157 kJ/mole.

11.7   APPLICATION OF RRKM THEORY TO THERMAL ACTIVATION

The RRKM rate constant for the overall thermal unimolecular reaction is found by inserting equation (11-54) into equation (11-32) and integrating over the Boltzmann distribution for E and J; i.e.

$$k_{\text{uni}} = \int_{E_0}^{\infty} dE \sum_{J=0}^{\infty} \frac{k(e, J)P(E, J)}{1 + k(E, J)/\omega} \tag{11-60}$$

TABLE 11-3   RRKM Calculation for $CH_3NC \rightarrow CH_3CN^a$.

Vibrational frequencies (cm^{-1})	
CH_3NC	Transition state
3014	3070
3014	3049
2966	2945
2166	1970
1467	1439
1429	1311
1129	975
1129	956
945	609
263	230
263	$412i$

Moments of inertia (amu $-$ Å^2)		
	CH_3NC	Transition state
I_x	3.20	11.80
I_y	50.12	32.87
I_z	50.12	41.42

E_v^b	$E^{\ddagger b}$	$N(E_v)^c$	$G(E^{\ddagger})$	$k(E_v)^d$
167	9.6	1.502×10^3	5.547	3.321×10^8
209	51.5	7.515×10^3	8.798×10^2	1.053×10^{10}
251	93.3	3.031×10^4	1.927×10^4	5.717×10^{10}
335	177	3.135×10^5	1.302×10^6	4.257×10^{11}

[a]Frequencies and moments of inertia from reference 40.
[b]Energy in kJ/mol.
[c]Density of states in states/cm^{-1}.
[d]Rate constant in sec^{-1}.

The distribution $P(E,J)$ is given by

$$P(E, J) = \frac{(2J + 1)N(E, J)\exp(-E/k_BT)}{Q_rQ_v} \tag{11-61}$$

where $N(E,J)$ is the density of states for the reactant, equation (11-58), and Q_r and Q_v are the reactant's external rotational and vibrational partition functions. Inserting equation (11-61) into equation (11-60) and replacing $k(E, J)$ with equation (11-54) gives

$$k_{uni} = \frac{1}{hQ_rQ_v} \int_{E_0}^{\infty} dE \sum_{J=0}^{\infty} \frac{(2J + 1)G^{\ddagger}(E, J)\exp(-E/k_BT)}{1 + k(E, J)/\omega} \tag{11-62}$$

At high pressures ($\omega \to \infty$),

$$k_{\text{uni}}^{\infty} = \frac{1}{hQ_rQ_v} \int_{E_0}^{\infty} dE \sum_{J=0}^{\infty} 2J + 1 \sum_{K=-J}^{J} G^{\ddagger}(E,J,K) \exp(-E/k_BT) \quad (11\text{-}63)$$

where $G^{\ddagger}(E,J)$ is represented by equation (11-57). Since the Laplace transform of $\Sigma(2J+1)\Sigma G^{\ddagger}(E,J,K)$ in equation (11-63) is the product of the external rotational and vibrational partition functions $Q_r^{\ddagger}Q_v^{\ddagger}$ for the transition state multiplied by k_BT,[28] k_{uni}^{∞} can be written as

$$k_{\text{uni}}^{\infty} = \frac{k_BT}{h} \frac{Q_r^{\ddagger}Q_v^{\ddagger} \exp(-E_0/k_BT)}{Q_rQ_v} \quad (11\text{-}64)$$

which is in agreement with transition state theory. At low pressures ($\omega \to 0$),

$$k_{\text{uni}}^{0} = \frac{\omega}{Q_rQ_v} \int_{E_0}^{\infty} dE \sum_{J=0}^{\infty} (2J+1) N(E, J) \exp(-E/k_BT) \quad (11\text{-}65)$$

which is the collision frequency ω multiplied by the ratio of the partition function for energized reactants with energy in excess of E and the partition function for all reactants.

To avoid the summation over J in equation (11-60), Marcus[20] suggested the use of the above simplified RRKM model, with rotational energy treated classically, and replacing $E_r - E_r^{\ddagger}$ in equation (11-52) by the difference of the average rotational energies of reacting molecules, $\langle E_r \rangle$, and the transition states $\langle E_r^{\ddagger} \rangle$. The quantity $\langle E_r^{\ddagger} \rangle$ may be found by averaging $E_r^{\ddagger}$ over $\exp(-E_r^{\ddagger}/k_BT)$; $\langle E_r^{\ddagger} \rangle = k_BT$. Because $E_r/E_r^{\ddagger} = I^{\ddagger}/I$, the result for two dimensions is

$$\Delta E_r = \langle E_r \rangle - \langle E_r^{\ddagger} \rangle = [(I^{\ddagger}/I) - 1]k_BT \quad (11\text{-}66)$$

If we now define

$$k(E^{\ddagger}) = \frac{1}{h} \frac{G(E^{\ddagger})}{N(E^{\ddagger} + E_0)} \quad (11\text{-}67)$$

and introduce the factor

$$F = N(E^{\ddagger} + E_0 - \Delta E_r)/(N(E^{\ddagger} + E_0)) \quad (11\text{-}68)$$

the RRKM rate constant for transition states with internal energy $E^{\ddagger}$ and rotational energy $E_r^{\ddagger}$ may be approximated by

$$k(E^{\ddagger}, E_r^{\ddagger}) = k(E^{\ddagger})/F \quad (11\text{-}69)$$

The thermal unimolecular rate constant may then be written as (see Refs. 20 and 41 for details)

$$k_{\text{uni}} = \frac{Q_r^{\ddagger}\exp(-E_0/k_BT)}{hQ_rQ_v} \int_0^{\infty} \frac{G(E^{\ddagger})\exp(-E^{\ddagger}/k_BT)dE^{\ddagger}}{1 + k(E^{\ddagger})/F\omega} \quad (11\text{-}70)$$

In the low-pressure limit ($\omega \to 0$),

$$k_{uni}^0 = \frac{F\omega Q_r^{\ddagger} \exp(-E_0/k_B T)}{Q_r Q_v} \int_0^{\infty} N(E^{\ddagger} + E_0) \exp(-E^{\ddagger}/k_B T) dE^{\ddagger} \quad (11\text{-}71)$$

In the high-pressure limit ($\omega \to \infty$), equation (11-70) gives equation (11-64).

By performing exact numerical integrations over $E^{\ddagger}$ and $E_r^{\ddagger}$, Waage and Rabinovitch[41] achieved a very good approximation for F, viz,

$$F = \{1 + [(s + d/2 - 1)((I^{\ddagger}/I) - 1)k_B T]/(E_0 + aE_z)\}^{-1} \quad (11\text{-}72)$$

where d is the number of internal rotational degrees of freedom. Other approximations have been made for F that are similar to, but not as accurate as, that of Waage and Rabinovitch.

Different experimental techniques, including static pyrolysis,[23] carrier (flow) techniques,[42] shock tube methods[43] (also see chapter 3), and very-low-pressure-pyrolysis (VLPP),[44] have been used to measure the thermal unimolecular rate constant k_{uni} as a function of temperature and pressure. One of the most significant achievements of RRKM theory is its ability to match measurements of k_{uni} with pressure. The unimolecular fall-off measurements made by Rabinovitch and co-workers of alkyl isocyanide isomerization[23,45-48] have provided some of the most thorough comparisons with the theory. The first reaction studied was that of equation (11-2). A comparison of the RRKM calculations and fall-off measurements for this reaction is shown in Figure 11-11. In addition to examining CH_3NC, Rabinovitch and co-workers also studied the isomerization of CD_3NC, CH_2DNC, and C_2D_5NC and have found that the same model transition state that fits CH_3NC isomerization gives a good fit to the fall-off behavior of these molecules as well. Extensive comparisons between RRKM theory and experimental fall-off curves have also been made for cyclopropane isomerization, and again, a good fit has been found between theory and experiment.[42]

Agreement has also been found with experimental unimolecular rate constants when transition state structures derived from *ab initio* calculations have been used in RRKM calculations. For example, when used in RRKM theory the *ab initio* CH_3NC transition state structure in Table 11-3 gives a good fit to the thermal unimolecular rate constant.[40] Similarly, for the decomposition reaction $C_2H_5 \to H + C_2H_4$, the *ab initio* transition state fits both thermal and chemical activation (see next) rate constants.[49]

11.8   MEASUREMENT OF *k(E)*

One of the most sensitive tests of RRKM unimolecular rate theory is the direct determination of the microcanonical unimolecular rate constant $k(E)$ as a function of energy. Some of the most extensive studies of this type have involved the unimolecular dissociation of ions. A powerful and generally applicable approach to the monoenergetic excitation of ions is that of photoelectron-photoion coincidence (PEPICO).[50] Ions excited in this manner include those of the halobenzenes, aniline, benzonitrile, and phenol. In all cases the measured $k(E)$ values agree with those of RRKM theory.

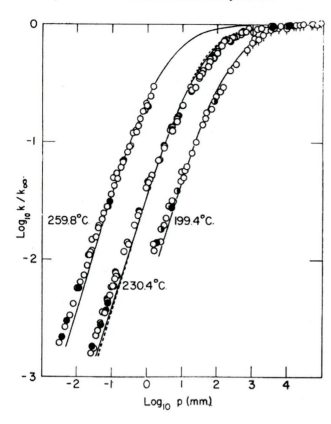

FIGURE 11-11 Pressure dependence of the unimolecular rate constant for CH_3NC isomerization. For clarity, the 260° and 200° curves are displaced one log unit to the left and right, respectively. [Adapted from F. W. Schneider and B. S. Rabinovitch, *J. Am. Chem. Soc. 84,* 4215 (1962).]

A comparison between the experimental and theoretical $k(E)$ for bromobenzene and deuterated bromobenzene ions[51] is given in Figure 11-12. Most purported examples of non-RRKM behavior for ionic reactions have been shown to involve direct dissociation from excited electronic, or repulsive, states.

Chemical activation and photoactivation techniques have been used to excite neutral molecules with much higher energies than is attained in thermal experiments and with narrow internal energy distributions. In chemical activation a chemical reaction is used to form molecules with sufficient energy to undergo unimolecular reactions.[52,53] Some of the more important chemical activation processes are H addition to olefins, CH_2 insertion and addition, and the radical recombination which is the reverse of equation (11-5). Photoactivation techniques, discussed in section 11.1, include internal conversion from an electronically excited state [equation (11-8)] and direct overtone pumping. These techniques are similar to chemical activation, in that energized molecules are formed with a narrow energy distribution. The distribution is not completely sharp, however, because the thermal energy distribution in the initial reactants (typically a few kJ/mol in width) is just translated upward by the energy of the absorbed photon.

The molecule A* formed by chemical activation or photoactivation may either undergo a unimolecular reaction or be collisionally stabilized. (In the PEPICO experi-

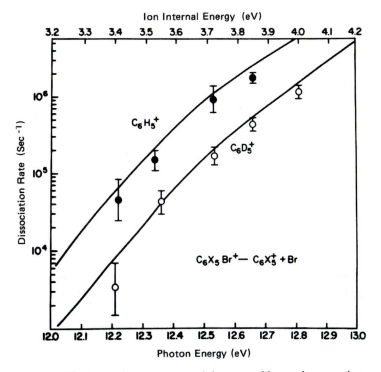

FIGURE 11-12 Bromobenzene and deuterated bromobenzene ion-dissociation rates as a function of ion internal energy. The points are the experimental results, and the solid lines are the statistical-theory results. [From T. Baer, *Adv. Chem. Phys. 74*, 111 (1986). Copyright © 1986 by John Wiley & Sons, Inc. Reprinted by permission of John Wiley & Sons, Inc.]

ments described, $k(E)$ is determined in the absence of collisions.) If the strong collision assumption is made and A* is assumed to be monoenergetically excited, we have

$$A^* \xrightarrow{k(E)} \text{decomposition products } (D)$$

$$A^* \xrightarrow{\omega} A_1 \qquad \text{stabilization } (S)$$

(11-73)

where ω is the collision frequency. From the equation, it is plain that the rate constant $k(E)$ equals $\omega D/S$. When A* is formed with a distribution of energies $f(E)$ the average rate constant $\langle k(E) \rangle$ equals $\omega D/S$. The ratio D/S may be found by averaging over $f(E)$, and the following expression is obtained for $\langle k(E) \rangle$:

$$\langle k(E) \rangle = \omega \frac{\int \{k(E)/[k(E) + \omega]\}f(E)de}{\int \{\omega/[k(E) + \omega]\}f(E)de}$$

(11-74)

At high pressures, i.e., where $\omega \gg k(E)$, we have

$$\langle k(E) \rangle_\infty = \frac{\int k(E)f(E)dE}{\int f(E)dE} \tag{11-75}$$

Similarly, at low pressures, i.e., where $\omega \ll k(E)$, equation (11-74) becomes

$$\langle k(E) \rangle_0 = \frac{\int f(E)dE}{\int [f(E)/k(E)]dE} \tag{11-76}$$

For some chemical activation processes it is possible to assume an equilibrium between the two reactants and A* in order to determine the form for the distribution function $f(E)$.

The most comprehensive applications of RRKM theory to chemical activation measurements have been for various homologous series. Agreement with the RRKM theory results when one transition state is suitable for the complete series. Rabinovitch and co-workers have made major contributions to studies involving H atom addition to olefins. Their most comprehensive studies have been of H + olefins to yield excited alkyl radicals that decompose by loss of CH_3, C_2H_5, C_3H_7, C_4H_9, and C_5H_{11} radicals. In all cases the results were found to be in satisfactory agreement with RRKM theory. [54] Agreement with RRKM theory has been found in many other chemical activation studies as well. [55-57]

In photoactivation the excited reactant is formed by

$$A + h\nu \xrightarrow{k_a} A^* \tag{11-77}$$

The rates for the overall loss of reactant A and the formation of products are equal and, according to the mechanism of equations (11-73) and (11-77), are given by

$$\frac{-d[A]}{dt} = k(E)A^* \tag{11-78}$$

Applying a steady-state treatment to A* leads to the rate law

$$\frac{-d[A]}{dt} = \frac{k(E)k_a\rho_\nu[A]}{k(E) + \omega} \tag{11-79}$$

where ρ_ν is the photon number density. Hence, the apparent first-order rate constant for loss of A is

$$k = \frac{k(E)k_a}{k(E) + \omega} \tag{11-80}$$

Inversion of equation (11-80) gives

$$k^{-1} = k_a^{-1} + \frac{\omega}{k_a k(E)} \tag{11-81}$$

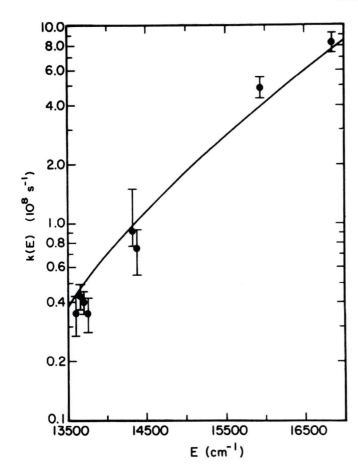

FIGURE 11-13   Plot of $k(E)$ vs. E for cyclobutene to 1,3-butadiene isomerization. Excitation is by a CH overtone transition. E is defined as the photon energy plus the average thermal energy (246 cm^{-1}). The solid line is the RRKM curve. [Adapted from J. M. Jasinki, et al. *J. Chem. Phys. 29*, 1312, (1983).]

which is the standard *Stern-Volmer* equation. This equation is used in most treatments of collisional quenching of excited species. A plot of k^{-1} vs. pressure (i.e., ω) should be linear with an intercept of k_a and a slope of $1/k_a k(E)$. The unimolecular rate constant $k(E)$ is then the intercept divided by the slope. Nonlinearity will arise in the plot if the energized molecules are not monoenergetic and/or if the strong-collision assumption for de-energization is invalid (see next section).

An example of a photoactivation experiment is the excitation of CH overtones in cyclobutene to induce isomerization to 1,3-butadiene.[58] The $k(E)$ values obtained in this experiment are compared with RRKM theory in Figure 11-13. Another sort of photoactivation is infrared multiple-photon excitation (IRMPE). In this case, however, a broad distribution of energies in the molecule is produced, rather than a nearly monoenergetic excitation, as discussed in section 11.13.

For energies slightly in excess of the unimolecular threshold E_0, the transition state energy is small and the transition state sum of states $G^{\ddagger}(E, J)$ versus E for fixed J is not continuous and, instead has "steps" (see Figure 11-5). Each unit increase in $G^{\ddagger}(E, J)$ corresponds to an additional internal state at the transition state becoming accessible. Since the reactant density of states is continuous for energies near E_0 (see

Figure 11-6), RRKM theory predicts $k(E, J)$ to have steps corresponding to the steps in $G^{\ddagger}(E, J)$. Such steps have been observed in energy-resolved photodissociation studies of $NO_2 \rightarrow NO + O$[59,60] and $CH_2CO \rightarrow CH_2 + CO$.[61,62] For each of these photodissociations, unimolecular reaction is thought to occur on the ground state potential energy surface and very good agreement is found between the experimental rate constants and the predictions of RRKM theory.[63,64]

11.9   INTERMOLECULAR ENERGY TRANSFER

The strong collision assumption can be tested experimentally for thermal unimolecular reactions by measuring k_{uni} in the second-order regime ($\omega \rightarrow 0$).[65] The low-pressure unimolecular rate is simply the rate at which the molecules are excited to vibrational levels having energy greater than E_0, and the rate constant is one for bimolecular reaction, viz., k_{bi}. If k_{bi} is measured for a bath gas M, the pressure-for-pressure activation efficiency β_p of M relative to reactant A can be found from

$$\beta_p = \frac{(k_{bi})_M}{(k_{bi})_A} \tag{11-82}$$

where (k_{bi}) is the slope of k_{uni} vs. the pressure of M with the reactant pressure kept constant, and $(k_{bi})_A$ is the slope of k_{uni} vs. the reactant pressure. This may be converted to a relative efficiency per collision by using the kinetic theory of gases to remove the M-A and A-A collision frequencies. The resulting expression for the relative activation efficiency of A is

$$\beta_c = \beta_p \left(\frac{\sigma_{AA}}{\sigma_{AM}} \right)^2 \left(\frac{\mu_{AM}}{\mu_{AA}} \right)^{1/2} \tag{11-83}$$

where the μ's are reduced masses and the σ's are the collision diameters. The most extensive measurements of β_p and thus β_c are those of Rabinovitch and co-workers for methyl isocyanide isomerization.[66] These researchers measured the relative efficiencies of over 100 bath gases, and some of their results are shown in Table 11-4.

For the noble gases and smaller molecules, β_c relative to β_c of CH_3NC is much less than 1.0. As the molecular complexity increases, the collisional efficiencies approach a limiting value. It seems reasonable that the limiting efficiency is unity and that the strong collision assumption of the Lindemann mechanism is valid for large molecules at not too high temperatures. However, for smaller molecules with β_c less than 1, a multistep activation and deactivation Lindemann mechanism leads to the following master equation for each energy level (see appendix to chapter 9):

$$\frac{dn_i}{dt} = \omega \sum_j P_{ij} n_j - \omega n_i - k_i n_i \tag{11-84}$$

In this equation, P_{ij} is the probability of conversion to the state indicated by i from the state indicated by j and the quantities k_i and n_i are the unimolecular rate constant for state i and the number of molecules in state i, respectively.

TABLE 11-4 Energy Transfer Efficiencies For
Methyl Isocyanide Isomerization In
The Second-Order Regime

Activating gas	Relative efficiency (β_c)
CH_3NC	1.00
$n\text{-}C_4H_{10}$	1.01
$n\text{-}C_4F_{10}$	1.00
C_3H_8	0.79
C_2H_6	0.76
CH_4	0.61
NH_3	0.93
CO	0.46
Xe	0.232
Ar	0.279
He	0.243

Source: Data from S. C. Chan, B. S. Rabinovitch, J. T. Bryant,
L. D. Spicer, T. Fujimoto, Y. N. Lin, and S. P. Pavlou, *J. Phys.
Chem.* **74**, 3160 (1970).

Exact solutions to the master equation are impossible because the individual P_{ij}'s and their time dependence are not known for complex molecules containing many quanta of vibrational energy. However, a formal approximate solution[67] to the equation indicates that after an initial short induction period the energy levels reach a steady state ($dn_i/dt = 0$). If this steady-state approximation is made, a stochastic solution is possible if a form for the P_{ij}'s is assumed.[68-70] The stochastic analysis will yield a value of $\langle \Delta E \rangle$, the average amount of energy transferred per collision. If any analytic form for P_{ij} is chosen an approximate analytic solution to the master equation is also possible.[29,71] The relative efficiency β_p at low pressures may be determined by calculating the total reaction rate from the steady-state distribution $\Sigma_i\, k_i n_i$, and dividing it by strong collision reaction rate given by equation (11-65). It is important to determine, for different bath gases, to what extent pressure-dependent thermal unimolecular rate constants differ from the prediction of RRKM theory, equation (11-62), which is based on the strong collision assumption described in Section 11.3. From an approximate analytic solution to the master equation, based on an exponential collision energy transfer probability between A and M, the relationship between $\langle \Delta E \rangle$ and β_c in the $\omega \to 0$ second-order regime is found to be[29,71]

$$\beta_c/(1 - \sqrt{\beta_c}) = -\langle \Delta E \rangle / F_E k_B T \qquad (11\text{-}85)$$

where F_E is the energy dependence factor of the vibrational density of states[71] and is given approximately by[72]

$$F_E \simeq \int_{E_0}^{\infty} \frac{dE}{k_B T} \frac{N(E)}{N(E_0)} \exp[-(E - E_0)/k_B T] \qquad (11\text{-}86)$$

Chemical activation experiments have also provided information about the dynamics of intermolecular energy transfer in unimolecular reactions.[73-75] The deactivation and unimolecular isomerization or decomposition of the chemically activated molecule A* may be regarded as the following cascade process proceeding through n energy levels which start at E and end at the first level below E_0:

$$A_1^* \xrightarrow{k(E_1)} \text{Products}$$
$$\downarrow \omega$$
$$A_2^* \xrightarrow{k(E_2)} \text{Products}$$
$$\downarrow \omega$$
$$A_3^* \cdots$$
$$\cdots$$
$$\xrightarrow{\omega} A_{n+1}^* \xrightarrow{k(E_{n+1})} \text{Products}$$
$$\downarrow \omega$$
$$\xrightarrow{\omega} A_n^*$$

Because A_n^* has energy less than E_0, it represents a stable molecule. The strong collision assumption means that the transition $A_1^* \rightarrow A_n^*$ occurs in one step. What is predicted by this cascade mechanism,[76] and what is found in many chemical activation experiments,[75] is decomposition or isomerization in excess of the strong collision prediction, because deactivation is a multistep process and there is a nonzero probability of unimolecular reaction after each of the deactivation steps until A* has energy less than E_0.

Extensive tables of $\langle \Delta E \rangle$ for different energized reactants and deactivating molecules are available from analyses of thermal and chemical activation studies.[66,77-79] Overall, $\langle \Delta E \rangle$ values for these two types of experiments are in agreement. However, these inferred values are considerably larger than values that are determined directly.[80-82] In one type of direct measurement, an ensemble of vibrationally excited molecules is prepared by laser excitation at a particular energy. Then, the average energy of the ensemble versus time in the presence of a deactivating bath is monitored by ultraviolet "hot band" absorption[81] or infrared fluorescence techniques.[80] Values for $\langle \Delta E \rangle$ are found by comparing the average vibrational energy of the ensemble versus time with the collision frequency. In the other type of direct measurement,[82] populations are determined for energy levels excited in the bath molecule M by A* + M collisions. The smaller $\langle \Delta E \rangle$ values obtained from the direct measurements are consistent with classical trajectory studies[83-88] and stochastic (random walk)[89] models of intermolecular energy transfer. Trajectory calculations[86,87] and the direct measurements of excited levels of M[82] indicate that a small fraction of the collisions may transfer a large amount of energy from A* to M.

The theory of unimolecular reactions still lacks a detailed microscopic picture of the collision dynamics of intermolecular energy transfer. What is needed is an understanding of how intermolecular energy transfer from highly energized molecules is influenced by the energizing level, the partitioning of the energy between vibration and rotation, and the relative translational energy between the colliding bath and energized molecules.

11.10   PRODUCT ENERGY PARTITIONING

Among the predictions of RRKM theory is that statistical distributions of relative translational, rotational, and vibrational energies will exist in the transition state. As an example, consider the relative translational energy distribution. If an ensemble of molecules is monoenergetically excited with a specific rotational energy, the rate constant for dissociation when the fragments have relative translational energy in the interval from $E_t^{\ddagger}$ to $E_t^{\ddagger} + dE_t^{\ddagger}$ is $k(E, E_t^{\ddagger})$ and, according to equation (10-97), is given by

$$k(E, E_t^{\ddagger})dE_t^{\ddagger} = \frac{N(E^{\ddagger} - E_t^{\ddagger})dE_t^{\ddagger}}{hN(E_v)} \tag{11-87}$$

where the N's are densities of states and the E's are as defined in Figure 11-9. The probability of a particular $E_t^{\ddagger}$ is then

$$P(E_t^{\ddagger})dE_t^{\ddagger} = \frac{k(E^{\ddagger}, E_t^{\ddagger})dE_t^{\ddagger}}{\displaystyle\int_0^{\infty} k(E^{\ddagger}, E_t^{\ddagger})dE_t^{\ddagger}} \tag{11-88}$$

After canceling terms and integrating, equation (11-88) becomes

$$P(E_t^{\ddagger}) = \frac{N(E^{\ddagger} - E_t^{\ddagger})}{G(E^{\ddagger})} \tag{11-89}$$

This equation applies to the situation where the energized molecules have one specified rotational energy. If, instead, the energized molecules have a distribution of rotational energies, equation (11-89) must be averaged over this distribution to obtain the final $P(E_t^{\ddagger})$.

If the unimolecular reaction does not have a barrier for the reverse recombination reaction (Figure 11-1(c)), the energy distributions at the transition state may be used to predict those for the products. This can be done since there is a larger internuclear separation between the dissociation fragments at the transition state, so that the transition state resembles the products. Under special conditions RRKM theory becomes the same as two versions of phase space theory (PST).[30,90–94] If the transition state is assumed to be located at the product asymptotic limit, the RRKM theory transition state energy distributions are equivalent to the product energy distributions for a version of PST which does no explicitly treat the conservation of orbital angular momentum as the products separate.[30,90,91] The total energy available to the products is

$$E_{\infty} = E - E_0 = E_t + E_r + E_V \tag{11-90}$$

where E_t, E_r, and E_V are the product's relative translational, rotational, and vibrational energies, respectively. For fixed E and J, the PST probability distribution for E_t is

$$P_{E,J}(E_t) \propto \int_0^{E_{\infty} - E_t} N_V(E - E_0 - E_t - E_r) \, N_{ro}(E_t, E_r, J) \, dE_r \tag{11-91}$$

where $N_V(E - E_0 - E_t - E_r$ and $N_{ro}(E_t, E_r, J)$ are the densities of state, respectively, for the product's vibrational and rotational-orbital degrees of freedom. If the transition state is assumed to be located at the dissociation fragment's centrifugal barrier

(Section 8.3.4) and if an isotropic interfragment potential is assumed, RRKM theory becomes the same as the orbiting transition state/phase space theory (OTS/PST) model.[30,92–94] Product energy distributions are determined for this model by assuming conservation of orbital angular momentum from the centrifugal barrier to products. OTS and OTS/PST have been used extensively to calculate product energy distributions for dissociation of energized ions.[30]

If there is a barrier for recombination, as in reactions (11-3) and 11-4), the product energy distributions are usually nonstatistical and different from those at the transition state. This is because the potential energy released in going from the transition state to products is channeled nonstatistically to translation, vibration, and rotation. (Recall the discussion in chapter 9, section 9.4.2, concerning the relationship between energy partitioning and potential energy surfaces for direct bimolecular reactions.) Two types of unimolecular dissociation potential energy curves and the resulting relative translational energy distributions are depicted in Figure 11-14. In

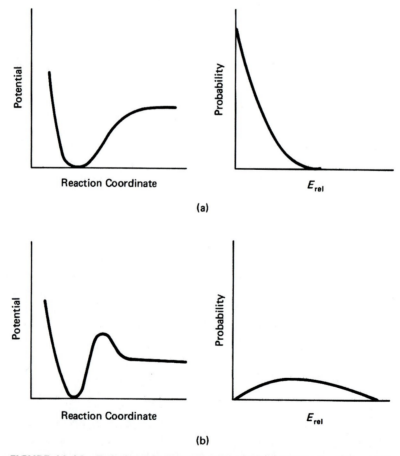

FIGURE 11-14   Relationship between the reaction path potential energy curve for a unimolecular dissociation reaction and the distribution of relative translational energy in the dissociation products.

(a) the product distribution is similar to that at the transition state; in (b) the barrier has broadened the distribution and shifted it to higher energies. These effects have been observed in numerous experimental and theoretical studies of unimolecular dissociation.

One of the predictions of RRKM theory is that in a ladder-climbing excitation process such as infrared multiphon dissociation (IRMPD) it is difficult to produce unimolecular fragments with large vibrational energies. This is because the unimolecular dissociation rate rises rapidly with energy in excess of the threshold energy E_0. Large amounts of vibrational energy could appear in the products if there is a large potential energy release in going from the transition state to products that is preferentially directed to vibration. However, for most unimolecular reactions, the potential energy release is preferentially partitioned to translation. Products with large amounts of vibrational energy can be formed if the reactant is highly energized, as can be accomplished in chemical activation (section 11.7) and photoactivation (section 11.8).

To interpret energy partitioning correctly in unimolecular reactions, angular momentum constraints must be treated correctly. This is illustrated by the chemical activation process

$$F + C_2H_4 \longrightarrow C_2H_4F^* \longrightarrow H + C_2H_3F \tag{11-92}$$

which has been studied in crossed molecular beam experiments.[95] The distribution of rotational energy in the energized $C_2H_4F^*$ molecules results from the distribution of impact parameters (see chapter 8) leading to reaction. The initial orbital angular momentum $l = \mu b v$ between the reactants is converted into rotational angular momentum J of $C_2H_4F^*$. However, since the dissociating H atom is very light, there is very little orbital angular momentum associated with the products H and C_2H_3F. Thus, conservation of angular momentum requires[96] that l be almost entirely converted into rotational excitation j of C_2H_3F, i.e., $l + j = J$. Extensive analyses[97] have been made of the role of angular momentum constraints in the partitioning of energy between unimolecular fragments. Such constraints should be included when using the RRKM theory to calculate energy distributions.

The adiabatic theory of chemical reactions[25,98,99] may also be used to calculate product energy distributions. In this theory, the reactive system is assumed to remain in the same vibrational/rotational quantum level as it moves along the reaction path. These levels are approximated by equation (7-48), which gives harmonic vibrational energy levels as a function of the reaction path. However, this equation is incomplete for describing unimolecular dissociation, since it does not include rotation in the energy levels and does not consider the transformation of vibrational degrees of freedom of the unimolecular reactant to rotational and translational degrees of freedom of the product fragments. Quantum mechanical calculations may be used to evaluate vibrational/rotational energy levels as a function of the reaction path.[100,101] Also, Quack and Troe[25] have given a very useful semi-empirical expression for constructing these energy levels. From the vibrational/rotational adiabatic curves connecting reactant and product states, product energy distributions may be determined if the population of the reactant states is known. In the statistical adiabatic channel model,[25] the energized reactant's energy levels are assumed to be populated statistically as assumed by RRKM theory.

11.11   APPARENT AND INTRINSIC NON-RRKM BEHAVIOR

In the preceding sections, applications of RRKM theory to thermal, photoactivation, and chemical activation experiments have been illustrated. One of the major achievements of unimolecular rate theory is the extent of agreement realized between RRKM theory and such experiments. However, there are situations where the RRKM theory does not appear to be successful. Thus, at this point it is useful to review the basic postulates of RRKM theory and consider situations where they may be violated.[102]

The central assumption of RRKM theory is that isolated molecules behave as if all their accessible states were occupied with equal probabilities. This means that the probability of $A^{\ddagger}$ [equation (11-46)] is due entirely to statistical considerations and the lifetime distribution is random, as given by equation (11-33). Figure 11-15(a) illustrates the situation schematically and shows random transitions among states at some energy high enough for eventual reaction (toward the right). In reality, transitions between quantum states (though coupled) are *not* equally probable: some are more likely than others. Therefore, the molecular motion must be disorderly enough for the RRKM assumption to be mimicked, as crudely illustrated in Figure 11-15(b).

The lifetime distribution will depend in part on the manner in which the energy needed for reaction is supplied. As discussed in section 11.3, the usual understanding of

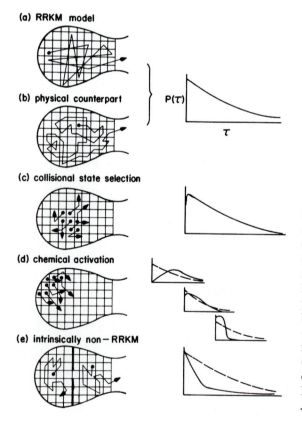

FIGURE 11-15   Relation of state occupation (schematically shown at constant energy) to lifetime distribution for the RRKM theory and for various actual situations. Dashed lines in lifetime distributions for (d) and (e) indicate RRKM behavior. (a) RRKM model. (b) Physical counterpart of RRKM model. (c) Collisional state selection. (d) Chemical activation. (e) Intrinsically non-RRKM. [Adapted from D. L. Bunker and W. L. Hase, *J. Chem. Phys.* 59, 4621 (1973).]

thermal unimolecular reactions assumes that the activating collisions randomly prepare the internal vibrational states of A*. However, it has been pointed out previously that thermal collisions surely at least approximately obey the Franck-Condon principle.[103] Thus, direct collisional preparation of a state with a large extension of some bond, in the direction of product formation, is suggested to be an implausible event. Consequently, one might expect some shortage of lifetimes less than one-fourth of a vibration period, as illustrated in Figure 11-15(c).

In many experiments, such as photoactivation and chemical activation, the molecular vibrational states are intentionally excited nonrandomly. Regardless of the pattern of initial energizing, the RRKM model in Figure 11-15(a) would require the distribution to become random in a negligibly short time. How well will the actual molecular motion approach this ideal? Three possibilities are represented by Figure 11-15(d). The lifetime distribution may be similar to that of RRKM theory. In other cases, the probability of a short lifetime with respect to reaction may be enhanced or reduced, depending on the location of the initial excitation within the molecule. These are examples of *apparent* non-RRKM behavior arising from state selection. If there are very strong internal couplings, $P(\tau)$ will become that of RRKM theory after rapid intramolecular vibrational energy redistribution.

The effect of apparent non-RRKM behavior is to make the experimental unimolecular rate constant time-dependent according to the formula

$$k(E, t) = \frac{1}{N(t)} \frac{dN(t)}{dt} \tag{11-93}$$

Obviously, if the excitation is preferentially localized in the reaction coordinate, the initial $k(E,t)$ will be larger than the RRKM value. However, if there is strong intramolecular coupling, $k(E,t)$ will rapidly approach $k_{RRKM}(E)$. There is widespread agreement between RRKM rate constants and those derived from photoactivation and chemical activation experiments, which illustrates that for many molecules and experimental conditions the rate of intramolecular vibrational energy redistribution (k_{IVR}) is larger than $k_{RRKM}(E)$.

To detect the initial apparent non-RRKM decay in a photoactivation or chemical activation experiment, one has to monitor the reaction at short times. This can be performed by studying the unimolecular decomposition at high pressures, where collisional stabilization competes with the rate of IVR. The first successful detection of apparent non-RRKM behavior was accomplished by Rabinovitch and co-workers,[104] who used chemical activation to prepare vibrationally excited hexafluorobicyclopropyl-d$_2$:

$$^1CD_2 + CF_2\!\!-\!\!CF\!\!-\!\!CF = CF_2 \rightarrow CF_2\!\!-\!\!CF\!\!-\!\!CF\!\!-\!\!CF_2^* \tag{11-94}$$
$$\underset{CH_2}{\diagdown\diagup} \qquad\qquad \underset{CH_2}{\diagdown\diagup}\quad\underset{CD_2}{\diagdown\diagup}$$

The molecule decomposes by elimination of CF_2, which should occur with equal probabilities from each ring when energy is randomized. However, at pressures in excess of 100 torr there is a measurable increase in the fraction of decomposition in the ring that was initially excited. Rabinovitch et al. also performed extensive studies on the series of chemically activated fluoroalkyl cyclopropanes:[105]

$$R = CF_3, C_3F_7, C_5F_{11}$$

(11-95)

The chemically activated molecules are formed by reaction of $^1CD_2$ with the appropriate fluorinated alkene. In all these cases apparent non-RRKM behavior was observed. As displayed in Figure 11-16, the measured unimolecular rate constants are strongly dependent on pressure. However, at low pressures each rate constant approaches the RRKM value.

A situation that is both more interesting and completely distinct from apparent non-RRKM behavior is *intrinsic* non-RRKM behavior. By this, it is meant that A* has a nonrandom $P(\tau)$ even if the internal vibration states of A* are prepared randomly. This situation arises when transitions between individual molecular vibrational states are slower than transitions leading to products. As a result, the vibrational states do not have equal dissociation probabilities. Slow transitions between the states occur when there is at least one bottleneck in the molecular phase space other than the one defining the transition state. An example of such a bottleneck with a possible resultant $P(\tau)$ is shown in Figure 11-15(e).

One kind of an intrinsically non-RRKM model is Slater's theory of unimolecular reactions.[15] A general discussion of this theory and the nature of the classical mechani-

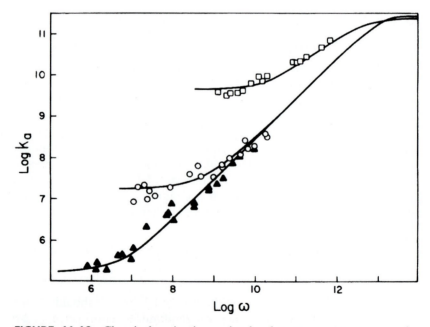

FIGURE 11-16 Chemical activation unimolecular rate constants vs. ω for fluoroalkyl cyclopropanes [see equation (11-95)]. [Adapted from J. F. Meagher, et al., *J. Phys. Chem. 78*, 2535 (1974). Copyright 1974 American Chemical Society.]

cal motion of highly vibrationally excited molecules is presented in the next section: experimental examples of intrinsic non-RRKM behavior are given in the last section of this chapter.

11.12   CLASSICAL MECHANICAL DESCRIPTION OF INTRAMOLECULAR MOTION AND UNIMOLECULAR DECOMPOSITION

The first classical mechanical model of unimolecular decomposition, developed by Slater,[15] was based on the normal-mode harmonic oscillator Hamiltonian. Although the Hamiltonian is rigorously exact only for small displacements from the molecular equilibrium geometry, Slater extended it to the situation where molecules are highly vibrationally energized, undergo large amplitude motions, and decompose. Since there are no couplings in the normal-mode Hamiltonian, the energies in the individual normal modes do not vary with time. This is the essential difference from the RRK and RRKM theories, which treat a molecule as a collection of coupled normal modes which freely exchange energy.

The harmonic oscillator classical Hamiltonian is

$$H = \sum_{i=1}^{s} \frac{(p_i^2 + \lambda_i q_i^2)}{2} \tag{11-96}$$

where $\lambda_i = 4\pi^2 \nu_i^2$. From the classical equations of motion [equation (8-20)], it is easily shown that each normal mode coordinate varies with time according to

$$q_i = q_i^0 \cos(2\pi\nu_i t + \delta_i) \tag{11-97}$$

where q_i^0 is the amplitude and δ_i is the phase of the motion. Thus, if an energy $E_i = (p_i^2 + \lambda_i q_i^2)/2$ and phase δ_i are chosen for each normal mode, the complete intramolecular motion of the energized molecule may be determined for this particular initial condition.

For most unimolecular reactions the reaction coordinate is not a particular normal mode, but instead is an internal coordinate displacement such as a bond extension. In the normal-mode approximation the internal coordinate displacements **d** are related to the normal modes through the linear transformation

$$\mathbf{d} = \mathbf{L}_d\mathbf{q} \tag{11-98}$$

The transformation matrix $\mathbf{L}_d$ is obtained from a normal-mode analysis performed in internal coordinates. Thus, as the development with time of the normal-mode coordinates is evaluated from equation (11-97), the value for the reaction coordinate $q^{\ddagger}$, a particular internal coordinate displacement, is found from equation (11-98). The variation in $q^{\ddagger}$ with time results from a superposition of the normal modes. At a particular time the normal-mode vibrations will phase together so that $q^{\ddagger}$ exceeds a critical extension, at which point decomposition is assumed to occur.

The preceding discussion gives the essential details of the Slater theory. Since energy does not flow freely within the molecule, the theory predicts intrinsic non-RRKM behavior. The attainment of the reaction coordinate critical extension is not a statistically random process as in RRKM theory, but depends on the energies and phases of the specific normal modes excited. To illustrate, consider the normal modes of the ethylene molecule, which consist of two separable groups, one for in-plane and one for out-of-plane vibrations. Thus, if in-plane normal vibrations are excited, only an in-plane dissociation path, such as C-H bond rupture, can occur. To promote an out-of-plane unimolecular path, such as rotation around the double bond (e.g., cis-trans isomerization in 1,2-dichloroethylene), the out-of-plane vibrations must be excited, but then displacements of the in-plane coordinates do not occur. Even if the molecule is excited randomly, in general its lifetime distribution will not be exponential and decay with the RRKM rate constant.

Overall, the Slater theory is unsuccessful in interpreting experiments. The theory is not compatible with photoactivation and chemical activation experiments where molecules are initially excited nonrandomly. Instead, the rate constants and unimolecular paths in these experiments are consistent with energy flowing randomly within the molecule. Nor does the theory adequately reproduce thermal unimolecular falloff curves. By an appropriate choice of parameters the high-pressure limit may be fit, but at lower pressures the unimolecular rate is severely underestimated. This is because certain patterns of normal-mode excitation prepared by collisional activation can never lead to reaction. The experimental results are consistent with ultimate dissociation if E is in excess of E_0, regardless of the excitation pattern.

If one considers the nature of classical Hamiltonians for actual molecules, at first glance it is not surprising that the Slater theory performs so poorly. The potential energy of actual molecules is anharmonic, and the modes are strongly coupled, not harmonic and separable as assumed in the normal-mode model. Also, for finite displacements of the atoms from their equilibrium positions, there will be couplings between the normal-mode momenta. Clearly, understanding the classical intramolecular motion of vibrationally excited molecules requires one to go beyond the normal-mode model.

The first classical trajectory study of unimolecular decomposition and intramolecular motion for realistic anharmonic molecular Hamiltonians was performed by Bunker.[14,21] Since this initial work, there have been numerous additional studies.[96,102,106–110] In some cases potential energy surfaces used in the trajectories have been derived from *ab initio* calculations. These studies have identified two distinct types of intramolecular motion: chaotic and quasiperiodic.[111,112] Chaotic motion is depicted in Figure 11-17(a). The molecule does not exhibit regular vibrational motion as predicted by the normal-mode model, and instead there is energy transfer between the normal modes. If all the normal modes of the molecule participate in the chaotic motion and energy flow is sufficiently rapid, a microcanonical ensemble is maintained as the molecule dissociates, and RRKM behavior is observed.[102] If the states in the molecular Hamiltonian are chosen randomly, exponential unimolecular decomposition occurs with a rate constant in agreement with that of RRKM theory. For nonrandom excitation initial apparent non-RRKM behavior is observed, but at longer times a microcanonical ensemble of states is formed and the probability of decomposition becomes that of RRKM theory.

The quasiperiodic motion depicted in Figure 11-17(b) is regular, like that of the normal-mode model. The vibrational motion can be represented by a superposition of individual vibrational modes, each containing a fixed amount of energy. For some cases

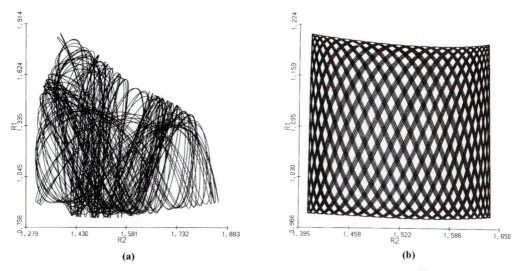

FIGURE 11-17   (a) Chaotic trajectory for a model HCC Hamiltonian. R_1 is the HC bond length, R_2 the CC bond length. Distance is in angstroms (Å). (b) Same as (a), but a quasi-periodic trajectory.

these modes resemble normal modes, but they may be identifiably different. Thus, although actual molecular Hamiltonians contain potential and kinetic energy coupling terms, they may still exhibit regular vibrational motion with no energy flow between modes as predicted by the Slater theory.

At low levels of excitation, the motion associated with the molecular Hamiltonian is quasiperiodic. However, as the energy is increased, in most cases there is a gradual transition from quasiperiodic to chaotic motion, with both types present at intermediate energies. In many cases the motion becomes totally chaotic before the unimolecular threshold energy is reached. However, in other cases quasiperiodic (or nearly quasiperiodic) motion exists above the unimolecular threshold, and intrinsic non-RRKM lifetime distributions result. This type of behavior has been found for Hamiltonians with low unimolecular thresholds, widely separated frequencies, and/or disparate masses.[14,108] Thus, classical trajectory simulations performed for realistic Hamiltonians predict that for some molecules the unimolecular rate constant may be strongly sensitive to the modes excited in the molecule, in agreement with the Slater theory. This property is called *mode specificity* and is discussed in section 11.14.

11.13 INFRARED MULTIPLE-PHOTON EXCITATION

In section 11.3, we noted that radiation could drive a molecular system to dissociation (the Perrin hypothesis). The molecule absorbs photons from the radiation field and thus acquires sufficient energy to react. The first such reaction to be identified was actually driven by visible rather than infrared photons,[113] and so was actually a multi-photon photochemical process. In this section, two approaches to promoting unimolecular dissociation by infrared excitation are described. One is laser-induced multiple-photon excitation. The other involves thermal unimolecular dissociation driven by blackbody radiation in the zero-pressure limit, when collisions don not contribute to the activation process.

The first observation of a reaction driven by multiple infrared photon absorption (IRMPA) was the decomposition of SiF_4.[114,115] Considerable attention was attracted by reports in the mid-1970s[116,117] of experiments in which the IRMPA-induced dissociation process was isotopically selective. These experiments used a high-powered CO_2 laser to dissociate SF_6; depending on the laser frequency employed, either $^{32}SF_6$ or the less abundant $^{34}SF_6$ could be dissociated. Such a reaction can be readily understood by noting that the excitation process occurs in the isolated molecule and that isotopic substitution results in frequency shifts of infrared absorption bands; if the pressure is kept low enough so that unimolecular dissociation proceeds faster than intermolecular collisions, the selectively excited isotopically labeled species will react preferentially. Following this early work, a large number of molecular species have been found to undergo IRMPD, and several comprehensive reviews of the subject are now available in the scientific literature.[118–121]

The details of the IRMPA process can be understood by reference to Figure 11-18. In the unexcited molecule (region I), the radiation from the laser source must be resonant, or nearly so, with a vibration frequency of the molecule in order for absorption to take place. In principle, if the frequency interval remained constant for successive transitions, a sufficiently intense field at $h\nu_0$ could drive the system through a sequence of levels to arbitrarily high excitation. In actual molecules, however, anharmonicity causes the overtone levels to shift to a lower frequency, so that the resonance condition is lost. Multilevel excitation is possible through mechanisms such as rotational and/or anharmonic compensation, two- or three-photon absorption, etc.,* but eventually the excitation process becomes too off-resonant to proceed further.

At this point, the *density of states* in the molecule becomes significant (see section 11.2). At an energy of several thousand cm^{-1}, or 40–50 kJ $mole^{-1}$, the vibrational state density $\rho(E)$ may be 10^3–10^5 or more (see Figure 11-6). This high density of states allows the laser-pumped vibrational levels to mix with a large number of other vibrational modes of the molecule, thus distributing the absorption strength over a range of transitions. In this dense manifold of states, called the *quasicontinuum* or QC (region II in Figure 11-18), photons of energy $h\nu_0$ can be absorbed by a range of states, and this absorption can proceed until sufficient energy has been acquired to reach the threshold of a dissociation or other chemical process. Above this threshold (region III in Figure 11-18), the reaction rate can generally be estimated quite well by RRKM-type expressions. Although non-RRKM behavior in such reactions has long been sought, there is little evidence to date for such behavior.

The rigorous mathematical treatment of IRMPD is known as the *optical Bloch equations*,[123] which represent a coupling of the Schrödinger equation for the molecule with the Maxwell equations for the radiation field. These equations, which treat the interaction between the complete set of molecular quantum states, the radiation field, and any dissipative processes such as collision or reaction, are generally far too detailed and difficult of solution for the desired objective, which is usually an ensemble-averaged quantity such as the total dissociation yield. The optical Bloch equations can be reduced to a more tractable master equation if certain assumptions

---

*For a discussion of the spectroscopic aspects of IRMPA, see references 121 and 122.

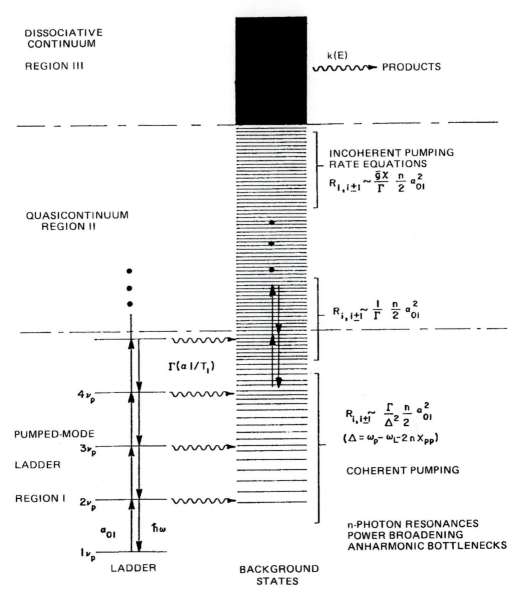

INFRARED MULTIPHOTON ABSORPTION:
POSTULATED MECHANISM

FIGURE 11-18 Schematic model for infrared multiple-photon absorption (IRMPA). Excitation occurs by resonant transitions at frequency $h\nu_0 = \hbar\omega_0$ in the discrete region (I), which connects to the quasi-continuum region (II). At sufficiently high energies, the true dissociative continuum (region III) is reached. [From J. I. Steinfeld, *Molecules and Radiation,* 2nd ed., (Cambridge, MA: MIT Press, 1985). Reproduced with permission.]

are made, the most important of which is that the molecular levels are characterized only by their total energy content E. This is a valid description if *intramolecular vibrational relaxation* (IVR), or coupling between the various vibrational modes of the molecule, is an extremely rapid process compared with laser pumping, relaxation, or reaction. The "rapid IVR" assumption has been the subject of controversy, but it seems to be appropriate for most cases. Using it, the equation of motion for the level populations $g(E,t)$ can be written in the form of a Master Equation (see chapter 9, Appendix):

$$\frac{dg(E,t)}{dt} = -[L_A(E) + L_S(E)]g(e) + L_A(E - h\nu_0)g(E - h\nu_0)$$

$$+ L_S(E + h\nu_0)g(E + h\nu_0) + \omega_{coll}\int [P(E, E')g(E') \qquad (11\text{-}99)$$

$$- P(E', E)g(E)]dE' - k(E)g(E)$$

Let us consider briefly each of the terms in this equation. The first four terms represent the interaction of the molecular energy levels with the radiation field. The absorption coefficients $L_A(E)$ are given by

$$L_A(E) = \sigma_{eff}(E)\frac{I}{h\nu_0} \qquad (11\text{-}100)$$

where the absorption coefficient σ_{eff} is assumed to be a slowly varying function of E in the quasicontinuum and the radiation intensity I is typically a time-dependent quantity, as for a pulsed laser. The stimulated-emission coefficients are related to the absorption coefficient by the Einstein coefficient relationships[122] as follows:

$$L_S(E)\rho(E) = L_A(E - h\nu_0)\rho(E - h\nu_0) \qquad (11\text{-}101)$$

Since $\rho(E)$ is a monotonically increasing function of E, as shown in Figure 11-6, the ratio $L_S(E)/L_A(E - h\nu_0)$ in less than unity, so that optical pumping tends to drive the system to higher energy, i.e., towards dissociation.

The collisional relaxation term $\omega P(E,E')$ is expressed as an integration over initial and final energies E', which makes analytical solution of equation (11-99) extremely difficult. The density of states $\rho(E)$ tends to bias the final-energy distribution to net excitation, which also drives the system toward dissociation. The final term in equation (11-99), which represents molecular dissociation or isomerization with rate $k(E)$, can usually be estimated quite satisfactorily by an RRKM expression.

Before proceeding with a solution of equation (11-99), it is instructive to consider the magnitudes of the terms involved. To estimate the radiative pumping rate $L = \sigma I/h\nu_0$, we shall assume an infrared absorption cross section of 10^{-20} cm^2 and a photon energy $h\nu_0 = 2 \times 10^{-20}$ J/photon for $\nu_0/c = 1000$ cm^{-1} (CO$_2$ laser frequency). This gives $L = 0.5I$ sec^{-1}, where the intensity I has units of W/cm^2. In order to drive a molecule to dissociation, it is first necessary to *saturate* the absorbing transition—that is, to cause the molecule to absorb photons at a rate which exceeds spontaneous emission $\tau_{rad}^{-1} \approx 10^3$ sec^{-1} for infrared transitions). This requires a saturation intensity I_{sat} greater than $10^3/0.5 \cong 2 \times 10^3$ W/cm^2. Saturation alone is not sufficient to induce multiple-photon absorption, however: in order to overcome the "anharmonic bottleneck," the intensity must be sufficiently high to cause power broadening of the resonance with a magnitude comparable to the spacing between vibrational levels, i.e., $\rho^{-1}(E)$. Since $\rho(E)$ may be on the order of several thousand states per cm^{-1} at energies corresponding to the anharmonic bottleneck, power broadening on the order of 10 MHz is required, corresponding to $L \cong 10^7 \ sec^{-1}$. Also, the pumping rate must exceed collisional relaxation, which occurs at a rate of about $10^7 \ sec^{-1}$ at 1 torr. Thus, the intensity required for IRMPD may be estimated as $I > 10^7 \ sec^{-1}/0.5 \approx 2 \times 10^7 \ W/cm^2$.

For weakly bound species at very low pressures, where only one or a few infrared photons can induce dissociation and collisional relaxation is unimportant, dissociation has been found to occur at intensities of $10^2 - 10^3 \ W/cm^2$, in accordance with the estimate from saturation. Some examples are van der Waals complexes in molecular beams[124] and ionic complexes in ion cyclotron resonance traps.[125] For most medium-sized polyatomic molecules, IRMPD threshold intensities are on the order of $10^6 - 10^7 \ W/cm^2$. Such intensities can be obtained from pulsed-discharge (TEA) CO_2 lasers, which typically generate 1-J pulses with 10^{-7} sec durations. The beams can be focused to $0.1 \ cm^2$ or less to obtain the requisite intensity. Dissociation yields are often reported as a function of time-integrated pulse intensity, or fluence, given by

$$\phi = \int I(t)dt \qquad J \ cm^{-2} \qquad (11\text{-}102)$$

because yields tend to correlate best with fluence for a variety of pulse peak intensities and durations. It is now also clear why IRMPD products generally possess little excess internal excitation: $k_{RRKM}(E)$ rises rapidly with excess energy (see Figure 11-7) and exceeds the pumping rate L after only one or two infrared photons in excess of $E_0/h\nu_0$ are absorbed.

Analytic solutions of equation (11-99) can be approximated when collisional relaxation can be neglected ($\omega = 0$) and the extent of reaction is small. Alternatively, post-pulse reaction and relaxation can be solved by setting $I = 0$, a treatment equivalent to that of photoactivated unimolecular reactions. Solution of the complete equation must be carried out numerically, using matrix techniques.[126] One difficulty encountered in carrying out such a solution is that the energy E is a continuous variable. In the absence of collisions, this can be dealt with by recognizing that energy is added to or removed from the molecule in units of $h\nu_0$, the infrared photon energy; this results in an *energy-grained* master equation, with grain size $h\nu_0$. Collisional relaxation will tend to redistribute molecules over a wide range of energies, but it has been found that solutions are still practical in this case with a grain size equal to one-third or one-fourth $h\nu_0$.

The object of solving the master equation is the prediction of an experimentally observed quantity, such as net dissociation yield. The yield can be found as the net fraction of molecules removed from populated energy states, given by

$$f = 1 - \lim_{t \to \infty} \frac{\int g(E,t)dE}{\int g(E,0)dE} \qquad (11\text{-}103)$$

The IRMPD yield typically has a threshold fluence on the order of 0.1 to 10 $J \ cm^{-2}$, below which dissociation does not occur; at high fluences, the yield saturates when all the molecules in the high-field region, which can interact with the laser, are dissociated.

Some experimental examples are shown in Figure 11-19. Only limited information about the details of the IRMPA process can be obtained from highly averaged measurements, such as net dissociation yield. Additional information can be obtained either by studying systems which possess two closely spaced competitive unimolecular dissociation channels,[127] studying the kinematics of the dissociated fragments using molecular-beam techniques,[128] or carrying out time-resolved spectroscopic measurements on IRMP-excited molecules.[121,129]

IRMPD is a useful technique for producing large densities of reactive free radicals in low-temperature environments, enabling the reaction rates of these radicals to be measured.[130] It can also be used to generate reactive species in the proximity of a surface, at which etching or deposition reactions may then occur.[131] In some recent experiments, thermal unimolecular dissociation has also been demonstrated in the zero-pressure limit, where collisions are not involved in the excitation, as assumed by the Lindemann mechanism (section 11.3). It is found that ions trapped in an ion cyclotron resonance cell may exchange energy with the surroundings by emission and absorption of infrared blackbody radiation.[125,132] Analogous to the rate constant for the Lindemann mechanism, equation (11-18), the thermal unimolecular rate constant in the zero-pressure limit is

$$k_{\mathrm{uni},zpl} = k_d \left(k_{1,\mathrm{rad}} / k_{-1,\mathrm{rad}} + k_d \right)$$

where k_d is the unimolecular rate constant and $k_{1,\mathrm{rad}}$ and $k_{-1,\mathrm{rad}}$ are the rate constants for absorption and emission of infrared photons. In a more detailed analysis, the rate

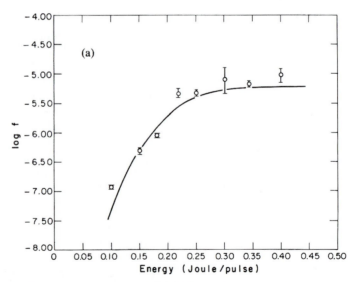

FIGURE 11-19 Examples of IRMPD yield vs. infrared fluence, as measured by several different experimental techniques. (a) Yield of total ethylene, measured by gas chromatographic techniques, in the dehydrohalogenation of chloroethane-d₃ by the 9P(14) CO₂ laser line. [From Francisco, Zhu, and Steinfeld, *J. Chem. Phys. 78,* 5339 (1983). Reproduced with permission.]

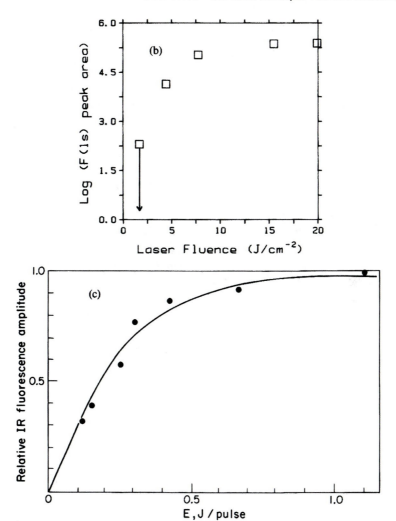

FIGURE 11-19   (b) Fluorine coverage on single-crystal silicon exposed to CF_3 radicals produced by IRMPD of C_2F_6 with the CO_2 9R(30) laser line. The quantity measured is the integrated F(ls) X-ray photoelectron peak intensity [from Roop, et al., *J. Chem. Phys. 83,* 6012 (1985). Reproduced with permission.] (c) In this experiment, bistrifluoromethyl peroxide (CF_3OOCF_3(BTMP)) is dissociated with the CO_2 9R(30) laser line to produce CF_3O radicals which are further dissociated to CF_2O and F atoms by continued absorption of infrared photons. The F atoms react with added HI to produce vibrationally excited HF, which in turn fluoresces at 2.8 μm, producing an infrared emission that is detected. The threshold fluence for this system appears to be very low or even zero, because the BTMP molecule is already in its vibrational quasi continuum under the conditions of the experiment. [From Zhang, Francisco, and Steinfeld, *J. Chem. Phys. 86,* 2402 (1982). Reproduced with permission.]

constants are energy dependent as in equations (11-29) to (11-32). There are three regimes for $k_{uni,zpl}$, defined by the relative values of $k_{-1,rad}$ and k_d. Ions activated above E_0, by the absorption of a single photon, dissociate promptly if $k_d \gg k_{-1,rad}$. Both master equation modeling and a truncated Boltzmann model may be used to interpret this regime.[133] Master equation modeling is needed for a detailed accounting of activation, deactivation, and dissociation in the regime where $k_{-1,rad} \approx k_d$.[133] Finally, when $k_{-1,rad} \ll k_d$, the ions equilibrate with the blackbody radiation field and their energies are described by a Boltzmann distribution. Arrhenius parameters measured in this regime are equivalent to those obtained in the traditional high-pressure limit using equation (11-63).[134]

One sought-after result of studying laser-driven reactions has been the stimulation of "bond-selective" chemistry, in which the functional group of the molecule initially excited by the laser is the one that selectively reacts. That this has not been obtained is due to the rapid IVR process discussed earlier, which effectively redistributes the absorbed energy among all the vibrational modes of the molecule. Reaction then proceeds in an essentially statistical manner. In the following section, we consider some cases where mode or bond selectivity is found to occur.

11.14   MODE SPECIFICITY

Mode specificity has been widely observed in the unimolecular decomposition of van der Waals molecules,[135,136] e.g.,

$$I_2 - He \longrightarrow I_2 + He$$

and

$$HF - HF \longrightarrow 2HF$$

A covalent bond (or particular normal mode) in the van der Waals molecule (e.g., the I_2 bond in I_2-He) can be selectively excited, and what is usually observed experimentally is that the unimolecular dissociation rate constant is orders of magnitude smaller than the RRKM prediction. This is thought to result from weak coupling between the van der Waals bond and the other modes in the molecule.[137] In the $(HF)_2$ dimer the lifetimes associated with the two HF vibrational modes differ by a factor of 24. Other van der Waals molecules studied include $(NO)_2$,[138] $(C_2H_2)_2$,[139] and $(C_2H_4)_2$.[139]

There are fewer documented examples of mode specificity for the unimolecular decomposition of covalently bound molecules. One example is the decomposition of the formyl radical HCO, viz.,

$$HCO \longrightarrow H + CO$$

for which mode-specific effects have been measured experimentally.[140,141] The formyl radical is excited above its unimolecular threshold, in individual vibrational/rotational states, by stimulated emission pumping (SEP) (see Figure 11-3). In the absence of unimolecular dissociation, the absorption spectrum for each of the excited vibrational/rotational states of formaldehyde would consist of a narrow line. As a result of the Heisenberg uncertainty principle, when dissociation occurs, the absorption line for each state is broadened. The relationship between the unimolecular lifetime of the excited vibrational/rotational state τ and the width at half-maximum Γ of its absorption line is given by[30]

$$\tau = \frac{\hbar}{\Gamma} \qquad (11\text{-}104)$$

These excited states which dissociate are called resonance states.[30] The unimolecular rate constant k for the resonance state is the inverse of its lifetime τ.

What is observed experimentally is that resonance states, with total energies that differ by only several cm^{-1}, may have rate constants that vary by more than an order of magnitude.[140,141] In fact, the rate constant may become significantly smaller if a state with a larger energy is excited. These results are contrary to those of RRKM theory, which predicts a *monotonic increase* in the rate constant with energy. To explain these experimental results for HCO, consider the idealized situation where unimolecular dissociation does not occur. The states prepared by SEP could not then dissociate and would simply be the highly excited vibrational/rotational states of bound HCO. The extent of mode specificity in HCO dissociation depends on the properties of these states. At low levels of excitation, it is known that wave functions ϕ_i for vibrational/rotational states are products of normal-mode vibration and rotational wave functions. If anharmonicity and vibrational-rotational coupling remain unimportant as the energy increased, such normal-mode rotational wave functions would also represent the high-energy states. Each state would then have a specific intramolecular vibrational property identified by the set of HC stretch, CO stretch, and HCO bend quantum numbers v_1, v_2, and v_3, respectively.

Well defined progressions are seen in the SEP spectrum[140,141] so that quantum numbers may be assigned to the HCO resonance states. Thus, lifetimes for these states may be associated with the degree of excitation in the HC stretch, CO stretch, and HCO bend vibrational modes. States with large v_1 and a large excitation in the HC stretch have particularly short lifetimes, while states with large v_2 and a large excitation in the CO stretch have particularly long lifetimes. These are mode-specific effects. Short lifetimes for states with a large v_1 might be expected, since the reaction coordinate for dissociation is primarily HC stretching motion. The mode specific effects are illustrated by the nearly isoenergetic (v_1, v_2, v_3) resonance states $(0, 4, 5)$, $(1, 5, 1)$, and $(0, 7, 0)$ whose respective energies (i.e. position in spectrum) are 12373, 12487, and 12544 cm^{-1} and whose respective linewidths Γ are 42, 55, and 0.72 cm^{-1}.

Quantum mechanical scattering calculations (see Section 8.3.2) have also been performed to study the HCO resonance states.[142–144] The resonance energies, linewidths, and quantum number assignments determined from these calculations are in excellent agreement with the experimental results.

In the preceding discussion, we have considered the situation where highly excited vibrational/rotational states exhibit mode-specific effects akin to the prediction of the Slater theory. The situation should also be considered where mode-specific effects would not be observed, even though dissociation occurs through resonance states. For many molecules, anharmonic and vibration/rotation coupling terms are important at high levels of excitation.[30] As a result of these couplings, the vibrational/rotational states of the molecule cannot be defined by a single ϕ_i which is a product of normal-mode and rotational wave functions. Instead, the wave function ψ_j for each state is a linear combination of the ϕ_i, i.e.,

$$\psi_j = \sum_i c_{ji}\phi_i \qquad (11\text{-}105)$$

If ψ_j is similar to a particular normal-mode rotational state, all except one of the coefficients c_{ji} are small. This is the situation for the above mode-specific dissociation of HCO. However, if the anharmonic and vibrational/rotational couplings are extensive, ψ_j will be a linear combination of many ϕ_i's. Thus, quantum numbers cannot be assigned to ψ_j for the resonance state, since it is not comprised of a predominant ϕ_i.[145] The spectrum for these states is irregular without patterns, and random fluctuations in the resonance states' rate constants k_j are related to the manner in which the ψ_j are randomly distributed in coordinate space. The coefficients c_{ji} in equation (11-105) are random variables.[146]

This type of unimolecular dynamics has been referred to as *statistical state-specific* behavior.[30,147] The average of the statistical state-specific rate constants within a small energy interval $E \rightarrow E + \Delta E$ is thought to approximate the RRKM rate constant $k(E)$.[30] Statistical state-specific behavior has been observed in experimental SEP studies of $H_2CO \rightarrow H_2 + CO$ dissociation[62,148] and quantum mechanical scattering calculations of $HO_2 \rightarrow H + O_2$ dissociation.[149] The state-specific rate constants for the latter calculation are plotted in Figure 11-20. The RRKM rate constant and the average of the state-specific quantum rate constants are in good agreement.

These observations of mode-specific kinetic processes have led to numerous suggestions for "molecular control" of a chemical reaction by specific excitation of chemical bonds.[150] As noted in the preceding section, this had been the hope for IRMPD, but in that case the molecules refused to cooperate. An early example of nonmonotonic (and therefore mode-specific) dependence of rate on excitation energy had been

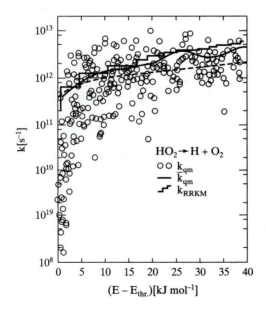

FIGURE 11-20 Comparison of the unimolecular dissociation rates for $HO_2 \rightarrow H + O_2$ as obtained from the quantum mechanical resonances (k_{qm} open circles) and from variational transition state RRKM (k_{RRKM}, step function). E_{thr} is the threshold energy for dissociation. Also shown is the quantum mechanical average (solid line) as well as the experimental prediction for J = 0 derived from a simplified SACM analysis of high pressure unimolecular rate constants [A. J. Dobbyn et al., *J. Chem. Phys.* **102**, 5869 (1995). Reproduced with permission.]

found by Pimentel,[151] for excitation of ethylene reacting with F_2 in a low-temperature matrix. Beginning in the early 1990's, selective overtone excitation was found to be effective in guiding the branching ratios of reactions such as unimolecular decomposition of DOOH or bimolecular reaction of HOD with H or Cl atoms, yielding {HD/H_2} or {HCl/DCl}, respectively.[2,152,153] More recently, this type of vibrational mode selectivity has been found to occur in ion-molecule reactions such as $NH_3^+(\nu) + ND_3$[154] and $C_2H_2(\nu) + CH_4$.[155]

A more general approach[156] to molecular control involves preparation of a wavepacket in a specified region of the potential energy surface using coherent picosecond pulses (see chapter 10, section 10.9 and reference 41). The wavepacket then propagates on that surface, and in principle can be "guided" into a specified product channel by judicious initial placement on the surface. The principle has been demonstrated[157] by controlling the photoion distribution from HI (viz., HI^+ or I^+) when irradiated with either three 355-nm photons or a single 118.5-nm photon. When using femtosecond pulses (see section 10.9) to initiate and probe unimolecular reactions, the time scale may be too fast for the IVR process to be completed, and nonstatistical (nonergodic) behavior may be observed. Zewail and co-workers[158] have observed such behavior in the laser-induced decomposition of a series of cyclic alkyl ketones. These ideas and observations have led to grandiose speculations on "controlling the future of matter"[159] by laser-selective initiation of reactions; but given the intricate and expensive equipment needed to generate the required laser pulse sequences, and the specialized conditions under which such reactions are found to occur, it seems very unlikely that such an approach will ever have a major impact on the way most chemicals are actually produced.

REFERENCES

[1] D. W. Lupo and M. Quack, *Chem. Rev. 87*, 181 (1987).

[2] L. J. Butler, T. M. Ticich, M. D. Likar, and F. F. Crim, *J. Chem. Phys. 85*, 233 (1986); A. Sinha, M. C. Hsiao, and F. F. Crim, *J. Chem. Phys. 92*, 6333 (1990).

[3] H.-L. Dai and R. W. Field, eds., *Molecular Dynamics and Spectroscopy by Stimulated Emission Pumping* (Singapore: World Scientific Publishing Pte Ltd, 1995).

[4] T. Beyer and E. F. Swinehart, *Commun. Assoc. Comput. Machin. 16*, 372 (1973); S. E. Stein and B. S. Rabinovitch, *J. Chem. Phys. 58*, 2438 (1973).

[5] R. A. Marcus and O. K. Rice, *J. Phys. and Colloid Chem. 55*, 894 (1951).

[6] G. Z. Whitten and B. S. Rabinovitch, *J. Chem. Phys. 41*, 1883 (1964); D. C. Tardy, B. S. Rabinovitch, and G. Z. Whitten, *J. Chem. Phys. 48*, 1427 (1968).

[7] J. Perrin, *Ann. Phys. 11*, 1 (1919).

[8] I. Langmuir, *J. Am. Chem. Soc. 42*, 2190 (1920).

[9] F. A. Lindemann, *Trans. Faraday Soc. 17*, 598 (1922).

[10] C. N. Hinshelwood, *Proc. Roy. Soc. (A) 113*, 230 (1927).

[11] N. B. Slater, *Proc. Camb. Phil. Soc. 35*, 56 (1939).

[12] I. E. Farquhar, *Ergodic Theory in Statistical Mechanics* (New York: John Wiley and Sons, 1964).

[13] D. Chandler, *Introduction to Modern Statistical Mechanics* (Oxford: University Press, 1987).

[14] D. L. Bunker, *J. Chem. Phys. 40*, 1946 (1964).

[15] N. B. Slater, *Theory of Unimolecular Reactions* (London: Methuen, 1959), Chapter 9.

[16] O. K. Rice and H. C. Ramsperger, *J. Am. Chem. Soc. 50,* 617 (1928).

[17] L. S. Kassel, *J. Phys. Chem. 32,* 1065 (1928).

[18] Slater, *op. cit.,* Chapter 2.

[19] R. A. Marcus and O. K. Rice, *J. Phys. Colloid Chem. 55,* 894 (1951); R. A. Marcus, *J. Chem. Phys. 20,* 359 (1952); H. M. Rosenstock, M. B. Wallenstein, A. L. Wahrhaftig, and H. Eyring, *Proc. Natl. Acad. Sci. USA 38,* 667 (1952).

[20] R. A. Marcus. *J. Chem. Phys. 43,* 2658 (1965); *52,* 1018 (1970).

[21] D. L. Bunker and M. Pattengill, *J. Chem. Phys. 48,* 772 (1968).

[22] D. L. Bunker, in *Proceedings of the International School of Physics,* Course XLIV, edited by C. Schlier (New York: Academic Press, 1970), p. 315.

[23] F. W. Schneider and B. S. Rabinovitch, *J. Am. Chem. Soc. 84,* 4215 (1962).

[24] J. H. Current and B. S. Rabinovitch, *J. Chem. Phys. 38,* 783 (1963).

[25] M. Quack and J. Troe, *Ber. Bunsenges. Phys. Chem. 78,* 240 (1974).

[26] W. H. Miller, *J. Am. Chem. Soc. 101,* 6810 (1979).

[27] P. J. Robinson and K. A. Holbrook, *Unimolecular Reactions* (New York: Wiley Interscience, 1972).

[28] W. Forst, *Theory of Unimolecular Reactions* (New York: Academic Press, 1973).

[29] R. G. Gilbert and S. C. Smith, *Theory of Unimolecular and Recombination Reactions* (London: Blackwell Scientific Publications, 1990).

[30] T. Baer and W. L. Hase, *Unimolecular Reaction Dynamics. Theory and Experiment* (Oxford: University Press, 1996).

[31] Baer and Hase, *ibid.,* Chapter 6.

[32] L. Zhu and W. L. Hase, *Chem. Phys. Letts. 175,* 117 (1990).

[33] L. Zhu, W. Chen, W. L. Hase, and E. W. Kaiser, *J. Phys. Chem. 97,* 311 (1993).

[34] J. Troe, *J. Phys. Chem. 83,* 114 (1979).

[35] L. B. Bhuiyan and W. L. Hase, *J. Chem. Phys. 78,* 5052 (1983).

[36] H. S. Johnston, *Gas Phase Reaction Rate Theory* (New York: Ronald, 1966), Chapter 5.

[37] H. E. O'Neal and S. W. Benson, *J. Phys. Chem. 71,* 2903 (1967).

[38] K. Dees and D. W. Setser, *J. Chem. Phys. 49,* 1193 (1968).

[39] W. L. Hase, *Acc. Chem. Res. 16,* 258 (1983).

[40] P. Saxe, Y. Yamaguchi, P. Pulay, and H. F. Schaefer, III, *J. Am. Chem. Soc., 102,* 3718 (1980).

[41] E. V. Waage and B. S. Rabinovitch, *Chem. Rev. 70,* 377 (1970).

[42] H. Furue and P. D. Pacey, *Can. J. Chem. 60,* 916 (1982).

[43] W. Tsang, *J. Chem. Phys. 41,* 2487 (1964); W. Tsang, in *Shock Tubes in Chemistry,* edited by A. Lifshitz (New York: Dekker, 1981), p. 59.

[44] G. N. Spokes, D. M. Golden, and S. W. Benson, *Angew. Chem. Int. Ed. Engl. 12,* 534 (1973).

[45] F. W. Schneider and B. S. Rabinovitch, *J. Am. Chem. Soc. 85,* 2365 (1963).

[46] B. S. Rabinovitch, P. W. Gilderson, and F. W. Schneider, *J. Am. Chem. Soc. 87,* 158 (1965).

[47] K. M. Maloney, S. P. Pavlou, and B. S. Rabinovitch, *J. Phys. Chem. 73,* 2756 (1969).

[48] K. M. Maloney and B. S. Rabinovitch, *J. Phys. Chem. 73,* 1652 (1969).

[49] W. L. Hase and H. B. Schlegel, *J. Phys. Chem. 86,* 3901 (1982).

[50] T. Baer, *Adv. Chem. Phys. 64,* 111 (1986).

[51] T. Baer and R. Kury, *Chem. Phys. Lett. 92,* 659 (1982).

[52] B. S. Rabinovitch and M. C. Flowers, *Quart. Rev. 18,* 122 (1964).

[53] B. S. Rabinovitch and D. W. Setser, *Advan. Photochem. 3,* 1 (1964).

[54] C. W. Larson and B. S. Rabinovitch, *J. Chem. Phys. 52,* 5181 (1970).

[55] W. L. Hase, R. L. Johnson, and J. W. Simons, *Int. J. chem. Kinet. 4,* 1 (1972).

[56] F. B. Growcock, W. L. Hase, and J. W. Simons, *Int. J. Chem. Kinet. 5,* 77 (1973).

[57] K. Dees, D. W. Setser, and W. G. Clark, *J. Phys. Chem. 75,* 2231 (1971).

[58] J. M. Jasinski, J. K. Frisoli, and C. B. Moore, *J. Chem. Phys. 79,* 1312 (1983).

[59] S. I. Ionov, G. A. Brucker, C. Jacques, Y. Chen, and C. Wittig, *J. Chem. Phys. 99,* 3420 (1993).

[60] J. Miyawaki, K. Yamanouchi, and S. Tsuchiya, *J. Chem. Phys. 99,* 254 (1993).

[61] E. R. Lovejoy, S. K. Kim, and C. B. Moore, *Science 256,* 1541 (1992); S. K. Kim, E. R. Lovejoy, and C. B. Moore, *J. Chem. Phys. 102,* 3202 (1995).

[62] W. H. Green, Jr., C. B. Moore, and W. F. Polik, *Ann. Revs. Phys. Chem. 43,* 591 (1992).

[63] S. J. Klippenstein and T. Radivoyevitch, *J. Chem. Phys. 99,* 3644 (1993).

[64] S. J. Klippenstein, A. L. L. East, and W. D. Allen, *J. Chem. Phys. 105,* 118 (1996).

[65] H. S. Johnston, *J. Am. Chem. Soc. 75,* 1567 (1953); D. J. Wilson and H. S. Johnston, *J. Am. Chem. Soc. 75,* 5673 (1953).

[66] S. C. Chan, B. S. Rabinovitch, J. T. Bryant, L. D. Spicer, T. Fujimoto, Y. N. Lin, and S. P. Pavlou, *J. Phys. Chem. 74,* 3160 (1970).

[67] W. G. Valance and E. W. Schlag, *J. Chem. Phys. 45,* 4280 (1966).

[68] J. Troe and H. G. Wagner, *Ber. Bunsenges. Phys. Chem. 71,* 937 (1967).

[69] D. C. Tardy and B. S. Rabinovitch, *J. Chem. Phys. 48,* 1282 (1968).

[70] D. C. Tardy and B. S. Rabinovitch, *Chem, Revs. 77,* 369 (1977).

[71] J. Troe, *J. Chem. Phys. 66,* 4745 (1977).

[72] J. Troe, *J. Chem. Phys. 66,* 4758 (1977).

[73] R. E. Harrington, B. S. Rabinovitch, and M. R. Hoare, *J. Chem. Phys. 33,* 744 (1960).

[74] G. H. Kohlmaier and B. S. Rabinovitch, *J. Chem. Phys. 48,* 1282 (1968).

[75] D. W. Setser, in *MTP International Review of Sciences, Physical Chemistry,* Vol. 9, edited by J. Polanyi (Baltimore: University Park Press, 1972), p. 1.

[76] M. Hoare, *J. Chem. Phys. 38,* 1630 (1963).

[77] D. W. Setser and E. E. Siefert, *J. Chem. Phys. 57,* 3623 (1972).

[78] Von S. H. Luu and J. Troe, *Ber. Bunsenges. Phys. Chem. 78,* 766 (1974).

[79] I. Oref and D. C. Tardy, *Chem. Revs. 90,* 1407 (1990).

[80] J. R. Barker, *J. Phys. Chem. 88,* 11 (1984).

[81] J. E. Dove, H. Hippler, and J. Troe, *J. Chem. Phys. 82,* 1907 (1985); H. Hippler, L. Lindemann, and J. Troe, *J. Chem. Phys. 83,* 3906 (1985).

[82] C. A. Michaels and G. W. Flynn, *J. Chem. Phys. 106,* 3558 (1997).

[83] A. J. Stace and J. N. Murrell, *J. Chem. Phys. 68,* 3028 (1978).

[84] N. Date, W. L. Hase, and R. G. Gilbert, *J. Phys. Chem. 88,* 5135 (1994).

[85] G. Lendvay and G. C. Schatz, *J. Chem. Phys. 96,* 4356 (1992).

[86] T. Lenzer, K. Luther, J. Troe, R. G. Gilbert, and K. F. Lim, *J. Chem. Phys. 103,* 626 (1995).

[87] V. Bernshtein and I. Oref, *J. Chem. Phys. 104,* 1958 (1996).

[88] T. Lenzer and K. Luther, *J. Chem. Phys. 105,* 10944 (1996).

[89] K. F. Lim and R. G. Gilbert, *J. Chem. Phys. 84,* 6129 (1986).

[90] C. E. Klots, *J. Chem. Phys. 58,* 5364 (1973).

[91] C. E. Klots, *J. Chem. Phys. 64,* 4269 (1976).

[92] W. J. Chesnavich and M. T. Bowers, *J. Chem. Phys. 66,* 2306 (1977).

[93] J. C. Light, *Disc. Faraday Soc. 44,* 14 (1967).

[94] W. J. Chesnavich and M. T. Bowers, in *Gas Phase Ion Chemistry, Vol. 3* (New York: Academic Press), p. 119.

[95] J. M. Parson and Y. T. Lee, *J. Chem. Phys. 56,* 4658 (1972); J. M. Farrar and Y. T. Lee, *J. Chem. Phys. 65,* 1414 (1976).

[96] W. J. Hase and K. C. Bhalla, *J. Chem. Phys. 75,* 2807 (1981).

[97] W. B. Miller, S. A. Safron, and D. R. Herschbach, *Disc. Faraday Soc. 44,* 108 (1967).

[98] R. A. Marcus, *J. Phys. Chem. 83,* 204 (1979).

[99] F. H. Mies, *J. Chem. Phys. 51,* 798 (1969).

[100] K. Song, G. H. Peslherbe, W. L. Hase, A. J. Dobbyn, M. Stumpf, and R. Schinke, *J. Chem. Phys. 103,* 8891 (995).

[101] A. J. Dobbyn, M. Stumpf, H. -M. Keller, and R. Schinke, *J. Chem. Phys. 104,* 8357 (1996).

[102] D. L. Bunker and W. L. Hase, *J. Chem. Phys. 59,* 4621 (1973).

[103] R. C. Baetzold and D. J. Wilson, *J. Phys. Chem. 68,* 3141 (1964).

[104] J. D. Rynbrandt and B. S. Rabinovitch, *J. Phys. Chem. 75,* 2164 (1971).

[105] J. F. Meagher, K. J. Chao, J. R. Barker, and B. S. Rabinovitch, *J. Phys. Chem. 78,* 2535 (1974).

[106] C. S. Sloane and W. L. Hase, *J. Chem. Phys. 66,* 1523 (1977).

[107] W. L. Hase and D. G. Buckowski, *J. Comput. Chem. 3,* 335 (1982); W. L. Hase, D. G. Buckowski, and K. N. Swamy, *J. Phys. Chem. 87,* 2754 (1983).

[108] R. J. Wolf and W. L. Hase, *J. Chem. Phys. 72,* 316 (1980); R. J. Wolf and W. L. Hase, *J. Chem. Phys. 73,* 3779 (1980).

[109] T. Uzer, J. T. Hynes, and W. P. Reinhardt, *J. Chem. Phys. 85,* 5791 (1986).

[110] T. A. Holme and J. S. Hutchinson, *J. Chem. Phys. 83,* 2860 (1985); J. S. Hutchinson, *J. Phys. Chem. 91,* 4495 (1987).

[111] M. C. Gutzwiller, *Chaos in Classical and Quantum Mechanics* (Berlin: Springer Verlag, 1991).

[112] A. J. Lichtenberg and M. A. Lichtenberg, *Regular and Chaotic Dynamics,* 2d ed. (Berlin: Springer Verlag, 1992).

[113] G. Porter and J. I. Steinfeld, *J. Chem. Phys. 45,* 3456 (1966).

[114] N. Isenor and M. C. Richardson, *Appl. Phys. Letts. 18,* 224 (1971).

[115] N. Isenor, V. Merchant, R. Hallsworth, and M. Richardson, *Can. J. Phys. 51,* 1281 (1973).

[116] R. V. Ambartzumian, Yu. A. Gorokhov, V. S. Letokhov, and G. N. Makarov, *J.E.T.P. Letts. 21,* 375 (1975).

[117] J. L. Lyman, R. J. Jensen, J. Rink, C. P. Robinson, and S. D. Rockwood, *Appl. Phys. Letts. 27,* 87 (1975).

[118] J. I. Steinfeld (ed.), *Laser-Induced Chemical Processes,* (New York: Plenum, 1981).

[119] M. Quack, in *Adv. Chem. Phys. Vol. 50: Dynamics of the Excited State,* edited by K. P. Lawley (Chichester, England: Wiley, 1982), pp. 395–473.

[120] D. W. Lupo and M. Quack, *Chem. Rev. 87,* 181 (1987).

[121] J. S. Francisco and J. I. Steinfeld, in *Multiphoton Processes and Spectroscopy, Vol. 2,* edited by S. H. Lin (Singapore: World Publishing Co., 1986), pp. 79–173.

[122] J. I. Steinfeld, *Molecules and Radiation,* 2d ed. (Cambridge, MA: M.I.T. Press, 1985).

[123] J. Ackerhalt and B. Shore, *Phys. Rev. A16,* 277 (1977).

[124] M. P. Casassa, D. S. Bomse, J. L. Beauchamp, and R. C. Janda, *J. Chem. Phys. 72,* 6805 (1980); M. A. Hoffbauer, K. Liu, C. F. Giese, and W. R. Gentry, *J. Chem. Phys. 78,* 5567 (1983).

[125] R. L. Woodin, D. S. Bomse, and J. L. Beauchamp, *J. Am. Chem. Soc. 100,* 3248 (1978); R. L. Woodin, D. S. Bomse, and J. L. Beauchamp, *Chem. Phys. Lett. 63,* 630 (1979); R. N. Rosenfeld, J. M. Jasinski, and J. L. Beauchamp, *Chem. Phys. Lett. 71,* 400 (1980); R. C. Dunbar and T. B. McMahon, *Science 279,* 194 (1998).

[126] W. D. Lawrance, A. E. W. Knight, R. G. Gilbert, and K. D. King, *Chem. Phys. 56,* 343 (1981).

[127] J. S. Francisco, Zhu Qingshi, and J. I. Steinfeld, *J. Chem. Phys. 78,* 5339 (1983).

[128] P. A. Schulz, Aa. S. Sudbø, D. J. Krajnovich, H. S. Kwok, Y. R. Shen, and Y. T. Lee, *Ann. Rev. Phys. Chem. 30,* 379 (1979).

[129] J. I. Steinfeld, in *Energy Storage and Redistribution in Molecules,* edited by J. Hinze (New York: Plenum Press, 1983), p. 1.

[130] Zhang Fumin, J. S. Francisco, and J. I. Steinfeld, *J. Phys. Chem. 86,* 2402 (1982).

[131] B. Roop, S. Joyce, J. C. Schultz, and J. I. Steinfeld, *J. Chem. Phys. 83,* 6012 (1985).

[132] D. Tholmann, D. S. Tonner, and T. B. McMahon, *J. Phys. Chem. 98,* 2002 (1994).

[133] R. C. Dunbar, *J. Phys. Chem. 98,* 8705 (1994).

[134] W. D. Price, P. D. Schnier, R. A. Jokusch, E. F. Strittmatter, and E. R. Evans, *J. Am. Chem. Soc. 118,* 10640 (1996); W. D. Price, P. D. Schnier, and E. R. Williams, *J. Phys. Chem. B101,* 664 (1997).

[135] K. E. Johnson, L. Wharton, and D. H. Levy, *J. Chem. Phys. 69,* 2719 (1978).

[136] Z. S. Huang, K. W. Jucks, and R. E. Miller, *J. Chem. Phys. 85,* 3338 (1986).

[137] G. E. Ewing, *J. Chem. Phys. 71,* 3143 (1979); G. E. Ewing, *J. Chem. Phys. 72,* 2096 (1980).

[138] M. P. Casassa, A. M. Woodward, J. C. Stephenson, and D. S. King, *J. Chem. Phys. 85,* 6235 (1986).

[139] G. Fischer, R. E. Miller, P. F. Vohralik, and R. O. Watts, *J. Chem. Phys. 83,* 1471 (1985).

[140] J. D. Tobiasen, J. R. Dunlap, and E. A. Rohlfing, *J. Chem. Phys. 103,* 1448 (1995).

[141] C. Stock, X. Li, H. -M. Keller, R. Schinke, and F. Temps, *J. Chem. Phys. 106,* 5333 (1997).

[142] H. -M. Keller, H. Floethmann, A. J. Dobbyn, H. -J. Werner, C. Bauer, and P. Rosmus, *J. Chem. Phys. 105,* 4983 (1996).

[143] R. N. Dixon, *J. Chem. Soc. Faraday Trans. 88,* 2575 (1992).

[144] D. Wang and J. M. Bowman, *J. Chem. Phys. 100,* 1021 (1994).

[145] G. Hose and H. S. Taylor, *J. Chem. Phys. 76,* 5356 (1982).

[146] W. F. Polik, D. R. Guyer, W. H. Miller, and C. B. Moore, *J. Chem. Phys. 92,* 3471 (1990).

[147] W. L. Hase, S. -W. Cho, D. -L. Lu, and K. N. Swamy, *Chem. Phys. 139,* 1 (1989).

[148] W. F. Polik, D. R. Guyer, and C. B. Moore, *J. Chem. Phys. 92,* 3453 (1990).

[149] A. J. Dobbyn, M. Stumpf, H. -M. Keller, W. L. Hase, and R. Schinke, *J. Chem. Phys. 104,* 8357 (1996).

[150] F. F. Crim, *Science 249,* 1387 (1990).

[151] H. Frei and G. C. Pimentel, *J. Chem. Phys. 78,* 3698 (1983).

[152] A. Sinha, M. C. Hsiao, and F. F. Crim, *J. Chem. Phys. 92,* 6333 (1990).

[153] F. Flam, *Science 266,* 215 (1994).

[154] R. D. Guettler, G. C. Jones, Jr., L. A. Posey, and R. N. Zare, *Science 266,* 259 (1994).

[155] Y. -H. Chiu, H. Fu, J. -T. Huang, and S. L. Anderson, *J. Chem. Phys. 101,* 5410 (1994).

[156] S. A. Rice, *Science 258,* 412 (1992); D. J. Tannor and S. A. Rice, *Adv. Chem. Phys. 70,* 441 (1988).

[157] L. Zhu, V. Kleiman, X. Li, S. P. Lu, K. Trentelman, and R. G. Gordon, *Science 270,* 77 (1995).

[158] I. Oref, *Science 279,* 820 (1998); E. W. -G. Diau, J. L. Herek, Z. H. Kim, and A. H. Zewail, *Science 279,* 847 (1998).

[159] B. Kohler, J. M. Krause, F. Raksi, K. R. Wilson, V. Y. Yakovlev, R. M. Whitnell, and YJ. Yan, *Accts. Chem. Research 28, 133* (1995).

BIBLIOGRAPHY

Theory of Unimolecular Reactions

BAER, T., and HASE, W. L. *Unimolecular Reaction Dynamics. Theory and Experiments.* New York: Oxford, 1996.

BEYNON, J. H., and GILBERT, J. R. *Application of Transition State Theory to Unimolecular Reactions.* New York: Wiley, 1984.

BUNKER, D. L. *Theory of Elementary Gas Reaction Rates.* New York: Pergamon Press, 1966.

CHESNAVICH, W. J., and BOWERS, M. T. In *Gas Phase Ion Chemistry, Vol. 1,* edited by M. T. Bowers. New York: Academic Press, 1979, p. 119.

FORST, W. *Theory of Unimolecular Reactions.* New York: Academic Press, 1973.

GARDINER, Jr., W. C. *Rates and Mechanisms of Chemical Reactions.* Menlo Park, California: W. A. Benjamin, 1972.

GILBERT, R. G., and SMITH, S. C. *Theory of Unimolecular and Recombination Reactions.* London: Blackwell Scientific Publications, 1990.

HASE, W. L. In *Modern Theoretical Chemistry, Dynamics of Molecular Collisions, Part B,* edited by W. H. Miller, New York: Plenum, 1976, p. 121.

HOLBROOK, K. A. *Chem. Soc. Rev. 12,* 163 (1983).

PRITCHARD, H. O. *The Quantum Theory of Unimolecular Reactions.* New York: Cambridge University Press, 1984.

QUACK, M., and TROE, J., In *Gas Kinetics and Energy Transfer, Vol. 2. Specialist Periodical Reports,* edited by P. G. Ashmore and R. J. Donovan. London: The Chemical Soc., Burlington House, 1977, p. 175.

QUACK, M., and TROE, J., *Int. Rev. Phys. Chem. 1,* 97 (1981).

ROBINSON, P. J., and HOLBROOK, K. A. *Unimolecular Reactions.* London: Wiley-Interscience, 1972.

SLATER, N. B. *Theory of Unimolecular Reactions.* Ithaca, NY: Cornell Press, 1959.

SMITH, I. W. M. *Kinetics and Dynamics of Elementary Gas Reactions.* London: Butterworths, 1980.

TROE, J. In *Physical Chemistry an Advanced Treatise, Vol. VI B,* edited by W. Jost. New York: Academic Press, 1975, p. 835.

Energy-Selected Dissociation

BAER, T. *Adv. Chem. Phys. 64,* 111 (1986).

CRIM, F. F. *Ann. Rev. Phys. Chem. 35,* 657 (1984).

LUPO, D. W., and QUACK, M. *Chem. Rev. 87,* 181 (1987).

Intermolecular Energy Transfer

TARDY, D. C., and RABINOVITCH, B. S. *Chem. Rev. 77,* 369 (1977).

Intramolecular Energy Transfer

BRUMER, P. *Adv. Chem. Phys. 47*, 201 (1981).

HASE, W. L. In *Potential Energy Surfaces and Dynamics Calculations,* edited by D. G. Truhlar. New York: Plenum, 1981, p. 1.

HASE, W. L., *J. Phys. Chem. 90*, 365 (1986).

HASE, W. L., and WOLF, R. J. In *Potential Energy Surfaces and Dynamics Calculations,* edited by D. G. Truhlar. New York: Plenum, 1981, p. 37.

NOID, D. W., KOSZYKOWSKI, M. L., and MARCUS, R. A., *Ann. Rev. Phys. Chem. 32*, 267 (1981).

OREF, I., and RABINOVITCH, B. S. *Acc. Chem. Res. 12*, 166 (1979).

PARMENTER, C. S. *Faraday Discuss. Chem. Soc. 75*, 7 (1983).

QUACK, M. In *Energy Storage and Redistribution in Molecules,* edited by J. Hinze. New York: Plenum, 1983, p. 493.

Mode Specificity

BISSELING, R. H., KOSLOFF, R., MANZ, J., MRUGALA, F., RÖMELT, J., and WEICHSEL-BAUMER, G. *J. chem. Phys. 86*, 2626 (1987).

HEDGES, R. M., JR., SKODJE, R. T., BORONDO, F., and REINHARDT, W. P. In *Resonances in Electron Molecule Scattering, Van der Waals Complexes, and Reactive Chemical Dynamics.* Kansas City, ACS Symp. Ser. No. 263, 1974.

HOSE, G., and TAYLOR, H. S. *J. Chem. Phys. 76*, 5356 (1982).

HOSE, G., TAYLOR, H. S., and BAI, Y. Y. *J. Chem. Phys. 80*, 4363 (1984).

MIES, F. H., and KRAUSS, M. *J. Chem. Phys. 45*, 4455 (1966).

MIES, F. H., *J. Chem. Phys. 51*, 798 (1969).

MILLER, W. H., *Chem. Rev. 87*, 19 (1987).

SWAMY, K. N., HASE, W. L., GARRETT, B. C., MCCURDY, C. W., and MCNUTT, J. F. *J. Phys. Chem. 90*, 3517 (1986).

WAITE, B. A., GRAY, S. K., and MILLER, W. H. *J. Chem. Phys. 78*, 259 (1983).

WAITE, B. A., and MILLER, W. H. *J. Chem. Phys. 73*, 3713 (1980).

WAITE, B. A., and MILLER, W. H. *J. Chem. Phys. 74*, 3910 (1981).

PROBLEMS

11.1 Consider the three-oscillator harmonic system with frequencies of 800, 1000, and 1400 cm^{-1}. By direct count, determine the sum of states at E = 5000 cm^{-1}.

11.2 Use the Whitten-Rabinovitch approximation to determine the sum of states for cyclopropane at 40, 200, and 400 kJ/mol. Compare with the values in Table 11-1.

11.3 Compare $G(E)$ in equations (11-11), (11-13) and (11-14) by applying these equations to CH$_4$ for E in the range of 0–400 kJ/mol. The harmonic vibrational frequencies for CH$_4$ are 3025, 1583(2), 3157(3), and 1367(3) cm^{-1}. Calculate the sum of states every 40 kJ/mol.

11.4 Using the Wien radiation law

$$E_\nu d\nu = \frac{2\pi h\nu^3}{c^2} e^{-h\nu/k_B T} \, d\nu$$

calculate the amount of infrared energy radiated in a 100 cm^{-1} bandwidth centered at 3000 cm^{-1} by a blackbody source at T = 1000 K. How does this compare with the amount of energy required to sustain a unimolecular reaction having a rate coefficient $k_{uni} = 10^{-2}$ sec^{-1} and an activation energy of 300 kJ/mole?

11.5 Using equations (11-14) through (11-17) derive the following density of states expression for $(1.0 < E' < 8.0)$

$$N(E_V) = \frac{(E_V + aE_Z)^{s-1}}{(s-1)!\Pi h\nu_i} [1 + 0.60478\,\beta w(E')^{-0.75}]$$

11.6 Show that if classical mechanical expressions are used for the sum and density of states given by equations (11-11) and (11-12), RRKM theory gives the classical mechanical RRK expression, equation (11-39) for the unimolecular rate constant.

11.7 The high pressure A-factor for the dissociation of methane, i.e., $CH_4 \rightarrow CH_3 + H$ is 10^{16} sec^{-1}, and the activation energy is 400 kJ/mole. Using the RRK theory calculate the unimolecular rate constant for methane at an excitation energy of 500 kJ/mole.

11.8 Silane (SiH$_4$) unimolecularly decomposes to SiH$_2$. The frequencies for SiH$_4$ are four stretches of 2,000 cm^{-1} and five bends of 900 cm^{-1}.

(a) Propose a structure for the transition state.

(b) Estimate values for the transition state frequencies.

(c) Calculate the difference in harmonic zero-point energies between the transition state and reactant (83 cm^{-1} = 1 kJ/mol.).

11.9 What is the expression for the harmonic sum of states for a diatomic harmonic oscillator at energy E? Is the sum of states for the Morse oscillator at this energy larger or smaller? Why?

*11.10 Consider the thermal unimolecular rearrangement of hydrogen isocyanide, which involves only three atoms:

$$HNC \rightarrow HCN$$

It is reasonable to assume that this reaction proceeds via a three-center transition state:

(a) Specify all the (non-translational) degrees of freedom of HNC and the transition state shown above. Note that HNC, being collinear, has two rotational degrees of freedom, while the transition state has three.

(b) Given the following values for the various degrees of freedom in HNC and the transition state, evaluate $Q^{\ddagger}$ and Q at T = 1000 K.

NOTE: HNC: $I = 11.3$ amu-Å^2, $\nu = 3300$ cm^{-1} (stretch); 713 cm^{-1} (doubly degenerate bend), 2100 cm^{-1} (stretch); Transition state: $I^{\ddagger} = 1.02, 10.5,$ and 12.0 amu-Å^2. $\nu = 1000$ cm^{-1} (bend), 2000 cm^{-1} (ring-breathing stretch), 3000 cm^{-1} (stretch). One of the reactant degenerate bends (corresponding to a bend of the transition state) we shall take to be the reaction coordinate.

*11.11 (a) Evaluate $k_B T/h$ at $T = 1000$ K, given $k_B = 1.38 \times 10^{-23}$ J K^{-1}, $h = 6.63 \times 10^{-34}$ J s.

(b) If we assume that $Q^{\ddagger}$ and Q are approximately the same for a particular reaction, give a numerical value for the frequency factor.

(c) Evaluate the frequency factor for the "tight" (three-center) transition state for the HNC + HCN reaction given above. Note: for a unimolecular reaction, the frequency factor is usually expressed notationally as A = $10^{13.1}$ s^{-1} (or whatever the value is).

(d) Would the frequency factor for a "loose" transition state (where one of the vibrational degrees of freedom is of lower frequency in the transition state than in the reactant) be larger or smaller than $k_B T/h$?

11.12 Calculate the densities-of-states and RRKM dissociation rates for the unimolecular dissociation of chloroethane-d_1:

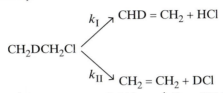

$$CH_2DCH_2Cl \Big\langle \begin{array}{l} \xrightarrow{k_I} CHD = CH_2 + HCl \\ \\ \xrightarrow{k_{II}} CH_2 = CH_2 + DCl \end{array}$$

(a) Calculate $k(E_v)$ [sec^{-1}] for $E_v = 0$ to 50,000 cm^{-1} (0 to 594 kJ/mole), at 1000 cm^{-1} intervals, for the two reaction paths I (elimination of HCl) and II (elimination of DCl). The Whitten-Rabinovitch approximation should be used for calculating the sum, $G(E^{\ddagger})$, and density, $N(E_v)$, terms. The critical energies (E_0), rotational symmetry numbers, and vibrational frequencies are given in the following table:

CH$_2$DCH$_2$Cl		T. S. I (HCl elimination)		T. S. II (DCl elimination)	
g_i	ν_i, cm^{-1}	g_i	ν_i, cm^{-1}	g_i	ν_i, cm^{-1}
4	2940	3	3000	4	3000
1	2160	2	2200	1	2088
4	1340	2	1380	1	1380
2	1270	1	1115	1	1100
3	960	5	960	4	960
2	720	1	850	3	850
1	330	1	820	1	750
1	200	1	645	1	570
		1	403	1	501
s = 18		$s^{\ddagger} = 17$		$s^{\ddagger} = 17$	
critical energies (kJ/mole)		232.7		237.9	
symmetry numbers	0.33	0.25		0.497	

Suggestions: Make sure to count each vibration for a number of times equal to its degeneracy g_i (some of the frequencies, e.g., the C—H stretches, have been "grouped" to reduce computation). The symmetry numbers for the transition state are based on a vibrational transition state model (with hindered internal rotation treated as a torsion) sue to Hassler and Setser; that for the reactant takes into account the entropy of mixing of *trans* and *gauche* forms. Assume that the moments of inertia of the transition states and of the parent molecule are approximately equal.

(b) Using transition state theory calculate k_∞, the net high-pressure limiting thermal unimolecular decomposition for C$_2$H$_5$Cl at 1000°K, and the predicted ratio $[HCl]_\infty / [DCl]_\infty$, assuming no secondary reactions.

11.13 The RRKM theory can be used to calculate the unimolecular rate constant as a function of E. The decomposition rate constant for $C_2H_6 \rightarrow 2CH_3$ is 5×10^9 sec^{-1} at $E = 481$ kJ/mole. E_0 for ethane dissociation is 355 kJ/mole and the frequencies in cm^{-1} for ethane are

$$
\left.\begin{array}{l}
2985 \\
2985 \\
2969 \\
2969 \\
2954 \\
2896
\end{array}\right\} \text{C—H stretches}
\qquad
\left.\begin{array}{l}
1190 \\
1190 \\
822 \\
822
\end{array}\right\} \text{CH}_3 \text{ rocks}
$$

955 C—C stretch

289 internal rotation

$$
\left.\begin{array}{l}
1469 \\
1469 \\
1468 \\
1468 \\
1388 \\
1379
\end{array}\right\} \text{H—C—H bends}
$$

Using computer program RRKM repeatedly, find a critical configuration that fits the above rate constant within a factor of two. Explain the frequency adjustments you made in your critical configuration.

11.14 Consider the simplest (i.e., 2-level) version of the Lindemann-Hinshelwood mechanism:

$$ A + M \underset{k_{-1}}{\overset{k_1}{\rightleftarrows}} A^* + M $$

$$ A^* \xrightarrow{k_2} \text{products} $$

Let

$c_1(t) \equiv [A]$, $c_2(t) \equiv [A^*]$ and $M = [M]$ is the bath gas concentration.

The rate equations are:

$$ \dot{c}_1(t) = -k_1 M c_1(t) + k_{-1} M c_2(t) $$
$$ \dot{c}_2(t) = k_1 M c_1(t) - (k_{-1}M + k_2)c_2(t) $$

(a) Solve the rate equations using the steady-state approximation (i.e., $\dot{c}_2 = 0$) to find the unimolecular rate constant

$$ k_{\text{uni}} \equiv -\dot{c}_1/c_1. $$

(b) Solve the rate equations exactly by assuming a solution of the form

$$ c_i(t) = A_i e^{-\lambda t}; $$

substitute into the rate equations, find λ, and thus k_{uni}:

$$ k_{\text{uni}} \equiv -\dot{c}_1/c_1 = \lambda $$

(c) Compare the unimolecular rate constants obtained in parts (a) and (b). In what limit are they equivalent, i.e., in what limit is the steady-state approximation valid?

11.15 The probability distribution of total thermal energy E is given by the Boltzmann distribution $P(E)$.

(a) Show that

$$P(E) = \frac{N(E)e^{-\beta E}}{Q(T)}$$

where Q is the canonical partition function and N the microcanonical density of states ($\beta = 1/k_B T$).

(b) For the case of s independent classical harmonic oscillators, show that

$$P(E) = \frac{1}{(s-1)!}\left(\frac{E}{k_B T}\right)^{s-1}\frac{e^{-\beta E}}{k_B T}$$

(c) For the $P(E)$ distribution in part (b), calculate the average energy,

$$\langle E \rangle \equiv \int dE\, P(E)E$$

and the width of the distribution about this average value,

$$\Delta E \equiv \sqrt{\langle E^2 \rangle - \langle E \rangle^2}$$

where

$$\langle E^2 \rangle \equiv \int dE\, P(E)E^2$$

11.16 Recall the following Lindemann expression for the thermal unimolecular rate constant, which assumes continuous energies:

$$k_{\text{uni}}(\omega) = \int_0^\infty \frac{k(E)\omega P(E)dE}{k(E) + \omega} \tag{1}$$

where

$$P(E) = N(E)\exp(-\beta E)/Q$$

and

$$Q = \int_0^\infty N(E)\exp(-\beta E)dE$$

Since $k(E) \equiv 0$ for $E < E_0$ the lower limit of the integral in Eq. (1) is actually E_0,

$$k_{\text{uni}}(\omega) = \int_{E_0}^E \cdots dE \tag{2}$$

(a) Show that equation (1) or (2) implies that a plot of $k_{\text{uni}}(\omega)$ versus ω (which is proportional to the pressure) has positive slope and negative curvature for all ω.

(b) For interpretational reasons it is useful to consider $k_{\text{uni}}(\omega)$ as the average of the quantity $k(E)\omega/[k(E) + \omega]$ over a normalized distribution function. The problem is that the function $N(E)\exp(-\beta E)/Q$ is *not* normalized over the interval (E_0, ∞). One thus introduces the constant C,

$$k_{uni}(\omega) = C \int_0^\infty dE \, \frac{N(E)\exp(-\beta E)}{CQ} \, \frac{k(E)\omega}{k(E) + \omega}, \tag{3}$$

so that the distribution function $W(E)$

$$W(E) \equiv \frac{N(E)e^{-\beta E}}{CQ}, \tag{4}$$

is normalized on (E_0, ∞); i.e.,

$$\int_{E_0}^\infty dE \, W(E) = 1;$$

show that this implies that

$$c = \int_{E_0}^\infty dE \, N(E)e^{-\beta E}/Q.$$

The expression for $k_{uni}(\omega)$ then reads

$$k_{uni}(\omega) = c \left\langle \frac{k(E)\omega}{k(E) + \omega} \right\rangle \tag{5a}$$

where

$$\langle \cdots \rangle \equiv \int_{E_0}^\infty dE \, W(E)(\cdots) \tag{5b}$$

Use equation (5) to show that in the high and low pressure limits, respectively, one has the limiting forms

$$k_{uni}(\omega) = C\langle k \rangle - C\langle k^2 \rangle(1/\omega) + \cdots$$
$$k_{uni}(\omega) = \omega C - \omega^2 C\langle 1/k \rangle + \cdots.$$

11.17 Some of the data of Schneider and Rabinovitch [*J. Am. Chem. Soc. 84*, 4215 (1962)] on the thermal isomerization of methyl isocyanide, CH_3NC, are given below.

(a) Compare the pressure dependence with that which would be expected if the simple Lindemann-Hinshelwood theory held.

(b) From suitable plots and calculations, derive approximate values for $\langle k \rangle$ and for the standard derivation $\sqrt{\langle k^2 \rangle - \langle k \rangle^2}$. [Hing: Convert the pressure P into the collision frequency ω by the relation

$$\omega = \bar{v}\pi d^2 C, \tag{A}$$

where C is the concentration, given from the ideal gas law as $C = P/k_B T$. Viscosity measurements imply a value for d of 4.5 Å. Show that equation (A) gives the conversion ω (in sec^{-1}) = $9.03 \times 16^6 \, P$ (in mm Hg).

Pressure mm Hg	$k \times 10^6$ sec^{-1}	Pressure mm Hg	$k \times 10^6$ sec^{-1}	Pressure mm Hg	$k \times 10^6$ sec^{-1}
5980	75.0	570	69.8	24.8	30.0
5010	74.1	337	66.2	15.7	21.6
2888	74.5	208	60.2	12.7	19.8
2248	73.7	155.2	55.6	10.0	18.0
1780	73.5	120.1	55.8	9.07	16.6
1500	72.5	101.5	52.0	7.25	14.2
1143	73.0	72.3	45.0	5.21	13.2
948	67.9	35.7	34.0		

11.18 The IRMPD of chloroethane-d$_1$ was investigated in a classic study of this subject [C. Reiser et al., *J. Am. Chem. Soc. 101*, 350 (1979)]. The RRKM dissociation rates $k(E_v)$ for the two dissociation channels

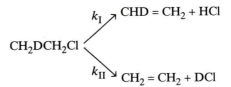

were calculated in problem 11.12 over the energy range 0 to 600 kJ/mole.

(a) Each CO_2 laser photon at 966 cm^{-1} carries 11.3 kJ/mole of energy. Make a plot of the expected lifetime of the multiphoton-activated CH_2DCH_2Cl as a function of the number of CO_2 laser photons absorbed. Also, calculate the predicted HCl/DCl branching ratio over the same range.

(b) At what level of excitation does the HCl/DCl branching ratio become equal to the thermal branching ratio calculated in part (b) of problem 11-12?

---

*These problems have been provided by Dr. R. G. Gilbert ("The Surfing Professor"), Department of Theoretical Chemistry, University of Sydney, Australia

CHAPTER 12

Dynamics Beyond the Gas Phase

Most reactions of interest to the chemical community occur either in solution or at the gas-surface interface. Much less is known about the reaction dynamics of these processes at the microscopic level than about the dynamics of gas phase reactions. For some situations, concepts derived from gas phase studies are helpful in interpreting the dynamics of reactions in solution and heterogeneous reactions. However, many dynamical characteristics of the latter processes are not prominent in the gas phase. An example is the manner in which solvent molecules can affect the molecular motion of a reactive chemical system in solution. Diffusion on a solid surface, a property important in many heterogeneous reactions, does not have a comparable analogue in gas phase reactions.

For reactions in solution, the solvent often has a dramatic effect on the potential energy surface of the reactive system. As an example, consider the generic S_N2 reaction

$$X^- + CH_3Y \longrightarrow XCH_3 + Y^-$$

Representative reaction path enthalpy profiles are shown in Figure 12-1 for reaction in the gas phase, and in dipolar aprotic and protic solvents. Due to the extensive delocalization of the electron in the transition state region of the potential, polar solvent molecules significantly stabilize the reactants and products in comparison to the transition state.[1] Indeed, rate constants for S_N2 reactions in aqueous solution can be up to 20 orders of magnitude smaller than in the gas phase!

Experimental and theoretical studies have provided much insight and detail into the dynamics of solution and gas-surface reactions. In this chapter, we shall consider models which have been developed for these two kinds of reaction.

12.1 TRANSITION STATE THEORY OF SOLUTION REACTIONS

12.1.1 Static Solvent Effects: Thermodynamic Formulation

In the phenomenological treatment of solution reactions presented in chapter 4, a bimolecular solution reaction is modeled by equation (4-30), i.e.,

$$A + B \underset{k_{uni}}{\overset{k_D}{\rightleftharpoons}} AB \overset{k_2}{\longrightarrow} products$$

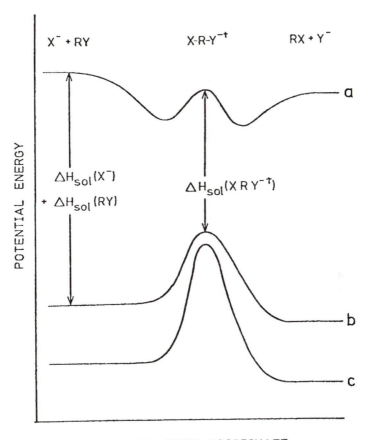

$X^- + RY$ $X\text{-}R\text{-}Y^{-\dagger}$ $RX + Y^-$

POTENTIAL ENERGY

$\Delta H_{sol}(X^-)$
$+ \Delta H_{sol}(RY)$

$\Delta H_{sol}(X\,R\,Y^{-\dagger})$

a

b

c

REACTION COORDINATE

FIGURE 12-1   Representative reaction coordinate diagrams for a nucleophilic displacement reaction in (a) the gas phase and in (b) dipolar aprotic and (c) protic solvents. [Reproduced with permission from W. N. Olmstead and J. I. Brauman. *J. Am. Chem. Soc.* *99*, 4219 (1977). Copyright 1977, American Chemical Society.]

Here, AB is the encounter complex of the reactants trapped inside a solvent cage. In contrast, transition state theory represents the reaction by the equation

$$A + B \longrightarrow AB^{\ddagger} \longrightarrow \text{products} \qquad (12\text{-}1)$$

where $AB^{\ddagger}$ is the transition state.

There are two ways the solvent cage enters into the transition state theory of solution reactions. First, for fast diffusion-controlled bimolecular reactions, the process of diffusional approach of the reactants, i.e., cage-to-cage jumping, may be the rate-limiting elementary step. Second, the inner solvation shell interacts with the chemical reactive system as the AB encounter complex forms the reaction products.

In principle, transition state theory can be applied to either the physical diffusion process or the chemical reaction inside the solvent cage. For the former, the transition state would be a molecule in the act of making a jump from one position to another. The more common application of transition state theory is to the chemical transformation

$$AB \longrightarrow AB^{\ddagger} \longrightarrow \text{products} \qquad (12\text{-}2)$$

It is this process which is treated here. If the chemical transformation is rate-determining, the transition state theory and experimental rate constants may be directly compared. The transition state theory of solution reactions can be developed so that the structure of the solvent and its interaction with the reactive system are explicitly considered. Another way to apply transition state theory is through its thermodynamic formulation. The latter approach is discussed here, while the former is considered in the next section.

From chapter 10, it is easily shown that the transition state theory rate for the reaction given by equation (12-2) is proportional to the concentration of transition states; i.e.,

$$\text{Rate} = \frac{k_B T}{h} [AB^{\ddagger}] \qquad (12\text{-}3)$$

In solution, the equilibrium constant for formation of the transition state should be expressed in terms of activities instead of concentrations, so that

$$K^{\ddagger} = \frac{a^{\ddagger}}{a_A a_B} \qquad (12\text{-}4)$$

where the a's are the activities. By introducing the concept of the activity coefficient γ (i.e., $a_A = \gamma_A[A]$) into equation (12-4), the concentration of transition states becomes

$$[AB^{\ddagger}] = K^{\ddagger} \frac{\gamma_A \gamma_B}{\gamma^{\ddagger}} [A][B] \qquad (12\text{-}5)$$

If this equation is inserted into equation (12-3), one obtains the transition state theory expression for the rate constant for a reaction in solution in thermodynamic form:

$$k = \frac{k_B T}{h} K^{\ddagger} \frac{\gamma_A \gamma_B}{\gamma^{\ddagger}} \qquad (12\text{-}6)$$

Note that both $K^{\ddagger}$ and the γ's refer to the same standard state. Equation (12-6) is often written as

$$k = k_0 \frac{\gamma_A \gamma_B}{\gamma^{\ddagger}} \qquad (12\text{-}7)$$

where k_0 is the rate constant in the limit in which the activity coefficients become unity.

An important application of equation (12-7) is to the reaction between the ions, i.e.,

$$A^{z_A} + B^{z_B} \longrightarrow [AB^{\ddagger}]^{z_A + z_B} \longrightarrow \text{products} \qquad (12\text{-}8)$$

From Debye-Hückel theory (see chapter 4, section 4.5) the activity coefficient for an ion in an aqueous solution at 25°C is given by

$$\log \gamma_i = -0.51 z_i^2 I^{1/2} \qquad (12\text{-}9)$$

where the ionic strength I is given by $I = 1/2 \Sigma m_j z_j$, in which the coefficient m_j is the molality of ionic species j in the solution. By inserting equation (12-9) into equation (12-7), one obtains

$$\log k = \log k_0 + \log \gamma_A + \log \gamma_B - \log \gamma^{\ddagger}$$
$$= \log k_0 - 0.51 I^{1/2}[z_A^2 + z_B^2 - (z_A + z_B)^2]$$

so that

$$\log k = \log k_0 + 1.02 z_A z_B I^{1/2} \tag{12-10}$$

which is the same as equation (4-55) derived by the phenomenological approach.

Equation (12-7) is particularly useful for comparing reactions to the gas and liquid phases if a reference state is chosen so that k_0 equals the gas phase rate constant k_g. For a unimolecular reaction, the equation becomes

$$k_s = k_g \frac{\gamma_A}{\gamma^{\ddagger}}$$

Rate constants for many unimolecular reactions are the same in the gas and liquid phases. This is because the transition state, with a few bonds elongated and/or shortened, is often similar to the reactant. Consider, for example, the rearrangement of bicycloheptene to 1,3-cycloheptadiene:

(12-11)

The reactant itself is one of the best possible models for the transition state, and very similar interactions with the solvent are expected for both the reactant and the transition state. Thus, the activity coefficient ratio, $\gamma_A/\gamma^{\ddagger}$ should be unity and independent of the solvent. Accordingly, very similar rate constants and Arrhenius parameters are found for this reaction in the gas phase and in solution.[2] Differences arise between gas and liquid phase unimolecular rate constants when the reactant and transition state charge distributions (e.g., polarities) are different, so that a polar solvent preferentially stabilizes either the reactant or the transition state. Also, for a unimolecular dissociation reaction such as

$$C_2H_6 \longrightarrow 2CH_3 \tag{12-12}$$

the gas and liquid phase kinetics may differ if the products have difficulty escaping from the solvent cage and, thus, re-form the reactant. If diffusion effects are important, transition state theory as developed by equations (12-2) through (12-7) would not be applicable.

To consider differences between gas and liquid phase kinetics more explicitly, the activity coefficient can be expressed in terms of thermodynamic functions. At equilibrium the free energies of a species in solution and in the gas are equal, so that

$$G_s = G_g = G^0 + RT \ln C_g \tag{12-13}$$

where C_g is the concentration in the gas phase and G^0 is the free energy for one mole/liter of the gas. The gas is assumed to behave ideally, so that the activity in the gas phase is equal to the concentration, i.e., $a_g = C_g$. In solution, the activity is given by $a_s = \gamma C_s$. Since a_s and a_g are equal, equation (12-13) may be written as

$$G_s = G_g = G^0 + RT \ln \gamma C_s \tag{12-14}$$

which, for one mole/liter in solution, becomes

$$G_s = G_g = G^0 + RT \ln \gamma$$

Recognizing that $G^0 - G_s$ is the free energy of vaporization ΔG_{vap} for one mole of the species in solution, we see that the activity coefficient is given by

$$\gamma = \exp(-\Delta G_{vap}/RT) = \exp(-\Delta H_{vap}/RT)\exp(\Delta S_{vap}/R) \tag{12-15}$$

From the free-volume theory of liquids,[3] the term $\exp(\Delta S_{vap}/R)$ is related to the molar volume V_s^0 of the solution and free volume V_f per mole of the species by the expression

$$\exp(\Delta S_{vap}/R) = \frac{V_s^0}{V_f} \tag{12-16}$$

The free volume of the solute species is defined as the molar volume minus the actual volume of the molecules. Empirically, V_s^0/V_f turns out to be approximately 100.

Equations (12-15) and (12-16) are very useful for comparing bimolecular reactions in the gas and liquid phases. Inserting these equations into equation (12-17) and identifying k_0 as k_s gives

$$\frac{k_s}{k_g} = \exp(\Delta \Delta H_{vap}/RT) \frac{V_s^0}{V_{fA}} \frac{V_s^0}{V_{fB}} \frac{V_f^\ddagger}{V_s^0} \tag{12-17}$$

where $\Delta \Delta H_{vap}$ is the difference in the heats of vaporization of the transition state and the reactants. Unless there is a significant polarity change between the transition state and the reactants, $\Delta \Delta H_{vap}$ is approximately zero. Thus, from equation (12-17), the free-volume theory of liquids predicts the bimolecular rate constant to be approximately 100 times larger in solution than in the gas phase. Unfortunately, few bimolecular rate constants are known in both the gas and liquid phases.

It is of interest to consider the effect of pressure on a solution rate constant. With the thermodynamic formulation of transition state theory (see chapter 10, section 10.5), the rate constant may be written as

$$\ln k = \ln \frac{k_B T}{h} - \frac{\Delta G^\ddagger}{RT} \tag{12-18}$$

where activity coefficients and other nonideal properties of the solution are absorbed in $\Delta G^\ddagger$. This can be done by choosing the appropriate reference state. The change in the rate constant with pressure at constant temperature is

$$\left(\frac{d \ln k}{dP}\right)_T = -\frac{1}{RT}\left(\frac{d \Delta G^\ddagger}{dP}\right)_T \tag{12-19}$$

$$= -\Delta V^\ddagger/RT$$

since $(dG/dP)_T = V$.

The term $\Delta V^\ddagger$ is the standard change in volume in going from the reactants to the transition state, and is usually referred to as the *volume of activation*. A rate constant increases with increasing pressure if $\Delta V^\ddagger$ is negative. Conversely, the rate constant

TABLE 12-1   Volumes of Activation for Some Reactions in Solution.

Reaction	Solvent	$\Delta V^{\ddagger} (cm^3/mol)$
$OH^- + N(C_2H_5)_4^+ \rightarrow CH_3OC_2H_5 + N(C_2H_5)_3$	methanol	$+20.0$
$(CH_3)_3COOC(CH_3)_3 \longrightarrow 2(CH_3)_3CO$	cyclohexane	$+6.7$
sucrose $+ H_2O \xrightarrow{H+}$ glucose $+$ fructose	water	$+2.5$
$CH_3COOCH_3 + H_2O \xrightarrow{H+} CH_3COOH + CH_3OH$	water	-8.7
$2C_5H_6(\text{cyclopentadiene}) \longrightarrow C_{10}H_{12}$	n-butyl chloride	-22.0

decreases with pressure if there is an increase in volume when the transition state is formed. For many cases $\Delta V^{\ddagger}$ is independent of pressure, so that integration of equation (12-19) gives

$$\ln k = \ln k_0 - \frac{\Delta V^{\ddagger}}{RT} P \qquad (12\text{-}20)$$

Some typical values of $\Delta V^{\ddagger}$ are listed in Table 12-1.

12.1.2   Static Solvent Effects: Formulation Based on Solvent Structure

In the foregoing application of transition state theory to solution reactions, the structures of the solvent and solvent-solute (i.e., reactant) interactions were not directly considered. More insight into the microscopic nature of the solution reaction is often revealed if solvent terms are explicit in the formulation of the transition state theory rate constant. Such an approach allows a direct calculation of the solution rate constant from the solvent-solvent, reactant-solvent, and reactant-reactant potential energies.

The relationship between the solution and gas phase transition state theory rate constants may be expressed as[4,5]

$$k_s = k_g e^{-(\Delta G_s^{\ddagger} - \Delta G_g^{\ddagger})/RT} \qquad (12\text{-}21)$$

where $\Delta G_s^{\ddagger}$ and $\Delta G_g^{\ddagger}$, calculated for the same standard-state convention, are the free energies of activation in solution and in the gas-phase, respectively. For the association process $A + B \rightarrow AB^{\ddagger}$, the free energy of activation in solution is given by

$$\Delta G_s^{\ddagger} = \Delta G_g^{\ddagger} + \Delta w^{\ddagger}(r^{\ddagger}) \equiv \Delta G_g^{\ddagger} + \Delta G_{solv}^{\ddagger} \qquad (12\text{-}22)$$

Here $\Delta w^{\ddagger}(r^{\ddagger}) = \Delta G_{solv}^{\ddagger}$ is the solvent contribution to the *potential of mean force*; i.e., the solvent contribution to the molar free energy change on bringing together A and B up to the transition state separation $r^{\ddagger}$ from infinite separation $r = \infty$ in the solvent.

In variational transition state theory (section 10.8), $\Delta G_s^{\ddagger}$ is the maximum in the free energy along the reaction path. Equation (12-22) shows that this free energy maximum arises from contributions from ΔG_g and the potential of mean force. Each of these free energy terms is temperature dependent so that both the position of the solution transition state and $\Delta G_s^{\ddagger}$ are temperature dependent.

From the equilibrium statistical mechanics of liquids[6,7] it is well-known that the above potential of mean force is related to the A-B pair distribution function $g_{A,B}(r)$ at any separation by

$$g_{A,B}(r) = e^{-U(r)/RT} e^{-\Delta w(r)/RT} \qquad (12\text{-}23)$$

where $U(r)$ is the gas-phase interaction potential at separation r. At the transition state equation (12-23) becomes

$$g_{A,B}(r^{\ddagger}) = e^{-E_0/RT} e^{-\Delta w^{\ddagger}(r^{\ddagger})/RT} \qquad (12\text{-}24)$$

To illustrate the evaluation of a radial distribution function, consider particles 1 and 2 separated by the distance r_{12} in a solution of N particles. The radial distribution function for particles 1 and 2 is defined by

$$g(r_{12}) = V^2 \int \cdots \int \frac{e^{-U(r_1, r_2, \cdots r_N)/k_B T} d^3\mathbf{r}_3 \cdots d^3\mathbf{r}_N}{Z} \qquad (12\text{-}25)$$

where V^2 is the square of the unit volumes corresponding to the coordinates for particles 1 and 2, $U(r_1, \ldots, r_N)$ is the potential energy for a particular configuration of the N particles, $d^3 r_N$ is the volume element $dx_N dy_N dz_N$ for particle N, and

$$Z = \int \cdots \int e^{-U(r_1, r_2, \cdots r_N)/k_B T} d^3\mathbf{r}_1 \cdots d^3\mathbf{r}_N \qquad (12\text{-}26)$$

is the complete configuration integral over all $3N$ coordinates. The integral in equation (12-25) is a restricted configuration integral. That is, particles 1 and 2 are fixed at a separation r_{12} and the integration is over all possible configurations for the remaining $N - 2$ particles. Accurate evaluation of the multidimensional integral can be accomplished by Monte Carlo methods. By varying r_{12} in equation (12-25), the complete radial distribution function may be determined. Then, from equation (12-23), the potential of mean force is found as a function of the distance between particles 1 and 2.

Determining $g(r^{\ddagger})$ and $\Delta w^{\ddagger}(r^{\ddagger})$ for a reactive chemical system follows the description just given, except the reaction coordinate $r^{\ddagger}$ replaces r_{12}. The potential of mean force (i.e., free energy) is an equilibrium quantity. It is calculated by fixing the reaction coordinate and averaging over an equilibrium distribution of the solvent interacting with the reactive system. The potential of mean force is therefore analogous to Born-Oppenheimer electronic energies, which are determined by fixing the nuclear coordinates and averaging over distributions of the electrons.

If equations (12-21) and (12-22) are combined, the difference between the gas phase and liquid phase rate constants may be expressed as

$$k_s = k_g \exp\{-\Delta G^{\ddagger}_{\text{solv}}/RT\} \qquad (12\text{-}27)$$

To illustrate the use of this equation, consider a bimolecular reaction in a high-density solvent. As shown in Figure 12-2, the solvent can favor the more compact transition state configuration as compared to the separated reactants. As a result, $\Delta G^{\ddagger}_s$ will be smaller than $\Delta G^{\ddagger}_g$. Calculations of potentials of mean force have shown that a nonpolar solvent can enhance bimolecular atom transfer reactions up to a factor of 35.[4] The diagram in Figure 12-2 is fairly typical for nonpolar reactants and a nonpolar transition state in a nonpolar solvent. In the thermodynamic formulation of transition

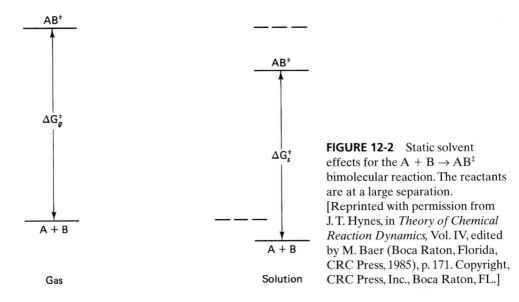

FIGURE 12-2   Static solvent effects for the $A + B \rightarrow AB^{\ddagger}$ bimolecular reaction. The reactants are at a large separation. [Reprinted with permission from J. T. Hynes, in *Theory of Chemical Reaction Dynamics,* Vol. IV, edited by M. Baer (Boca Raton, Florida, CRC Press, 1985), p. 171. Copyright, CRC Press, Inc., Boca Raton, FL.]

state theory, the potential of mean force is absorbed in γ, ΔH_{vap}, and ΔS_{vap} [see equation (12-15)].

Calculations of potentials of mean force have been applied to the S_N2 nucleophilic substitution reaction[8]

$$Cl^- + CH_3Cl \longrightarrow ClCH_3 + Cl^- \tag{12-28}$$

As illustrated earlier in Figure 12-1, a polar solvent is expected to have a pronounced effect on the kinetics of this reaction. Calculations of $\Delta G_s^{\ddagger}$ are shown in Figure 12-3 for reactions in aqueous and dimethylformanide (DMF) solutions and compared with the gas phase reaction coordinate potential energy. The Cl^- anion is strongly stabilized in solution in comparison to the transition state. Also, in solution, the pronounced ion-dipole complex is not observed and the overall potential energy barrier is significantly higher. The dipolar aprotic solvent DMF shows less anion solvating ability than does water.

12.1.3   Electron Transfer Reactions

The direct transfer of an electron from one molecule to another, without the participation of an intermediate solvated electron, is important in many oxidation-reduction reactions, particularly those involving transition metals; e.g.,

$$Fe^{3+} + Cr^{2+} \longrightarrow Fe^{2+} + Cr^{3+} \tag{12-29}$$

Such an electron transfer reaction is identified as cross-exchange in contrast to self-exchange reactions where the reactants and products are chemically the same. The latter reactions may be studied by using radio-isotopes as labels, e.g.,

$$*Fe^{3+} + Fe^{2+} \longrightarrow *Fe^{2+} + Fe^{3+} \tag{12-30}$$

For self-exchange reactions the standard $\Delta G°$ of reaction is zero. The free energy of activation is then associated with the rearrangement of solvent molecules around the

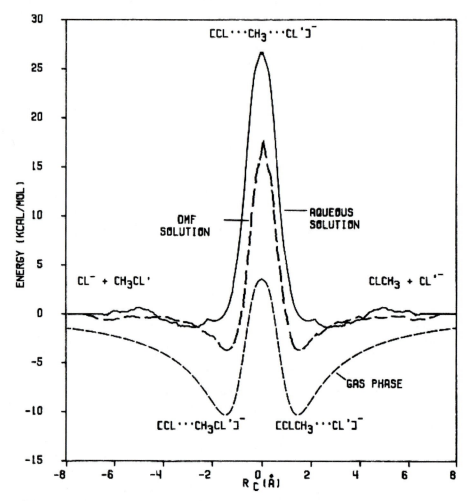

FIGURE 12-3 Calculated internal energies in the gas phase (short dashed and the potential of mean force in DMF (long dashes) and in aqueous (solid curve) for the reaction of Cl^- with CH_3Cl as a function of the reaction coordinate r_c in angstroms. [Adapted from J. Chandrasekhar and W. L. Jorgensen, *J. Am. Chem. Soc. 107*, 2974 (1985)].

ions, changes in intramolecular degrees of freedom of any ligands attached to the ions, and changes in ionic sizes which accompany electron transfer.

There are two principal mechanisms for electron transfer reactions: inner sphere and outer sphere. Each may be viewed as the interaction of the ions to form a transition state for electron transfer. For inner sphere electron transfer there are marked changes in the coordination shells of the ions in forming the transition state. Reaction may involve bond breaking and making and the actual transfer of a ligand; e.g.,

$$Cr^{2+}(aq) + Co(NH_3)_5Cl^{2+} \xrightarrow[H^+]{excess} CrCl^{2+} + Co(NH_3)_4(aq)^{2+} + NH_4^+$$

The Cl^- ligand is transferred in this reaction and a possible representation of the transition state is

$$[(NH_3)_5Co - Cl - Cr(H_2O)_5]^{4+}$$

Though there may be some distortion, the coordination shells of the reactant ions remain intact during outer sphere electron transfer. An electron is transferred from one ion to another without any bond breaking or making, as for

$$*Fe(phen)_3^{2+} + Fe(phen)_3^{3+} \rightarrow *Fe(phen)_3^{3+} + Fe(phen)_3^{2+}$$

where phen is phenanthraline

Different criteria may be used to determine whether the electron transfer is inner or outer sphere. The formation of reaction products, that indicate bond breaking and making has occurred, is indicative of inner sphere electron transfer. More direct evidence for the inner sphere mechanism, is the observation (e.g., by spectroscopy) of a bridged intermediate in equilibrium with reactants and products.

A theoretical model for outer sphere electron transfer was developed by Marcus,[9-11] for which he received the Nobel prize in 1993. This model envisages the solvent around the reactant ions first arranging to a configuration favorable for electron transfer, at which electron transfer occurs. There is a solvent configuration around each reactant ion for which the Gibbs free energy G is a minimum and changes in the solvent structure from this configuration increases the free energy. To attain the transition state structure for electron transfer the separation between the two reactant ions decreases and reorganization of the solvent structure about each ion occurs. A reaction coordinate for electron transfer may be conceived as a combination of these ion-ion separation and solvent reorganization coordinates. Without loss of generality, the Gibbs free energy of the reactants versus this reaction coordinate may be depicted as a parabola as shown in Figure 12-4. There is a similar curve for the products and the transition state is at the point the two curves cross. Following the discussion in section 7.10, this picture of crossing free energy curves for the reactant and product states is a diabatic model. In an adiabatic model the free curves will interact and not cross, leading to adiabatic states with upper and lower curves as shown in Figure 7-13.

The Marcus expression for the electron transfer rate constant is formulated by considering the reaction

$$A^{z_A} + B^{z_B} \rightleftharpoons A^{z_A + \Delta z} + B^{z_B - \Delta z} \tag{12-31}$$

where Δz is the charge transferred. The rate constant for the forward reaction can be expressed as

$$k_{AB} = Z_{AB}\, e^{-\Delta G^*_{AB}/RT} \tag{12-32}$$

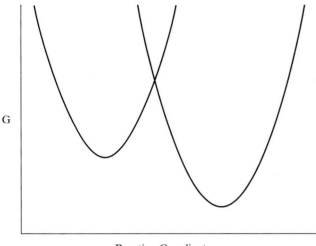

G

Reaction Coordinate

FIGURE 12-4   Depiction of Gibbs free energy versus reaction coordinate for reactants (left curve) and products (right curve). The reaction coordinate is a combination of ion-ion separation and solvent reorganization.

where Z_{AB} is the collision frequency in solution (chapter 4) and ΔG^{*}_{AB} is the free energy required to form the transition state from the reactants in their equilibrium solvation configuration. Note that ΔG^{*}_{AB} is different than $\Delta G^{\ddagger}$ of transition state theory (chapter 10), since Z_{AB} is used for the preexponential factor instead of $k_B T/h$.

The free energy change ΔG^{*}_{AB} is a sum of two terms:

$$\Delta G^{*}_{AB} = w_{AB} + \Delta G'_{AB} \tag{12-33}$$

The first term is the work required to bring the reactants together, with the solvent in dielectric equilibrium, so that the ions separation r_{AB} is that for the transition state. This quantity is given by

$$w_{AB} = \frac{z_A z_B e^2}{\varepsilon_s r_{AB}} \tag{12-34}$$

and is the same as that in equation (4-27). Here, ε_s is used to explicitly denote the static equilibrium dielectric constant. The second term $\Delta G'_{AB}$ is the work required to transfer the charge Δz, with the solvent in the nonequilibrium configuration corresponding to the transition state structure.

Electron transfer between the ions occurs on a time scale much shorter than that required for solvent reorientations, which maintain equilibrium configurations about the ions and screen the ionic charges. Thus, the effective dielectric constant for electron transfer is not the static equilibrium dielectric constant $\varepsilon_s = 78$ for water, but the much smaller optical dielectric constant ε_o, which is approximately 2 for water. The effect of using ε_o, instead of ε_s, is to increase $\Delta G'_{AB}$ for electron transfer.

To evaluate $\Delta G'_{AB}$ it is necessary to determine the free energy required to charge an ion from 0 to $\Delta z e$. The differential free energy for charging an ion the amount de is given by $dG = E de$, where E is the electrical potential (see section 4.5). From equation (4-49), E for a solution of negligible ionic strength is $e/\varepsilon r$, so ΔG for charging an ion $\Delta z e$ is

$$\Delta G = \int_0^{\Delta ze} \frac{e}{\varepsilon r} \, de = \frac{(\Delta ze)^2}{2\varepsilon r} \tag{12-35}$$

As shown by equation (12-31), charge Δze is transferred in the electron transfer reaction. At the transition state some fraction of this charge $m\Delta ze$ has been transferred. Marcus has shown that $\Delta G'_{AB}$ only depends on $m\Delta ze$ and equals $G_o - G_s$, where G_o is the free energy of formation of the pair of charges $m\Delta ze$ and $-m\Delta ze$ on the ions at the transition state separation r_{AB} in the medium of dielectric constant ε_o, and G_s is the corresponding free energy for the medium with dielectric constant ε_s. The free energy G_o is the sum of the work done to charge the ions A and B the amounts $m\Delta ze$ and $-m\Delta ze$, respectively, and the work required to bring the ions from infinite separation to their separation r_{AB} at the TS; i.e.,

$$G_o = \frac{(m\Delta ze)^2}{2\varepsilon_o r_A} + \frac{(-m\Delta ze)^2}{2\varepsilon_o r_B} + \frac{(m\Delta ze)(-m\Delta ze)}{\varepsilon_o r_{AB}} \tag{12-36}$$

The expression for G_s is the same, except ε_o is replaced by ε_s, so that

$$\Delta G'_{AB} = G_o - G_s = m^2 (\Delta ze)^2 \left(\frac{1}{2r_A} + \frac{1}{2r_B} - \frac{1}{r_{AB}} \right) \left(\frac{1}{\varepsilon_o} - \frac{1}{\varepsilon_s} \right)$$

$$= m^2 \lambda_o \tag{12-37}$$

where λ_o, known as the reorganization energy, is

$$\lambda_o = (\Delta ze)^2 \left(\frac{1}{2r_A} + \frac{1}{2r_B} - \frac{1}{r_{AB}} \right) \tag{12-38}$$

The free energy of activation for electron transfer may then be expressed as

$$\Delta G^*_{AB} = w_{BA} + m^2 \lambda_o \tag{12-39}$$

The value of m is found by deriving an expression for ΔG^*_{AB} for the reverse reaction. Since the fractional charge transferred for the reverse reaction is $-(1 - m)\Delta ze$,

$$\Delta G^*_{BA} = w_{BA} + (1 + m^2)\lambda_o \tag{12-40}$$

Combining the reaction standard free energy change

$$\Delta G^o_{AB} = \Delta G^*_{AB} - \Delta G^*_{BA} \tag{12-41}$$

with equations (12-39) and (12-40) gives

$$m = -\left(1/2 + \frac{\Delta G^o_{AB} + w_{BA} - w_{AB}}{2} \right)$$

Substituting this expression for m into equation (12-39) gives

$$\Delta G^*_{AB} = 1/2 \, (w_{AB} + w_{BA}) + \frac{\lambda_o}{4} + \frac{\Delta G^o_{AB}}{2} + \frac{(\Delta G^o_{AB} + w_{BA} - w_{AB})^2}{4\lambda_o} \tag{12-42}$$

The electron transfer rate constant is obtained by inserting this expression for ΔG^*_{AB} into equation (12-32). The above derivation of ΔG^*_{AB} does treat any changes in the potential energy function for the inner coordination shells of the reactant ions

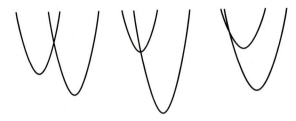

FIGURE 12-5 Three possible free energy profiles for electron transfer in solution. The first case, on left, is the normal free energy region. The middle curve depicts an activation-less process. The inverted region is depicted on the right.

upon electron transfer. Marcus has derived a factor for treating this effect, which is added to ΔG^*_{AB} in equation (12-42).

Several simplifications of the above expression for ΔG^*_{AB} may be made. If $z_A z_B = (z_A + \Delta z)(z_B - \Delta z)$, $w_{AB} = w_{BA}$ and equation (12-42) becomes

$$\Delta G^*_{AB} = w_{AB} + \frac{1}{4\lambda_o}(\Delta G^o_{AB} + \lambda_o)^2 \qquad (12\text{-}43)$$

If the work term is assumed to be negligible, ΔG^*_{AB} is further simplified and, when put in equation (12-32), gives

$$k_{AB} = z_{AB}\, e^{-(\Delta G^o_{AB} + \lambda_o)^2/4\lambda_o RT} \qquad (12\text{-}44)$$

This equation identifies three different cases, depicted in Figure 12-5, for the electron transfer rate constant. The first case, illustrated by the left diagram in Figure 12-5, is termed the normal free energy region. Here ΔG^o_{AB} is negative, with a magnitude smaller than λ_o, to give a positive free energy of activation ΔG^*_{AB}. In this region, the activation free energy is decreased as ΔG^o_{AB} becomes more negative until the situation in the middle diagram of Figure 12-5 occurs, where $\lambda_o = -\Delta G^o_{AB}$ and the activation free energy is zero. In this case the electron transfer rate constant is Z_{AB}, the diffusion rate constant. The third case is depicted in the right-most diagram of Figure 12-5 and is known as the "inverted" region. Here ΔG^*_{AB} increases as ΔG^o_{AB} becomes more negative. Evidence for the "inverted" region has come from experiments by Miller, Calcaterra, and Closs,[12] where the electron transfer rate constant was found to decrease for large values of $-\Delta G^o_{AB}$.

Finally, for a self-exchange reaction, such as equation (12-30), k_{AB} in equation (12-44) becomes

$$k_{AB} = Z_{AB}\, e^{-(\lambda_o/4RT)} \qquad (12\text{-}45)$$

The activation energy only depends on the reorganization energy λ_o, consistent with the discussion following equation (12-30).

12.2   KRAMERS' THEORY AND FRICTION

Most modern investigations of the validity of transition state theory in solutions are based on the classic paper by Kramers[13] on reactions in solution. Kramers viewed a reaction as a barrier crossing influenced by interactions with surrounding solvent molecules. Passage across the barrier then appears similar to Brownian motion, which is described via the Fokker-Planck (or Langevin)[14] equation. To account for the dynamical influence of the solvent, a damping force is introduced which is proportional to both the velocity along the reaction coordinate and a *friction constant* β. Although this friction constant is not clearly defined, it is expected to increase with solvent density, pressure, and viscosity.

The friction constant can have different influences on the solution rate constant, and Kramers identified several major kinetic regimes. (a) In one regime the requirement for validity of transition state theory is the same as for normal gas phase reactions. Thus, in the absence of intrinsic recrossing of the transition state, the equilibrium conditions assumed in the standard formulation of transition state theory are satisfied. Collisions with solvent molecules are sufficient to maintain an equilibrium distribution of energized reactants, but are not sufficiently frequent to perturb free passage over the potential energy barriers. (b) As the friction is increased, the second regime is reached. Here, collisions with solvent molecules lead to recrossing of the potential barrier before a stable product molecule is formed. As a result, the rate is less than the transition state theory prediction, and the difference depends on the solvent friction. For broad, flat barriers and high friction, the rate becomes diffusion-controlled and is inversely proportional to the friction. Barrier recrossings are so frequent that the rate may become more than an order of magnitude less than the transition state theory value. (c) A third regime, which occurs at low friction, represents a different breakdown of transition state theory. Because the reactive system is weakly coupled to the solvent, an equilibrium distribution of reactant states cannot be maintained. This phenomenon is similar to that responsible for the first-to-second-order transition in thermal unimolecular reactions (chapter 11). For extremely low friction, reactant activation becomes rate limiting and the rate is proportional to the solvent friction. In what follows, Kramers's treatment of the effects of solvent friction on solution reactions rates is outlined.

Kramers idealized the reactive system as a particle with effective mass m moving in the one-dimensional potential shown in Figure 12-6. The reaction is regarded as a Markov process, and the Fokker-Planck equation

$$\frac{\partial \rho}{\partial t} + v \frac{\partial \rho}{\partial t} + \frac{1}{m} \frac{\partial V}{\partial r} \frac{\partial \rho}{\partial v} = \beta v \frac{\partial \rho}{\partial v} + \beta \rho + \beta \frac{k_B T}{m} \frac{\partial^2 \rho}{\partial v^2} \qquad (12\text{-}46)$$

is solved for the distribution $\rho(r, v, t)$ of the particle's position and velocity vs. time. In this equation, $V = V(r)$ is the potential and β is the friction constant. To solve equation (12-46), $V(r)$ is assumed to be of the form

$$V(r) = \frac{1}{2} m \omega_0^2 (r - r_0)^2 \qquad (12\text{-}47)$$

in the vicinity of the reactant, and of the form

$$V(r) = E_0 - \frac{1}{2} m \omega^{\ddagger 2} (r - r^{\ddagger})^2 \qquad (12\text{-}48)$$

near the transition state dividing the reactant from the product.

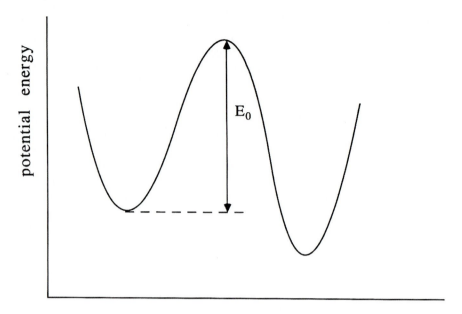

reaction   coordinate

FIGURE 12-6   One-dimensional potential model for Kramers' theory.

With the preceding potentials, Kramers solved the steady-state version of equation (12-46) for medium to high friction where energy equilibration is guaranteed. The resulting expression for the rate constant is

$$k = k^{\text{TST}} \kappa^{\text{KR}} \tag{12-49}$$

where k^{TST} is the transition state theory rate constant and $\kappa^{\text{KR}} \leq 1$, the Kramers transmission coefficient, is given by

$$\kappa^{\text{KR}} = \left[1 + \left(\frac{\beta}{2\omega^{\ddagger}} \right)^2 \right]^{1/2} - \frac{\beta}{2\omega^{\ddagger}} \tag{12-50}$$

Deviation from transition state theory depends on the parameter $\beta/2\omega^{\ddagger}$. For intermediate friction, $\beta \ll 2\omega^{\ddagger}$, κ^{KR} is unity, and transition state theory results. In the high-friction limit, $\beta \gg 2\omega^{\ddagger}$ and equation (12-49) becomes

$$k = k^{\text{TST}} \left(\frac{\omega^{\ddagger}}{\beta} \right) \tag{12-51}$$

so that the rate constant decreases with increasing friction. Equation (12-51) can also be derived by a steady-state solution of the Smoluchowski diffusion equation.[5] Thus, k in that equation is a rate constant for diffusion-controlled passage across the potential energy barrier and may be expressed as

$$k_D = k^{\text{TST}} \left(\frac{\omega^{\ddagger} D}{k_B T} \right) \tag{12-52}$$

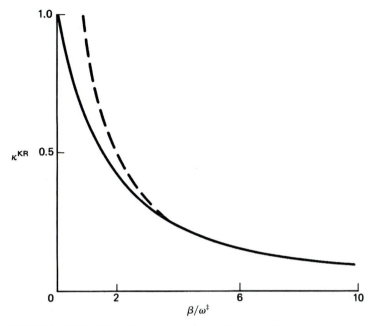

FIGURE 12-7 The Kramers transmission coefficient, as given by equation (12-50), vs. the frictional parameter $\beta/\omega^{\ddagger}$. The dashed curve is the high-friction diffusive limit expressed in equation (12-52). Transition state theory is approached for small values of friction. [Adapted from J. T. Hynes, in *Theory of Chemical Reaction Dynamics.* Vol. IV. Edited by M. Baer (Boca Raton, FL: CRC Press, 1985).]

where $D = k_B T/\beta$ is the diffusion constant (see chapter 4) determined by the Einstein relation.[15]

The preceding equations show that substantial differences may exist between actual solution rate constants and those predicted by transition state theory. A plot of the Kramers transmission coefficient versus the frictional parameter $\beta/\omega^{\ddagger}$ is given in Figure 12-7. Transition state theory is approached for small values of friction, as predicted by equation (12-50).

The foregoing approach is incomplete for a unimolecular reaction: at sufficiently low solvent friction, a thermal equilibrium no longer exists for the reactant energy states. The same situation occurs for unimolecular reactions in the falloff regime (see chapter 11, sections 11.2 and 11.3). It has been suggested[16] that for the unimolecular decomposition of a polyatomic molecule in solution the rate constant should be expressed as

$$k = \frac{k_1 k_2 [M]}{k_{-1}[M] + k_2} \kappa^{KR} \tag{12-53}$$

The first term is the Lindemann-Hinshelwood expression (see chapter 11, section 11.3) for the unimolecular rate constant vs. pressure (i.e., [M]), while the second is the frictional transmission coefficient from equation (12-50). A high-pressure limit plateau in

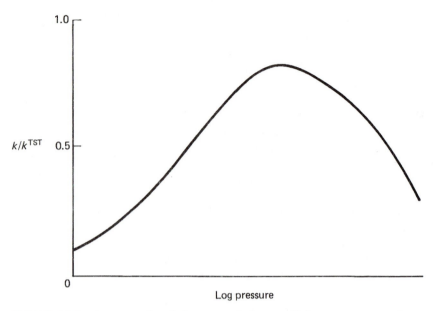

FIGURE 12-8 An example of the transmission coefficient vs. pressure for a unimolecular reaction. The plot shows that the transition state theory high-pressure limit is not attained. [Adapted from J. T. Hynes, in *Theory of Chemical Reaction Dynamics.* Vol. IV. Edited by M. Baer (Boca Raton, FL: CRC Press, 1985).]

the thermal unimolecular rate constant equal to $k_\infty = k_1 k_2 / k_{-1}$ is predicted by the Lindemann-Hinshelwood theory. For molecules with few internal degrees of freedom, there may not be a plateau in k vs. pressure. Since the frictional constant β increases with gas density and, thus, κ^{KR} decreases, a high-pressure limiting rate constant does not result from equation (12-53). Indeed, as illustrated in Figure 12-8, the rate constant decreases on approaching the high-pressure limit. However, with more internal degrees, there usually is a significant plateau, since the high-pressure limit for the first term in equation (12-53) is reached before solvent friction suppresses the rate.

If as predicted by Kramers' theory, there is a breakdown in transition state theory, then less significance should be placed on thermodynamic kinetic parameters (section 12.1) determined by inappropriately fitting rate constants to transition state theory rate expression.[5] This is readily apparent for the parameter $\Delta V^\ddagger$, which could have a nontransition-state contribution if the transmission coefficient has a pressure (i.e., density) dependence. If the correct thermodynamic $\Delta V^\ddagger$ is negative, then, according to equation (12-20), the rate should increase with pressure. However, for the high-friction limit, where the rate decreases with pressure, the apparent $\Delta V^\ddagger$ would be positive.

Trajectory calculations as discussed in section 8.3 for gas-phase reactions have also been applied to reactions in solution.[17,18] Collinear models of the atom transfer reaction $A + BC \rightarrow AB + C$ and the S_N2 reaction $Cl^- + CH_3Cl \rightarrow ClCH_3 + Cl^-$ have been studied.

12.3   GAS-SURFACE REACTION DYNAMICS

Heterogeneous reactions, i.e., those that occur at the interface of two phases, are typical examples of catalysis. The field of heterogeneous catalysis has enormous practical applicability, and, therefore, heterogeneous chemical reaction kinetics has been extensively studied. The interest here is in a molecular-level (microscopic) description of gas-surface dynamics, to complement the more macroscopic approach in chapter 5.

Chemical reactions at a gas-surface interface involve the following sequence of steps:

1. Transport of gas-phase reactants to the surface
2. Adsorption of the reactants on the surface
3. Diffusion of adsorbed reactants into proximity
4. Reaction on the surface
5. Desorption of products
6. Transport of products away from the surface

Any one or a combination of these steps may be slow and, therefore, rate determining. Measurements of transport rates allow steps 1 and 6 to be characterized separately. Also, it may be possible to measure individual adsorption and desorption rates. Thus, it can often be established whether the actual reaction on the surface is rate limiting.

12.3.1   Adsorption and Langmuir Isotherms

Adsorption occurs on a solid surface and is due to the attractive forces between the surface and the adsorbed species (adsorbate). Two types of adsorption can be distinguished. Heterogeneous catalysis is more intimately related to adsorption processes in which the adsorbed species forms a chemical bond with the surface (adsorbent). Such a process is called chemical adsorption, or *chemisorption,* as contrasted to the much weaker adsorption of gases due to van der Waals forces. This second kind of adsorption is more like a condensation process and is called physical adsorption, or *physisorption.* The dividing line between the two adsorption processes is not always sharp.

The enthalpy changes for chemisorption are usually substantially greater than those for physical adsorption. For chemisorption, once a monolayer of adsorbed gas molecules covers the solid's surface, no further chemical reaction between the gas and the solid can occur. For physical adsorption, once a monolayer has formed, intermolecular interactions between adsorbed molecules in the monolayer and gas-phase molecules can lead to the formation of a second layer of adsorbed gas. Thus, the enthalpy change for formation of the first layer of physically adsorbed molecules is determined by solid-molecule intermolecular forces, whereas the enthalpy change for formation of the second, third, etc., physically adsorbed layers is determined by molecule-molecule intermolecular forces and is about the same as ΔH in the condensation of a gas to a liquid. Although only one layer can be chemically adsorbed, physical adsorption of further layers on top of a chemisorbed monolayer sometimes occurs.

It should be recognized that there may be exceptions to the above general properties. In some cases the surface can influence the adsorption of the second layer. It is also possible for the second layer to be chemisorbed to the first layer.

The difference between physical and chemical adsorption is illustrated by the behavior of nitrogen on iron. At $-190°C$, nitrogen is liquid and nitrogen molecules are physisorbed on iron. The equilibrium amount of N_2 adsorbed in this manner decreases rapidly as the temperature rises until, at room temperature, nitrogen does not adsorb on iron at all. At high temperature, approximately $500°C$, nitrogen is chemisorbed on the iron surface. Physisorption is thus a very weak gas-surface interaction with no potential energy barrier and only occurs at low temperatures. In contrast, there may be activation energy for chemisorption, and a high temperature often is required. Since an appreciable activation energy may be associated with chemisorption, it is often called *activated adsorption.*

Adsorption and desorption may be represented by the chemical equation

$$A(g) + S \rightleftharpoons AS \qquad (12\text{-}54)$$

where $A(g)$ is the gaseous adsorbate, S is a vacant site on the surface, and AS represents an adsorbed molecule of A. A simple model can be developed for physisorption, since no chemical reaction is involved; in the case of chemisorption, a similar model is applicable for chemically stable adsorbates. The amount of adsorbed A is measured as a function of the temperature and pressure of $A(g)$; it is convenient to express this amount as the equivalent volume of adsorbed A, per gram of adsorbent, at the gas temperature and pressure. A plot of this equivalent volume V as a function of pressure is known as an *adsorption isotherm* and is illustrated in Figure 12-9. If only a monolayer is formed on the surface, as is the normal situation for chemisorption, V will approach a limiting value V_m as illustrated in the figure.

A simple model of adsorption was given by Langmuir[19] in 1916. In the model, all adsorption sites on the surface are the same, there are no interactions between one

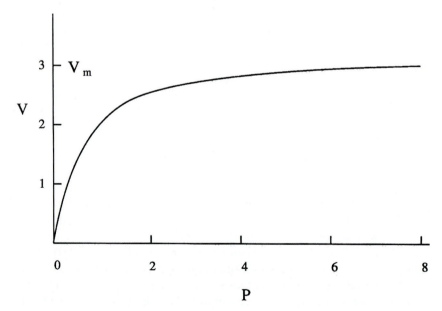

FIGURE 12-9   Langmuir adsorption isotherm. Units of V and P are arbitrary.

adsorbed molecule and another (i.e., sites are independent), and only a monolayer can be adsorbed. Now consider a surface with N adsorption sites. If N_1 sites are occupied by adsorbate, then $N_0 = N - N_1$ are bare. At equilibrium, the rate of adsorption is equal to the rate of desorption and is proportional to the rate of collisions of gas-phase molecules with unoccupied adsorption sites, which in turn is proportional to the pressure P (see chapter 8). Thus,

$$\text{Rate of adsorption} = k_a P N_0 = k_a P(N - N_1)$$

The rate of desorption is proportional to the number N_1 of occupied sites. So

$$\text{Rate of desorption} = k_d N_1$$

Setting the two rates equal to each other and solving for N_1/N, we obtain equation (5-27)

$$\frac{N_1}{N} = \theta = \frac{bP}{1 + bP}$$

where θ is the fraction of surface covered and $b = k_a/k_d$. The limiting value V_m of Figure 12-9 corresponds to $\theta = 1$; hence, $V/V_m = \theta$, and we can write

$$V = \frac{V_m bP}{1 + bP} \tag{12-55}$$

These two equations are known as the *Langmuir adsorption isotherm.*

The Langmuir isotherm may be interpreted in the context of statistical mechanics. The reaction in equation (12-54) has an equilibrium constant

$$K = \frac{[AS]}{[A][S]} \tag{12-56}$$

For an ideal gas, $[A] = P/k_B T$. Also, for $[S]$ and $[AS]$, we have $(1 - \theta)N$ and θN, respectively. Thus, by substituting those quantities into equation (12-56) and rearranging terms, we obtain

$$\theta = \frac{(K/k_B T)P}{1 + (K/k_B T)P} \tag{12-57}$$

which is equivalent to the Langmuir adsorption isotherm if

$$b = \frac{K}{k_B T} \tag{12-58}$$

The statistical mechanical expression for the equilibrium constant (chapter 10) is

$$K = \frac{Q_{AS}}{Q_A Q_S} e^{-E_{ad}/RT} \tag{12-59}$$

where the Q's are the partition functions and E_{ad} is the energy of adsorption, given by the difference in ground state energy between the molecules in the gaseous and adsorbed states. Thus, equations (12-58) and (12-59) provide a molecular interpretation of the constant b in terms of adsorption energy and the molecular quantities that determine the partition function, viz., vibrational and rotational energies, temperature, and masses.

The Langmuir model oversimplifies gas-surface interactions. In reality, the surfaces of most solids are not uniform, and the desorption rate depends on the location of the adsorbed molecule. Also, the force between adjacent adsorbed molecules is frequently substantial, as evidenced by changes in the heat of adsorption with increasing θ. To complicate matters further, there is substantial experimental and theoretical evidence that adsorbed molecules can move about on the surface and that this mobility is much greater for physically adsorbed molecules than for chemisorbed ones and increases as T increases. In addition, multilayer adsorption is quite common in physical adsorption. To account for all these effects, different adsorption isotherms may be used. The Freundlich isotherm[20] can be derived by modifying the Langmuir assumptions to allow for several kinds of adsorption sites on the solid, each having a different heat of adsorption.[21] The Brunauer, Emmett, and Teller isotherm[22] is a modification of Langmuir's model that allows for multilayer adsorption.

12.3.2 Adsorption/Desorption Kinetics and Transition State Theory

Two important kinetic parameters in gas-surface adsorption and desorption are the sticking probability s and the desorption rate constant k_d. The sticking probability is the fraction of the incident flux impinging on the surface which is completely accommodated (i.e., adsorbed). It is related to the adsorption rate R_a via

$$s = \frac{R_a}{N_g v} \tag{12-60}$$

where N_g is the gas concentration and v is the component of the gas velocity that is perpendicular to the surface. For a thermal velocity distribution, s is a thermally averaged quantity. The sticking probability of NO on a Pt(111) surface is shown in Figure 12-10 as a function of surface coverage. Note that it decreases with coverage, since there is less probability that the NO molecule will find a chemisorption site before it desorbs.

Equation (12-60) represents the sticking probability per adsorption site. The adsorption rate per site is $R_a = k_a N_g$. According to transition state theory, k_a is given by equation (10-20), i.e.,

$$k_a = \frac{k_B T}{h} \frac{Q^{\ddagger}}{Q_A Q_B} e^{-E_0/k_B T}$$

In the limit of a loose transition state which resembles the reactants A and S (i.e., the separation between A and S approaches infinity in the transition state), it is straightforward to show that the sticking probability is unity. For a loose transition state, the only difference between the partition function $Q^{\ddagger}$ and the partition function product $Q_A Q_S$ is that $Q^{\ddagger}$ lacks the reaction coordinate degree of freedom which is a translation perpendicular to the surface. Thus,

$$\frac{Q^{\ddagger}}{Q_A Q_S} = \frac{1}{\left(\frac{2\pi m k_B T}{h^2}\right)^{1/2}} \tag{12-61}$$

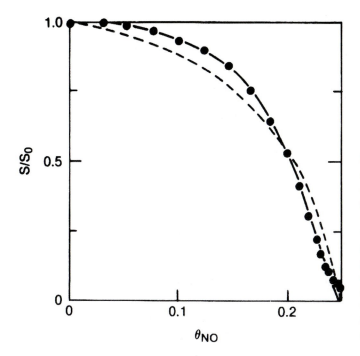

FIGURE 12-10   Sticking probability of NO on a Pt(111) surface vs. surface coverage. The surface temperature is 300 K, and S_o, the sticking probability for a clean surface, is 0.85. The solid curve is passed through the experimental points. The dashed curve is a model calculation. [Adapted from R. D. Levine and R. B. Bernstein, *Molecular Reaction Dynamics and Chemical Reactivity* (New York: Oxford, 1987).]

Since E_0 is equal to zero for a loose transition state, we have, upon substituting equation (12-61) into equation (10-20) reproduced above,

$$k_a = \left(\frac{k_B T}{2\pi m} \right)^{1/2} \qquad (12\text{-}62)$$

This value for k_a is the same as the average thermal velocity of particles moving perpendicular to the surface. Thus, from equations (12-60) and (12-62), together with the relationship $R_a = k_a N_g$, the sticking coefficient is seen to equal unity.

As the transition state becomes tighter and acquires properties close to those of the adsorbed state AS, the sticking probability becomes less than unity. This results from configurational (or entropic) effects which cause the gas to "turn back" before it crosses the transition state.

The desorption rate constant is given by

$$k_d = \frac{R_d}{N_a}$$

where R_d is the desorption rate and N_a is the concentration of the adsorbed species. If the adsorbed molecule is localized at one surface site, desorption is analogous to a gas phase unimolecular decomposition reaction. In Figure 12-11, two types of desorption potential energy curves are shown. Since the physisorption potential energy curve has no saddle point, variational transition state theory is needed to determine the temperature-dependent transition state location. For chemisorption the transition state may be tight, with the adsorbed species still strongly interacting with the surface. Thus, there

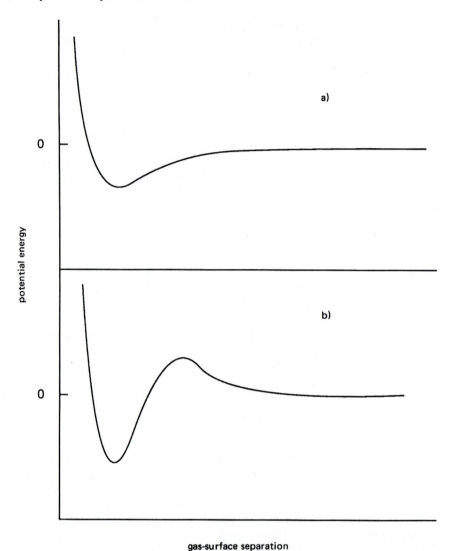

FIGURE 12-11 Two possible adsorption/desorption potential energy curves: (a) physisorption, (b) chemisorption.

may be a well-identified saddle point on the potential curve, at which the transition state will be positioned.

In Arrhenius form, the desorption rate constant is written as

$$k_d = Ae^{-E_a/RT}$$

For a physisorbed molecule or atom, the normal situation is a small activation energy and a large A-factor in the range of 10^{13} to 10^{15} sec^{-1}, similar to that for gas phase bond dissociations (e.g., $C_2H_6 \rightarrow 2CH_3$). The reason for a large A-factor is that in the transition state the desorbing molecule is free to rotate. The A-factor for the desorption of a chemisorbed molecule may be smaller if the desorbing molecule is not free to rotate in the transition state.

The thermodynamic transition state theory expression for the desorption rate constant is

$$k_d = \frac{k_B T}{h} e^{-\Delta G^{\ddagger}/RT}$$

where $\Delta G^{\ddagger}$ is the difference in free energy between the transition state and the adsorbed species. By combining this equation with equations (12-60) through (12-62), k_d can be written in terms of the sticking probability as

$$k_d = s \left(\frac{k_B T}{2\pi m} \right)^{1/2} e^{-\Delta G/RT} \tag{12-63}$$

where ΔG is the difference in free energy between the gaseous and adsorbed states.

The inverse of the desorption rate constant is often referred to as the *average residence time* of the molecule on the surface.

The concept of detailed balance (section 6.2) may be used to understand the relationship between the sticking probability and the dynamics of adsorption and desorption.[23,24] At steady state, the thermal equilibrium distribution of kinetic energies of all molecules striking the surface (and leaving the surface) is Maxwell-Boltzmann at temperature T and given by[25]

$$P(E) \, dE = \frac{E}{(RT)^2} e^{-E/RT} \, dE \tag{12-64}$$

The average value of E for this distribution is $2k_B T$. The Maxwell-Boltzmann angular distribution $P(\theta)$ for the gaseous molecules is proportional to $\cos\theta$,[26] where the θ is the angle between the gaseous molecule's velocity vector and the surface normal.

At equilibrium the rate $R_{eq}(E, \theta; T)$ at which molecules strike the surface is the same as that at which they leave the surface. Thus, the desorption rate for fixed E and θ may be expressed as

$$R_d(E, \theta; T) = s(E, \theta; T) R_{eq}(E, \theta; T) \tag{12-65}$$

where $s(E, \theta; T)$ is the sticking probability versus E and θ. From this equation it is seen that, if the sticking is unity, the desorption rate versus E and θ is the same as the equilibrium rate. Therefore, with s of unity $P(E)$ and $P(\theta)$ for the desorbing molecules will be the above Maxwell-Boltzmann distributions.

If the sticking probability is not unity, the angular and kinetic energy distributions of desorbing molecules may not be Maxwell-Boltzmann.[27] For chemisorption processes, with a potential energy barrier as shown in Figure 12-11(b), gaseous molecules impinging from the gas phase with a component of kinetic energy normal to the surface less than the barrier height will simply bounce off. Thus, only atoms impinging with high velocity in the normal direction are adsorbed. By detailed balance, desorbing molecules will have average velocities higher than Maxwell-Boltzmann and will be focused toward the surface normal. The combined distribution of molecules leaving the surface, i.e. those desorbing and those bouncing off, is Maxwell-Boltzmann under equilibrium conditions.

The sticking probability may also be less than unity for the potential energy shown in Figure 12-11(a). Since the kinetic energy of the incident gas molecule must be transferred to the surface for association to occur, incident molecules with high velocities might be expected to have a lower probability of sticking than low energy incident molecules. If this is the case, the atoms which stick to the surface and the desorb will be "colder" than the surface and have a mean kinetic energy less than $2k_BT$.

12.3.3   Dissociative Adsorption and Precursor States

A process that is particularly important for catalysis is *dissociative adsorption,* in which a molecule does not adsorb intact but undergoes bond rupture. An example is the adsorption of H_2 on a Cu(100) surface, depicted by

$$H_2(g) + \cdots \underset{\displaystyle CuCu}{\overset{\displaystyle \begin{array}{cc} H\text{–}H \\ | \quad | \end{array}}{}} \cdots \rightarrow \cdots CuCu \cdots \rightarrow \cdots \underset{\displaystyle CuCu}{\overset{\displaystyle \begin{array}{cc} H \quad\ H \\ | \quad\ | \end{array}}{}} \cdots Cu \cdots$$

In the dissociative adsorption of a polyatomic molecule there can be total bond rupture so that only atoms are adsorbed, or there may be partial bond rupture as in the dissociative chemisorption of CH_4 on Ni(111) to an adsorbed methyl radical and a hydrogen atom[28].

Mechanisms for dissociative chemisorption often include a *precursor state,* the notion of which is derived from the idea that before chemisorbing a molecule becomes trapped temporarily in a weakly bound molecular state which arises from long-range physisorption interactions. The existence of a precursor state means that there is an intermediate barrier (ΔE_b) to the chemisorbed state, as depicted in Figure 12-12. Two families of curves are illustrated that at large distances describe possible physisorbed states. The different curves represent different gas-surface systems, or different gas-surface adsorption sites for the same system. Only one curve is drawn for the short-range, strongly bound chemisorbed state. The barrier between the precursor and chemisorbed states allows the molecule to become temporarily trapped on the surface in the precursor state before chemisorbing. The curve for the chemisorbed state depicts dissociative chemisorption. The asymptote ($r \rightarrow \infty$) of this curve gives the potential energy of the fragments, which are formed by chemisorption.

For some of the potential curves in the figure, there is an overall barrier ΔE_b for chemisorption. If the molecule is to become adsorbed to the surface, this barrier must be surmounted, or else the molecule will reside in the precursor state for a short time before desorbing. One way to detect this energy barrier, as shown in Figure 12-13, is to measure the sticking probability as a function of the component of the molecular translational energy normal to the surface, i.e., $E_n = E_1 \cos^2 \theta_i$, where θ_i is the angle of incidence. An increase in the sticking probability with E_n is indicative of a barrier for dissociative chemisorption.

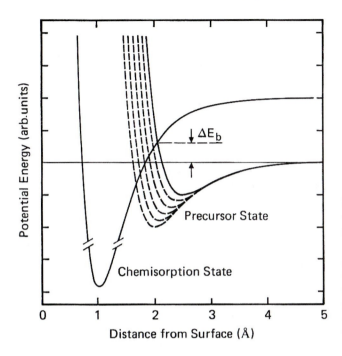

FIGURE 12-12   Schematic representation of potentials for mechanisms of dissociative chemisorption which include a precursor state. [Adapted from D. J Auerbach and C. T. Rettner, in *Kinetics of Interface Reactions,* edited by M. Grunze and H. J. Kreuzer (New York: Springer-Verlag, 1987), p. 125.]

As is found for gas phase reactions (see chapter 9), the effectiveness of translational and vibrational energy in surmounting the barrier for dissociative chemisorption may differ for different reactions. An example is the dissociative chemisorption of CO on Ni(111), for which a barrier of 30 kcal/mol is estimated. However, no dissociative adsorption is observed at a translational energy of 45 kcal/mol.[29] The conclusion is that CO must be vibrationally excited for chemisorption to occur.

12.3.4   Heterogeneous Reactions

Experimental studies of gas-surface heterogeneous reactions have provided considerable dynamical details of the elementary mechanisms of catalytic activity. An important aspect of these reactions is that a catalytic surface is *not* flat! Some examples of surface imperfections are shown in Figure 12-14. Kinks and monatomic ledges (or steps) are particularly important in catalysis. Active adsorption sites which promote bond breaking are usually found at these irregular regions of the surface. Steps and kinks may have different specificities for bond rupture. For example, a step on a particular surface may preferentially rupture the C—H bond of a hydrocarbon, while C—C bond rupture may occur only at a kink.

Adatoms can modify surface reactivity by blocking binding at an adsorption site and/or altering the binding at sites not directly blocked by through-surface electronic effects. These electronic effects may alter the potential energy barriers for dissociative chemisorption and surface migration. If the adatom slows down the catalytic behavior of a surface, it is called a *poison;* if it speeds up catalytic behavior, it is called a *promoter.*

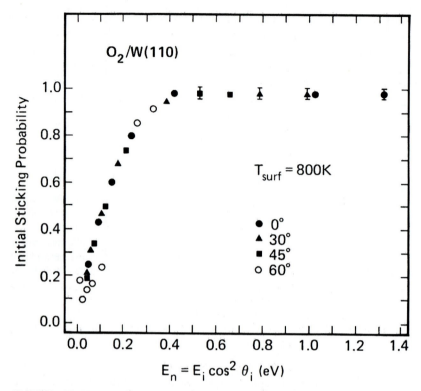

FIGURE 12-13 Initial dissociative chemisorption probability of O_2 on W(110) at a surface temperature of 800 K as a function of the kinetic energy of the incident molecules for angles of incidence between 0° and 60°. [Adapted from D. J. Auerbach and C. T. Rettner, in *Kinetics of Interface Reactions,* edited by M. Grunze and H. J. Kreuzer (New York: Springer-Verlag, 1987), p. 125.]

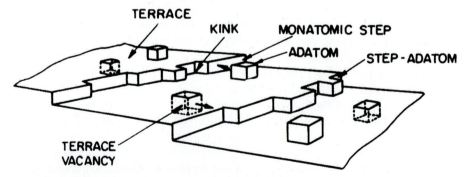

FIGURE 12-14 Model of the heterogeneous surface depicting the different surface sites: terrace, kink, ledge, vacancy, ledge-adatom, and adatom. [Adapted from G. A. Somorjai, *Principles of Surface Chemistry* (Englewood Cliffs, NJ: Prentice-Hall Inc., 1972).]

A poison may physically block binding sites, increase potential barriers for dissociative chemisorption, or increase barriers of lateral motion (i.e., migration) on the surface. Lateral motion on the surface is important for the bimolecular recombination of adsorbed species. A promoter adatom might lower the potential barriers for chemisorption and/or migration.

An important practical catalytic process is the oxidation of CO to CO_2. We may characterize the process by the mechanism[30]

$$O_2 + * \rightleftharpoons O_{2,ad} \xrightarrow{k_1} 2O_{ad}$$

$$CO + * \underset{k_{-2}}{\overset{k_2}{\rightleftharpoons}} CO_{ad}$$

$$CO_{ad} + O_{ad} \xrightarrow{k_3} CO_2 + 2*$$

$$CO + O_{ad} \xrightarrow{k_4} CO_2 + *$$

(12-66)

where * denotes a free adsorption site. The first step is the dissociative chemisorption of O_2, which proceeds through a molecular precursor state. In the second step carbon monoxide adsorbs molecularly. Next, there are two possible reaction steps for CO_2 formation. In the Langmuir-Hinshelwood[31] mechanism (the third step), chemisorbed CO and O recombine to form CO_2. In the fourth step, the Eley-Rideal mechanism,[32] CO_2 is formed by direct collision of CO from the gas phase with a chemisorbed O-atom. Experiments of Pd, Pt, Ir, and Rh surfaces have demonstrated the dominance of the Langmuir-Hinshelwood mechanism. A schematic potential energy diagram is given in Figure 12-15. All of the individual rate constants k_1, k_2, k_{-2}, and k_3 depend on temperature and surface coverage. The catalytic formation of CO_2 is only slightly influenced by the surface structure. Step sites apparently enhance O_2 dissociative chemisorption

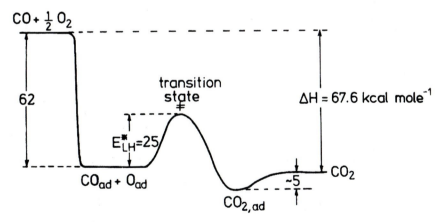

FIGURE 12-15 Schematic potential diagram illustrating the energetic changes associated with the individual reaction steps of equation (12-66) on Pd(111) at low coverages. [Adapted from T. Engel and G. Ertl, in *The Chemical Physics of Solid Surfaces and Heterogeneous Catalysis*, ed. D. A. King and D. P. Woodruff (New York: Elsevier, 1982).]

(step 1), but decrease the probability of CO_2 formation (step 3). Both effects nearly compensate each other.

Ammonia synthesis on an iron catalyst, $N_2 + 3H_2 \rightarrow 2NH_3$, (the Haber process), illustrates the importance of promoters.[33] The rate of ammonia formation is determined by the rate of N_2 dissociative chemisorption on the iron surface. Introducing potassium as an adatom promotes the reaction by decreasing the barrier for dissociative adsorption of N_2.

One of the most carefully studied heterogeneous reactions on a single crystal is the methanation reaction[34]

$$3H_2 + CO \longrightarrow CH_4 + H_2O \tag{12-67}$$

The mechanism for this reaction on a Ni(100) surface is represented by

$$H_2 + 2* \longrightarrow 2H_{ad}$$

$$CO + * \longrightarrow CO_{ad} \longrightarrow C_{ad} + O_{ad} \tag{12-68}$$

$$C_{ad} + 4H_{ad} \longrightarrow CH_4$$

where the last step represents a series of surface hydrogenation steps leading to the desorption of CH_4. Sulfur adatoms are a catalytic poison for this reaction as well as for others. The poisoning for this methanation reaction results from an inhibition in chemisorption of both hydrogen and CO. One sulfur atom deactivates approximately ten nickel atom adsorption sites.[34]

For many heterogeneous reactions, dissociative chemisorption is the rate-determining step. In these cases it is important to know how the dissociative chemisorption is affected by reactant translation, rotational, and vibrational energy. In some catalytic processes activated adsorption leads to unimolecular isomerization. The catalytic isomerization of cyclopropane to propylene has been studied on mica surfaces.[35] A model that explains the experimental results assumes a high barrier for adsorption that is surmounted most effectively by vibrational energy in the incident cyclopropane molecule.

12.3.5 Trajectory Studies

The classical trajectory theoretical approach (see chapter 8, section 8.3) has been particularly effective in simulating the dynamics of adsorption/desorption and of reactive gas-surface collisions. A potential energy function (chapter 7) which depends on the positions of all atoms is chosen to represent molecular and surface degrees of freedom, as well as molecule-surface and adsorbate-adsorbate interactions. Solving the classical equations of motion then gives a microscopic "picture" of the gas-surface event.

An important application of classical trajectories is the interpretation of angle-resolved secondary-ion-mass spectrometric (SIMS) measurements. As is evidenced in Figure 12-15, the surface structure can be a crucial factor in catalytic activity. A trajectory simulation[36] performed for a microcrystallite of approximately 240 atoms has been used to calculate the angular distribution of Ni^+ and Ni_2^+ ions ejected from a Ni(100) surface by 2-keV Ar^+ bombardment. Anisotropies in the experimental angular distributions are well reproduced by the dynamical calculations. A particularly significant discovery from the trajectories is that there is a dominant ejection mechanism for dimer formation, as shown in Figure 12-16. Atoms 1 and 3 are ejected through four-fold holes in the Ni surface and are moving parallel to each other. As such, they are in close proximity and susceptible to dimer formation. For this situation, the constituent atoms in the ejected dimer are known to form from surface atoms with a specific location relative to the primary impact point. Extrapolated to alloy surfaces such as CuNi, this finding would allow the determination of the relative placement of the alloy components on the surface.

In this trajectory simulation, a large number of atoms are treated explicitly to model the surface layers because a large number of atoms are directly involved in the gas-surface collision. For processes such as the scattering of gases from surfaces at

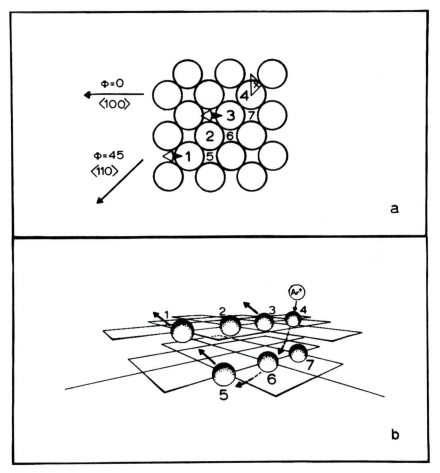

FIGURE 12-16 Elementary mechanism for dimer formation in the SIMS experiment of Ar^+ ion bombardment of a Ni(100) surface. The shaded spheres represent Ni atoms in the first and second layers. [Adapted from S. P. Holland, et al., *Phys. Rev. Lett. 44,* 756 (1980).]

thermal energies a method called *stochastic classical trajectory approach*[37] is often used to reduce the number of surface degrees of freedom treated explicitly. Only atoms in the "primary" surface zone (usually a few upper layers of the surface) are treated explicitly and allowed to interact with gas molecules. Energy transfer between the "primary" surface zone and the rest of the lattice is included via the generalized Langevin technique, which introduces two additional terms into the classical equations of motion. The first, a generalized friction, describes the dissipation of energy from the primary zone to the rest of the lattice. The second, a fluctuation force, accounts for energy imparted to the primary atoms through thermal vibrations of the secondary atoms. The generalized Langevin technique has been used to simulate the effect of translational, rotational, and vibrational energies on the dissociative chemisorption of diatoms,[38] and to calculate adsorption/desorption probabilities.[39]

REFERENCES

[1] W. N. Olmstead and J. I. Brauman, *J. Am. Chem. Soc. 99,* 4219 (1977).

[2] G. R. Branton, H. M. Frey, D. C. Montague, and I. D. R. Stevens, *Trans. Faraday Soc. 62,* 659 (1966).

[3] J. F. Kincaid, H. Eyring, and A. E. Stearn, *Chem. Rev. 28,* 301 (1941).

[4] B. M. Ladanyi and J. T. Hynes, *J. Am. Chem. Soc. 108,* 585 (1986).

[5] J. T. Hynes, in *Theory of Chemical Reactions Dynamics.* Vol. IV, edited by M. Baer (Boca Raton, FL.: CRC Press, 1985), p. 171.

[6] T. L. Hill, *Statistical Mechanics* (New York: McGraw-Hill, 1956).

[7] N. Davidson, *Statistical Mechanics* (New York: McGraw-Hill, 1962).

[8] J. Chandrasekhar, S. F. Smith, and W. L. Jorgensen, *J. Am. Chem. Soc. 106,* 3049 (1984); J. Chandrasekhar and W. L. Jorgensen, *J. Am. Chem. Soc. 107,* 2974 (1985).

[9] R. A. Marcus, *J. Chem. Phys. 24,* 966, 979 (1956).

[10] R. A. Marcus, *J. Chem. Phys. 43,* 679 (1965).

[11] R. A. Marcus and N. Sutin, *Biochim. Biophys. Acta 811,* 265 (1985).

[12] J. R. Miller, T. Calcaterra, and G. L. Closs, *J. Am. Chem. Soc. 106,* 3047 (1984).

[13] H. A. Kramers, *Physica* (The Hague) 7, 284 (1940).

[14] D. A. McQuarrie, *Statistical Mechanics* (NY: Harper and Row, 1976); S. Chandrasekhar, *Rev. Mod. Phys. 15,* 1 (1943).

[15] F. Daniels and R. A. Alberty, *Physical Chemistry* (New York: John Wiley, 1963).

[16] J. Troe in *Physical Chemistry, An Advanced Treatise, Vol VI B,* edited by W. Jost (New York: Academic Press, 1975), p. 835.

[17] J. P. Bergsma, J. R. Reimers, K. R. Wilson, and J. T. Hynes, *J. Chem. Phys. 85,* 5625 (1986).

[18] J. P. Bergsma, B. J. Gertner, K. R. Wilson, and J. T. Hynes, *J. Chem. Phys. 86,* 1356 (1987).

[19] I. Langmuir, *J. Am. Chem. Soc. 38,* 2221 (1916); *40,* 1361 (1918).

[20] H. M. F. Freundlich, *Kappilarchemie,* Leipzig, 1909.

[21] R. Sips, *J. Chem. Phys. 18,* 1024 (1950); F. C. Tompkins, *Trans. Faraday Soc. 46,* 569 (1950).

[22] S. Brunauer, P. H. Emmett, and E. Teller, *J. Am. Chem. Soc. 60,* 309 (1938).

[23] J. C. Tully, *Surface Sci. 111,* 461 (1981).

[24] J. C. Tully, *Surface Sci. 299/300,* 667 (1994).

[25] E. K. Grimmelmann, J. C. Tully and M. J. Cardillo, *J. Chem. Phys. 72,* 1039 (1980).

[26] W. H. Weinberg and R. P. Merrill, *J. Chem. Phys. 56,* 2881 (1972).

[27] M. J. Cardillo, M. Baloach and R. E. Stickney, *Surface Sci. 50,* 263 (1975).

[28] M. B. Lee, Q. Y. Yang, and S. T. Ceyer, *J. Chem. Phys. 87,* 2724 (1987).

[29] M. B. Lee, J. D. Beckerle, S. L. Tang, and S. T. Ceyer, *J. Chem Phys. 87,* 723 (1987).

[30] T. Engel and G. Ertl, in *The Chemical Physics of Solid Surfaces and Heterogeneous Catalysis, Vol 4,* edited by D. A. King and D. P. Woodruff (New York: Elsevier, 1982), p. 73.

[31] C. N. Hinshelwood, *Kinetics of Chemical Change in Gaseous Systems* (Oxford: Clarendon Press, 1926), p. 145.

[32] D. D. Eley and E. K. Rideal, *Proc. R. Soc. London A 178,* 429 (1941).

[33] M. Grunze, in *The Chemical Physics of Solid Surfaces and Heterogeneous Catalysis, Vol. 4,* edited by D. A. King and D. P. Woodruff (New York: Elsevier, 1982), p. 143.

[34] R. D. Kelly and D. W. Goodman, in *The Chemical Physics of Solid Surfaces and Heterogeneous Catalysis, Vol. 4,* edited by D. A. King and D. P. Woodruff (New York: Elsevier, 1982), p. 427.

[35] L. Tsou, G. L. Haller, and J. B. Fenn, *J. Phys. Chem. 91,* 2654 (1987).

[36] S. P. Holland, B. J. Garrison, and N. Winograd, *Phys. Rev. Letters 44,* 756 (1980).

[37] S. Adelman and J. D. Doll, *J. Chem Phys. 61,* 4242 (1974); *Acc. Chem. Res. 10,* 378 (1977); J. C. Tully, *J. Chem. Phys. 73,* 1975 (1980).

[38] C.-Y. Lee and A. E. DePristo, *J. Chem. Phys. 87,* 1401 (1987).

[39] J. C. Tully, *Ann. Rev. Phys. Chem. 31,* 319 (1980); J. C. Tully, *Acc. Chem. Res. 14,* 188 (1981); J. C. Tully and M. J. Cardillo, *Science 223,* 445 (1984); C. W. Muhlhausen, L. R. Williams, and J. C. Tully, *J. Chem. Phys. 83,* 2594 (1985).

BIBLIOGRAPHY

Thermodynamic Formulation of Transition State Theory

HAMMES, G. C. *Principles of Chemical Kinetics.* New York: Academic Press, 1978.

LAIDLER, K. J. *Chemical Kinetics.* New York: Harper and Row, 1987.

MOORE, J. W., and PEARSON, R. G. *Kinetics and Mechanism.* New York: Wiley, 1981.

WESTON, R. E., JR., and SCHWARZ, H. A. *Chemical Kinetics.* Englewood Cliffs, NJ: Prentice-Hall, 1972.

Electron Transfer Reactions

WESTON, R. E., JR., and SCHWARZ, H. A. *Chemical Kinetics.* Englewood Cliffs, NJ: Prentice-Hall, 1972.

HAMMES, G. G. *Principles of Chemical Kinetics.* New York: Academic Press, 1978.

LOGAN, S. R. *Fundamentals of Chemical Kinetics.* Essex, England: Longman House, 1996.

BILLING, G. D., and MIKKELSEN, K. V. *Molecular Dynamics and Chemical Kinetics.* New York: John Wiley, 1996.

Kramers' Theory

HYNES, J. T. In *Theory of Chemical Reaction Dynamics, Vol. IV,* edited by M. Baer. Boca Raton, FL: CRC Press, 1985, p.171.

KAPRAL, R. *Adv. Chem. Phys. 48,* 71 (1981).

TROE, J. In *Physical Chemistry, An Advanced Treatise, Vol. VI B,* edited by W. Jost. New York: Academic Press, 1975, p. 835.

TROE, J. *Ann. Rev. Phys. Chem. 38,* 163 (1987).

Gas-Surface Reaction Dynamics

GRUNZE, M., and KREUZER, H. J. *Kinetics of Interface Reactions.* New York: Springer-Verlag, 1987.

JORDAN, P. C. *Chemical Kinetics and Transport.* New York: Plenum, 1979.

KING, D. A., and WOODRUFF, D. P. *The Chemical Physics of Solid Surfaces and Heterogeneous Catalysis, Vol. 4, Fundamental Studies of Heterogeneous Catalysis.* New York: Elsevier, 1982.

LAIDLER, K. J. *Chemical Kinetics.* New York: Harper and Row, 1987.

PULLMAN, B., JORTNER, J., NITZAN, A., and GERBER, B. *Dynamics on Surfaces.* Boston: D. Reidel, 1984.

SOMORJAI, G. A. *Principles of Surface Chemistry.* Englewood Cliffs, NJ: Prentice-Hall, 1972.

PROBLEMS

12.1 A study of the rate of alkaline hydrolysis of propionamide as a function of hydrostatic pressure at 300 K gave the following results:

$\log k/k_0$	0.10	0.21	0.31
P, lb per sq. in.	5,000	10,000	15,000

What is the volume of activation, $\Delta V^{\ddagger}$, for this reaction in units of liters/mole?

12.2 Outline the different assumptions made in deriving equation (12-10) as compared with equation (4-57).

12.3 A strong correlation has been found between volumes and entropies of activation for reactions in aqueous solution [K. J. Laidler and D. T. Y. Chen, *Can. J. Chem. 37,* 599 (1959)]. What is the nature of this correlation? What is its chemical significance?

12.4 The radial distribution function, $g_{ab}(r^{\ddagger})$, at the transition state for an association reaction in solution has a value of 5.69×10^{-3} at 300 K. The potential energy barrier for the association reaction in the gas-phase is 21 kJ/mole. What is the ratio between the gas-phase and solution association rate constants at 300 K? What is $\Delta G_{solv}^{\ddagger}$?

12.5 What is the mathematical relationship between the ω_0 and $\omega^{\ddagger}$ terms in equations (12-47) and (12-48) and the vibrational frequency?

12.6 The velocities of molecules desorbing from a surface are usually characterized by a Maxwell-Boltzmann distribution at the temperature of the surface. It might seem that the distribution of molecular velocities should be colder than the distribution of velocities of the molecules adsorbed on the surface by just the well depth of the interaction, since the molecules have to overcome a barrier to desorption. This problem will show that the former statement is correct.

Consider a one-dimensional square well of depth U. Let c represent the velocities inside the well (measured from the bottom of the well) and v represent the experimentally determinable velocities (the velocities of the desorbed particles).

(a) Draw a diagram showing the potential and the probability density of molecules adsorbed in the well.

(b) Write the mathematical expression for the probability density, and covert it to a quantity proportional to the probability *flux* of molecules in the well with velocities between c and $c + dc$.

(c) Using the diagram from part (a), transform the above expression to one in terms of the experimentally measurable velocities. Make sure to do the differentials (the Jacobian of the transformation) correctly.

(d) The answer to the original problem is clear from the expression found in part (c). The mean of the experimentally observed velocity distribution is

$$\left(\frac{\pi k_B T}{2m} \right)^{1/2} = \langle |v| \rangle$$

Derive this expression. Is the temperature of the desorbing molecules equal to or less than that of the surface? Why?

(e) The flux of molecules desorbing from a surface is characterized by a cosine angular distribution where the angle is measured from the surface normal. From the expression in part (b) or (c), identify what determines the rate at which the molecules escape from the surface. How does the rate relate to the angular distribution?

12.7 In order to study the properties of a metal-semiconductor interface one approach is to evaporate a monolayer of metal atoms onto the surface of the semiconductor in a vacuum chamber. Such an apparently simple procedure is, however, fraught with experimental difficulties. In particular, if the substrate is at room temperature, the metal adatoms will sometimes form "islands" rather than a continuous uniform layer. To circumvent this, the substrate is often cooled, the rationale being that on a sufficiently cold surface an atom will stick where it lands on the surface and a uniform coverage will result. We wish to examine this rationale quantitatively. Assume for simplicity that the substrate and adsorbate atoms are virtually the same size and that the adsorbed metal has the same crystalline arrangement as the semiconductor.

(a) If enough metal atoms are evaporated to give an average number of metal atoms/ surface site $\overline{N}$ and there are b surface sites, what is the probability that a metal atom is adsorbed at a particular site b_0? What is the order of magnitude of b for a typical crystal surface area of 1 cm^2?

(b) The probability that a site is covered by a layer of N atoms deep can be calculated from the binomial distribution

$$P_N = \left(\frac{b(b-1)\cdots(b-N+1)}{N!} \right) \left(\frac{\overline{N}}{b} \right)^N \left(1 - \frac{\overline{N}}{b} \right)^{b-N} \tag{1}$$

However, this form is cumbersome as it includes b explicitly. Show that in the limit as $b \to \infty$ $(b \gg N, b \gg \overline{N})$,

$$P_N \approx \frac{(\overline{N})^N e^{-\overline{N}}}{N!}$$

Hints: This derivation involves making approximations which are good in the limit $b \to \infty$. (Justify briefly the ones used.) Eliminate b from the first two factors of (1) alone. Then recognize a truncated Taylor series in the third factor.

(c) Suppose enough metal is evaporated to give an average surface coverage of 1 monolayer. What fraction of the substrate atoms are not covered by metal? By one metal atom exactly? By five metal atoms?

CHAPTER 13

Information-Theoretical Approach to State-to-State Dynamics

13.1 INTRODUCTION

The information-theoretical approach to the reaction dynamics of systems possessing a large number of internal levels was first introduced in the early 1970s by R. D. Levine, R. B. Bernstein, and J. L. Kinsey.[1] This approach is designed to deal with large amounts of data in a systematic manner. In the macroscopic, phenomenological approach with which we began, we averaged over internal state distributions in reactants and products. That approach was simply not detailed enough to describe experiments which employed state-selected reactants and/or state-resolved product detection, such as those described in chapter 9. Many instances are known of selective energy consumption in endothermic reactions and specific energy release in exothermic reactions which do not follow a simple, Boltzmann-like statistical distribution. A microscopic theory of reaction dynamics has been developed to deal with such experiments, but this turns out to be in many ways *too* detailed: the theory frequently requires the solution of many closely coupled equations and calculates far more pieces of information than could ever be observed. A middle ground between these extremes would be a statistical thermodynamic approach, analogous to that used for prediction of equilibrium thermodynamic properties. In such a theory, all of the dynamical variables are considered in principle, but suitable averages are taken over those variables that are not observed in any particular measurement. For reasons which will become apparent shortly, this approach has come to be called *information-theoretic*. Details of the early development of this theory, together with its practical implementations, may be found in several excellent reviews.[2,3]

13.2 THE MAXIMAL-ENTROPY POSTULATE

13.2.1 Basic Definitions

Consider an experiment having n possible distinguishable outcomes, e.g.,

$$F + H_2 \rightarrow HF(v) + H$$

Since the energy available in the reaction, $-\Delta H^0 + E_{act} + \langle E_{thermal} \rangle \approx 140$ kJ/mole, the vibrational states $v = 0, 1, 2$, and 3 are energetically accessible. Thus, $n = 4$ for this system.* After N repetitions of this molecular process, we get a number of possible sequences representing the v level observed for each individual reaction. Such a sequence might be, for example,

$$\{v\} = 0\,2\,1\,3\,2\,2\,3\,1\,2\,0\,2\,1\,2\,2\,3\ldots$$

which is distinct from, say,

$$\{v\}' = 0\,2\,1\,3\,2\,3\,2\,1\,2\,0\,2\,1\,2\,2\,3\ldots$$

even though the first 15 members of the sequence include two reactions giving $v = 0$, three with $v = 1$, seven with $v = 2$, and three with $v = 3$. The number of possible sequences is clearly n^N, which can be expressed conveniently as $\exp(N \ln n)$. This is in general an enormous number; even in the present trivial case, $n^N = 4^{15} \approx 10^9$!

The complete specification of a sequence is entirely too detailed for practical use. What we are really interested in is the *set* of outcomes for a particular experiment, denoted by $\{N_i\}$, with $i = 1, 2, \ldots, n$, and normalized so that

$$\sum_{i=1}^{n} N_i = N \tag{13-1}$$

For example, for both of the preceding sequences, $\{N_i\} = (2, 3, 7, 3)$. The thermodynamic weight W associated with a set $\{N_i\}$ is the number of possible sequences which correspond to a distinct distribution or set $\{N_i\}$. This is equivalent to the combinatorial problem of placing N balls in n boxes, which gives[4]

$$W(\{N_i\}) = \frac{N!}{\prod_{i=1}^{n}(N_i)!} \tag{13-2}$$

with the normalization

$$\sum_{\text{all sets}} W(\{N_i\}) = n^N \tag{13-3}$$

The unbiased or *a priori* probability of finding a particular set is then

$$p^0(\{N_i\}) = \frac{W(\{N_i\})}{\sum W(\{N_i\})} = \frac{W(\{N_i\})}{n^N} \tag{13-4}$$

and clearly,

$$\sum_{\text{all distributions}} P^0\{(N_i)\} = 1 \tag{13-5}$$

---

*For F + D$_2$, $v = 0, 1, 2, 3$, and 4 can be populated, and $n = 5$.

Since typical experiments involve large N (on the order of 10^{23} or so), evaluating these expressions is facilitated considerably by use of the *Stirling approximation*[4,5]

$$\ln N! \equiv \log_e(N!) = N \ln N - N + \mathbb{O}\{\ln N\} \tag{13-6}$$

For $N \approx 10^{23}$, the contribution of $\mathbb{O}(\ln N)$ relative to that of N is approximately

$$\frac{2.3 \times 23}{10^{23}} \approx \frac{50}{10^{23}} \approx 1 \text{ in } 10^{21}$$

so that $\mathbb{O}(\ln N)$ can be safely neglected. Using the Stirling approximation, the *a priori* probability can be evaluated from equation (13-4) as

$$\ln P^0(\{N_i\}) = N \ln N - N - \sum_i (N_i \ln N_i - N_i) - N \ln n$$

$$= -N \ln n - \sum_i (N_i \ln N_i - N_i \ln N)$$

$$= -N\left(\ln n + \sum_i X_i \ln X_i \right)$$

where

$$X_i = \frac{N_i}{N}, \text{ with } \sum_{i=1}^{n} X_i = 1 \tag{13-7}$$

The *statistical entropy* associated with this distribution is given by

$$S(\{N_i\}) = \frac{1}{N} \ln W(\{N_i\})$$

$$= \frac{1}{N} \ln [n^N P^0(\{N_i\})]$$

$$= \frac{1}{N}\left(N \ln n - N \ln n - N \sum_i X_i \ln X_i \right)$$

so that

$$S(\{N_i\}) = - \sum_i X_i \ln X_i \tag{13-8}$$

Equation (13-8) will look familiar, as the *entropy of mixing* in equilibrium thermodynamics, if X_i is the mole fraction and the entropy is expressed in entropy units (e.u.), or ergs mole^{-1} degree^{-1}, using the coefficient $S = (R/N_0) \ln W = k \ln W$.

We shall shortly prove that the entropy is a maximum for the *uniform distribution*, in which $X_i = 1/n$ for all i—that is, all possible outcomes are equally probable. From equation (13-8), this is

$$\max(S) = - \sum_{i=1}^{n} \frac{1}{n} \ln\left(\frac{1}{n}\right) = -\frac{n}{n} \ln\left(\frac{1}{n}\right) = \ln n \tag{13-9}$$

For any other distribution, we may define an *entropy deficiency*

$$\Delta S = -\frac{1}{N} \ln\left(P\{N_i\}\right) = \ln n + \sum_i X_i \ln X_i \qquad (13\text{-}10)$$

The entropy S is the difference between the maximum entropy of equation (13-9) and the entropy deficiency of equation (13-10), or

$$S = \max(S) - \Delta S$$

$$= \ln n - \left(\ln n + \sum_i X_i \ln X_i\right)$$

$$= -\sum_i X_i \ln X_i$$

so that

$$S = \frac{1}{N} \ln W(\{N_i\}) \qquad (13\text{-}11)$$

as before. For $X_i = 1/n$, i.e., $P(\{N_i\}) = P^0$, obviously $\Delta S = 0$. In communication theory,[6] this quantity is also called the *information content* of the distribution; the rationale is that a completely flat, unmodulated signal conveys no information, and only the departures from uniform amplitude [the entropy deficiency, according to equation (13-10)] provide useful information to the receiver. A further important consequence of the expression given for the entropy in equations (13-8) and (13-11) can be seen from writing the statistical weight as $W(\{N_i\}) = \exp(NS)$. For large N (i.e., in bulk experiments), this means that a distribution possessing even a slightly smaller value of S than any other distribution will have an *overwhelmingly* smaller statistical weight; so, in effect, the experimentalist observes only a single outcome, the maximal-entropy outcome, every time the measurement is repeated. This is the basis of the *maximal-entropy postulate*, which is central to the development of the information-theoretic approach. It is also expressed, much more succinctly, by the Talmudic quotation with which Levine begins his 1978 review article.[7]

13.2.2   Constraints and the Variational Principle

While unique distributions are indeed observed in experiments, such as the vibrational distribution of the HF product in the $F + H_2$ reaction (chapter 9), these are seldom the uniform distribution with $X_i = 1/n$. This is because of the existence of constraints on the distribution.

What is meant by "constraints"? Constraints are conditions imposed on the distribution, e.g., *normalization*, given by

$$\sum_{i=1}^{n} X_i = 1 \qquad (13\text{-}12)$$

which is of course equivalent to both equation (13-1) and equation (13-5). Let us begin by finding the distribution which maximizes the entropy, subject only to the normalization constraint of equation (13-12). The technique for doing this is the *calculus of variations,* which is described in any standard calculus textbook.[8] We define a variational function

$$F = S - (\lambda_0 - 1) \sum_i X_i \qquad (13\text{-}13)$$

where the first term, S, is the function to be maximized (in this case, the entropy), and the second term is the statement of the constraint (normalization). The coefficient of the latter term is the *Lagrange multiplier,* which we write here as $\lambda_0 - 1$ for convenience, as we shall shortly see. We require that $\delta F / \delta X_i = 0$ for all variations δX_i, i.e., that F is an *extremum* (in this case, a maximum) in multidimensional $\{X_i\}$ space.

Substituting the definition of the statistical entropy, equation (13-8), into the variational function, equation (13-13), gives

$$F = \sum (-\ln X_i - \lambda_0 + 1) X_i$$

The variation δF is found by simple differentiation:

$$\delta F = \sum [-\ln X_i (\delta X_i) - X_i (\delta \ln X_i) - \lambda_0 \delta X_i + \delta X_i]$$

Recalling that $\delta \ln X_i = \delta X_i / X_i$, we have

$$\delta F = \sum_i (-\ln X_i - 1 - \lambda_0 + 1) \delta X_i$$

$$= \sum_i (-\ln X_i - \lambda_0) \delta X_i$$

Note the usefulness of writing the Lagrange multiplier as $\lambda_0 - 1$ here. In order for $\delta F / \delta X_i$ to be zero for all i, we require that the coefficient of δX_i be equal to zero, i.e.,

$$-\ln X_i - \lambda_0 = 0$$

or

$$X_i = e^{-\lambda_0}$$

But the normalization equation (13-12) requires that

$$\sum_{i=1}^{n} X_i = \sum_{i=1}^{n} e^{-\lambda_0} = n e^{-\lambda_0} = 1$$

Therefore, $n = e^{\lambda_0}$, $\lambda_0 = \ln n$, and $X_i = e^{-\lambda_0} = e^{-\ln n} = 1/n$, which is the uniform distribution defined on page 426. This means that normalization provides us with no "new information" about the distribution, which can be stated as a theorem:

When the only constraint is normalization of probabilities, the most probable distribution is the uniform one.

A lemma can also be stated:

Furthermore, the entropy of such a distribution is an absolute maximum.

Now suppose we have one more piece of information, such as the first moment of the distribution, defined by

$$\sum_{i=1}^{n} iX_i = \langle i \rangle \tag{13-14}$$

This furnishes an additional constraint on the distribution, and the variational function now becomes

$$F = S - (\lambda_0 - 1) \sum_{i=1}^{n} X_i - \lambda_1 \sum_{i=1}^{n} iX_i$$

or

$$F = - \sum_i X_i \ln X_i - (\lambda_0 - 1) \sum_i X_i - \lambda_1 \sum_i iX_i \tag{13-15}$$

The variational quantity is thus

$$\delta F = \sum_i (-\ln X_i - 1 - \lambda_0 + 1 - \lambda_1 i) \delta X_i$$

and requiring that $\delta F / \delta X_i = 0$ gives

$$\ln X_i = -\lambda_0 - \lambda_1 i$$

or

$$X_i = e^{-\lambda_0 - \lambda_1 i} \tag{13-16}$$

To find the Lagrange multipliers λ_0 and λ_1, we use a *sum rule* based on the constraint given by equation (13-14). Factoring out the common element e^{λ_0}, we can rewrite this equation as

$$\langle i \rangle = \frac{\sum_i i \exp(-\lambda_1 i)}{\sum_i \exp(-\lambda_1 i)} \tag{13-17}$$

If $\langle i \rangle$ is known, equation (13-17) is sufficient to determine λ_1; however, because of the sums over exponentials, such determination in general must be done by numerical fitting. Once λ_1 is known, the normalization parameter λ_0 can be easily determined using equation (13-12) as follows:

$$\sum_i X_i = \sum_i \exp(-\lambda_0 - \lambda_1 i)$$

$$= \sum_i \exp(-\lambda_0) \exp(-\lambda_1 i) = 1$$

Therefore, $\exp(\lambda_0) = \Sigma_i \exp(-\lambda_1 i)$, and if the sum is taken over $n \to \infty$, this has the simple closed form

$$\exp(\lambda_0) = \frac{1}{1 - e^{-\lambda_1}} \tag{13-18}$$

13.2.3 Example: Harmonic-Oscillator Levels at Thermodynamic Equilibrium

As an illustration of the variational principle discussed in the preceding section, let us consider $P(v)$, the equilibrium distribution among harmonic-oscillator levels. The variable i is the oscillator energy

$$E_v = v\hbar\omega_0 \tag{13-19}$$

We take the zero of energy to be the zero-point level $1/2\ \hbar\omega_0$ and note that v can take any integer value between zero and infinity. (The sums then have convenient closed forms.) The constraints we specify are normalization, viz.,

$$\sum_v P(v) = 1 \tag{13-20a}$$

and fixed total energy, i.e.,

$$\sum_v E_v P(v) = \langle E_v \rangle = \text{constant} \tag{13-20b}$$

The second condition is that which specifies a canonical ensemble.[9] From the result previously obtained [equation (13-16)],

$$P(v) = \exp[-\lambda_0 - \beta v] \tag{13-21}$$

with the Lagrange multiplier $\lambda_1 = \beta\hbar\omega_0$. This quantity is determined by application of equation (13-17), the sum rule:

$$\langle E_v \rangle = \frac{\displaystyle\sum_v (v\hbar\omega_0)\exp(-\beta v\hbar\omega_0)}{\displaystyle\sum_v \exp(-\beta\hbar\omega_0)}$$

$$= \frac{\hbar\omega_0}{e^{\beta\hbar\omega_0} - 1}$$

The normalization constant is determined from

$$\exp(\lambda_0) = \frac{1}{1 - \exp(-\beta\hbar\omega_0)}$$

If we make the further connection with equilibrium thermodynamics, then β^{-1} must be proportional to a temperature T and $\beta = 1/k_B T$. Equation (13-21) is then the familiar Boltzmann distribution,

$$P(v) = [1 - e^{-\hbar\omega_0/k_BT}]e^{-v\hbar\omega_0/k_BT}$$

The purpose of this example was to illustrate the application of entropy maximization by means of the variational principle to a familiar situation, namely, the calculation of an equilibrium distribution. In the information-theoretical approach to chemical kinetics, the same procedures are applied to nonequilibrium systems, i.e., to calculation of rate coefficients, branching ratios, and reaction probabilities. While this approach cannot really be called a dynamic theory, since it cannot actually predict the values of these quantities, numerous investigations have demonstrated its utility in describing reactive systems possessing a very large number of possible channels in terms of a much smaller number of Lagrange multipliers. In many instances, only a single such parameter is sufficient to represent all of the available information about such a system. It is by no means axiomatic, however, that a distribution possess only a single constraint. In general, there may be r independent constraints having the form

$$\sum_{i=1}^{n} g_m(i)X_i = \langle g_m \rangle \tag{13-22}$$

We have already considered constraints with

$$m = 0 \text{ (normalization):} \qquad g_0(i) = 1 \quad \langle g_0 \rangle = 1$$

and

$$m = 1 \text{ (first moment):} \qquad g_1(i) = i \quad \langle g_1 \rangle = \langle i \rangle$$

For $m = 2$, it is usually most useful to take $\langle g_2 \rangle = (\langle i^2 \rangle - \langle i \rangle^2)$, i.e., the *dispersion* of the distribution; but any form can be specified, as long as it is independent of all other constraints. The maximal-entropy distribution has the form

$$X_i = \exp\left[-\lambda_0 - \sum_{m=1}^{r} \lambda_m g_m(i)\right] \tag{13-23}$$

with

$$S = -\sum_{i} X_i \ln X_i$$

i.e.,

$$S = \lambda_0 + \sum_{m=1}^{r} \lambda_m \langle g_m \rangle \tag{13-24}$$

In the applications to be discussed in this chapter, we shall see numerous examples with $r \ll n$, which means that it is possible to describe a distribution over many possible outcomes with a very small number of moments of the distribution. An equivalent way of saying this is that there are only r independent pieces of information required to describe the experiment in full.

13.3 SURPRISAL ANALYSIS AND SYNTHESIS: PRODUCT STATE DISTRIBUTION IN EXOTHERMIC REACTIONS

We shall now proceed to apply these concepts to a nonequilibrium situation, specifically, the product state distribution in an exothermic reaction. The reaction we shall consider is one which has been discussed previously in the context of reaction dynamics, namely, the hydrogen + fluorine atom reaction

$$F + H_2 \longrightarrow HF(v = n') + H$$

along with its isotopic variant,

$$F + D_2 \longrightarrow DF(v = n'') + D$$

We seek the *branching ratios* for this reaction, given by

$$P(n') = \frac{k(n')}{\displaystyle\sum_{n'} k(n')} \tag{13-25}$$

which takes the place of the probability densities discussed in the context of equilibrium distributions. While in principle we may make a complete specification of product states, for the time being we shall consider just the total production rate into a specified vibrational state:

$$k(n') = \sum_{J'} \sum_{M'_j} k(n', J', M'_j) \tag{13-26}$$

The set $\{k(n')\}$ may incorporate one or more dynamic constraints, but we are generally not aware of that prior to carrying out the experiment. Therefore, we shall take the *prior rate* $k^0(n')$ to be simply proportional to the density of states available to the products. This is equivalent to the uniform distribution introduced earlier, and the result will be dominated by the translational density of states. (We shall take care of vibrational and rotational state degeneracies shortly.) The equation is shown in Appendix 2 to have the form

$$\rho_T(E'_T) = \frac{(\mu')^{3/2}}{2^{1/2}\pi^2\hbar^3}(E'_T)^{1/2} \tag{13-26a}$$

or

$$\rho_T(E'_T) = A_T(E'_T)^{1/2} \tag{13-26b}$$

We use the symbol $\rho(E)$ here, rather than $N(E)$ as in chapter 11, to facilitate reference to the extensive literature on information theory. Also, μ' is the reduced mass of the products, and E'_T is the translational energy available to the products, which is equal to the total reaction energy

$$E = -\Delta H^0(\text{reaction}) + E_{\text{act}} + \langle E_{\text{thermal}}(\text{reactants})\rangle$$

less the energy in the product internal states, $E_n = E'_{\text{vib}} + E'_{\text{rot}}$. We assert that

$$k^0(n') = R\rho_T(E'_T)\rho_n(E'_n) = R\rho_T(E - E'_n)\rho_n(E'_n) \tag{13-27}$$

with R a coefficient that provides the correct units for the rate coefficient. Note that $\rho_T(E_T') \propto (E_T')^{1/2}$, which is in turn proportional to the relative velocity of separation of the products. Thus, equation (13-27) can also be interpreted as stating that the prior rate coefficient is proportional to the net flux of products in a specified internal state out of the reaction zone.

By analogy with equation (13-25), we define

$$k^0 = \sum_{n'} k^0(n')$$

$$= R \sum_{n'} \rho_T(E - E_n')\rho_n(E_n')$$

so that

$$k^0 = R\rho(E) \tag{13-28}$$

and the prior branching ratio

$$P^0(n') = \frac{k^0(n')}{k^0} = \frac{\rho_T(E - E_n')\rho_n(E_n')}{\rho(E)} \tag{13-29}$$

The $P^0(n')$ are, of course, normalized to unity.

The observed distribution can be written in the maximal-entropy form [cf. equation (13-23)]

$$P(n') = P^0(n')\exp\left[-\sum_{m=1}^{r} \lambda_m g_m(r')\right] \tag{13-30}$$

We then define the "departure from the prior expectation,"

$$I(n') = -\ln\frac{P(n')}{P^0(n')} \tag{13-31}$$

The quantity $I(n')$ has been called the *surprisal* of the product state distribution. Although the term "surprisal" has attracted a degree of opprobrium, its usage is quite reasonable: if you don't get what you expected beforehand, then by definition you are surprised. On the other hand, if $P(n') = P^0(n')$ for all n', then $I = 0$; in other words, the observed and prior distributions are identical, and that's no surprise at all! Indeed, the zero-surprisal situation implies no constraints beyond normalization, which is implicit in the definition of P (or P^0). In general, however, one finds nonzero surprisals which have the maximal-entropy form

$$I(n') = \sum_{m=1}^{r} \lambda_m g_m(n') \tag{13-32}$$

Now let us apply this analysis to the special case of the HF product distribution. The most detailed specification includes both the vibrational state v' and the rotational state J', so that

$$P^0(v', J'|E) = \frac{(2J' + 1)\rho_T(E - E_{v'J'})}{\rho(E)} \tag{13-33}$$

The factor $2J' + 1$ comes from having summed over $M_{J'}$ states in equation (13-26), while the quantity $P^0(v' J' | E)$ is a kind of *conditional* probability[4] which can be read as "the probability of finding a product molecule in state $v'J'$ given that a total energy E is available." We shall come across conditional probabilities often in the rest of this chapter.

The desired quantity, the net branching ratio into state v', is found by summing equation (13-33) over all accessible states J' up to the maximum $J^*(v')$ which is compatible with the energy $E_{v'}$ in vibration and the total energy E:

$$P^0(v' | E) = \sum_{J=0}^{J^*(v')} P^0(v' J' | E) \tag{13-34}$$

In order to evaluate equation (13-34), we need to choose a model for the vibration-rotation energy levels of the molecule. The most convenient model is the rigid rotor-harmonic oscillator (RRHO), given by

$$E_{v'J'} = E_R + E_v = BJ'(J' + 1) + \hbar\omega_0 v' \tag{13-35}$$

We thus have, for the RRHO,

$$
\begin{aligned}
(2J' + 1)\rho(v'J'E) &= (2J' + 1)\rho_T(E_T) \\
&= (2J' + 1)\rho_T(E - E_{v'J'}) \\
&= (2J' + 1)\left(\frac{(\mu')^{3/2}}{2^{1/2}\pi^2\hbar^3}\right)(E - E_{v'J'})^{1/2}
\end{aligned}
$$

and

$$
\begin{aligned}
\rho(E) &= \sum_{v'=0}^{v^*(E)} \left[\sum_{J'=0}^{J^*(v')} \rho(v'J'E) \right] \\
\rho(E) &= \frac{4}{15}\left(\frac{1}{\hbar\omega_0 B}\right)\left(\frac{(\mu')^{3/2}}{2^{1/2}\pi^2\hbar^3}\right)E^{5/2}
\end{aligned}
\tag{13-36}
$$

In equation (13-36), $v^*(E)$ is the *exothermic limit*—the vibrational state of the product in which all the energy E appears in vibration.

Combining equations (13-33) and (13-36), we have

$$P^0(v'J'E) = (2J' + 1)\hbar\omega_0 B \frac{15}{4}(E - E_v - E_R)^{1/2}E^{-5/2} \tag{13-37}$$

and

$$P^0(v' | E) = \sum_{J'=0}^{J^*(v')} P^0(v' J' | E)$$

so that

$$P^0(v' | E) = (\text{const})\frac{5}{2}(E - E_{v'})^{3/2}E^{-5/2} \tag{13-38}$$

It is convenient to rewrite these quantities in terms of the fraction of energy in product vibration.

$$f_v = \frac{E_{v'}}{E} \tag{13-39}$$

Thus,

$$P^0(f_v) \propto \frac{5}{2} E^{3/2} (1 - f_v)^{3/2} E^{-5/2}$$

or

$$P^0(f_v) \propto \frac{5}{2} \frac{(1 - f_v)^{3/2}}{E} \tag{13-40}$$

In order to perform the calculation, we need to specify E. For $F + H_2$, $-\Delta H° = 130$ kJ mole^{-1}, $E_{act} = 4$ kJ mole^{-1}, and $\langle E_{thermal} \rangle \approx 6$ kJ mole^{-1}, for a total of $E = 140$ kJ mole^{-1}. Because of the large value of $-\Delta H°$, we can use a single value of E, i.e., we can treat the system as a microcanonical ensemble. In other situations, we have to carry out an average over energy distributions at some temperature (see section 13.4.1).

The analysis proceeds as follows for HF. We calculate the vibrational energy levels from the vibrational constants[10] and note the experimental values of $P(v')$ (see chapter 9). Then the quantities f_v, $P^0(v')$, and $I(v')$ are readily calculated from equations (13-39), (13-40), and (13-31), respectively. The results are shown in Table 13-1(a). When $-I(v')$ is plotted against f_v, a straight line is obtained, as shown in Figure 13-1. If the calculation is now repeated for the reaction $F + D_2 \rightarrow DF(v') + D$ (Table 13-1(b)), the surprisal points are found to lie on the same straight line as previously found. Therefore, a *single dynamic constraint*, independent of isotopic substitution, is sufficient to determine energy disposal in this reaction. The $F + (H_2, D_2)$ reaction system was among the first to be analyzed in this way,[11] and its success led to many further

TABLE 13-1 Surprisal Analysis for HF and DF Product Vibrational Distributions.

v'	$E_v - E_0$, kJ/mole	f_v	$(1 - f_v)^{3/2}$	$P(v')$, experimental	$I(v')$
(a) $F + H_2 \rightarrow HF(v') + H$					
0	0	0	1.00	(0.02)	(3.9)
1	47	0.355	0.54	0.15	1.28
2	92	0.657	0.20	0.57	−1.04
3	132	0.943	0.014	0.26	−2.9
(b) $F + D_2 \rightarrow DF(v') + D$					
0	0	0	1.00	(0.01)	(4.6)
1	34.7	0.247	0.65	0.10	1.8
2	68.5	0.489	0.365	0.23	0.46
3	101.1	0.722	0.146	0.38	−0.95
4	132	0.944	0.013	0.28	3.0

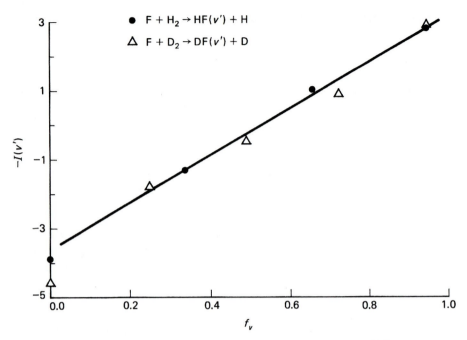

FIGURE 13-1 Surprisal plot for the reaction $F + H_2 \rightarrow HF(v') + H[\bullet]$ and $F + D_2 \rightarrow DF(v') + D[\triangle]$.

applications. It is worth pointing out that if we had possessed this key piece of information before actually doing the experiment, we could have justifiably incorporated it into the analysis together with our prior branching ratios $P^0(v)$; in that case we would have had $P(v) = P^0(v)$ for all v with $I(v) = -\ln [P^0(v)/P^0(v)] = 0$, and there would have been no surprise.

The obvious utility of the information-theoretical approach to this problem is that nine data points, giving the product branching ratios in the $F + H_2$ and $F + D_2$ reactions, can be reduced to a single parameter. In subsequent sections we shall see even more spectacular compaction of molecular collision data. In the meantime, let us consider how to obtain the single parameter from the data. Reducing equation (13-30) to a single term and writing the normalization separately gives

$$k(v) = k_{\text{total}} P(v')$$
$$= \exp(-\lambda_0 - \lambda_v f_v)$$

so that

$$k(v') = k^0(v') \exp(-\lambda_v f_v) \tag{13-41}$$

and the surprisal is given by

$$I(v') = -\ln \frac{k(v')}{k^0(v')} = \lambda_v f_v \tag{13-42}$$

For this system, we obtain λ_v from the slope of the straight line in Figure 13-1:

$$\lambda_v = \frac{\Delta(I_v)}{\Delta(f_v)} = -7.2$$

A large value for $|\lambda_v|$ indicates a considerably nonstatistical distribution; moreover, when the value of λ_v is negative, as in the present case, it indicates preferential disposition of the energy in product vibration. Values of λ close to zero always indicate a near-statistical outcome.

Other useful measures of the reaction can be derived from the preceding. The average amount of the total available energy appearing as product vibration is

$$\langle f_v \rangle = \sum_{v'} f_{v'} P(v'|E) \tag{13-43}$$

For HF, $\langle f_v \rangle = 0.66$. This quantity is the first moment of the distribution, but by itself it cannot specify $P(v'|E)$ until a linear-surprisal, maximal-entropy form is assumed for P. If such a form is assumed, then $\langle f_v \rangle$ can be used to find λ_v by means of a sum rule, such as that in equation 13-17.

By analogy with the discussion of equilibrium distributions in section 13.2.3, we can define an effective *temperature* characterizing the HF or DF product distribution:

$$T_v = \frac{E}{R\lambda_v} = \frac{1.4 \times 10^5 \text{ J mole}^{-1}}{(8.3 \text{ J mole}^{-1} \text{ K}^{-1}) \times (-7.2)} \approx -2{,}400 \text{ K}$$

A large negative temperature, such as that just found, indicates a *population inversion* in the product distribution. An inverted population can be used to produce a laser,[12] and indeed, the F + (H$_2$, D$_2$) reaction has been used to power high-energy chemical laser systems. Such a system can do thermodynamic work on any positive-temperature system to which it is coupled.

Many other applications of surprisal analysis to vibrational and rotational product distributions have appeared in the literature on kinetics[2,3]. Surprisal analysis of the rotational distributions of the OH product of the H + NO$_2$ reaction has been carried out by Kinsey and co-workers.[13] Faist and Levine[14] have applied the method to the distribution of electronic states of excited metal atoms in the exothermic reaction, M$_2$ + X → MX + M*. The prototypical H + H$_2$ reaction, discussed in chapter 9, has also been analyzed using these techniques.[15-17]

13.4   INFORMATION-THEORETICAL ANALYSIS OF ENERGY TRANSFER PROCESSES

We shall describe one further application of information theory, to *energy transfer processes,* and then mention several others briefly, giving references to the literature on the subject. Energy transfer processes are state-to-state events in which no net chemical change occurs, but in which energy is exchanged among vibrational, rotational, and translational degrees of freedom as a result of *inelastic* collisions (see chapter 8, Section 8.2). The most common types of energy transfer include the following:

	Example
Vibration-translation (V-T)	$HF(v = 1) + He \rightarrow HF(v = 0) + He$ The amount of energy ΔE_T transferred into translation is equal to $E_{vib}(v = 1) - E_{vib}(v = 0) = 47$ kJ/mole.
Rotation-translation (R-T)	$HF(J = 9) + He \rightarrow HF(J = 8) + He$ $\Delta E_T = 2$ BJ ≈ 5 kJ/mole.
Vibration-(rotation & translation) (V-R, T)	$HF(v = 1, J = 0) + He \rightarrow HF(v = 0, J = 8) + He$ In this case, $\Delta E_T = -\Delta E_{vib} + \Delta E_{rot} \approx (47 - 45)$ $= 2.5$ kJ/mole.
Vibration-vibration (V-V)	For harmonic oscillators, $\Delta E_T = 0$; but for real molecules, $\Delta E_T = 2\omega_e x_e$, which for HF ≈ 150 cm$^{-1} \approx 2$ kJ/mole.

13.4.1   Thermally Averaged Prior Rates

The new feature introduced in analyzing energy transfer processes is that we must now *average* a priori rates and probabilities over a distribution of initial energies. This becomes particularly important for *near-resonant* energy transfer processes, such as V-R, T and V-V, in which ΔE_T is on the order of or less than $k_B T$. In other words, upon specifying an initial temperature, we will have a Boltzmann distribution of collision energies at that temperature.*

The procedure for carrying out the desired averaging is quite straightforward.[18,19] Consider the general energy transfer process

$$AB(n) + M \rightarrow AB(n') + M$$

where n and n' refer to the internal state (vibrational, rotational, electronic, etc.) of the molecule before and after the inelastic collision. We define the "state-to-state" rate coefficient $k(n \rightarrow n'; T)$. Then, to find the total rate of collision removing molecules from state n, we sum over final states:

$$k(n; T) = \sum_{n'} k(n \rightarrow n'; T) \tag{13-44}$$

The transition probability is a conditional probability [cf. equation (13-33)]:

$$P(n'|n; T) = \frac{k(n \rightarrow n'; T)}{k(n; T)} \tag{13-45}$$

The mean energy transfer out of state n is simply the first moment of the P distribution:

$$\langle \Delta E_n \rangle = \sum_{n'} (E_{n'} - E_n)P(n'|n; T) \tag{13-46}$$

---

*This is not the case, however, in experiments with velocity-selected molecular beams.

To find the total inelastic collision rate, we must average the microscopic rates over all initial states of the molecule:

$$k(T) = \sum_n \rho(n, T) \sum_{n'} k(n \rightarrow n'; T)$$

or

$$k(T) = \sum_n \rho(n, T) k(n; T) \tag{13-47}$$

We can also define a bulk-averaged energy transfer:

$$\langle\langle \Delta E \rangle\rangle = \sum_n \rho(n, T) \langle \Delta E_n \rangle \tag{13-48}$$

This quantity is related to relaxation properties of the system, such as acoustic dispersion.

If the probability densities $\rho(n, T)$ in equations (13-47) and (13-48) are the Boltzmann equilibrium probabilities $\rho^{eq}(n, T)$, then the important *detailed balancing* condition applies:

$$\rho^{eq}(n, T) k(n \rightarrow n'; T) = \rho^{eq}(n'T) k(n' \rightarrow n; T) \tag{13-49}$$

This is just a consequence of the definition of equilibrium, $\dot{\rho}^{eq}(n, T) = 0$ for all n, T. Equation (13-49) can be written

$$\frac{k(n \rightarrow n'; T)}{k(n' \rightarrow n; T)} = \frac{\rho^{eq}(n'T)}{\rho^{eq}(n, T)} = \frac{g(n')}{g(n)} e^{-(E_{n'} - E_n)/k_B T}$$

For a simple V-T process in a diatomic molecule, this reduces to

$$\frac{k(0 \rightarrow 1)}{k(1 \rightarrow 0)} = e^{-\hbar \omega_0 / k_B T}$$

As for the reactive case, we first find the prior rate for the energy transfer process at a specified total energy $E = E_n + E_r$, which is proportional to the total density of states at the final translational energy:

$$k^0(n \rightarrow n'; E) = R \rho(n', E)$$

$$= R \rho_T(E_{T'})$$

$$= R \left(\frac{\mu^{3/2}}{2^{1/2} \pi^2 \hbar^3} \right) (E - E_{n'})^{1/2}$$

or

$$k^0(n \rightarrow n'; E) = R A_T (E - E_{n'})^{1/2} \tag{13-50}$$

The amount of energy transferred from internal degrees of freedom into translational energy is

$$\Delta E = E_n - E_{n'} = E_{T'} - E_T \tag{13-51}$$

We now have to carry out the thermal average over a distribution of collision energies. This can be expressed as

$$k^0(n \to n'; T) = \frac{\int k^0(n \to n'; E) e^{-E/k_B T} \rho(n, E) dE}{\int e^{-E/k_B T} \rho(n, E) dE}$$

or

$$k^0(n \to n'; T) = \frac{R \int \rho(n, E) \rho(n', E) e^{-E/k_B T} dE}{\int \rho(n, E) e^{-E/k_B T} dE} \tag{13-52}$$

To evaluate the foregoing integrals, we have to consider several specific cases. The most detailed possibility is full knowledge of the molecular vibrational and rotational states before and after the collision, as in the V-R, T case noted earlier:

$$AB(v, J) + M \to AB(v', J') + M$$

Making use of the RRHO approximation of equation (13-35) gives

$$E_{vJ} = \hbar\omega_0\left(v + \frac{1}{2}\right) + BJ(J + 1)$$

and we have

$$\rho(v, J; E) = A_T(2J + 1)(E - E_{vJ})^{1/2} \tag{13-53}$$

In many experiments, the rotational states are thermalized or not individually resolved. For such a V-T process.

$$AB(v) + M \to AB(v') + M$$

and we have, as previously discussed,

$$\rho(v; E) = A_T \sum_{J=0}^{J^*(v)} (E - E_{vJ})^{1/2}(2J + 1)$$

or

$$\rho(v; E) = \frac{A_T}{B\hbar c}\left(\frac{2}{3}\right)(E - E_v)^{3/2} \tag{13-54}$$

For the V-V case,

$$AB(v = n) + AB(v = m) \to AB(v = n') + AB(v = m')$$

and we have

$$\rho(nJ, mL; E) = A_T(2J + 1)(2L + 1)(E - E_{nJ} - E_{mL})^{1/2}$$

where J and L are the rotational angular momenta of the two molecules. Thus,

$$\rho(n, m; E) = \sum_{L=0}^{L^*(m)} \sum_{J=0}^{J^*(n)} \rho(nJ, mL; E)$$

$$= A_T\left(\frac{4}{15}\right)(E - E_n - E_m)^{5/2} \qquad (13\text{-}55)$$

Calculation of $k^0(T)$ from equation (13-52) involves integrals over products of $\rho(E)$ having the form of equation (13-53), (13-54), or (13-55), depending on the case under consideration. All of these integrals can be expressed in terms of modified Bessel functions.[20] The general form is

$$\int_0^\infty x^{\nu-1}(2y + x)^{\nu-1}e^{-x}dx = \pi^{-1/2}(2y)^{\nu-1/2}e^y\Gamma(\nu)K_{\nu-1/2} \qquad (13\text{-}56)$$

On examination, this can be seen to cover all of the possibilities noted if we let $x = E/kT$, $y = \Delta \equiv (E_n - E_{n'})/2k_BT$, and $\nu = 3/2$ for V-R, T, 5/2 for V-T, and 7/2 for V-V. $K_n(y)$ is called the *modified Bessel function of the second kind of order n*; the coefficient $\Gamma(\nu)$ is the *gamma function,* which is the analytic continuation of the factorial function, i.e., $\Gamma(n) = (n - 1)!$ for integer n.

The results for the three cases we have been considering are as follows:

$$V\text{-}R, T: k^0(vJ \to v,' J'; T) = RA_T(k_BT)^{1/2}\left(\frac{2}{\pi^{1/2}}\right)(2J' + 1)\Delta e^\Delta K_1(\Delta) \qquad (13\text{-}57)$$

$$V\text{-}T: \qquad k^0(v \to v'; T) = \frac{8RA_T(k_BT)^{3/2}}{3Bhc\pi^{1/2}} \Delta^2 e^\Delta K_2(\Delta)$$

$$\text{with } \Delta = \frac{E_v - E_{v'}}{2k_BT} \qquad (13\text{-}58)$$

$$V\text{-}V: \qquad k^0(nm \to n'm'; T) = \frac{32}{15} RA_T(k_BT)^{5/2}\pi^{-1/2}\Delta^3 e^\Delta K^3(\Delta)$$

$$\text{with } \Delta = \frac{(E_n + E_m) - (E_n - E_{m'})}{2k_BT} \qquad (13\text{-}59)$$

Let us examine the properties of these prior rates, with reference to Figure 13-2, which shows the modified Bessel functions $K_\nu(x)$. First, we see that $K_\nu(x)$ is a monotonically decreasing function of x for $x > 0$ and any ν. If we look at the behavior of the V-T prior rate, we see that, after dividing through by the Δ-independent factor, we have $k^0(v \to v'; T)/A^0(T) = \Delta \to 0$ ($v' = v$, or $k_BT \gg \hbar\omega_0$), and, using the asymptotic properties of Bessel functions,[20] we have

$$k^0(0; T) \to 2A^0(T) = T^{3/2}$$

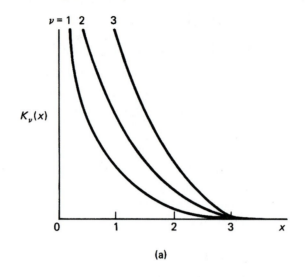

(a)

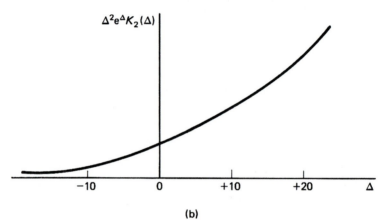

(b)

FIGURE 13-2   (a) Modified Bessel function $K_\nu(x)$ as a function of x, for $\nu = 1, 2,$ and 3. (b) Dependence of $k^0(v \to v'; T)/A^0(T) = \Delta^2 e^\Delta K_2(\Delta)$ on $\Delta = (E_v - E_{v'})/2kT$.

which is independent of Δ. Similarly, for $\Delta \to \infty$ $(k_B T \ll \hbar\omega_0)$, we have

$$k^0(\Delta; T) \to \left(\frac{\pi}{2}\right)^{1/2} A^0(T)\Delta^{3/2}$$

In other words, as is clear from Figure 13-2(b), the prior rate favors transferring as much energy as possible from internal motions into translation.

It is instructive to substitute $-\Delta$ for Δ in equation (13-58). Physically, this corresponds to vibrational activation $(T \to V)$ as compared with deactivation $(V \to T)$. Since K_2 is an even function,*

---

*In general,[20] $K_\nu(-\Delta) = (-1)^\nu K_\nu(\Delta)$.

$$\Delta^2 K_2(\Delta) = (-\Delta)^2 K_2(-\Delta) = |\Delta|^2 K_2(|\Delta|)$$

We can rewrite equation (13-58) in terms of the absolute value of Δ as

$$k^0(v \to v'; T) = A^0(T)|\Delta|^2 K_2(|\Delta|)\exp[|\Delta| + (\Delta - |\Delta|)]$$
$$= k^0(|\Delta|; T)\exp(\Delta - |\Delta|)$$

The purpose of doing this becomes evident if we examine the last factor for $\Delta > 0$ ($v > v'$, i.e., deactivation $V \to T$) and for $\Delta < 0$ ($v < v'$, i.e., activation $T \to V$). For $\Delta > 0$, $\Delta = |\Delta|$ and $\exp(\Delta - |\Delta|) = \exp(0) = 1$. For $\Delta < 0$, $\Delta = -|\Delta|$ and $\exp(\Delta - |\Delta|) = \exp(-2|\Delta|)$. So for processes in which $\Delta v = 1$,

$$\frac{k^0_{act}}{k^0_{deact}} = \frac{k^0(v \to v+1)}{k^0(v+1 \to v)} = \exp\left(-2\frac{E_{v+1} - E_v}{2k_BT}\right)$$
$$= e^{-\hbar\omega_0/k_BT}$$

and the correct detailed balancing relationship [equation (13-49)] is obtained. The same holds true for all the other prior rates as well.

13.4.2 Surprisal Analysis of Energy Transfer Processes

We next make the same maximal-entropy assertion as we did for the reactive case, namely,

$$k(n \to n') = k^0(n \to n')\exp\left[-\lambda_0 - \sum_{m=1}^{r} \lambda_m g_m(n, n')\right] \tag{13-60}$$

where r may indeed be equal to one (linear surprisal), but in principle may incorporate more than one independent piece of information. Now, what piece of information (i.e., what measure of the energy transfer) should we choose? The obvious one (for the V-T case) is $E_v - E_{v'} = 2k_BT\Delta$; since k^0 is symmetric in $|\Delta|$, the absolute value is especially convenient. Equation (13-60) then becomes

$$k(v \to v'; T) = k^0(v \to v'; T)\exp[-\lambda_0 - \lambda_v|E_v - E_{v'}|/k_BT - \ldots]$$
$$= k^0(\Delta; T)e^{-\lambda_0 - 2\lambda_v|\Delta|} - \ldots \tag{13-61}$$

and the surprisal is

$$I(v, v'; T) = -\ln\frac{k(v \to v'; T)}{k^0(v \to v'; T)}$$
$$= I_0 + \frac{\lambda_v|E_v - E_{v'}|}{k_BT} + \ldots$$
$$= I_0 + \frac{\lambda_v|v - v'|\hbar\omega_0}{k_BT} + \ldots \tag{13-62}$$
$$= I_0 + \Lambda_v|v - v'|$$

References 21 through 29 provide examples of a number of systems for which surprisal analysis has been applied to V-T and V-V energy transfer data. An example

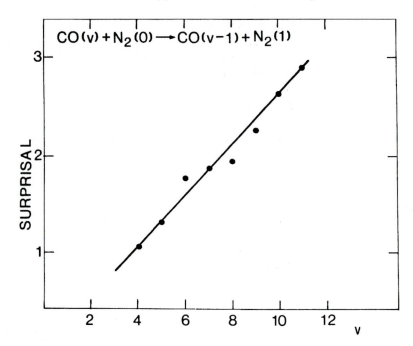

FIGURE 13-3 Surprisal plots for $CO(v) + N_2(0) \rightarrow CO(v-1) + N_2(1)$: experimental results from reference 29. The energy gap increases with v because of the increasing mismatch between the anharmonic CO vibrational levels and the N_2 fundamental [From Procaccia and Levine, *J. Chem. Phys. 63*, 4261 (1975)].

of such an analysis is shown in Figure 13-3, for the system $CO(v) + N_2(0) \rightarrow CO(v-1) + N_2(1)$.[19,29] A conclusion from these and numerous additional examples is that an "exponential gap law" [EGL, equations (13-61) or (13-62)] provides a good representation of vibrational energy transfer propensities. A large value of λ_v indicates a persistence of the original vibrational state, as is found in collisions with rare-gas atoms; small values of λ_v, on the other hand, are characteristic of a more nearly statistical distribution of final states, as may occur in reactive collisions. The reduced surprisal parameter, $\Lambda_v = \hbar\omega_v\lambda_v/k_BT$, is found to have values between 0.4 and 1.2 for many systems; this suggests that a simple scaling behavior can be applied to V-T and V-V processes.

Application of surprisal analysis to systems more complex than diatomic molecules presents additional difficulties, since the more complex energy-level expressions lead to correspondingly more complex expressions for the prior rate coefficients. The case of V-T deactivation of large polyatomic molecules is important in many areas of kinetics, such as activation and deactivation in unimolecular and infrared multiple-photon dissociation (see chapter 11). The central quantity to be determined is $P(E, E')$, the conditional probability that a molecule possessing initial energy E will still contain energy E' following an inelastic collision. This is equivalent to a "collision step size distribution", $P(\Delta E)$ with $\Delta E = E - E'$, from which an average collision step size $\langle \Delta E \rangle$ may be determined. Application of information theory to this problem, together with some simplifying approximations, yields closed-form expressions which are in reasonable agreement with other models for $P(E, E')$.[30] However, as noted in chapter 11, much work remains to be done on establishing the form of $P(E, E')$ and its dependence on collision partner, initial E, and temperature.

State-to-state rotational energy transfer (R—T) has also been analyzed using information-theoretical methods.[31] Application of the maximum-entropy principle, following the same procedure as for vibrational energy transfer, would also lead to an exponential-gap law,

$$k(J \to J'; T) = k^0(J \to J'; T) \exp\left[-\lambda_0 - \lambda_{rot}\left(\frac{\Delta E_{rot}}{k_B T} \right) \right] \qquad (13\text{-}63)$$

Extensive studies of R—T transfer show, however, that an angular-momentum-based scaling law, such as can be derived from the sudden approximation in scattering theory[3,32,33] gives a much better representation of state-to-state R—T rates. An example of such a comparison[34] is shown in Figure 13-4.

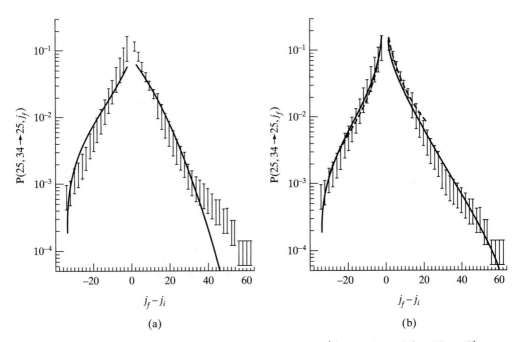

FIGURE 13-4   R—T energy transfer probabilities for $I_2^*(v_i = 25, j_i = 34) + \text{He} \to I_2^*$ $(v_f = 25, j_f) + \text{He}$ compared with (a) an exponential gap law (EGL), equation (13-63), and (b) a scaling law based on the energy-corrected sudden (ECS) approximation. It is clear that the EGL underpredicts the low-Δj and high-Δj probabilities, while the ECS provides a good representation throughout the entire range of Δj. Experimental data are from J. I. Steinfeld, and W. Klemperer, *J. Chem. Phys. 42*, 3475 (1965). Figure adapted with permission from J. I. Steinfeld, P. Ruttenberg, G. Millot, G. Fanjoux, and B. Lavorel, *J. Phys. Chem. 95*, 9638 (1991).

13.4.3   Additional Applications of Information Theory

Several additional applications of information theory—surprisal analysis will serve to demonstrate the versatility of the technique. Thus far we have shown that information theory can be used to reduce a large number of rate coefficients to a much smaller number of parameters which reflect constraints on the final-state distribution. This suggests an approach to simplifying the master equation previously discussed in chapter 9: namely, if individual rate coefficients are not arbitrary or independent, then it should be possible to bypass the rate coefficients and coupled differential equations entirely, using a maximum-entropy principle to compute time-dependent populations directly.

To carry out this prescription, we make use of a sum rule, and consider some bulk property A of a system evolving in time with time-dependent populations $P_v(t)$. The average value of A is

$$\langle\langle A \rangle\rangle = \sum_v A_v P_v \tag{13-64}$$

and both $\langle\langle A \rangle\rangle$ and P_v may be time-dependent. One example of A is the energy and v the harmonic-oscillator level number, but we are by no means restricted to that example. The evolution with time of $\langle\langle A \rangle\rangle$ can be written in terms of the normalized, time-dependent probabilities as

$$\frac{d}{dt}\langle\langle A(t) \rangle\rangle = \sum_v A_v \frac{dP_v(t)}{dt} \tag{13-65a}$$

$$= -\sum_v A_v \sum_{v'} P_v(t)k(v \rightarrow v') + \sum_v A_v \sum_{v'} P_{v'}(t)k(v' \rightarrow v) \tag{13-65b}$$

$$= -\sum_v P_v(t) \sum_{v'} A_v k(v \rightarrow v') + \sum_v P_v(t) \sum_{v'} A_{v'} k(v' \rightarrow v) \sum_{v'} \tag{13-65c}$$

$$= \sum_v P_v(t) \sum_{v'} [A_{v'} - A_v]k(v \rightarrow v') \tag{13-65d}$$

$$= \sum_v P_v(t) \frac{d\langle A_v \rangle}{dt} \tag{13-65e}$$

In going from equation (13-65a) to equation (13-65b), we used the basic master equation $dP_v(t)/dt = N^{-1}dn_v(t)dt$ (cf. chapter 9); equation (13-65c) is obtained by exchanging v and v' in the second set of terms. Equation (13-65e) serves to define $d\langle A \rangle/dt$ as $\sum_{v'}[A_{v'} - A_v]k(v \rightarrow v')$.

We now introduce a time-dependent surprisal, in analogy with equations (13-31) and (13-32):

$$I(v, t) = -\ln \frac{P_v(t)}{P_v^0(t)} \tag{13-66a}$$

$$= \sum_{m=1}^{r} \lambda_m(t)\langle A_v^m \rangle \tag{13-66b}$$

Here, $P_v^0(t)$ is the Boltzmann equilibrium distribution with a time-dependent temperature, i.e.,

$$P_v^0(t) = P(v|T(t))$$

$$= \frac{g_v e^{-E_v/k_B T(t)}}{\sum_v g_v e^{-E_v/k_B T(t)}} \qquad (13\text{-}67)$$

The time-dependent *entropy deficiency* of the distribution [cf. equation (13-10)] is

$$\Delta S[P(t)] = -R \sum_v P_v(t) I(v, t)$$

$$= \sum_v P_v(t) \ln \frac{P_v(t)}{P_v^0(t)} \qquad (13\text{-}68)$$

As before, we seek to find a set $\{P_v(t)\}$ which is consistent with the available information (in this case, bulk relaxation of an observable $\langle\langle A \rangle\rangle$) and which minimizes the entropy deficiency $\Delta S(t)$, i.e., which maximizes the rate of entropy production in the relaxing system. Note that all reference to individual state-to-state rate coefficients or transition probabilities has now disappeared.

We can do this by specifying a set of constraints, which may now be explicitly time-dependent:

$$\langle\langle A^r(t) \rangle\rangle = \sum_j A_v^r P_v(t)$$

The first, as always, is normalization:

$$A^0 = 1, \quad \langle\langle A^0 \rangle\rangle = 1 \qquad \text{(this is not time-dependent)}$$

The first nontrivial constraint may just be the average energy:

$$A^1 = E_v, \quad \langle\langle A^1 \rangle\rangle = \langle\langle V(t) \rangle\rangle$$

Higher moments of the distribution may be introduced if needed. We use the now familiar calculus of variations, setting up a variational function

$$\mathcal{L} = \Delta S + (\lambda_0 - 1)(1) + \frac{\lambda_1}{k_B T} \langle\langle V(t) \rangle\rangle + \cdots, \qquad \delta\mathcal{L} = 0$$

The Lagrange multipliers $\lambda_0(t)$ and $\lambda_1(t)$ are now time-dependent. The solution is clearly

$$P_v(t) = P_v^0(t) \exp[-\lambda_0(t) - \lambda_1(t) E_v/k_B T - \cdots] \qquad (13\text{-}69)$$

with the surprisal given by

$$I(v, t) = \lambda_0(t) + \lambda_1(t) \frac{E_v}{k_B T} + \cdots \qquad (13\text{-}70)$$

The temperature T must be specified. For a system of harmonic oscillators in contact with a "bath gas" at a bulk temperature T^{eq}, $T = T^{eq}$ should be used.

To find the $\lambda(t)$'s, we employ the same procedure as for individual rate coefficients; that is, we make use of a sum rule for the experimentally known quantity, to obtain

$$\frac{d}{dt}\langle\langle V(t)\rangle\rangle = \sum_v P_v(t) \sum_j (E_{v'} - E_v)k(v \to v')$$

$$= \sum_v P_v(t) \frac{d}{dt}\langle V\rangle$$

$$= \sum_v P_v(t)(\beta - \alpha E_v)$$

where $\beta/\alpha = \langle\langle V^{eq}\rangle\rangle = \langle\langle V(\infty)\rangle\rangle$ and α^{-1} is proportional to an experimental relaxation time. This can be clearly seen from

$$\frac{1}{\alpha}\frac{d}{dt}\langle\langle V(t)\rangle\rangle = \frac{\beta}{\alpha}\sum_v P_v(t) - \sum_v P_v(t)E_v$$

or

$$\frac{1}{\alpha}\frac{d}{dt}\langle\langle V(t)\rangle\rangle = \langle\langle V(\infty)\rangle\rangle - \langle\langle V(t)\rangle\rangle \qquad (13\text{-}71)$$

Integrating the simple first-order equation (13-71) gives

$$\int_{t_0}^t \frac{d\langle\langle V(t)\rangle\rangle}{\langle\langle V(t)\rangle\rangle - \langle\langle V(\infty)\rangle\rangle} = -\alpha \int_{t_0}^t dt$$

$$\ln\left[\langle\langle V(t)\rangle\rangle - \langle\langle V(\infty)\rangle\rangle\right]_{t_0}^t = -\alpha(t - t_0)$$

or

$$\langle\langle V(t)\rangle\rangle = \langle\langle V(\infty)\rangle\rangle + [\langle\langle V(t_0)\rangle\rangle - \langle\langle V(\infty)\rangle\rangle]e^{-\alpha(t-t_0)} \qquad (13\text{-}72)$$

But we also have the sum rule

$$\langle\langle V(t)\rangle\rangle = \frac{\sum_v E_v P_v(t)}{\sum_v P_v(t)} = \frac{\sum_v E_v e^{-\lambda_1(t)E_v/k_BT}}{\sum_v e^{-\lambda_1(t)E_v/k_BT}} \qquad (13\text{-}73)$$

Equating the right-hand sides of equations (13-72) and (13-73) now furnishes an algorithm for finding $\lambda_1(t)$ and thus $P_v(t)$. Every piece of information is now specified: we have the initial $\langle\langle V(t_0)\rangle\rangle$, the bulk temperature determines $\langle\langle V(\infty)\rangle\rangle$, and the experimental relaxation time determines α. Consequently, we can find $\lambda_1(t)$ at any time between $t = t_0$ and $t \to \infty$. The parameter $\lambda_0(t)$ is determined by normalization, i.e.,

$$\lambda_0(t) = \ln\left[\sum_v e^{-\lambda_1(t)E_v/k_BT}\right]$$

and $P_v(t)$ is calculated from equation (13-69). For a distribution of harmonic oscillators, this is identical to the exact analytic solution of the master equation; for other systems, this information-theoretical prescription for $P_v(t)$ is generally found to be indistinguishable from a numerical solution to the master equation. An example of the use of information theory to model V—V relaxation of an ensemble of harmonic oscillators is given in reference 35.

One further example may be mentioned, not involving collision-induced processes. This is spontaneous radiative decay of an electronically excited molecule. By considering the coupling of the excited states to a random distribution of background levels, Levine and co-workers have shown[36-38] that the distribution of radiative lifetimes τ is given by

$$P\left(\frac{\tau}{\tau(E)}\right) = \left(\frac{\tau(E)}{\tau}\right)^{(\nu/2)+1} \exp\left[\frac{\nu}{\tau} \bigg/ \frac{2}{\tau(E)}\right] \bigg/ \left(\frac{2}{\nu}\right)^{\nu/2} \Gamma\left(\frac{\nu}{2}\right) \qquad (13\text{-}74)$$

where $\tau(E) = 1/\langle 1/\tau \rangle$, ν is the number of decay channels, and Γ is the gamma function, as before. Equation (13-74) provides a very good representation of the measured decay times[39] for 356 rovibronic levels in the A^1B_1 state of SiH_2, for which lifetimes vary over nearly three orders of magnitude, from a few nsec to nearly 1 μsec.

13.5 CONCLUSION

We have seen how the judicious use of information theory, in the form of surprisal analysis and synthesis, can systematize a vast amount of molecular dynamics data and uncover correlations between seemingly unrelated phenomena. In many cases, though not all, selective energy consumption and specific energy release in state-to-state chemical reactions can be well represented by a single dynamical constraint superimposed on a statistical prior distribution. It is thus tempting to conclude that much of the vast amount of experimental and theoretical work in molecular dynamics described in earlier chapters has been unnecessary: since the results can be described by one- or two-parameter distributions, all that would have been necessary would be to measure one or two microscopic cross sections very accurately and then use surprisal synthesis to generate the rest of the data. It must be remembered, though, that information theory is not predictive, but is rather a methodology for systematizing and interpreting data. The experimental and theoretical results were essential for empirically establishing the validity of linear-surprisal representations, and instances where such a representation fails are certainly known. On the basis of this prior work, we can use information theory to organize and even predict dynamical results with a good degree of confidence, and innovative applications of the theory continue to be found. The bibliography lists several additional examples of such applications.

REFERENCES

[1] J. L. Kinsey, *J. Chem. Phys.* 54, 1206 (1971); A. Ben-Shaul, R. D. Levine, and R. B. Bernstein, *J. Chem. Phys.* 57, 5427 (1972).
[2] R. D. Levine, *Ann. Rev. Phys. Chem.* 29, 59 (1978).

[3] R. D. Levine and J. L. Kinsey, "Information Theory Approach to Molecular Collisions," in *Atom-Molecule Collision Theory: A Guide for the Experimentalist,* edited by R. B. Bernstein (New York: Plenum, 1978), pp. 693–750.

[4] W. Feller, *An Introduction to Probability Theory and its Applications,* 2d ed. (New York: Wiley, 1957).

[5] P. Franklin, *Methods of Advanced Calculus* (New York: McGraw-Hill, 1944), pp. 263–266.

[6] C. E. Shannon, *Bell Syst. Tech. J. 27,* 379 (1948); N. Wiener, *Cybernetics* (Cambridge, Mass.: M.I.T. Press, 1948).

[7] "Everything is determined even though freedom of choice exists." *Ethics of the Fathers* 3:19, Tractate *Nezikin,* Babylonian Talmud (ca. 200 B.C.E.).

[8] F. B. Hildebrand, *Methods of Applied Mathematics* (Englewood Cliffs, N.J.: Prentice-Hall, 1952), pp. 120–226.

[9] D. A. McQuarrie, *Statistical Thermodynamics* (Mill Valley, CA: University Science Books, 1985).

[10] K. P. Huber and G. Herzberg, *Constants of Diatomic Molecules* (New York: Van Nostrand Reinhold, 1979), pp. 304–305.

[11] A. Ben-Shaul, R. D. Levine, and R. B. Bernstein, *Chem. Phys. Letts. 15,* 160 (1964).

[12] J. Steinfeld, *Molecules and Radiation,* 2d ed. (Cambridge, MA: M.I.T. Press, 1984), pp. 308–313, 318–320.

[13] J. A. Silver, W. L. Dimpfl, J. H. Brophy, and J. L. Kinsey, *J. Chem. Phys. 65,* 1811 (1976).

[14] M. B. Faist and R. D. Levine, *Chem. Phys. Letts. 47,* 5 (1997).

[15] E. E. Marinero, C. T. Rettner, and R. N. Zare, *J. Chem. Phys. 80,* 4142 (1984).

[16] D. P. Gerrity and J. J. Valentini, *J. Chem. Phys. 81,* 1298 (1984).

[17] S. H. SuckSalk, C. K. Lutrus, and D. A. Reago, Jr., *Phys. Rev. A35,* 1074 (1987).

[18] M. Rubinson and J. I. Steinfeld, *Chem. Phys. 4,* 467 (1975).

[19] I. Procaccia and R. D. Levine, *J. Chem. Phys. 63,* 4261 (1975).

[20] M. Abramowitz and I. A. Stegun, *Handbook of Mathematical Functions* (New York: Dover Publications, 1965), pp. 374–379, 417–429.

[21] M. Rubinson, B. Garetz, and J. I. Steinfeld, *J. Chem. Phys. 60,* 3082 (1974).

[22] J. I. Steinfeld and W. Klemperer, *J. Chem. Phys. 42,* 3475 (1965).

[23] D. L. Thompson, *J. Chem. Phys. 60,* 455 (1974).

[24] J. M. White and D. L. Thompson, *J. Chem. Phys. 61,* 719 (1974).

[25] R. L. Wilkins, *J. Chem. Phys. 59,* 698 (1973).

[26] R. L. Wilkins, *J. Chem. Phys. 58,* 3038 (1973).

[27] G. Ennen and C. Ottinger, *J. Chem. Phys. 3,* 404 (1974).

[28] I. W. M. Smith, "Vibrational Relaxation in Small Molecules," in *Molecular Energy Transfer,* edited by R. D. Levine and J. Jortner (London: Wiley, 1975), pp. 85–113.

[29] G. Hancock and I. W. M. Smith, *Appl. Opt. 10,* 1827 (1971).

[30] C. C. Jensen, J. I. Steinfeld, and R. D. Levine, *J. Chem. Phys. 69,* 1432 (1978).

[31] R. D. Levine, R. B. Bernstein, P. Kabana, I. Procaccia, and E. T. Upchurch, *J. Chem. Phys. 64,* 796 (1976).

[32] A. DePristo, S. D. Augustin, R. Ramaswamy, and H. Rabitz, *J. Chem. Phys. 71,* 850 (1979).

[33] D. Secrest, "Rotational Excitation I: The Quantal Treatment," in *Atom-Molecule Collision Theory: A Guide for the Experimentalist,* edited by R. B. Bernstein (New York: Plenum, 1978), pp. 265–299.

[34] J. I. Steinfeld, P. Ruttenberg, G. Millot, G. Fanjoux, and B. Lavorel, *J. Phys. Chem. 95*, 9638 (1991).

[35] M. Tabor, R. D. Levine, A. Ben-Shaul, and J. I. Steinfeld, *Mol. Phys. 37*, 141 (1979).

[36] Y. Alhassid and R. D. Levine, *Phys. Rev. Letts. 57*, 2879 (1986).

[37] R. D. Levine, *Adv. Chem. Phys. 70*, 53 (1987).

[38] Y. M. Engel, R. D. Levine, J. W. Thoman, Jr., J. I. Steinfeld, and R. I. McKay, *J. Chem. Phys. 86*, 6561 (1987).

[39] J. W. Thoman, Jr., J. I. Steinfeld, R. I. McKay, and A. E. W. Knight, *J. Chem. Phys. 86*, 5909 (1987).

BIBLIOGRAPHY

Information Theory: Reviews, Fundamentals and Mathematical Techniques

ALHASSID, Y., and LEVINE, R. D., *J. Chem. Phys. 67*, 4321 (1977).

ALHASSID, Y., AGMON, N., and LEVINE, R. D., *Chem. Phys. Letts. 53*, 22 (1978).

AGMON, N., ALHASSID, Y., and LEVINE, R. D., *J. Comput. Phys. 30*, 250 (1979).

LEVINE, R. D., and BEN-SHAUL, A., "Thermodynamics of Molecular Disequilibrium," in *Chemical and Biochemical Applications of Lasers*, edited by C. B. Moore (New York: Academic Press, 1977), pp. 145–97.

Application to Reactive Processes

LEVINE, R. D., and BERNSTEIN, R. B., *Chem. Phys. Letts. 22*, 217 (1973).

LEVINE, R. D., and KOSLOFF, R., *Chem. Phys. Letts. 28*, 300 (1974).

LEVINE, R. D., and MANZ, J., *J. Chem. Phys. 63*, 4280 (1975).

POLLAK, E., and LEVINE, R. D., *Chem. Phys. Letts. 39*, 199 (1976).

LEVINE, R. D., "Energy Consumption and Energy Disposal in Elementary Chemical Reactions," in *The New World of Quantum Chemistry*, edited by B. Pullman and R. Parr (Dordrecht: Reidel, 1976), pp. 103–129.

POLANYI, J. C., and SCHREIBER, J. L., *Chem. Phys. 31*, 113 (1978).

Application to Electronically Nonadiabatic Processes

LEVINE, R. D., "The Formation of Electronically Excited Products in Chemical Reactions: An Information Theoretic Analysis," in *Electronic Transition Lasers*, edited by J. I. Steinfeld (Cambridge, Mass.: M.I.T. Press, 1976), pp. 251–267.

Application to Inelastic Collision Processes

ALEXANDER, M. H., *Chem. Phys. Letts. 40*, 267 (1976).

NANBU, K., *J. Chem. Phys. 66*, 136 (1977).

Application to Addition Reactions, Dissociation, and Ionization

ZAMIR, E., HAAS, Y., and LEVINE, R. D., *J. Chem. Phys. 73*, 2680 (1980).

ZAMIR, E., and LEVINE, R. D., *Chem. Phys. 52*, 253 (1980).

LAWRANCE, W. D., SILVERSTEIN, J., ZHANG, Fu-min, ZHU, Qing-Shi, FRANCISCO, J. S., and STEINFELD, J. I., *J. Phys. Chem. 85,* 1961 (1981).

SILBERSTEIN, J., and LEVINE, R. D., *Chem. Phys. Letts. 74,* 6 (1980).

Application to Nuclear Physics

LEVINE, R. D., STEADMAN, S. G., KARP, J. S., and ALHASSID, Y., *Phys. Rev. Letts. 41,* 1537 (1978).

PROBLEMS

13.1 In V-V processes such as

$$H + HF(v) \longrightarrow HF(v-1) + H$$

and

$$CO(v) + AB(0) \longrightarrow CO(v-1) + AB(1)$$

where AB = CO, N_2, or NO, linear surprisal plots are obtained, as shown in Figure 13-3. Using the equation

$$E_v = \left(\frac{v+1}{2}\right)\hbar\omega_e - \left(\frac{v+1}{2}\right)^2 \hbar\omega_e x_e$$

for energy levels of an anharmonic oscillator, show that the exponential gap law [equation (13-61)] predicts that a plot of the surprisal $I(v, v-1)$ vs. v will be linear. Find the effective λ_v values for the given examples. Is the use of the preceding energy-level equation inconsistent with the use of the RRHO approximation to derive equation (13-53)?

13.2 Derive the following expression for the prior rate for pure rotation-translation energy transfer where

$$\Delta = E_J - E_{J'}/2k_B T.$$

13.3 Show that the statistical power-gap scaling law for R—T energy transfer, given by the following equation, can be obtained by applying entropy maximization to $\ln|\Delta E_{\text{rot}}|$:

$$k(J' \to J'_f; E) \propto (2J' + 1)\left(\frac{E_{T'}}{E_T}\right)^{1/2} |\Delta E_{\text{rot}}|^{-\gamma}$$

C H A P T E R   1 4

Kinetics of Multicomponent Systems: Combustion Chemistry

14.1 INTRODUCTION

Thus far, we have considered individual elementary kinetic processes and mechanisms consisting of a small number of such processes. In the latter case, the mechanism was often simplified to the point at which closed-form analytic solutions for the species concentrations could be obtained. The emphasis throughout was on the details of the individual steps, such as energy consumption and disposal.

For many kinetically controlled systems, however, the number of coupled reaction steps is so large that simple analytic solutions are not possible. One source of such complexity arises from keeping track of internal states of the reacting species, resulting in the *master equation* discussed in chapter 9. Even more complex kinetic equations arise when a large number of interacting chemical species are involved. The objective of this and the following chapter is to illustrate how the fundamental principles of kinetics apply to important real-world systems. In this chapter, we will consider the kinetics of two relatively simple combustion systems, the hydrogen-oxygen reaction and the oxidation of methane. In chapter 15, we will discuss an important example from atmospheric chemistry.

14.2 THE HYDROGEN-OXYGEN REACTION, AN EXPLOSIVE COMBUSTION PROCESS

An example of a multicomponent kinetic system, consider the reaction between hydrogen and oxygen. Although starting with only two reactants, the complete mechanism involves eight major species and at least 16 reactions for its description. It is also an important example of a *branched-chain reaction* leading to an explosion; we shall analyze this feature using a simplified reaction set which includes five reactions. Further discussion of the hydrogen-oxygen system may be found in the texts by Espenson[1] and Emanuel' and Knorre.[2]

The overall reaction

$$2H_2 + O_2 = 2H_2O \tag{14-1}$$

is exothermic by 228.7 kJ/(mol H_2O); but mixtures of gaseous hydrogen and oxygen are quite stable, because the rate of the four-center $H_2 + O_2$ reaction, or any other conceivable direct reaction between the two gases, is zero. (The reaction half-time has been estimated to be much larger than the age of the universe.) If, however, the reaction is initiated by some free-radical species, reaction (14-1) proceeds very rapidly and indeed violently. The radicals are typically H and O atoms arising from the dissociation of H_2 and O_2, respectively; since these species are consumed in the reaction, the radicals should be considered not catalysts (see chapter 5), but rather initiators of the reaction.

A simplified mechanism which represents the behavior of the $H_2 + O_2$ system at low pressures is the following:

$$
\begin{array}{lll}
\textit{Initiation:} & & H_2 \xrightarrow{w_0} 2H \\[2mm]
 & & O_2 \xrightarrow{w_0} 2O \\[2mm]
\textit{Chain branching:} & (2) & H + O_2 \xrightarrow{k_2} OH + O \\[2mm]
 & (3) & O + H_2 \xrightarrow{k_3} OH + H \\[2mm]
\textit{Chain propagation:} & (1) & OH + H_2 \xrightarrow{k_1} H_2O + H \\[2mm]
\textit{Chain termination:} & (4) & H + \text{wall} \xrightarrow{k_4} 1/2\, H_2 \\[2mm]
 & (5) & H + O_2 + M \xrightarrow{k_5} HO_2 + M
\end{array}
\tag{14-2}
$$

In this mechanism, the initiation step is the dissociation of some amount of molecular species by a spark, flame, electric discharge, or other means; the step is characterized by a rate w_0 of free-radical production. To assess the contributions of steps 1 through 3, we must look at the enthalpy changes for these reactions. We can take the $O{=}O$ bond energy to be approximately 5.1 eV, while for $H{-}H$ it is 4.5 eV and for an $O{-}H$ bond it is 4.4 eV. Thus, reaction 2, which involves breaking an $O{=}O$ bond and making an $O{-}H$ bond, is 0.7 eV endothermic (about 70 kJ/mol) and progresses slowly. Reactions 3 and 1, on the other hand, involve breaking an $H{-}H$ bond and making an $O{-}H$ bond, so the enthalpy change is +0.1 eV, or about 9 kJ/mol endothermic. This energy is on the order of a few $k_B T$, and so these reactions are relatively fast. The OH and O radicals are therefore rapidly consumed by the reactions, and the principal *chain carrier* is H atoms, the number of which we can set equal to the total free-radical density n. Reactions 2 and 3 are called *chain-branching* reactions, because two radical species are produced for every one consumed; reaction 1 is a propagation reaction which generates one radical for each one consumed. The principal termination steps are loss of H atoms at the wall (reaction 4) and three-body recombination of H and O_2 to form HO_2 (reaction 5). The latter is a termination step because of HO_2 species (which has been identified spectroscopically in hydrogen-oxygen flames) is unreactive with either H_2 or O_2, and therefore cannot propagate the chain reaction.

The analysis proceeds by finding the rate equation for the free-radical density n, which is taken to be the same as that for [H] atoms:

$$\frac{dn}{dt} = \frac{d[\text{H}]}{dt}$$

$$= w_0 - k_2[\text{H}][\text{O}_2] + k_3[\text{O}][\text{H}_2] + k_1[\text{OH}][\text{H}_2] - k_4[\text{H}] - k_5[\text{H}][\text{O}_2][\text{M}] \quad (14\text{-}3)$$

The other free-radical species are governed by the equations

$$\frac{d[\text{OH}]}{dt} = k_2[\text{H}][\text{O}_2] + k_3[\text{O}][\text{H}_2] - k_1[\text{OH}][\text{H}_2] \quad (14\text{-}4)$$

and

$$\frac{d[\text{O}]}{dt} = k_2[\text{H}][\text{O}_2] - k_3[\text{O}][\text{H}_2] \quad (14\text{-}5)$$

Since [O] and [OH] are both much lower than [H], we can make the steady-state assumption for these species; we must *not* do so for [H], however, if we are to obtain a solution. Accordingly, we set

$$\frac{d[\text{OH}]}{dt} = \frac{d[\text{O}]}{dt} = 0$$

giving

$$[\text{O}]_{ss} = \frac{k_2[\text{H}][\text{O}_2]}{k_3[\text{H}_2]}$$

and

$$[\text{OH}]_{ss} = \frac{k_2[\text{H}][\text{O}_2] + k_3[\text{O}][\text{H}_2]}{k_1[\text{H}_2]}$$

$$= \frac{k_2[\text{H}][\text{O}_2] + k_3[\text{H}_2]k_2[\text{H}][\text{O}_2]/k_3[\text{H}_2]}{k_1[\text{H}_2]}$$

$$= \frac{2k_2[\text{H}][\text{O}_2]}{k_1[\text{H}_2]}$$

Combining all these relationships, we find, for the free-radical production rate,

$$\frac{dn}{dt} = w_0 - k_2[\text{H}][\text{O}_2] + k_3 \frac{k_2[\text{H}][\text{O}_2]}{k_3[\text{H}_2]}[\text{H}_2]$$

$$+ k_1[\text{H}_2] \cdot \frac{2k_2[\text{H}][\text{O}_2]}{k_1[\text{H}_2]} - k_4[\text{H}] - k_5[\text{H}][\text{O}_2][\text{M}]$$

$$= w_0 + (2k_2[\text{O}_2] - k_4 - k_5[\text{O}_2][\text{M}])n$$

so that

$$\frac{dn}{dt} w_0 + (f - g)n \quad (14\text{-}6)$$

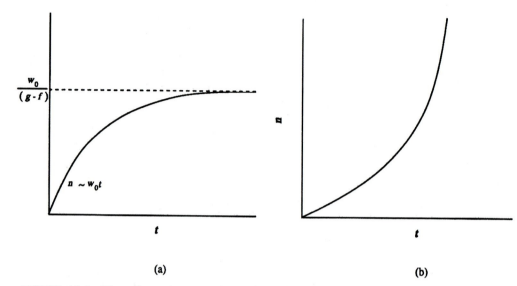

FIGURE 14-1 Time dependence of free-radical density n in the hydrogen-oxygen reaction. (a) below explosion limit, $g > f$. (b) above explosion limit, $g < f$.

In equation (14-6), the term $f = 2k_2[O_2]$ represents chain branching, which leads to an increase in the free-radical concentration, while the term $g = k_4 + k_5[O_2][M]$ represents chain termination, which removes free radicals.

Two classes of solutions of equation (14-6) for $n(t)$ are possible, depending on whether $g > f$ (termination exceeds branching) or, conversely, $g < f$. The former condition implies that $k_4 > (2k_2 - k_5[M])[O_2]$ so $g > f$ is guaranteed at sufficiently low O_2 pressures. The solution is then

$$n = \frac{w_0}{g - f}\left\{1 - e^{-(g-f)t}\right\} \tag{14-7}$$

At short times, n increases linearly with slope $w_0 t$ and reaches a steady-state value $n_{ss} = w_0/(g - f)$, as shown in Figure 14-1(a). At higher O_2 pressures $g < f$, and the solution in this case, viz.,

$$n = \frac{w_0}{f - g}\left\{e^{+(f-g)t} - 1\right\} \tag{14-8}$$

diverges, as shown in Figure 14-1(b). The free-radical concentration then increases exponentially, and, since the overall rate depends on the radical concentration, the reaction velocity also increases rapidly. This constitutes an explosion.

The hydrogen-oxygen reaction behaves quite differently in different pressure regimes, as shown in Figure 14-2. At low pressures (region I), the reaction is wall-recombination limited, and a steady reaction occurs. At a critical pressure p_1^*, at which f begins to exceed g, the branched-chain reaction takes over and explosion ensues (region II). This pressure, known as the *first explosion limit,* depends on the tempera-

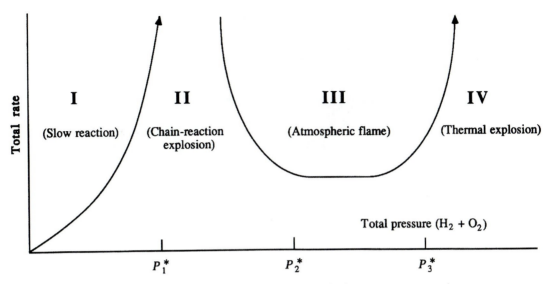

FIGURE 14-2   Explosion limits in the hydrogen-oxygen reaction.

ture, initial $[H_2][O_2]$ ratio, presence of free-radical scavengers, and other parameters. As the pressure is further increased, the explosion is quenched and another region of steady reaction is encountered (region III, atmospheric oxy-hydrogen flame). In this region, the kinetics are dominated by the relatively unreactive hydroperoxyl free radical.[3] Other reactions that are important in this region, in addition to those in the mechanism given by equations (14-2), include the following:

$$2HO_2 \longrightarrow H_2O_2 + O_2$$
$$H_2 + HO_2 \longrightarrow H_2O_2 + H$$
$$H + HO_2 \longrightarrow 2OH$$
$$\longrightarrow H_2 + O_2$$
$$\longrightarrow H_2O + O$$
$$H_2O_2 + M \longrightarrow 2OH + M$$
$$H + H_2O_2 \longrightarrow H_2O + OH$$
$$\longrightarrow O_2 + H_2$$
$$OH + H_2O_2 \longrightarrow H_2O + HO_2$$
$$H + OH + M \longrightarrow H_2O + M$$
$$H + H + M \longrightarrow H_2 + M$$

At a still higher pressure p_3^*, the amount of heat liberated in the exothermic steps of the mechanism becomes larger than can be dissipated by conduction and other thermal transport processes, and the temperature of the reaction mixture rises. This, in

FIGURE 14-3   (Top) Explosion of the hydrogen-filled dirigible *Hindenburg* in Lakehurst, New Jersey, May 1937. [UPI/Bettman Newsphotos.] (Bottom) Explosion of the space shuttle *Challenger* at Cape Canaveral, Florida, January 1986. [AP/Wide World Photos.]

turn, causes the rates of initiation and of the endothermic reactions to increase, liberating more heat, and so on, resulting in a *thermal explosion* (region IV).

We have noted that the hydrogen-oxygen reaction, once ignited, can proceed violently and explosively. This in fact has been the cause of several well-known major disasters. The hydrogen-filled dirigible *Hindenburg* was destroyed in 1937 by an atmospheric flame reaction (region III); newsreel photographs (see Figure 14-3(a)) show the ship burning but not actually exploding. The loss of the space shuttle *Challenger* in January, 1986 (Figure 14-3(b)), was most likely a thermal explosion (region IV), which occurred when the hydrogen- and oxygen-filled fuel tanks were ruptured and ignited by rocket exhaust penetrating a faulty O-ring seal.

14.3   THE METHANE COMBUSTION PROCESS

14.3.1   General Methane Oxidation Mechanism

The kinetics of the methane-oxygen system have been studied extensively over the years, because methane is the simplest hydrocarbon fuel and details of its combustion kinetics can be helpful in understanding the combustion of heavier hydrocarbon fuels. Further, methane is an important practical fuel which constitutes about 90% of the composition of natural gas. Like the hydrogen-oxygen system, the methane-oxygen system is complicated, involving at least 22 reaction steps and at least 12 species, depending on temperature and pressure. Also like the hydrogen-oxygen system, the methane-oxygen system is a branched chain reaction. However, before any branched chain reaction can take place, the reactions that generate radical intermediates must occur. This is not a facile process, since bonds must be broken. Methane oxidation is initiated by essentially two reactions, a hydrogen abstraction reaction from methane by oxygen, given by

$$CH_4 + O_2 \longrightarrow CH_3 + HO_2 \tag{14-9}$$

and a collision-induced dissociation of methane, represented by

$$CH_4 + M \longrightarrow CH_3 + H + M \tag{14-10}$$

The thermal $C-H$ bond fission described by equation (14-10) is slower for CH_4 than for any other hydrocarbon because the $C-H$ bond in CH_4 is stronger than that of most other hydrocarbons (e.g., 435 kJ/mol for methane as compared with 409 kJ/mol for n-propane or 385 kJ/mol for t-butane).

Following initiation, the reaction is propagated as a result of the following hydrogen abstraction reactions from CH_4 by radicals such as H, O, OH, and HO_2:

$$CH_4 + H \longrightarrow CH_3 + H_2$$
$$CH_4 + O \longrightarrow CH_3 + OH$$
$$CH_4 + OH \longrightarrow CH_3 + H_2O \tag{14-11}$$
$$CH_4 + HO_2 \longrightarrow CH_3 + H_2O_2$$

The methyl radicals produced by the reactions in the mechanism given by equations (14-11) are oxidized to formaldehyde (CH_2O) and formyl radical (HCO) and eventually form carbon monoxide (CO). This is accomplished by the reaction of methyl radicals with O and O_2:

$$CH_3 + O_2 \longrightarrow CH_3O + O$$
$$CH_3 + O_2 \longrightarrow HCO + H_2O$$
$$CH_3 + O \longrightarrow CH_2O + H$$

These reactions form three new species, HCO, CH_2O, and CH_3O, which also participate in the methane-oxygen reaction by the following steps:

$$CH_3O + O \longrightarrow CH_3 + O_2$$

$$CH_3O + O_2 \longrightarrow CH_2O + HO_2$$

$$CH_3O + M \longrightarrow CH_2O + H + M$$

$$CH_2O + O \longrightarrow HCO + OH$$

$$CH_2O + H \longrightarrow HCO + H_2$$

$$CH_2O + M \longrightarrow HCO + H + M$$

$$HCO + O \longrightarrow CO + OH$$

$$HCO + HO_2 \longrightarrow CO + H_2O_2$$

Clearly, once methyl radicals are produced, they react to oxidize to the final product, carbon monoxide (CO). The oxidation process is terminated by the recombination reactions

$$H + OH + M \longrightarrow H_2O + M$$

$$HCO + OH \longrightarrow CO + H_2O$$

and

$$HCO + H \longrightarrow CO + H_2$$

The net reaction mechanism is

$$CH_4 + \frac{3}{2}O_2 \longrightarrow CO + 2H_2O$$

The exothermicity of this reaction is -543.8 kJ/mol methane.

14.3.2   Low-Temperature Oxidation of Methane

The mechanism for low-temperature combustion of methane differs from the general mechanism just given in that a number of reactions decrease in importance because reactions with high activation energies become less significant at lower temperatures. For example, of the initiation reactions, the collision-induced dissociation given by equation (14-10) is two orders of magnitude faster than equation (14-9) at a temperature of 600 K. Consequently, the rate-limiting step at lower temperatures is the collision-induced dissociation of methane. After the initiation process, reaction is accelerated by radicals that enhance the rate of oxidation of methane to CO and H_2O. Among the ensuing radical-methane reactions, those which are most significant are

$$CH_4 + O \longrightarrow CH_3 + OH$$

and

$$CH_4 + OH \longrightarrow CH_3 + H_2O$$

since the activation energies for these two processes are about 8.4 and 31.9 kJ/mol, respectively. The latter reaction is the most important water-producing reaction in the methane combustion process. The other abstraction reactions produce little water at lower temperatures due to the fact that they have high activation energies. The principal reactions involving CH_3 are

$$CH_3 + O_2 \longrightarrow CH_3O + O$$

and

$$CH_3 + O \longrightarrow CH_2O + H$$

The former is more important in this mechanism than the reaction

$$CH_3 + O_2 \longrightarrow CH_2O + OH$$

Methoxyl radicals (CH_3O) react primarily by

$$CH_3O + M \longrightarrow CH_2O + H + M$$

with decomposition providing the major fraction of CH_2O. At lower temperatures the decomposition of CH_2O via the reaction

$$CH_2O + M \longrightarrow HCO + H + M$$

becomes insignificant because of the high activation energy about (339 kJ/mol). The most significant reaction of formaldehyde at lower temperatures is

$$CH_2O + O \longrightarrow HCO + OH$$

which has an activation energy of about 18 kJ/mol. This is followed by the production of CO by the reaction

$$HCO + O \longrightarrow CO + OH$$

which occurs at nearly gas-kinetic rates at low temperatures. The reaction responsible for the production of CO_2 is

$$CO + OH \longrightarrow CO_2 + H$$

This reaction has a low activation energy of 33 kJ/mol compared to 96 kJ/mol for the reaction of carbon monoxide with hydroperoxy radicals via

$$CO + HO_2 \longrightarrow CO_2 + OH$$

Consequently, the reaction of CO with OH is the most significant source of CO_2 in the combustion of methane at low temperatures.

Termination of the process results from the radical-radical reactions

$$HCO + H \longrightarrow CO + H_2$$

and

$$HCO + OH \longrightarrow CO + H_2O$$

Thus, the number of reactions that take place is significantly reduced from that of the initial simplified mechanism when combustion occurs at lower temperatures. The reduced lower temperature mechanism can be summarized as follows:

$$CH_4 + O_2 \longrightarrow CH_3 + HO_2$$

$$CH_4 + O \longrightarrow CH_3 + OH$$

$$CH_4 + OH \longrightarrow CH_3 + H_2O$$

$$CH_3 + O_2 \longrightarrow CH_3O + O$$

$$CH_3O + M \longrightarrow CH_2O + H + M$$

$$CH_2O + O \longrightarrow HCO + OH$$

$$HCO + O \longrightarrow CO + HO$$

$$CO + OH \longrightarrow CO_2 + H$$

$$O_2 + M \longrightarrow 2O + M$$

$$H + O_2 \longrightarrow OH + O$$

$$H + H_2O \longrightarrow H_2 + OH$$

$$H + OH + M \longrightarrow H_2O + M$$

$$HO_2 + O \longrightarrow OH + O_2$$

$$HO_2 + OH \longrightarrow O_2 + H_2O$$

$$HCO + H \longrightarrow CO + H_2$$

$$HCO + OH \longrightarrow CO + H_2O$$

As mentioned earlier in this section, combustion of methane at low temperatures is slow. Usually there is an induction period lasting several minutes, or even hours, depending on temperature and pressure. Following the induction period, there is an increase in the reaction rate, but under conditions of low pressure and temperature the reaction rate may reach a maximum and then decrease. Typical pressure-vs.-time curves, illustrated in Figure 14-4, show that for methane there is an induction period followed by acceleration and then a slowdown in the oxidation rate. During this induction period, the reactions which are most significant are the following:

$$CH_4 + M \longrightarrow CH_3 + H + M$$

$$H + O_2 \longrightarrow OH + O$$

$$CH_4 + O \longrightarrow CH_3 + OH$$

$$OH + CH_4 \longrightarrow CH_3 + H_2O \qquad (14\text{-}12)$$

$$CH_3 + O_2 \longrightarrow CH_3O + O$$

$$CH_3O + M \longrightarrow CH_2O + H + M$$

$$CH_3 + O \longrightarrow CH_2O + H$$

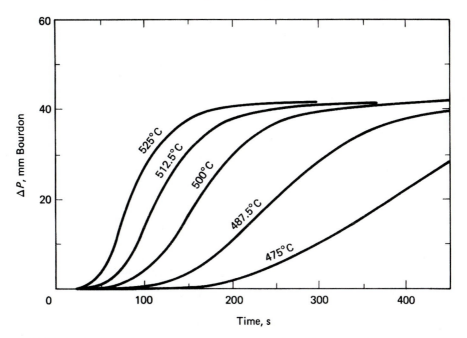

FIGURE 14-4 Typical pressure-vs.-time curves in the oxidation of CH_4, starting with 200 torr CH_4 and 100 torr O_3 in an HF-treated silica vessel (1.8 mm bourdon-1 torr). [From D. E. Hoare, "Low Temperature Oxidation," in *Combustion of Methane* (W. Jost, ed.) p. 125 (New York: Gordon and Breach, 1968). Reproduced with permission.]

Since the concentrations of CH_4, O_2, and OH are approximately constant during the induction period, the rate equations describing the kinetics reduce to a set of solvable linear differential equations. Also, since the species CH_3O, H, O, and OH are produced in small concentrations, the steady-state approximation may be used. The rate equations for the mechanism given by equations (14-12) may be written as follows:

$$\frac{d[CH_3O]}{dt} = k_5[CH_3][O_2] - k_6[CH_3O][M] \tag{14-13a}$$

$$\frac{d[H]}{dt} = k_1[CH_4][M] - k_2[H][O_2] + k_6[CH_3O][M] + k_7[CH_3][O] \tag{14-13b}$$

$$\frac{d[O]}{dt} = k_2[H][O_2] - k_3[CH_4][O] + k_5[CH_3][O_2] - k_7[CH_3][O] \tag{14-13c}$$

$$\frac{d[OH]}{dt} = k_2[H][O_2] + k_3[CH_4][O] - k_4[OH][CH_4] \tag{14-13d}$$

$$\frac{d[CH_3]}{dt} = k_1[CH_4][M] + k_3[O][CH_4] + k_4[CH_4][OH] - k_5[CH_3][O_2]$$
$$- k_7[CH_3][O] \tag{14-13e}$$

Applying the steady-state approximation to equations (14-13a–d), and simplifying the expression, we find that the rate equation for the methyl radical is

$$\frac{d[CH_3]}{dt} = 4k_1[CH_4][M] + 4k_5[CH_3][O_2] \tag{14-14}$$

and if $[CH_4]$, $[O_2]$, and $[M]$ are approximately constant, this expression can be integrated to give

$$CH_3]_{ss} = \frac{k_1[CH_4][M]}{k_5[O_2]}(e^{4k_5[O_2]t} - 1) + [CH_3]_0 e^{4k_5[O_2]t} \tag{14-15}$$

The rate of consumption of methane can be written as

$$\frac{d[CH_4]}{dt} = -4k_1[CH_4][M] - 5k_5[CH_3][O_2] - k_7[CH_3][O] \tag{14-16}$$

Assuming that $[CH_3][O]$ is negligible during the induction period, the last term in equation (14-16) can be ignored. Consequently, equation (14-16) simplifies to

$$\frac{d[CH_4]}{dt} = -4k_1[CH_4][M] - 5k_5[CH_3][O_2] \tag{14-17}$$

Substituting the steady-state expression for $[CH_3]$ into this equation, assuming that $[CH_3]_0 = 0$, and simplifying we obtain

$$\frac{d[CH_4]}{dt} = -k_1[CH_4][M] + \{5(e^{4k_5[O_2]t} - 1)\} \tag{14-18}$$

Integration of this expression gives the fractional methane loss as

$$\frac{[CH_4]_0 - [CH_4]}{[CH_4]_0} = 1 - \exp\left\{\frac{5k_1[M]}{4k_5[O_2]}\left(1 - e^{4k_5[O_2]t} + \frac{4}{5}k_5[O_2]t\right)\right\} \tag{14-19}$$

Defining the induction time t_i to be the time period during which 10% of the methane is consumed, we obtain the result

$$t_i = \frac{1}{4k_5[O_2]}\ln\left\{\frac{0.08k_5[O_2]}{k_1[M]} + 1 + \frac{5}{4}k_5[O_2]t_i\right\} \tag{14-20}$$

This equation may be solved iteratively for t_i. The expression is valid until combustion has progressed to a point at which radical concentrations become significant, so that the steady-state approximation is no longer valid.

14.3.3   Higher Temperature Oxidation of Methane

At elevated temperatures ($T > 700$ K), the combustion mechanism for methane differs considerably from the low-temperature oxidation process described. The process becomes rapid, resulting in the following major changes in the reaction mechanism that lead to an accelerated rate of oxidation of methane:

1. The rates of the initiation processes increase.
2. Chain branching steps become more important. Because of the elevated temperatures, reactions with high activation energies, such as

$$CH_3 + O_2 \longrightarrow CH_3O + O$$

$$CH_3O + M \longrightarrow CH_2O + H + M$$

and

$$CH_3 + O \longrightarrow CH_2O + H$$

accelerate the oxidation process by increasing the radical pool of [H] and [O] species that propagate branched chain oxidation. Another important chain-branching step is the abstraction of hydrogen from methane by HO radicals, given by

$$HO + CH_4 \longrightarrow H_2O + CH_3$$

3. At the higher temperatures, CO plays an important role as an intermediate in the oxidation process. The reaction of interest is

$$CO + HO \rightleftharpoons CO_2 + H$$

The left-to-right reaction enhances chain branching by increasing the source of H atoms; conversely, the right-to-left reaction acts to reduce chain branching. The additional CO also promotes termination via the reaction

$$O + CO + M \longrightarrow CO_2 + M$$

At higher temperatures, the radical concentrations can be quite large. The rate constants for radical-radical reactions, which have virtually zero activation energies, do not themselves change significantly, but because of the high radical concentrations the rates of these reactions may become appreciable. In fact, it is these reactions which give rise to C_2H_6 by recombination, i.e.,

$$CH_3 + CH_3 + M \longrightarrow C_2H_6 + M$$

The ethane, in turn, is oxidized by H, O, and OH radicals. A summary of the high-temperature mechanism for the oxidation of methane is shown in Figure 14-5.

14.3.4   Sensitivity Analysis

Sensitivity analysis was introduced in chapter 3 as a tool for analyzing complex kinetic systems. To demonstrate the usefulness of sensitivity analysis in combustion kinetics, consider a moderate-size model for the combustion of CH_4 in O_2–Ar mixtures.[4] In this system there are 13 species present and 23 rate constants; therefore, there is a total of 24 input parameters, 23 rate constants and one initial condition. Of interest is an analysis of the system near ignition, i.e., when $T \cong 2000°K$. The reactions and their rates used in this study are shown in Table 14-1. Integrating the system of kinetic rate equations

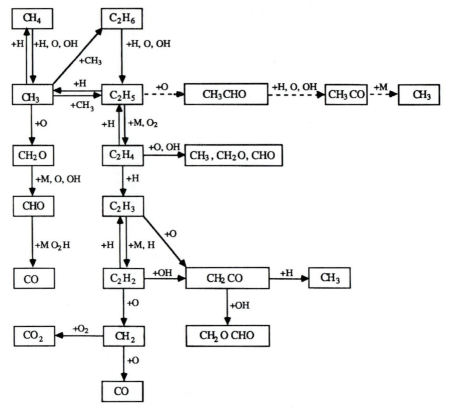

FIGURE 14-5   Mechanism of combustion of C_1/C_2 hydrocarbon species at high temperature. [After J. Warnatz, *Eighteenth Symposium (International) on Combustion* (Pittsburgh, PA: The Combustion Institute, 981), pp. 369–384, and "Chemistry of Stationary and Nonstationary Combustion," in *Modeling of Chemical Reaction Systems,* edited by K. H. Ebert, P. Deuflhared, and W. Jager (Berlin: Springer-Verlag, 1981), pp. 162–188.]

for the set of reactions given in Table 14-2 up to 10^{-2} sec, we obtain the concentration profiles shown in Figure 14-6 or the initial concentrations

$$[CH_4] = 2.265 \times 10^{18} \text{ molecules/cm}^3$$
$$[O_2] = 4.530 \times 10^{18} \text{ molecules/cm}^3$$
$$[Ar] = 1.790 \times 10^{19} \text{ atoms/cm}^3$$
$$[CO_2] = [H_2] = 1.0 \text{ molecules/cm}^3$$

The concentrations of all other intermediates and products are set to zero. As in the example discussed previously in chapter 3, the sensitivity coefficients vary with time, so in this example we look at sensitivity information at $t = 2.5 \times 10^{-5}$ sec; this corresponds to the time at which most transient species in the reaction have reached their maximum values (see Figure 14-6). Using the Green's Function Method and solving the simultaneous differential equations (see chapter 3, section 3.3.2), we find the reactions which

TABLE 14-1 Methane Oxidation Reaction Kinetics.

Reaction	Forward rate constant[a]	Forward rate constant at 2,000 K	Reverse rate constant at 2,000 K
(1) $CH_4 + M \longrightarrow CH_3 + H + M$	$3.32 \times 10^{-7} \exp(-44,500/T)$	7.2	3.05
(2) $CH_4 + OH \longrightarrow CH_3 + H_2O$	$9.96 \times 10^{-10} \exp(-6,290/T)$	4.3	1.74
(3) $CH_4 + H \longrightarrow CH_3 + H_2$	$3.72 \times 10^{-20} T^2 \exp(-4,400/T)$	3.3	1.34
(4) $CH_4 + O \longrightarrow CH_3 + OH$	$3.49 \times 10^{-11} \exp(-4,560/T)$	3.6	1.1
(5) $CH_3 + O \longrightarrow CH_2O + H$	1.66×10^{-10}	1.66	8.8
(6) $CH_3 + O_2 \longrightarrow CH_2O + OH$	3.32×10^{-14}	3.32	7.2
(7) $CH_2O + O \longrightarrow CHO + OH$	$8.30 \times 10^{-11} \exp(-2,300/T)$	2.63	1.5
(8) $CH_2O + OH \longrightarrow CHO + H_2O$	$8.97 \times 10^{-10} \exp(-3,170/T)$	1.84	1.42
(9) $CH_2O + H \longrightarrow CHO + H_2$	$2.24 \times 10^{-11} \exp(-1,890/T)$	8.7	7.0
(10) $CH_2O + M \longrightarrow CHO + H + M$	$6.64 \times 10^{-12} \exp(-18,500/T)$	6.38	5.4
(11) $CHO + O \longrightarrow CO + OH$	1.66×10^{-10}	1.66	1.9
(12) $CHO + OH \longrightarrow CO + OH$	1.66×10^{-10}	1.66	2.1
(13) $CHO + H \longrightarrow CO + H_2$	3.32×10^{-10}	3.32	4.6
(14) $CHO + M \longrightarrow CO + H + M$	$8.30 \times 10^{-12} \exp(-9,570/T)$	7.0	1.0
(15) $CO + OH \longrightarrow CO_2 + H$	$6.64 \times 10^{-12} \exp(-4,030/T)$	8.9	3.8
(16) $H_2 + OH \longrightarrow H + H_2O$	$4.82 \times 10^{-10} \exp(-5,530/T)$	3.0	3.0
(17) $H_2 + O \longrightarrow H + OH$	$5.31 \times 10^{-10} \exp(-7,540/T)$	1.2	1.1
(18) $H + O_2 \longrightarrow O + OH$	$3.65 \times 10^{-10} \exp(-8,450/T)$	5.34	2.3
(19) $H + OH + Ar \longrightarrow H_2O + Ar$	$2.31 \times 10^{-26} T^{-2}$	5.8	5.14
(20) $H + OH + H_2O \longrightarrow H_2O + H_2O$	$3.86 \times 10^{-25} T^{-2}$	9.65	8.56
(21) $H + HO_2 \longrightarrow OH + OH$	$4.15 \times 10^{-10} \exp(-950/T)$	2.58	1.0
(22) $H + O_2 + M \longrightarrow HO_2 + M$	$4.13 \times 10^{-33} \exp(500/T)$	5.32	3.6
(23) $OH + OH \longrightarrow H_2O + O$	$9.13 \times 10^{-11} \exp(-3,520/T)$	1.57	2.0

[a]Units: no. particles, cm^3, sec, K.
Source: A. A. Boni and R. C. Penner, *Combustion Sci. Tech.* 15, 99 (1976).

TABLE 14-2 Methane Oxidation Sensitivity Analysis, Time = 2.5×10^{-5} Sec.

Component	Most sensitive rate constants[a]									
CH_4	1	1	18	14	10	3	13	4	8	2
CH_3	6	1	14	5	8	2	13	4	18	10
H	1	6	14	5	8	2	13	4	18	10
OH	1	18	6	14	2	8	10	13	8	7
H_2O	1	6	18	3	10	14	4	5	13	7
H_2	1	6	14	10	8	18	2	5	13	12
CO	1	6	18	8	10	14	3	2	8	13
O	18	1	6	5	14	10	8	3	4	2
CO_2	1	6	18	15	2	10	14	3	8	5
O_2	1	6	18	10	3	14	13	4	5	12
CH_2O	6	8	1	2	5	18	7	14	10	13
CHO	6	1	18	14	13	8	10	12	3	4
HO_2	1	6	14	22	10	21	13	3	18	8

[a]Reaction numbers are given in Table 14-1. Reactions are listed in order of decreasing sensitivity from left to right.
Source: A. A. Boni and R. C. Penner, *Combustion Sci. Tech.* 15, 99 (1976).

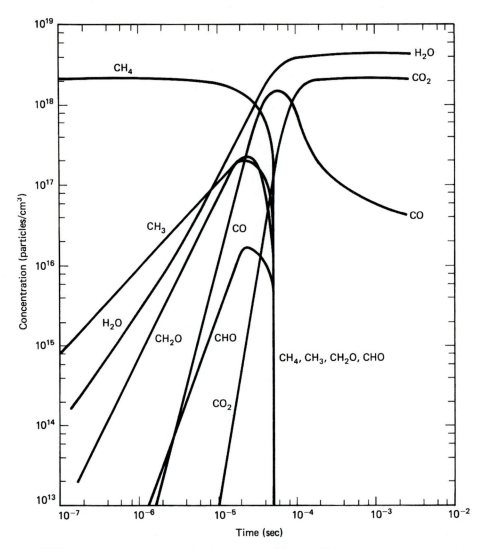

FIGURE 14-6 Calculated concentration profile for CH_4—O_2—Ar system at 2,000 K and 1 atm total pressure. (a) Profiles for CH_4, H_2O, CO, and hydrocarbon species. (*Continued*)

show significant sensitivities (see Table 14-2). The conclusion that can be drawn is that reactions 1, 6, 14, and 18 are the four most important reactions for all the species. Reaction 15 is important for the production of CO_2, and some reactions of lesser significance are 10, 3, 8, 13, 2 and 5; all other reactions are insensitive. As is plain from this example, the number of reactions that are needed to model the process can be reduced considerably. This becomes very useful in studying atmospheric and combustion processes, which may involve over 200 elementary reactions.

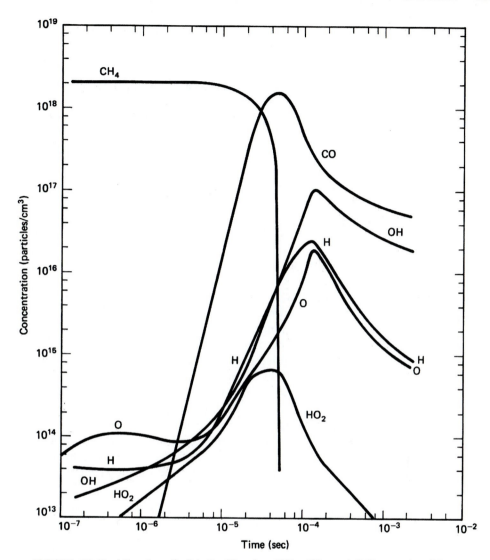

FIGURE 14-6 (*Continued*) (b) Profiles for CH_4, CO, and HO_x species. [From *Combustion Science and Technology 15*, 99 (1976).]

REFERENCES

[1] J. H. Espenson, *Chemical Kinetics and Reaction Mechanisms* (New York: McGraw-Hill, 1981), section 7-7.

[2] N. M. Emanuel' and D. G. Knorre, *Chemical Kinetics: Homogeneous Reactions,* transl. by R. Kondor (New York: J. Wiley and Sons/Halsted Press, 1973), section 8–4.

[3] M. Kaufman and J. Sherwell, *Progr. Reaction Kinetics 12,* 1 (1983).

[4] A. A. Boni and R. C. Penner, *Combustion Sci. Tech. 15*, 99 (1976).

CHAPTER 15

Kinetics of Multicomponent Systems: Atmospheric Chemistry

In the preceding chapter, we considered an example of a system comprised of a large number of coupled chemical reactions, namely, hydrogen and hydrocarbon combustion kinetics. In this chapter, we shall look at an even more complex coupled system, which is of considerable current interest: the kinetics of the Earth's atmosphere. Because of the vital importance of the atmosphere to the well-being—indeed the survival—of ourselves and the other species with whom we share our planet, a great deal of effort has been expended in recent years in understanding the processes which control its composition and how our activities are affecting this system. While we cannot, in a single chapter, provide a complete account of atmospheric chemistry (for that, see the excellent texts listed in the Bibliography), the quest to understand the atmosphere and predict its behavior provides an excellent example of the interaction between laboratory measurements, theory, environmental monitoring, and numerical models, and it is this aspect on which we will focus here. First, however, it is necessary to understand something about the physical processes governing the atmosphere, and this is described in the next section.

15.1 PHYSICAL STRUCTURE OF THE ATMOSPHERE

The earth's atmosphere is divided into regions primarily on the basis of temperature gradients, as shown in Figure 15-1. The *troposphere,* also referred to as the lower atmosphere, is characterized as the region where the air temperature falls with increasing altitude above the earth's sea level. The rate of temperature decrease in the troposphere with increasing altitude is given by the *adiabatic lapse rate,*

$$dT/dz \simeq -9.8°\text{C km}^{-1} \tag{15-1}$$

for dry air. The temperature reaches a minimum of around 210 K at the *tropopause.* The height of the troposphere varies from 8 km (25,000 ft) near the earth's poles to about 16 km (50,000 ft) in the tropics. Lying above the troposphere is the *stratosphere,* or upper atmosphere, which begins approximately 8 to 15 km above the earth's surface, depending on latitude and time of year, and extends to a height of around 50 km;

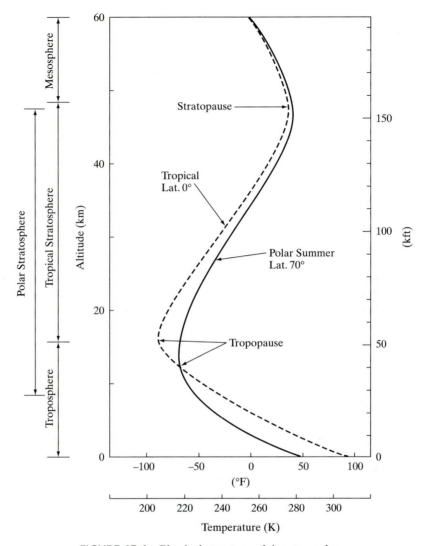

FIGURE 15-1   Physical structure of the atmosphere.

here the temperature rises to a maximum value of about 280 K. The reason for this increase in temperature will be given shortly. Both of these regions are illustrated in Figure 15-1. The stratospheric region is virtually cloudless and is characterized by slow vertical circulation because the dense cooler air at the lower altitudes does not readily rise. Atmospheric pressure in this region decreases almost exponentially with increasing altitude; in fact, the pressure falls by about a factor of 100 from the interface of the stratosphere and troposphere to the top of the stratosphere. At the stratopause the temperature passes through a maximum of about 270 K, while the temperature in the stratosphere increases to this maximum. Gases within the stratosphere are transported and mixed mainly by convection rather than by molecular diffusion. Consequently, the rate of transport of gases in this region becomes independent of the nature of the molecule, and settling out of heavier molecules due to gravity is negligible.

The chemistry involving most neutral species that enter the atmosphere takes place in two distinct regions: the troposphere and the stratosphere. Two important factors arise from the physical structure of these two regions which have important consequences for chemical reactions occurring in the atmosphere: first, the exponentially decreasing pressure generally results in a decrease in third-order recombination reaction rates; and second, the lower temperature in the upper troposphere and in the lower stratosphere reduces the rates of reactions that possess significant activation energies. Consequently, reactions with low activation energies are favored—in particular, recombination reactions with little or no activation energy.

15.2 CHEMICAL COMPOSITION OF THE ATMOSPHERE

The major constituents of the atmosphere, in terms of per cent composition, are molecular nitrogen and oxygen. Molecular nitrogen accounts for approximately 78% and molecular oxygen, 21%. The remainder consists of several inert gases such as argon and neon, and several trace gases. Water vapor is present in the lower regions of the atmosphere at highly variable concentrations, ranging from nearly zero in cold, arid regions to nearly 4% by volume in saturated, humid atmospheres. A summary of atmospheric composition is given in Table 15-1.

These abundances have remained remarkably stable over time. Despite the overwhelming abundance of the major species in the atmosphere, it is the minor constituents that play the major role in atmospheric chemistry. These include a number of species that are present in small but highly varying amounts, such as carbon dioxide, water vapor, and ozone. Because the atmosphere is one component of our planet's ecosystem, there is a continual exchange of gaseous constituents between the atmosphere, oceans, lakes, soil, and vegetation, as illustrated in Figure 15-2. As a consequence, gases are introduced into the atmosphere by emission from biological organisms, volcanic eruptions, and industrial processes. In addition to these exchanges, gases are consumed and produced by chemical transformations within the atmosphere itself. Most minor constituents arise from both natural and man-made (sometimes called anthropogenic) sources.

There are other minor atmospheric constituents that are produced in turn from the source gases. Many of these species are free radicals (for a discussion of free-radical kinetics, see sections 2.3.1 and 14.2) and play a disproportionately important role in the chemistry of the atmosphere. Radicals dominate the chemistry of the troposphere, even though their concentrations are typically only in the range of 10^6–10^7 molecules/cm^3, or 10^{-13} to 10^{-12} atm. The concentrations of some free-radical atmospheric species, such as OH, HO_2, H, and O, are shown in Figure 15-3.

15.3 PHOTOCHEMISTRY IN THE ATMOSPHERE

Most of the species present in the atmosphere are stable gases such as N_2, O_2, Ar, CO_2, and H_2O. The temperature in the lower region of the atmosphere varies between 210K and 280K. To a first approximation, then, there is no chemistry taking place in the atmosphere: the species listed in Table 15-1 are all mutually nonreactive at these temperatures.

TABLE 15-1   Major and Minor Constituents of the
                Lower Atmosphere.

Major components	Average abundance in the troposphere
N_2	78.1%
O_2	20.9%
Ar	0.93%

Minor components	
H_2O	10–40,000 ppm$_v$[a]
CO_2	332 ppm$_v$
Ne	18 ppm$_v$
CH_4	1.6 ppm$_v$
CO	0.06–0.2 ppm$_v$
H_2	0.5 ppm$_v$
N_2O	0.2 ppm$_v$
O_3	0.02–0.1 ppm$_v$

[a] Abundances of minor constituents are expressed in parts per million by volume (ppm$_v$), which is a *mixing ratio* of a particular species (ratio of moles of species to total number of moles of all gaseous species present, multiplied by 10^6), assuming ideal-gas behavior. 1 ppm$_v$ at a total pressure of 1 atmosphere corresponds to ca. 0.1 pascal. A mole fraction of 1% = 10,000 ppm$_v$.

How, then, does chemistry occur and how are free radicals produced? As we saw in chapter 3, the initiation step is a critical step in reaction chemistry, particularly among those involving free radicals. In the laboratory reactions may be initiated thermally, but at the low temperatures in the atmosphere, there is insufficient energy available to initiate free radical processes. The major process that initiates chemistry in the atmosphere is photodissociation of stable atmospheric species into reactive fragments upon absorption of solar radiation. The sun's output can be represented by a blackbody emission curve at T = 5900 K, with most of the emission lying between 300 nm and 2500 nm. The solar constant is the integral of this emission over all wavelengths,

$$\int_{all\,\lambda} (radiance)\ d\lambda = 1368\ W/m^2$$

$$= 2.0\ cal\ min^{-1}\ cm^{-2}$$

(15-2)

Species present in the atmosphere absorb solar radiation at various wavelengths. The absorption of solar radiation by species in the atmosphere follows the well-known Beer-Lambert absorption law,

$$I(\lambda) = I_0(\lambda)\exp(-\tau(\lambda))$$

(15-3)

where the intensity of solar radiation at wavelength λ at any point in the atmosphere is described by the solar irradiance at the top of the atmosphere, $I_0(\lambda)$, and the optical

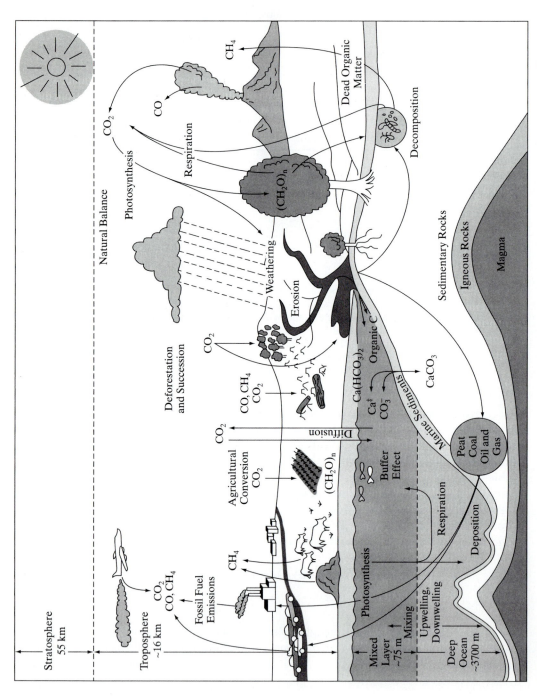

FIGURE 15-2   Exchange of carbon-containing gases between the biosphere, hydrosphere, geosphere, and atmosphere. (Redrawn with permission from Dr. Marguerite Schere, Jet Propulsion Laboratory, Pasadena, California).

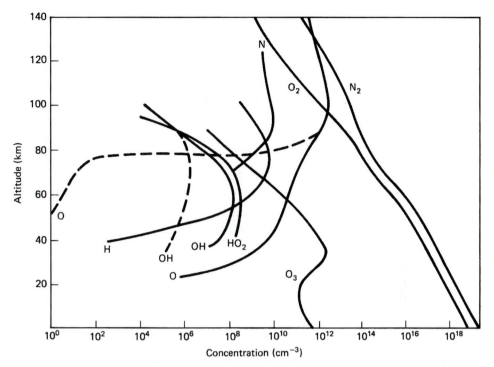

FIGURE 15-3 Concentration profiles of some atmospheric free radicals, along with major species (O_2, N_2) for reference. The solid line represents concentration profiles during the day, the dashed lines profiles at night.

depth, $\tau(\lambda)$. The optical depth is a function of the absorption characteristics of all the species absorbing at wavelength λ, viz. their absorption cross sections $\sigma_i(\lambda)$, and the species concentration at a given altitude $c_i(z)$:

$$\tau(\lambda) = \sum_i \sigma_i(\lambda) \int_{z_0}^{\infty} c_i(z) \, dz \tag{15-4}$$

In the process of absorbing solar radiation, photodissociation can cause the species to produce atoms or reactive free radicals. For example, photodissociation of molecular oxygen accounts for the production of oxygen atoms, since photons at wavelengths less than 242 nm are effective in the process

$$O_2 + h\nu(\lambda < 242 \text{ nm}) \longrightarrow O + O$$

Photodissociation can also lead to electronically excited products, e.g., in the photodissociation of ozone at wavelengths below 320 nm:

$$O_3 + h\nu(\lambda < 320 \text{ nm}) \longrightarrow O_2(^1\Delta_g) + O(^1D)$$

These processes will be discussed further in the following section.

Just as in the thermal initiation of a free-radical chain mechanism, this process occurs at some rate governed by a rate coefficient. This rate coefficient is customarily denoted by j, rather than k, for photodissociation processes. In the example for photodissociation of molecular oxygen, the rates of O_2 loss and O atom production are given by

$$d[O_2]/dt = -(1/2)\, d[O]/dt = -j_{O_2}[O_2] \tag{15-5}$$

j_{O_2} is the total oxygen photodissociation rate, given by

$$j_{O_2}\,(\text{sec}^{-1}) = \int \phi_{O_2}(\lambda)\, I(\lambda)\, \sigma_{O_2}(\lambda)\, d\lambda \tag{15-6}$$

where $\phi_{O_2}(\lambda)$ is the quantum efficiency for photodissociation, $I(\lambda)$ is the intensity of solar radiation at the point in the atmosphere where photodissociation is occurring, and $\sigma_{O_2}(\lambda)$ is the absorption cross section at wavelength λ. Figure 15-4 shows the variation of photodissociation rate with altitude for molecular oxygen. As one can see, photodissociation makes a much larger contribution to the chemistry at higher altitudes than at lower altitudes, because of the attentuation of the short-wavelength u.v. radiation as it passes through the atmosphere.

15.4 CATALYTIC CYCLES INVOLVING STRATOSPHERIC OZONE

The most important minor constituent from the point of view of the present discussion is ozone, which shields the earth from ultraviolet radiation that otherwise would reach the earth's surface. As shown in Fig. 15-3, the mixing ratio of ozone in the stratosphere is a strong function of altitude. If the ozone in a column of the atmosphere were collected and compressed to standard temperature and pressure,* it would form a layer only 3 to 4 mm thick. Most of the ozone is contained in a layer between 20 and 50 km above the earth's surface. This ozone layer is maintained as a result of a dynamic bal-

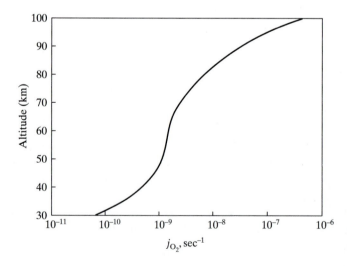

FIGURE 15-4 Total photodissociation rate as a function of altitude for molecular oxygen.

---

*1 atm at 25°C (STP).

ance between both chemical and photochemical formation and destruction processes for ozone.

15.4.1.   Chapman Mechanism for Ozone

The general mechanism for ozone formation and destruction derived from only oxygen species was originally suggested by Chapman.[1] In this mechanism, ozone formation occurs predominantly at altitudes above 30 km (see Figure 15-5), where solar ultraviolet radiation with wavelengths less than 242 nm dissociates molecular oxygen into oxygen atoms. The oxygen atoms rapidly react with molecular oxygen to form ozone in the presence of another molecule represented by M (this could be O_2 or N_2). The presence of M acts to stabilize ozone by removing its excess energy. The essentials of the process are represented by the following set of reactions:

$$O_2 + h\nu(\lambda < 242 \text{ nm}) \longrightarrow O + O \tag{15-7}$$

$$2[O + O_2 + M \longrightarrow O_3 + M]$$

$$\text{net}:3O_2 \longrightarrow 2O_3 \tag{15-8}$$

This mechanism accounts for the formation of ozone in the stratosphere; in the mesosphere (above 80 km), where three-body recombination processes are slower, additional mechanisms may be significant. The thermalization of absorbed u.v. energy, along with the exothermic recombination reaction (15-8) accounts for the increasing temperature in the stratosphere.

Ultraviolet radiation can also dissociate ozone in the longer wavelength region of 240–320 nm according to the reaction

$$O_3 + h\nu \longrightarrow O_2 + O \tag{15-9}$$

It is this reaction that is responsible for shielding the earth from harmful ultraviolet radiation. The photolysis of ozone is not, however, a true destruction mechanism, since almost all the oxygen atoms produced by this process rapidly combine with molecular oxygen to re-form ozone [equation (15-8)].

Several ozone destruction processes balance the formation of ozone [equations (15-7) and (15-8)]. One such process is the reaction of ozone with atomic oxygen to produce molecular oxygen:

$$O + O_3 \longrightarrow O_2 + O_2 \tag{15-10}$$

The ozone and oxygen are referred to as "odd oxygen species" in this scheme, and the reactions of both equation (15-10) and equation (15-7) are slower than the processes represented by equations (15-8) and (15-9), which merely interconvert odd oxygen. This can be illustrated by considering a simplified model using equations (15-7) through (15-10). At altitudes between 30 and 50 km, the stratosphere can be treated as a steady-state system of chemical reactions. Using a steady-state treatment leads to an estimation of the local concentrations of oxygen atoms and ozone. The rate of formation of [O] is given by

$$\frac{d[O]}{dt} = 2j_1[O_2] + j_3[O_3] - [O](k_2[O_2][M] + k_4[O_3]) \tag{15-11}$$

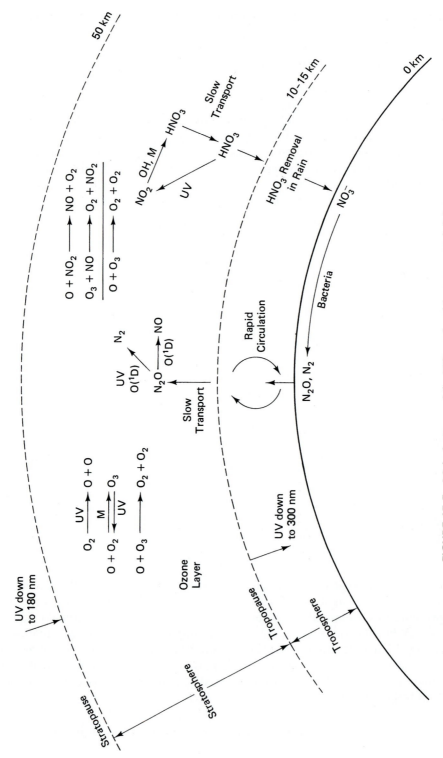

FIGURE 15-5   Main features of the NO_x cycle with sources and sinks.

where k_i is the chemical kinetic rate constant and j_i is the first-order rate constant for photolysis.

The rate of formation of ozone can be written as

$$\frac{d[O_3]}{dt} = k_2[O][O_2][M] - j_3[O_3] - k_4[O_3][O] \tag{15-12}$$

In the steady-state approximation,

$$\frac{d[O]}{dt} = \frac{d[O_3]}{dt} = 0 \tag{15-13}$$

Solving equation (15-11) for [O] gives

$$[O] = \frac{2j_1[O_2] + j_3[O_3]}{k_2[O_2][M] + k_4[O_3]} \tag{15-14}$$

and solving equation (15-12) for $[O_3]$ gives

$$[O_3] = \frac{k_2[O][O_2][M]}{j_3 + k_4[O]} \tag{15-15}$$

Substituting equation (15-14) into equation (15-15) and solving the resulting quadratic equation for $[O_3]$ gives

$$[O_3] = [O_2]\left(\frac{j_1}{2j_3}\right)\left\{\left(1 + 4\frac{j_3}{j_1}\frac{k_2}{k_4}[M]\right)^{1/2} - 1\right\} \tag{15-16}$$

The magnitudes of $[O_3]$ and $[O]$ may then be calculated from data for rate constants and values for j_1 and j_3. Values[†] for these parameters at 30 km are $j_1 = 10^{-12}$ sec^{-1}, $j_3 = 3 \times 10^{-4}$ sec^{-1}, and $k_4 = 10^{-15}$ cm^3 molecule^{-1} s^{-1}. The three-body recombination rate constant may be evaluated from the tabulated function $k_2 = 6 \times 10^{-34}$ $(T/300)^{-2.3}$, giving $k_2 = 2 \times 10^{-33}$ cm^6 molecule^{-2} s^{-1} at the temperature of the stratosphere. From the concentration profile shown in Figure 15-3, $[O_2]$ is estimated to be 10^{17} molecule cm^{-3}, and the total $[M] = [O_2] + [N_2] = 5 \times 10^{17}$ molecule cm^{-3}. With these estimates, we have $k_2[M]/k_1 = 0.6$ and $j_3/j_1 = 10^8$, which leads to the following simplification of equation (15-10):

$$[O_3] \approx [O_2] \cdot \left(\frac{j_1}{j_3}\frac{k_2}{k_4}[M]\right)^{1/2} \tag{15-17}$$

Evaluating equation (15-17) gives $[O_3] \approx 6 \times 10^{13}$ molecules/cm^3, and substitution of this into equation (15-14) gives $[O] \approx 10^{7.4}$ atoms/cm^3. Rearranging equation (15-15) then gives the following expression for $[O_3]$:

$$[O_3] = \frac{k_2[O_2][M]/k_4}{((j_3/k_4[O]) + 1)} \tag{15-18}$$

[†]Rate coefficients and photodissociation rates are taken from the compilation "Chemical Kinetics and Photochemical Data for Use in Stratospheric Modeling", Report JPL-97 (Jet Propulsion Laboratory, Pasadena, Calif.). A discussion of data sources in chemical and atmospheric kinetics may be found in Appendix 3 of this book.

and since

$$\frac{j_3}{k_4[O]} \approx \mathcal{O}(10^4) \gg 1$$

we have the result

$$\frac{[O_3]}{[O]} \approx \frac{k_2[O_2][M]}{j_3} \tag{15-19}$$

It can be seen from equation (15-19) that the reactions (15-8) and (15-9) serve merely to interconvert O and O_3. Furthermore, only reactions that destroy either O or O_3 without reproducing oxygen atoms or ozone will result in a direct reduction in $[O_3]$. In fact, the slowness of reaction (15-7) makes stratospheric ozone vulnerable to other removal processes. More precise estimates of the rate constants and more detailed calculations indicate that reaction (15-10) accounts for only 20% of the "odd oxygen" removal in the stratosphere, while other physical processes, such as the transport of ozone to the earth's surface, contribute only 0.5%. It can be seen from Figure 15-3 that the local density of ozone in the stratosphere is substantially less than the 6×10^{13} molecules cm^{-3} we estimated above. Therefore, other chemical processes must exist that contribute to more significant removal of ozone.[2,3] Several of these are discussed next.

15.4.2   HO$_x$ Cycle

In the upper stratosphere hydrogen-containing species such as H, OH, and HO_2 participate in ozone removal processes. The importance of these species in reactions for ozone removal was first recognized by Bates and Nicolet.[4] The primary source of the species, which forms the basis of the HO$_x$ cycle, is from the reaction of excited atomic oxygen atoms ($O(^1D)$) with hydrogen-containing species such as naturally occurring water vapor (H_2O) and methane (CH_4). $O(^1D)$ atoms are produced by the photolysis of ozone at wavelengths less than 310 nm according to the following reactions:

$$O_3 + h\nu(\lambda < 310 \text{ nm}) \longrightarrow O(^1D) + O_2 \tag{15-20}$$

$$O(^1D) + H_2O \longrightarrow OH + OH \tag{15-21}$$

$$O(^1D) + CH_4 \longrightarrow CH_3 + OH \tag{15-22}$$

The chief loss mechanism for $O(^1D)$ atoms is not by either of reactions (15-21) or (15-22), because water and methane are only minor components of the stratosphere, but rather through quenching by N_2 and O_2 via the reaction

$$O(^1D) + M \longrightarrow O(^3P) + M \tag{15-23}$$

Methane transported from the troposphere is chiefly responsible for water vapor in the stratosphere, which results from the oxidation of methane by hydroxyl radicals (OH·) according to

$$OH + CH_4 \longrightarrow H_2O + CH_3 \tag{15-24}$$

Other minor sources of OH radicals involve the photolysis of H_2O:

$$H_2O + h\nu(\lambda \le 190 \text{ nm}) \longrightarrow H + OH \tag{15-25}$$

The removal of HO_x species from the stratosphere is by downward transport of H_2O to the troposphere, where they are involved in precipitation—usually cloud droplets and aerosol particles.

The major catalytic cycles for ozone destruction involving HO_x species is attributed to the two-step reaction of hydroxy radicals with ozone and abstraction of oxygen atoms, the pair of which reactions regenerates oxygen as follows:

$$OH + O_3 \longrightarrow HO_2 + O_2 \qquad (15\text{-}26)$$

$$\underline{HO_2 + O \longrightarrow OH + O_2} \qquad (15\text{-}27)$$

$$\text{net:} \quad O + O_3 \longrightarrow 2O_2$$

In order for this overall reaction scheme to be effective, each reaction must be exothermic; rates for ozone destruction may then exceed those of the elementary reaction step (15-10), either because [OH] exceeds [O], or because the rate coefficient for reaction (15-26) exceeds that for reaction (15-10).

15.4.3 NO_x Cycle

The catalytic cycle involving nitric oxide (NO) and nitrogen dioxide (NO_2), collectively called NO_x, provides another important destruction process for ozone. Crutzen[5] first highlighted the importance of naturally occurring nitrogen oxides in the ozone removal process due to the following reactions:

$$NO + O_3 \longrightarrow NO_2 + O_2 \qquad (15\text{-}28)$$

$$\underline{NO_2 + O \longrightarrow NO + O_2} \qquad (15\text{-}29)$$

$$\text{net:} \quad O + O_3 \longrightarrow 2O_2$$

Clearly, this process produces the same net effect as the reaction (15-27), but the overall rate for the NO_x cycle is faster.

The main source of stratospheric NO_x is the reaction of $O(^1D)$ atoms with nitrous oxide:

$$O(^1D) + N_2O \longrightarrow 2NO \qquad (15\text{-}30)$$

N_2O is produced at the earth's surface by bacteria in soil and water. Although most of the nitrous oxide is photodissociated into N_2 and O, about 5% reacts with excited oxygen atoms ($O(^1D)$) to yield nitric oxide, thereby activating the NO_x cycle. The role of NO_x species is illustrated in Figure 15-5. There are several other minor sources of NO_x. Cosmic rays which enter the earth's atmosphere and ionization reactions involving N_2 also contribute to production of NO at high altitudes. NO_x is removed by reacting with hydroxyl radicals via the reactions

$$OH + NO_2 + M \longrightarrow HNO_3 + M \qquad (15\text{-}31)$$

and

$$HNO_3 + h\nu \longrightarrow OH + NO_2 \qquad (15\text{-}32)$$

Reaction (15-31) binds about 40% of the NO_x into HNO_3, which is inactive in regard to ozone removal and is transported downwards into the troposphere where it is rapidly removed by rain, thereby returning the nitrate to the soil.

Let us consider the effect of HO_x and NO_x species on the Chapman mechanism. To assess the role of these reactions, we are primarily interested in those reactions which will change the total odd oxygen concentration, $[O] + [O_3]$. The significant reactions are as follows for the simple mechanism envisaged here:

$$O_2 + h\nu \longrightarrow 2O \qquad (1)$$

$$O + O_2 + M \longrightarrow O_3 + M \qquad (2)$$

$$O_3 + h\nu \longrightarrow O + O_2 \qquad (3)$$

$$O + O_3 \longrightarrow 2O_2 \qquad (4)$$

$$NO + O_3 \longrightarrow NO_2 + O_2 \qquad (5)$$

$$NO_2 + O \longrightarrow NO + O_2 \qquad (6)$$

$$NO_2 + h\nu \longrightarrow NO + O \qquad (7)$$

$$OH + O_3 \longrightarrow HO_2 + O_2 \qquad (8)$$

$$O + HO_2 \longrightarrow OH + O_2 \qquad (9)$$

$$O + OH \longrightarrow H + O_2 \qquad (10)$$

The rate expressions for [OH], [H], and [NO] are

$$\frac{d[OH]}{dt} = -k_8[OH][O_3] + k_9[O][HO_2] - k_{10}[O][OH]$$

$$\frac{d[H]}{dt} = k_{10}[O][OH]$$

$$\frac{d[NO]}{dt} = -k_5[NO][O_3] + k_6[NO_2][O] + j_7[NO]$$

and the total odd oxygen concentration is given by

$$\frac{d([O] + [O_3])}{dt} = 2j_1[O_2] - 2k_4[O][O_3]$$

$$-k_5[NO][O_3] - k_6[NO_2][O] + j_7[NO_2]$$

$$-k_8[OH][O_3] - k_9[O][HO_2] - k_{10}[O][OH]$$

If steady-state conditions for [OH], [H], and [NO] are assumed, the total oxygen concentration can be rewritten as

$$\frac{d([O] + [O_3])}{dt} = 2j_1[O_2] - 2[O](k_4[O_3] + k_6[NO_2] + k_9[HO_2])$$

The magnitude of $k_4[O_3]$ is calculated to be on the order of the sum $k_6[NO_2] + k_9[HO_2]$.

15.4.4   ClO$_x$ Cycle

Many other substances can catalyze the destruction of ozone, but the HO$_x$ and NO$_x$ cycles constitute the main chemical and photochemical processes which determine the ozone concentration in the stratosphere. It is now known that there exists an additional mechanism for the depletion of the ozone layer, namely, photodissociation of halomethanes,[3,6] which are artificially produced substances such as CFCl$_3$ (Freon 11) and CF$_2$Cl$_2$ (Freon 12), as well as naturally occurring substances such as methyl chloride (CH$_3$Cl). These materials are inactive to reactions occurring in the troposphere and are transported into the stratosphere, where they are photolyzed according to the reaction

$$CCl_2F_2 + h\nu(180 \text{ nm} < \lambda < 220 \text{ nm}) \longrightarrow CF_2Cl + Cl \qquad (15\text{-}33)$$

or they react with electronically excited oxygen atoms (O(^{1}D)) to yield chlorine oxides (ClO$_x$):

$$CCl_2F_2 + O(^1D) \longrightarrow CF_2Cl + ClO \qquad (15\text{-}34)$$

These products then participate in catalytic cycles analogous to the HO$_x$ and NO$_x$ cycles whose net effect is the removal of ozone and oxygen atoms:

$$Cl + O_3 \longrightarrow ClO + O_2 \qquad (15\text{-}35)$$

$$\underline{ClO + O \longrightarrow Cl + O_2} \qquad (15\text{-}36)$$

$$\text{net:} \quad O + O_3 \longrightarrow 2O_2$$

The major removal process for Cl atoms is reaction with methane to form HCl, which is only slowly attacked by hydroxyl radicals, allowing transport of HCl to the troposphere where it is removed by rain via the reactions

$$Cl + CH_4 \longrightarrow HCl + CH_3 \qquad (15\text{-}37)$$

and

$$OH + HCl \longrightarrow H_2O + Cl \qquad (15\text{-}38)$$

15.4.5   Coupling of HO$_x$, NO$_x$, and ClO$_x$ Families

In the preceding sections, we introduced the individual catalytic cycles for the HO$_x$, NO$_x$, and ClO$_x$ families of radicals. These catalytic cycles, involving different families, do not operate in isolation, but interact with each other. The coupling between families can have catalytic consequences which can have important impacts on ozone chemistry in different regions of the atmosphere. This, in turn, can affect the distribution of each of the species within the atmosphere. Examples of reactions which couple the different radical families are presented in this section.

Important coupling reactions between the HO$_x$ and NO$_x$ families which impact the atmospheric distribution of HO$_x$ species (i.e., OH and HO$_2$) are:

$$HO_2 + NO \longrightarrow OH + NO \qquad (15\text{-}39)$$

$$OH + NO_2 + M \longrightarrow HNO_3 + M \qquad (15\text{-}40)$$

$$HNO_3 + h\nu \longrightarrow OH + NO_2 \qquad (15\text{-}41)$$

The reaction between HO_2 and NO generates OH radicals and plays a major role in the interconversion of HO_2 and OH. This in turn impacts the relative importance of the HO_x cycle in ozone destruction. Another important coupling involves the reaction between HO_2 and NO_2 via the reactions,

$$HO_2 + NO_2 + M \longrightarrow HO_2NO_2 + M \tag{15-42}$$

$$HO_2NO_2 + h\nu \longrightarrow HO_2 + NO_2 \tag{15-43}$$

$$HO_2NO_2 + OH \longrightarrow NO_2 + H_2O + O_2 \tag{15-44}$$

This reaction sequence impacts the NO_x cycle by sequestering NO_2 through its reaction with HO_2. The release of NO_2 results from photolysis of HO_2NO_2, but the photolysis lifetime is sufficiently long to permit the dispersion of HO_2NO_2 laterally throughout the stratosphere.

There are two important reactions that couple the HO_x and ClO_x families. These are

$$OH + HCl \longrightarrow H_2O + Cl \tag{15-45}$$

and

$$HO_2 + ClO \longrightarrow HOCl + O_2 \tag{15-46}$$

As mentioned above, a major removal process for atomic chlorine is the reaction of Cl with atmospheric CH_4 to yield HCl and methyl radicals. This reaction sequesters atomic chlorine in the form of hydrogen chloride. Because the reaction of HCl with OH radicals is slow, and HCl is water-soluble, HCl can be removed by rainout after being transported into the troposphere. However, as the OH concentrations increase at higher altitudes, the $OH + HCl$ reaction becomes effective in converting the chlorine tied up as HCl back into atomic chlorine. The net effect is to amplify the impact of the ClO_x cycle. At lower altitudes, the ClO_x cycle can link with the HO_x cycle via the following reactions[3,6]:

$$OH + O_3 \longrightarrow HO_2 + O_2 \tag{15-47}$$

$$Cl + O_3 \longrightarrow ClO + O_2 \tag{15-48}$$

$$HO_2 + ClO \longrightarrow HOCl + O_2 \tag{15-49}$$

$$\underline{HOCl + h\nu \longrightarrow OH + Cl} \tag{15-50}$$

$$\text{net:} \qquad 2O_3 \longrightarrow 3O_2$$

The reaction of HO_2 with ClO is central to this coupling chemistry.

Rapid reaction of ClO with nitric oxide is an important reaction cycle which links the NO_x to the ClO_x cycle. Two reactions play key roles in this coupling. These are:

$$ClO + NO \longrightarrow Cl + NO_2 \tag{15-51}$$

and

$$ClO + NO_2 + M \longrightarrow ClONO_2 + M \tag{15-52}$$

The first reaction influences the impact of the ClO_x catalytic cycle in the upper stratosphere. Other important reactions are:

$$Cl + O_3 \longrightarrow ClO + O_2 \tag{15-53}$$

$$ClO + NO \longrightarrow Cl + NO_2 \tag{15-54}$$

and
$$\underline{NO_2 + O \longrightarrow NO + O_2} \tag{15-55}$$

$$\text{net:} \quad O + O_3 \longrightarrow 2O_2$$

The second reaction, i.e., the reaction of ClO with NO_2, has a very interesting role. It terminates with the formation of a stable species: chlorine nitrate, $ClONO_2$. This species temporaily ties up ClO radicals in a nonreactive form. The photolysis of $ClONO_2$ regenerates the reactive species, hence the term, *temporary reservoir.* Once initiated, the longer catalytic cycle that results in ozone destruction involving NO_x and ClO_x radicals occurs by the following chain of reactions:

$$Cl + O_3 \longrightarrow ClO + O_2 \tag{15-56}$$

$$NO + O_3 \longrightarrow NO_2 + O_2 \tag{15-57}$$

$$ClO + NO_2 \longrightarrow ClONO_2 \tag{15-58}$$

$$ClONO_2 + h\nu \longrightarrow NO_3 + Cl \tag{15-59}$$

$$\underline{NO_3 + h\nu \longrightarrow NO + O_2} \tag{15-60}$$

$$\text{net:} \quad 2O_3 \longrightarrow 3O_2$$

In order to evaluate the combined effects of all these coupled cycles, a detailed atmospheric kinetic model must be run. In order to validate the model results, a coordinated suite of measurements of as many trace species as possible must be carried out and compared with model predictions. Section 15.7.1 describes the results of such a modeling and measurement campaign, which has provided the most detailed understanding to date of the contributions of the various catalytic families to ozone destruction in the stratosphere.

15.4.6   Homogeneous Reactions in the Polar Stratosphere

In the polar stratosphere, there is little atomic oxygen present. This is because at the two poles, the elevation of the sun is low, so the solar radiation flux is weak, resulting in a low photodissociation rate for molecular oxygen. A consequence of this physical constraint is a diminished role of catalytic cycles involving atomic oxygen. In the early 1980's, however, a significantly diminished abundance of ozone was detected over Antarctica. Early satellite observations of this reduction in ozone concentration were initially dismissed as instrumental artifacts, but when corroborating measurements were made using ground-based u.v. radiometry, it was realized that a substantial loss of ozone column density was occurring.[7] Balloon measurements subsequently indicated that ozone was being depleted at all altitudes. As shown in Figure 15-6, in mid-August 1987, the ozone partial pressure in the atmosphere over Antarctica peaked at ca. 170 nbar at 20 km altitude. By early October (the Antarctic spring), it had dropped to less than 40 nbar, resulting in a decrease in overall ozone

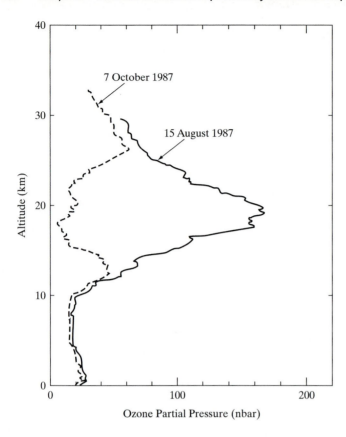

FIGURE 15-6 Balloon measurement of ozone concentration profiles in Halley Bay, Antarctica, during 1987. [From B. G. Malmstrom (ed.), *Nobel Lectures in Chemistry, 1991–1995* (Singapore: World Scientific, 1997), pp. 179–296. Reprinted with permission.]

column density of more than a factor of three. This allowed an enormous increase in the amount of damaging u.v. radiation from the sun to reach the surface; fortunately, the geographical area over which this was occurring was nearly devoid of human habitation.

It was clear that the sudden loss of ozone was associated with the increased solar flux during the Antarctic spring. However, the set of reactions discussed thus far was unable to reproduce such a large effect. In order to account for the chemistry, several additional ozone destruction mechanisms were proposed. One, suggested by Molina and co-workers,[8] involved the formation of a ClO dimer:

$$2\{Cl + O_3 \longrightarrow ClO + O_2\} \tag{15-61}$$

$$ClO + ClO + M \longrightarrow Cl_2O_2 + M \tag{15-62}$$

$$\underline{Cl_2O_2 + h\nu \longrightarrow 2Cl + O_2} \tag{15-63}$$

$$\text{net:} \quad 2O_3 + h\nu \longrightarrow 3O_2$$

McElroy and coworkers[9] suggested an additional mechanism similar to the reaction between two ClO radicals to produce ClO dimer, which involved bromine coupling with chlorine as follows:

$$Cl + O_3 \longrightarrow ClO + O_2 \tag{15-64}$$

$$Br + O_3 \longrightarrow BrO + O_2 \tag{15-65}$$

$$\underline{ClO + BrO \longrightarrow Cl + Br + O_2} \tag{15-66}$$

$$\text{net:} \qquad 2O_3 \longrightarrow 3O_2$$

The sources of atmospheric bromine are primarily man-made compounds such as halons (used to control electrical and aircraft fires) and methyl bromide used for fumigating agricultural produce; there is also a small "natural" methyl bromide contribution from marine organisms. The evidence that this process occurs is the observation of OClO in the Antarctic polar regions. The only known source of OClO in the atmosphere is the competing reaction channel from the ClO + BrO reaction, viz.:

$$ClO + BrO \longrightarrow Br + OClO \tag{15-67}$$

Molina's ClO dimer mechanism is believed to account for about 70% of the ozone loss, while McElroy's BrO + ClO mechanism accounts for the remainder. The measurement of a strong negative correlation between local ClO and ozone concentrations in the Antarctic polar vortex, described in section 15.6, was widely regarded as incontrovertible proof of the correctness of the basic ozone catalytic depletion mechanism. Even with these additional gas-phase reactions, however, it was not possible to completely describe the extent and time evolution of ozone destruction over Antarctica. Additional new chemistry had to be introduced, as described in the following section.

15.4.7   Heterogeneous Reactions in the Atmosphere

As mentioned in chapters 5 and 12, an important part of chemical kinetics involves reactions occurring at a gas-solid or gas-liquid interface. The two aspects of heterogeneous reactions with important applications to the chemistry of the atmosphere are reactions involving liquid cloud droplets and those involving stratospheric aerosols. The former is important because it is the principal means by which gas-phase species are removed from the atmosphere by rainout. How fast the atmosphere is able to cleanse itself is governed in large measure by how fast the gas-liquid interfacial chemistry can take place.

In the search for an understanding of the chemistry of the polar stratosphere, it was discovered that heterogeneous processes are critical and, in fact, are a main driver of the homogeneous chlorine catalytic chemistry.[10] In this case, gas-solid reactions are important. The solid surface is that of a crystalline ice particle containing mixtures of water and nitric acid which forms at temperatures above 195K. These ice crystals can form in polar stratospheric clouds at altitudes of about 20 km. The freezing out of nitric acid implies a loss of NO and NO_2, termed "denoxification". But the most important effect is that the surface of the ice crystals formed in polar stratospheric clouds can serve as reaction sites for heterogeneous gas-surface reactions. The key reaction is that between chlorine nitrate and hydrogen chloride. Both of these species are inert chlorine reservoirs. In principle, these species could react with each other in the gas phase via the homogeneous reaction,

$$ClONO_2 + HCl \longrightarrow HONO_2 + Cl_2 \tag{15-68}$$

Molecular chlorine formed from this reaction can then photodissociate with long-wavelength u.v. light to produce chlorine atoms:

$$Cl_2 + h\nu(\lambda < 450 \text{ nm}) \longrightarrow 2Cl \tag{15-69}$$

Hence chlorine is reactivated by this process. The homogeneous reaction (15-68) does not occur, however, because the reaction possesses a high activation energy barrier, leading to a small rate coefficient (an upper limit of 10^{-19} cm^3 molecule^{-1}sec^{-1} was measured at 298 K; at stratospheric temperatures of 230–250 K, the reaction will be negligibly slow). The presence of the ice surface, however, allows this reaction to occur efficiently:

$$ClONO_2 + HCl \xrightarrow{\text{ice surface}} HONO_2 + Cl_2 \tag{15-70}$$

It is the occurrence of reactions (15-69) and (15-70), as light becomes available during the onset of the Antarctic spring, that accounts for the rapid depletion of ozone in the polar stratosphere.

15.5   MODELING STUDIES OF THE ATMOSPHERE

One of the tasks of atmospheric modeling is to understand to what extent gases produced at the earth's surface affect the stratospheric ozone. To accomplish this, several pieces of information are needed: (1) the identity and amounts of these gases which are released into the atmosphere; (2) their rates of transport to the stratosphere; (3) the absorption and photolytic properties of these compounds in the stratosphere; (4) the rates of reaction in ozone destruction catalytic cycles; and (5) the location and nature of sinks for removal of these species and their end products from the atmosphere.

Atmospheric models deal with over 200 chemical reactions and more than 90 reactive species. Each time a reaction rate is remeasured, or a new reaction or new species is discovered, the whole interlocking system is affected and must be remodeled. Atmospheric models cover a wide range of sophistication and complexity. The simplest such model, called a "box model", is a spatially uniform, homogeneous volume of gas having no composition gradients, but with a net flow of species into and out of the box. The analysis is very similar to that presented for the flow reactor in chapter 3, namely, the variation of concentration in time for each species i is given by the continuity equation

$$\left(\frac{\partial c_i}{\partial t}\right)_{\text{box}} = \frac{u}{l}(c_i^o - c_i) + S_i - D_i + \left\{ \sum_{jl} k_{jl}c_jc_l + \sum_{l,\lambda} j_{l,\lambda} I(\lambda) \right.$$
$$\left. - \sum_{l'} k_{il'} c_i c_{l'} - j_{i,\lambda} I(\lambda) \right\} \tag{15-71}$$

where the first term represents the linear flow of material through the box, S_i and D_i are "source" and "deposition" terms representing exchange of gaseous species between the atmosphere and the Earth's surface, and the last set of terms incorporate the coupled chemical reactions (with rate coefficients k_{jl} and photodissociation rates $j_{l,\lambda}$. Pressure, temperature, flow velocity, initial concentrations, and source and deposition rates are specified, and an effective solar flux $I(\lambda)$, corresponding to the latitude, time of day, and season of interest, is used to estimate the photodissociation. The set of coupled differential equations is integrated forward in time using the methods described in chapter 2.

We know, however, that species concentrations vary strongly with altitude. This is addressed by using a "one-dimensional" model, which is essentially a stack of box models with vertical transport between the boxes, but no horizontal gradients. The equation for the 1-D model can be written as

$$\left(\frac{\partial c_i}{\partial t}\right)_{1-D} = \left(\frac{\partial c_i}{\partial t}\right)_{box} - \frac{\partial}{\partial t}\left(K_z\left(\frac{\partial c_i}{\partial z}\right)\right) \tag{15-72}$$

where z is the altitude and the transport terms K_z are complicated functions of species, altitude, etc. "Two-dimensional" models incorporate latitude variation as well, by means of additional terms:

$$\left(\frac{\partial c_i}{\partial t}\right)_{2-D} = \left(\frac{\partial c_i}{\partial t}\right)_{1-D} - \frac{\partial}{\partial t}\left(K_y\left(\frac{\partial c_i}{\partial y}\right)\right) \tag{15-73}$$

The most detailed models are "three-dimensional" models which include altitude, latitude, and longitude variation of the species of interest. An idea of the complexity of such models may be gained from the following. Let's say we want to resolve the lowest 50 km of the atmosphere into 5-km layers, which requires 10 vertical segments. To describe the 180° latitude variation from the North Pole (90° North) to the South Pole (90° South) in 20 degree sections would require 9 latitude segments. A similar resolution for the 360° of longitude would require 18 segments. Such a 3-D model would consist of $10 \times 9 \times 18 = 1,600$ "cells", each 5 km thick and 1300 miles on each side. In each of these cells, the 200 coupled kinetic equations all have to be solved, along with the transport terms. The variation of solar flux with altitude, latitude, longitude, and atmospheric composition also has to be modeled correctly in order to obtain accurate photodissocation terms. Calculations of this complexity are feasible for only the largest supercomputers; nevertheless, they are being carried out in order to gain an understanding of atmospheric chemistry and dynamics at the level necessary to make intelligent decisions about the consequences of emission of a wide range of substances into the Earth's atmosphere.

15.6 ATMOSPHERIC MEASUREMENTS

Although atmospheric chemistry models of great sophistication now exist, they rely for their validation on measurement of the species concentrations predicted by the model and the response of these concentrations to changing conditions. In assessing the models proposed for various reaction cycles, an essential step is the detection of species implicated in the mechanisms. Detecting the presence of various radicals in the atmosphere is the key to establishing its chemical structure. In parallel with measurements in the atmosphere itself, laboratory measurements of the elementary reaction rates for the many species implicated in proposed cycles makes it possible to assess the

TABLE 15-2 Measurement Methods for Atmospheric Trace Species.

Species	Expected abundance mixing ratio	Expected abundance number density	Measurement method
OH	1 ppt	$3 \times 10^6 \, \text{cm}^{-3}$	LIF
HO_2	3 ppt	5×10^6	$HO_2 + NO \rightarrow OH + NO_2$, then LIF of OH
NO	0.3 ppb	5×10^8	$NO + O_3 \rightarrow NO_2^* + O_2$ then chemiluminescence of NO_2^*
NO_2	0.5 ppb	8×10^8	modulated TDLAS cross checked by $NO_2 + h\nu \rightarrow NO + O$, then NO chemiluminescence detection
ClO	10 ppt	10^7	$ClO + O \rightarrow Cl + O_2$, Cl atom resonance fluorescence
BrO	a few ppt	10^6	$BrO + O \rightarrow Br + O_2$, Br atom resonance fluorescence
O_3			u.v. absorption
CO_2, CO, H_2O			IR absorption (TDLAS or FTIR)

reasonableness of complicated mechanisms describing certain regions of the atmosphere, which in turn allows concentration trends in the column levels of O_3 to be predicted. Studies aimed at species detection and rate constant measurement under laboratory conditions do, in turn, stimulate the development of experimental techniques for *in situ* measurements. Many of the techniques used in the laboratory and for *in situ* measurements are based on those discussed in chapter 3, such as laser induced fluorescence (LIF, section 3.2.2.4), chemiluminescence (section 3.2.2.3), and tunable diode laser absorption (TDLAS, section 3.2.2.2). Table 15-2 summarizes some of the measurement techniques employed for atmospheric trace gases, along with estimates of their expected abundance in the stratosphere. It can be seen that many of the important species are present at extremely low levels, placing severe demands on the sensitivity and reliability of the measurement method.

In order to obtain real-time, *in situ* concentration measurements, these instruments must be mounted on a suitable platform which travels through the air parcel to be monitored. High-altitude balloons, low-earth-orbit satellites, and most notably ER-2 aircraft[‡] are used for this purpose. Simultaneous measurements of ClO and ozone by instruments carried on an ER-2 flying though the polar vortex during the Antarctic Spring of 1987 provided clear and incontrovertible evidence[11] that stratospheric ozone was indeed being destroyed by a ClO_x catalytic cycle according to the mechanism discussed in sections 15.4.6 and 15.4.7; the strong anti-correlation of ozone and ClO concen-

---

[‡] The ER-2 aircraft is a non-military version of the U-2 "spy plane" which is operated by NASA for atmospheric observations.

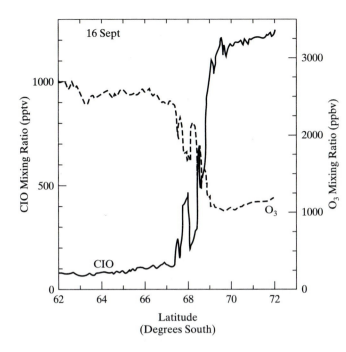

FIGURE 15-7   Mixing ratios of ClO and ozone measured in the Antarctic Polar vortex in September, 1987. [From J. G. Anderson, W. H. Brune, and M. H. Proffitt, *J. Geophys. Research 94,* 11465 (1989). Reprinted with permission.]

trations, called the "smoking gun" of the ozone depletion theory, is shown in Figure 15-7. These observations, along with model predictions, resulted in the international agreement to limit and eventually ban production of CFC's, discussed in section 15.7.2.

15.7   CURRENT UNDERSTANDING OF ATMOSPHERIC KINETICS

15.7.1   Relative Importance of the HO_x, NO_x, and ClO_x cycles

In the preceding sections, we have described (in simplified form) the chemistry which contributes to the global ozone balance. For many years, atmospheric scientists agreed that the NO_x catalytic cycle was the predominant ozone loss process in the middle and lower stratosphere, followed by the HO_x cycle and an uncertain (but crucially important) contribution from ClO_x and BrO_x acting together. A growing set of inconsistencies between observations and model predictions, along with the demand for improved scientific understanding in order to shape policy decisions, resulted in a coordinated modeling and measurement campaign, the Stratospheric Photochemistry Aerosols and Dynamics Expedition (SPADE) in 1993, with a follow-up campaign (Airborne Southern Hemisphere Ozone Experiment, or ASHOE) in 1994. The key feature of these campaigns was simultaneous, correlated measurements of trace atmospheric species using the techniques listed in Table 15-2, with all instruments packaged on the NASA ER-2 flying between 15° and 60° North latitude at altitudes up to 21 km. The results[12] were surprising: it was found that the HO_x cycle was the dominant mechanism, accounting for nearly 50% of the ozone removal rate in this region of the atmosphere. Halogen-radical photochemistry (most of which resulted from man-made, rather than natural halogen sources) was responsible for approximately one-third of

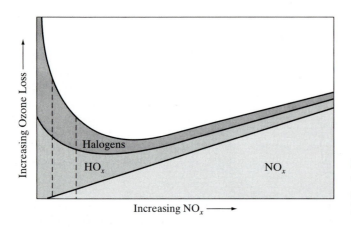

FIGURE 15-8   Ozone removal rate versus [NO_x]. [From P. O. Wennberg et al., *Science 266*, 398 (1994). Reproduced with permission.]

the ozone loss, with catalytic destruction by NO_x accounting for less than 20% of the ozone removal. The results of this study are summarized in Figure 15-8. In addition to demonstrating the importance of coordinated modeling and measurements, this study showed that in the lower stratosphere, at the present level of chlorine loading, additional NO_x input (as, for example, from supersonic aircraft flying at these altitudes) would actually decrease the net rate of ozone loss by tying up some of the active species participating in the HO_x and ClO_x cycles as inactive reservoir species.

15.7.2   Response to the Catalytic Ozone Destruction Issue

The idea that the release into the environment of relatively small amounts of seemingly benign compounds such as the CFC's could have profound effects on the chemical composition of the atmosphere was a bold and unproven hypothesis when first put forward by Molina and Rowland in 1973.[3] At about the same time, Harold Johnston[2] had warned of the possible deleterious effects of nitrogen oxide emissions from supersonic aircraft flying in the stratosphere, a concern which became generally recognized much later.[13] Since that time, an enormous effort in laboratory measurements, chemical-kinetic modeling, and atmospheric measurements has shown that the chemistry described in this chapter does indeed take place. As one result of these efforts, the international community agreed in 1987 to limit and eventually phase out completely the production, use, and release of compounds such as the CFC's. This agreement is known as the "Montreal Protocol on Substances that Deplete the Ozone Layer." Subsequent observations, most notably the Antarctic Ozone Hole and suggestions of a similar phenomenon occurring in the densely populated Northern Hemisphere, led to strengthening and accelerating this agreement in the "London" and "Copenhagen" amendments and implementations. The net result is that, as of the time this chapter is being written in the closing years of the 20th Century, emission of chlorine-containing compounds into the Earth's atmosphere has been sharply curtailed, the total chlorine loading in the atmosphere has leveled off and is expected to decrease during the next 50 to 100 years, and it is anticipated that stratospheric ozone levels will eventually return to their "natural", i.e., pre-CFC levels. Figure 15-9 shows the expected time development of this process.[14] The contributions of Professors Mario Molina, Sherry

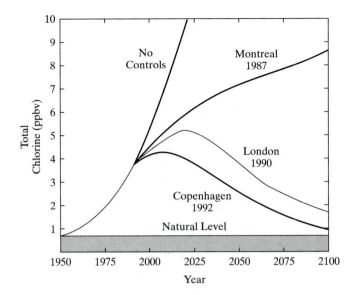

FIGURE 15-9 Total chlorine loading of the Earth's atmosphere, 1950–2100. The "natural level" corresponds to that resulting from methyl chloride generated in biological and hydrological processes. "No controls" is an extrapolation based on the CFC production rates prior to 1987. The "Montreal", "London", and "Copenhagen" curves correspond to the CFC emission caps mandated by each of these protocols, treaties, and amendments. [Adopted, with permission, from J. P. D. Abbatt and M. J. Molina, Ann. *Revs. Energy Environment 18,* 1(1993).]

Rowland, and Paul Crutzen to our understanding of the chemistry of the Earth's atmosphere, and more broadly for the insight that human activities can profoundly affect the global environment, were recognized by their sharing the Nobel Prize in Chemistry in 1995.[7]

15.8 CONCLUSION

In this chapter, we have tried to show how chemical kinetics can be used to understand an extremely complex problem of global importance. Many other kinetics-dominated problems in atmospheric chemistry will engage our concern in the years ahead. The consequences of flying high-speed aircraft in the lower stratosphere, an issue first raised by Johnston,[2,13] provided the motivation for the SPADE and ASHOE campaigns described in section 15.7.1. An even more complex problem is that of tropospheric photochemistry and air pollution, which directly affects the health of billions of people. Many more chemical species are involved in tropospheric chemistry, local variations occur on a much finer scale than in the stratosphere, and the sources of this problem (which involve energy production, transportation, and other large sectors of the economy) are much more difficult to solve than simply finding substitutes for CFC's. Nevertheless, the methods we have described here will continue to find use as these problems are addressed.

The atmosphere, just as all parts of the natural environment, operates as a complex system of coupled physical and chemical processes, with its own characteristic set of time constants. We have seen many instances in which such a system, when perturbed by a forcing function having a very different time constant, displays unexpected or even chaotic response. With the aid of the methods of chemical kinetics described in this book, scientists have been able to reach an understanding of how the atmosphere operates, and obtain reliable predictions of its response to human interference. Many questions still remain to be answered, however, before we achieve an understanding of the Earth's coupled ecosystem, including the atmospheric, biological, and hydrological systems illustrated in Figure 15-2, at the level which will be required to identify all of the potentially harmful effects we may be having on the environment, and to establish sustainable alternatives more consistent with our ecosystem's natural time constants.

REFERENCES

[1] S. Chapman, *Mem. Roy. Meteorol. Soc. 3,* 103 (1930).

[2] H. S. Johnston, *Ann. Rev. Phys. Chem. 26,* 315 (1975).

[3] M. J. Molina and F. S. Rowland, *Nature 249,* 810 (1974).

[4] D. R. Bates and M. Nicolet, *J. Geophys. Res. 55,* 30 (1950).

[5] J. P. Crutzen, *Quart. J. Roy. Meteorol. Soc. 96,* 320 (1970).

[6] F. S. Rowland and M. J. Molina, *Rev. Geophys. Space Phys. 13,* 1 (1975).

[7] B. G. Malmstrom (ed.), *Nobel Lectures in Chemistry, 1991–1995* (Singapore: World Scientific, 1997), pp. 179–296.

[8] L. T. Molina and M. J. Molina, *J. Phys. Chem. 91,* 433 (1987).

[9] M. B. McElroy, R. J. Salawitch, S. C. Wofsy, and J. A. Logan, *Nature 321,* 759 (1986).

[10] M. J. Molina, L. T. Molina, and C. E. Kolb, *Ann. Revs. Phys. Chem. 47,* 327 (1996).

[11] J. G. Anderson, W. H. Brune, and M. H. Proffitt, *J. Geophys. Research 94,* 11465 (1989).

[12] P. O. Wennberg, R. C. Cohen, R. M. Stimpfle, J. P. Koplow, J. G. Anderson, R. J. Salawitch, D. W. Fahey, E. L. Woodbridge, E. R. Keim, R. S. Gao, C. R. Webster, R. D. May, D. W. Toohey, L. M. Avallone, M. H. Proffitt, M. Loewenstein, J. R. Podolske, K. R. Chan, and S. C. Wofsy, *Science 266,* 398 (1994).

[13] H. S. Johnston, *Ann. Rev. Phys. Chem. 43,* 1 (1992).

[14] J. P. D. Abbatt and M. J. Molina, *Ann. Revs. Energy Environment 18,* 1 (1993).

BIBLIOGRAPHY

FINLAYSON-PITTS, B. J., and PITTS, J. N., JR. *Atmospheric Change: Fundamentals and Experimental Techniques.* New York: J. Wiley and Sons, 1986.

GOODY, R. *Principles of Atmospheric Physics and Chemistry.* Oxford: University Press, 1995.

GRAEDEL, T. E., and CRUTZEN, P. J. *Atmospheric Change: An Earth System Perspective.* New York: W.H. Freeman and Co., 1992.

MAKHIJANI, A., and GURNEY, K. R. *Mending the Ozone Hole.* Cambridge, Mass.: M.I.T. Press, 1995.

MCEWAN, M. J., and PHILLIPS, L. F. *Chemistry of the Atmosphere.* New York: J. Wiley and Sons, 1975.

NEWMAN, L. (Ed.) *Measurement Challenges in Atmospheric Chemistry. ACS Symposium Series No. 232.* Washington, D.C.: American Chemical Society, 1993.

SEINFELD, J. H., and PANDIS, S. N. *Atmospheric Chemistry and Physics.* New York: J. Wiley and Sons, 1998.

WAYNE, R. P. *Chemistry of Atmospheres.* Oxford: Clarendon Press, 1985.

PROBLEMS

15.1 Some reactions interconvert O and O_3, while other reactions carry out a net change in the total "odd oxygen" concentration, expressed as the sum $[O] + [O_3]$. Express $d([O] + [O_3])/dt$ in terms of the relevant rates of reactions in the mechanism, and evaluate the relative importance of the reaction $O + O_3 \rightarrow 2O_2$ and the reactions of the NO_x, HO_x, and CIO_x catalytic cycles in the destruction of O and O_3.

15.2 In this problem we shall examine the importance of various reaction steps in the dynamic atmospheric ozone balance. Consider the Chapman mechanism given by

$$O + O_2 + M \longrightarrow O_3 + M \qquad k_1 = 4.41 \times 10^{-33}\,[M]\ cm^6\ molecule^{-2}\ s^{-1}$$

$$O + O_3 \longrightarrow 2O_2 \qquad k_2 = 4.66 \times 10^{-16}\ cm^3\ molecule^{-1}\ s^{-1}$$

$$O_2 \longrightarrow 2O \qquad k_3 = 5.00 \times 10^{-11}\ s^{-1}$$

$$O_3 \longrightarrow O + O_2 \qquad k_4 = 2.50 \times 10^{-4}\ s^{-1}$$

At time $t = 0$, the initial concentrations are as follows:

$$[O] = [O]_0 = 1.0 \times 10^6\ atoms\ cm^{-3}$$

$$[O_3] = [O_3]_0 = 1.0 \times 10^{12}\ molecule\ cm^{-3}$$

$$[O_2] = [O_2]_0 = 3.7 \times 10^{16}\ molecule\ cm^{-3}$$

We assume that $[O_2]$ remains constant.

(a) Plot the evolution with time of $[O]$ and $[O_3]$ over the time range 1.0×10^{-2} to 10^4 hours. Comment on the changes in $[O]$ and $[O_3]$ over this time range.

(b) Determine sensitivity coefficients for this system and plot the evolution with time of the coefficients. What is the sensitivity of $[O]$ and $[O_3]$ to uncertainties in the rate constants? What are the most important rate constants?

(c) Suppose that the experimental uncertainties for each of the four independent rate constants are one order of magnitude, so that

$$k_i e^{-2.303} \leq k_i \leq k_i e^{2.303}, \quad i = 1, \ldots, 4$$

What is the sensitivity of $[O]$ and $[O_3]$ to these uncertainties in the rate coefficients? Suppose there is a two-order of magnitude uncertainty in k_2. What would your conclusions be then?

15.3 When NO and NO_2 are present in sunlight, ozone formation occurs as a result of the photolysis of NO_2:

$$NO_2 + h\nu \longrightarrow NO + O \qquad J_1 = \int_{sunlight} \sigma(\lambda)\phi(\lambda)I(\lambda)d\lambda \approx 0.0075\ sec^{-1}$$

$$O + O_2 + M \longrightarrow O_3 + M \qquad k_2 = 6.0 \times 10^{-34}(T/300)^{-2.3}\ cm^6\ molecule^{-2}\ sec^{-1}$$

Once formed, ozone reacts with NO to regenerate NO_2,

$$O_3 + NO \longrightarrow NO_2 + O_2 \qquad k_3 = 2.0 \times 10^{-12}\ exp(-1400/T)\ cm^3\ molecule^{-1}\ sec^{-1}$$

(a) Consider the dynamics of a system in which only these three reactions are taking place. Assume the initial concentrations of NO, NO_2 and ozone are $[NO]_0$, $[NO_2]_0$, and $[O_3]_0$, respectively. Write expressions for the rate of change of the concentrations of NO_2, ozone, and oxygen atoms once irradiation begins at constant temperature (300 K) and pressure (1 atm).

(b) Make a steady-state (ss) approximation for O atom concentration and find an expression for $[O]_{ss}$.

(c) Find a steady-state expression for the ozone concentration.

(d) Using the expression you found in (c), nitrogen atom conversion, and the stoichiometry of the O_3–NO reaction, find a relation giving the ozone concentration formed at steady state by irradiating an initial concentration $[NO_2]_0$ in the presence of excess O_2 (i.e., air); that is, assume $[NO]_0$ and $[O_3]_0$ are both equal to zero. Your final expression should include only $[NO_2]_0$ and the rate coefficients given above.

15.4 The chemistry of HONO in the atmosphere is not well understood, but there are some aspects which are known. The principal sources of HONO are the reactions

$$HO + NO + M \xrightarrow{k_1} HONO + M$$

and

$$HO_2 + NO_2 \xrightarrow{k_2} HONO + O_2$$

The processes for the removal of HONO are

$$HONO + h\nu \xrightarrow{j_1} HO + NO$$

and

$$HO + HONO \xrightarrow{k_3} H_2O + NO_2$$

(a) Calculate the steady-state concentration of HONO from the preceding mechanism.

(b) Couple the reactions given with the ozone-NO_x cycle discussed in the text, and determine the total odd oxygen concentration. What effect could this cycle have on the net depletion of ozone in the stratosphere?

15.5 In addition to describing the balance between formation and destruction of stratospheric ozone, the Chapman Mechanism [equations (15-7)–(15-10)] also predicts the altitude dependence of the ozone concentration, i.e., the ozone layer. Using the steady-state chemical model, the barometric formula for pressure vs. altitude, and the variation of solar flux with latitude, derive the Chapman formula for P, the rate of photon absorption in the atmosphere due to the oxygen–ozone cycle, viz.:

$$P = I_0 N_0 \sigma \cos\theta \exp[-(z/H) - N_0 \sigma H \sec\theta \, e^{-z/H}]$$

where I_0 is the solar flux at the top of the atmosphere, z is altitude, θ is latitude, N_0 is the sea-level atmospheric density, σ is the photon absorption cross section, and $H = RT/Mg$ is the barometric constant. Plot this function for several values of θ.

15.6 Bromine, similar to chlorine, is believed to play an important role in stratospheric ozone depletion. The largest anthropogenic source of bromine is methyl bromide (CH_3Br) which is used as an agricultural fumigant. Methyl bromide is broken down by u.v. light in the stratosphere, releasing active bromine atoms. Consider the following very simple model of how bromine reacts with ozone (O_3) in the stratosphere:

$$Br + O_3 \longrightarrow BrO + O_2 \qquad k_1 = 1.7 \times 10^{-11} \exp(-800/T) \, cm^3 \, molecule^{-1} s^{-1}$$

$$BrO + HO_2 \longrightarrow HOBr + O_2 \qquad k_2 = 2.5 \times 10^{-12} \exp(521/T) \, cm^3 \, molecule^{-1} s^{-1}$$

$$HOBr + h\nu \longrightarrow OH + Br \qquad j_3 = 1.0 \times 10^{-1} \, s^{-1}$$

$$OH + O_3 \longrightarrow HO_2 + O_2 \qquad k_4 = 1.6 \times 10^{-12} \exp(-940/T) \, cm^3 \, molecule^{-1} s^{-1}$$

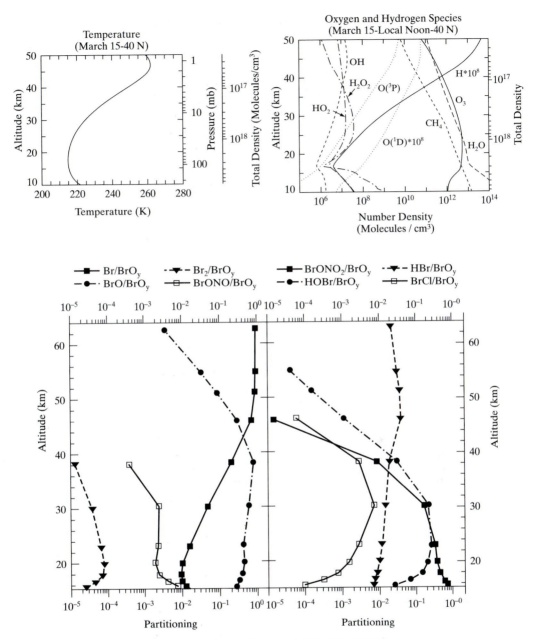

Source: Lary, D. J. *Journal of Geophysical Research 101*, 1505, (1996).
Note: BrO_y = total inorganic bromine. BrO_y = 20 ppt.

(a) Simulate these reactions at 25 km using the graphs shown above to determine the initial concentrations of all the species involved. Run the simulation for 1×10^6 seconds.
   (i) What is the net reaction if you add up reactions 1–4?
   (ii) Do reactions 1–4 lead to *catalytic* ozone depletion? Explain using the results of the simulation. (How many ozone molecules are destroyed per Br atom lost?)

       (iii) What percent ozone loss occures over 1×10^6 seconds? Using this amount, estimate what percent ozone loss would occur over a year? Over ten years? Discuss reasons why this predicted ozone loss is unlikely. What is missing from our simple chemical model?

       (iv) What is the rate limiting step in the model? Explain using the results of the simulation. Calculate the rates of all 4 reactions to confirm your answer.

(b) The U.S. is currently considering phasing our/banning methyl bromide. By how much would it be necessary to reduce the concentrations of all the bromine compounds in order to cut ozone depletion in half according to our simple model?

(c) Is the model an important source of ozone depletion at night? Explain. (Run the night simulation on a much shorter time scale, 20 seconds).

(d) Most reactions speed up at higher temperatures. Run the simulation assuming that T = 298 K. Letting the simulation run for 1×10^6 seconds, determine the percent of ozone depleted. Is more or less ozone depleted compared to the results from part a)? If there is less ozone depletion at 298 K explain this apparent contradiction of the first statement.

APPENDIX 1

Quantum Statistical Mechanics

According to the quantum mechanical Boltzmann distribution, the probability that a molecule has energy E_i is

$$P(E_i) = \frac{g_i \exp(-E_i/k_B T)}{\sum\limits_{i} g_i \exp(-E_i/k_B T)} \tag{A1-1}$$

where g_i is the degeneracy at energy E_i. The denominator in equation (A1-1) is the partition function Q, so $P(E_i)$ becomes

$$P(E_i) = \frac{g_i \exp(-E_i/k_B T)}{Q} \tag{A1-2}$$

If the energy is assumed to be continuous, then $W(E)$, which represents the number of states in the energy interval $E \rightarrow E + dE$, is equivalent to g_i. $W(E)$ is related to the density of states $N(E)$ via the equation $W(E) = N(E)\, dE$. Using this relationship, the probability that a molecule has energy in the interval $E \rightarrow E + dE$ is given by

$$P(E)\, dE = \frac{N(E) \exp(-E/k_B T)\, dE}{Q} \tag{A1-3}$$

where

$$Q = \int_0^\infty N(E) \exp(-E/k_B T)\, dE \tag{A1-4}$$

Besides $W(E)$ and $N(E)$, another important term is $G(E)$, which represents the total number of states having energy in the range 0 to E. Calculation of $G(E)$, $N(E)$, and $W(E)$ for quantum harmonic oscillators is discussed in chapter 11.

APPENDIX 2

Classical Statistical Mechanics

A2.1 SUM AND DENSITY OF STATES

The classical sum of states, $G(E)$, is based on a theorem in classical statistical mechanics which says that the number of states for one degree of freedom with momentum p and coordinate q is

$$\frac{dpdq}{h} \tag{A2-1}$$

where $dpdq$ is the phase space volume associated with p and q, and h is Planck's constant. Thus, for one degree of freedom, the volume of phase space associated with a single state is h. The number of states, $G(E)$, is then the total phase space volume V divided by h, i.e.,

$$G(E) = \frac{V_1}{h} \tag{A2-2}$$

where the subscript 1 indicates that one degree of freedom is under consideration.

This example may be extended to any multidimensional classical system with coordinates q_i with conjugate momenta p_i. The total energy for the system is given by the classical Hamiltonian $H(p_1 \cdots p_s q_1 \cdots q_s)$. To find the total phase space volume for the Hamiltonian at energy E, the integral

$$V_s = \int_{H=0}^{H=E} \cdots \int dp_1 \cdots dp_s dq_1 \cdots dq_s \tag{A2-3}$$

is solved by considering possible combinations of p_i and q_i with the restriction that H lies between 0 and E. The classical sum of states is then

$$G(E) = \frac{V_s}{h^s} = \int_{H=0}^{H=E} \cdots \int \frac{dp_1 \cdots dp_s dq_1 \cdots dq_s}{h^s} \tag{A2-4}$$

The number of states for H in the range $E \rightarrow E + dE$, which is the analog to quantum mechanical degeneracy, is denoted by $W(E) = G(E + dE) - G(E)$ and is written as

$$W(E) = \int_{H=E}^{H+dH=E+dE} \cdots \int \frac{dp_1 \cdots dp_s dq_1 \cdots dq_s}{h^s} \tag{A2-5}$$

This integral is a multidimensional volume of phase space enclosed by the two hyper-surfaces defined by H and $H + dH$. Another way to express this multidimensional volume is to multiply the area of the hypersurface for $H = E$ by dH, so that $W(E)$ becomes

$$W(E) = \left[\int_{H=E} \cdots \int \frac{dp_1 \cdots dp_s dq_1 \cdots dq_s}{h^s} \right] dH \tag{A2-6}$$

This density of states $N(E) = [G(E + dE) - G(E)]/dE$ is simply $W(E)/dH$ and is given by

$$N(E) = \int_{H=E} \cdots \int \frac{dp_1 \cdots dp_s dq_1 \cdots dq_s}{h^s} \tag{A2-7}$$

Thus, the density of states is determined by a surface integral, while the sum of states [equation (A2-4)] is found from a volume integral.

A greater understanding of the expressions for the classical sum and density of states may be attained by considering translation, harmonic oscillators, and the rigid rotor.

A2.1.1   Three-Dimensional Translation

The Hamiltonian for the free translation of a particle inside a cube with sides of length ℓ is

$$H = \frac{(p_x^2 + p_y^2 + p_z^2)}{2m} \tag{A2-8}$$

The potential energy V is zero inside the container and infinite at the walls. To obtain the translational sum of states, equation (A2-4) must be solved for the Hamiltonian. The integration limits for each coordinate are 0 and ℓ. In contrast, the limits for the momenta are more complex, since the maximum value of a momentum term at a particular energy E is determined by the values for the remaining momenta. Thus, the sum of states is given by

$$G(E) = \int_{H=0}^{H=E} \cdots \int dp_x dp_y dp_z \frac{\int_0^\ell dx \int_0^\ell dy \int_0^\ell dz}{h^3} \tag{A2-9}$$

The product of the integrals over the coordinates in equation (A2-9) is simply ℓ^3, the volume of the container. Although the integral over the momentum terms may appear complex, it is in fact quite simple. As discussed, the integral represents the momentum phase space volume for H having values between 0 and E. However, since the translation Hamiltonian in equation (A2-8) is the equation for a sphere, this volume is that for a sphere with a radius of $(2mE)^{1/2}$. Thus, the translational sum of states per unit volume ℓ^3 is

$$G(E) = \frac{4\pi(2mE)^{3/2}}{3h^3} \tag{A2-10}$$

To obtain the density of states $N(E)$, the sum $G(E)$ is differentiated with respect to E. The resulting $N(E)$ is usually written in terms of $\hbar$ instead of h, and is given by

$$N(E) = (2^{1/2} \pi^2 \hbar^3)^{-1} m^{3/2} E^{1/2} \qquad (A2\text{-}11)$$

This result is used extensively in chapter 13.

A2.1.2   Harmonic Oscillators

As a further illustration of the classical sum and density of states, consider a single harmonic oscillator for which the classical Hamiltonian is

$$H(p, q) = \frac{p^2}{2} + \frac{\lambda q^2}{2} \qquad (A2\text{-}12)$$

where $\lambda = 4\pi^2 \nu^2$. The phase space volume is that enclosed by energy E and is formally given by equation (A2-3), which for this example is

$$V_1 = \int_{H=0}^{H=E} \int dp\,dq \qquad (A2\text{-}13)$$

This is simply the area of an ellipse, logical enough since the Hamiltonian for the harmonic oscillator is the equation for an ellipse. The area of an ellipse is $\pi a b$, where a and b are the semiaxes. For the harmonic oscillator Hamiltonian, the semiaxes equal $(2E)^{1/2}$. Thus, the phase space volume for the harmonic oscillator is

$$V_1 = \frac{E}{\nu} \qquad (A2\text{-}14)$$

and the classical number of states is

$$G(E) = \frac{E}{h\nu} \qquad (A2\text{-}15)$$

The harmonic oscillator energy levels and their associated phase space volumes are illustrated in Figure A2-1.

For s harmonic oscillators, the classical number of states is given by equation (A2-4). The Hamiltonian for s harmonic oscillators is

$$H(p_i q_i) = \sum_{i=1}^{s} \left(\frac{p_i^2}{2} + \frac{\lambda_i q_i^2}{2} \right) \qquad (A2\text{-}16)$$

Evaluation of the phase space volume for s harmonic oscillators is straightforward if one recognizes that equation (A2-16) is the equation for a $2s$-dimensional ellipsoid with semiaxes along p_i and q_i. The volume for this ellipsoid with $H(p_i, q_i) = E$ is given by

$$V_s = \frac{E^s}{s! \displaystyle\sum_{i=1}^{s} \nu_i} \qquad (A2\text{-}17)$$

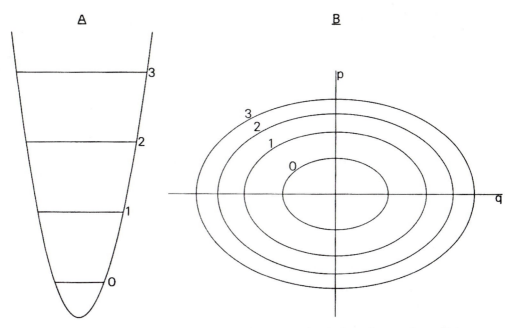

FIGURE A2-1   (a) The $n = 0, 1, 2,$ and 3 energy levels for a harmonic oscillator. (b) Phase space volumes for the $n = 0, 1, 2,$ and 3 harmonic oscillator energy levels.

The sum of states is then

$$G(E) = \frac{E^s}{s! \displaystyle\prod_{i=1}^{s} h\nu_i} \tag{A2-18}$$

and the similarity with equation (A2-15) should be noted. The density of states $dG(E)/dE$ is

$$N(E) = \frac{E^{s-1}}{(s-1)! \displaystyle\prod_{i=1}^{s} h\nu_i} \tag{A2-19}$$

A2.1.3   Rigid Rotor

The last example is a rigid rotor with two degrees of freedom, for which the Hamiltonian is

$$H(p_\theta, p_\phi, \theta, \phi) = \frac{p_\theta^2}{2I} + \frac{p_\phi^2}{2I \sin^2\theta} \tag{A2-20}$$

where θ varies from 0 to π and ϕ varies from 0 to 2π. This Hamiltonian is also the equation for an ellipse with semiaxes $a = (2IE)^{1/2}$ and $b = (2IE \sin^2\theta)^{1/2}$. Therefore, for fixed values of θ and ϕ, the phase space volume is $V_2(\theta, \phi) = 2IE \sin \theta$. Integration over θ and ϕ is necessary to find the total phase space volume

$$V_2 = 8\pi^2 EI \tag{A2-21}$$

Thus, the expression for the classical number of states for a rigid rotor is

$$G(E) = \frac{8\pi^2 EI}{h^2} \tag{A2-22}$$

and the classical density of states is

$$N(E) = \frac{8\pi^2 I}{h^2} \tag{A2-23}$$

A2.2   PARTITION FUNCTION AND BOLTZMANN DISTRIBUTION

As shown in equation (A1-4) of Appendix 1, for the situation where the energy is continuous, as in classical mechanics, the partition function is

$$Q = \int_0^\infty N(E)\, e^{-E/k_BT}\, dE$$

If the classical expression for $N(E)$ in equation (A2-7) is used, the partition function becomes

$$Q = \int_0^\infty \left[\int \cdots \int_{H=E} \frac{dp_1 \cdots dp_s dq_1 \cdots dq_s}{h^s} \right] e^{-H/k_BT}\, dH \tag{A2-24}$$

This multiple integral is over all possible combinations of coordinates and momenta, with each combination weighted by e^{-H/k_BT}. Equation (A2-24) is usually written as

$$Q = \frac{1}{h^s} \int dp_1 \cdots dp_s dq_1 \cdots dq_s e^{-H/k_BT} \tag{A2-25}$$

where the single integral denotes integration over all possible values of coordinates and momenta. The integral in equation (A2-25) is called the *phase integral*.

The classical expression for the Boltzmann distribution is

$$P(H)dH = \frac{\left[\int \cdots \int_{H=E} dp_1 \cdots dp_s dq_1 \cdots dq_s / h^s \right] e^{-H/k_BT}\, dH}{\int dp_1 \cdots dp_s dq_1 \cdots dq_s e^{-H/k_BT} / h^s} \tag{A2-26}$$

where the denominator is the partition function and the integral in the numerator is the density of states. As an illustration of equation (A2-26), consider the rigid rotor equations (A2-20) through (A2-23). The partition function for the rigid rotor is

$$Q = \int_0^\infty \frac{8\pi^2 I}{h^2} e^{-H/k_B T} \, dH$$

$$= \frac{8\pi^2 I k T}{h^2} \tag{A2-27}$$

so that the Boltzmann distribution becomes

$$P(H)dH = \frac{1}{k_B T} e^{-H/k_B T} \, dH \tag{A2-28}$$

This distribution may be used to calculate the average energy for the rigid rotor:

$$\langle E \rangle = \int_0^\infty H P(H) dH \tag{A2-29}$$

Using equation (A2-28) for $P(H)dH$, we find that

$$\langle E \rangle = \frac{1}{k_B T} \int_0^\infty H e^{-H/k_B T} \, dH = k_B T \tag{A2-30}$$

A particularly important classical Boltzmann distribution is that for s harmonic oscillators. By inserting equation (A2-19) for the classical density of states into equations (A2-24) through (A2-26), the classical probability that s harmonic oscillators contains energy E is found to be

$$P(E)dE = \frac{1}{(s-1)!} \left(\frac{E}{k_B T} \right)^{s-1} e^{-E/k_B T} \left(\frac{dE}{k_B T} \right) \tag{A2-31}$$

A molecule must possess some minimum energy before a reaction can be expected to occur. The classical mechanical value for the fraction of molecules which have energy greater than E_0 is found by integrating equation (A2-31) between the limits E_0 and ∞, i.e.,

$$F(E \geq E_0) = \int_{E_0}^\infty \frac{1}{(s-1)!} \left(\frac{E}{k_B T} \right)^{s-1} e^{-E/k_B T} \left(\frac{dE}{k_B T} \right) \tag{A2-32}$$

For the special case where $E_0/k_B T \gg 1$, this fraction becomes

$$F(E \geq E_0) = \frac{1}{(s-1)!} e^{-E_0/k_B T} \left(\frac{E_0}{k_B T} \right)^{s-1} \tag{A2-33}$$

BIBLIOGRAPHY

BLINDER, S. M. *Advanced Physical Chemistry.* London: Macmillan, 1969.

COURANT, R., and JOHN, F. *Introduction to Calculus and Analysis.* New York: Wiley, 1974.

DAVIDSON, N. *Statistical Mechanics.* New York: McGraw-Hill, 1962.

FORST, W. *Theory of Unimolecular Reactions.* New York: Academic Press, 1973.

HASE, W. L. *J. Chem. Ed. 60*, 379 (1983).

KAUFMAN, E. D. *Advanced Concepts in Physical Chemistry.* New York: McGraw-Hill, 1966.

MCQUARRIE, D. A. *Statistical Thermodynamics.* New York: Harper and Row, 1973.

PROBLEMS

A2-1 The probability distribution $P_T(E)$ of the total energy E in a Boltzmann distribution is the Boltzmann of $\delta(E - H)$.

(a) Show that

$$P_T(E) = \frac{N(E)e^{-E/k_B T}}{Q(T)}$$

where Q is the canonical partition function and N is the microcanonical density of states.

(b) For the case of F independent harmonic oscillators, show that

$$P_T(E) = \frac{1}{(F-1)!} \left(\frac{E}{k_B T} \right)^{F-1} \frac{e^{-E/k_B T}}{k_B T}$$

Sketch a graph of this function.

(c) For the system of part (b), calculate the average energy

$$\langle E \rangle \equiv \int dE \, P_T(E) E$$

and the width

$$\Delta E \equiv \sqrt{\langle E^2 \rangle - \langle E \rangle^2}$$

of the distribution about this average value, where

$$\langle E^2 \rangle \equiv \int dE \, P_T(E) E^2$$

A P P E N D I X   3

Data Bases in Chemical Kinetics

In order to carry out chemical kinetics modeling and calculations, it is necessary to have available accurate values of rate coefficients for a large number of elementary reactions over a wide range of temperatures. Until recently, the chemical kinetics data base has been rather inadequate. This is now being remedied, with the appearance of critically evaluated compilations of kinetic data. In this Appendix, we provide some leads to both printed and electronic kinetics data bases that are now available.

   1. Literature through 1987 (somewhat out-of-date, but useful for accessing the earlier literature):

HOCHSTIM, A. R. (ed.), "Bibliography of Chemical Kinetics and Collision Processes", IFI/Plenum, New York (1969).

TROTMAN-DICKINSON, A. F., and G. S. MILNE, "Tables of Bimolecular Gas Reactions", NSRDS-NBS-9, National Bureau of Standards, Washington, D.C. (1967).

KERR, J. A., and S. J. MOSS, "CRC Handbook of Bimolecular and Termolecular Gas Reactions", CRC Press, Boca Raton, Fla. Volume I (1981), Volume II (1981), Volume III (1987).

   2. Data Surveys for modeling atmospheric chemistry:

   For carrying out the atmospheric models described in chapter 15, it is necessary to have accurate rate parameters for the hundreds of reactions that may participate in atmospheric chemistry. Periodic compilations have been issued as Special Publications of the National Bureau of Standards (now NIST—see No. 4), but the most authoritative, and regularly updated source, is that prepared by the Jet Propulsion Laboratory. The most recent version is:

DEMORE, W. B., S. P. SANDER, D. M. GOLDEN, R. F. HAMPSON, M. J. KURYLO, C. J. HOWARD, A. R. RAVISHANKARA, C. E. KOLB, and M. J. MOLINA, "Chemical Kinetics and Photochemical Data for use in Stratospheric Modeling", JPL Publication 97-4, Jet Propulsion Laboratory, Pasadena, Calif. (1997).

3. Additional compilations regularly appear in the *Journal of Physical and Chemical Reference Data*. Examples include:

COHEN, N., and K. R. WESTBERG, "Chemical Kinetic Data Sheets for High-Temperature Chemical Reactions", *J. Phys. Chem. Ref. Data 12,* 531–590 (1983).

STEINFELD, J. I., S. M. ADLER-GOLDEN, and J. W. GALLAGHER, "Critical Survey of Data on the Spectroscopy and Kinetics of Ozone in the Mesosphere and Thermosphere", *J. Phys. Chem. Ref. Data 16,* 911–951 (1987).

BAULCH, D. L., C. J. COBOS, R. A. COX, P. FRANK, G. HAYMAN, TH. JUST, J. A. KERR, T. MURRELLS, M. J. PILLING, J. TROE, R. W. WALKER, and J. WARNATZ, "Evaluated Kinetic Data for Combustion Modeling", *J. Phys. Chem. Ref. Data 21,* 411–734 (1992); "Supplement I", *ibid. 23,* 847–1033 (1994).

4. A limitation of all printed databases is that, once published, they are unable to incorporate new information on kinetic parameters which is being continually generated. This is remedied in the computer-searchable, regularly updated databases now available from the National Institute of Standards and Technology (formerly NBS). Rate parameters are stored in the form $k(T) = A(T/298)^n \exp(-(E_a/R)/T)$, and can be displayed either in tabular form or as Arrhenius plots. Literature citations for each measurement include pressure and temperature range, experimental method, and experimental uncertainties. Two databases now available are:

NIST Chemical Kinetics (gas-phase reactions). Version 6.0 includes data on 29,200 rate coefficients for 9,200 reactions involving 4,400 compounds.

NDRL/NIST Solution Kinetics (NIST Standard Reference Database 40) includes data on 10,300 rate coefficients for 7,800 reactions involving 6,400 chemical species.

Both are available from Standard Reference Data, National Institute of Standards and Technology, Gaithersburg Md. 20899, Telephone (301) 975-2208, Fax (301) 926-0416, e-mail SRDATA@enh.nist.gov.

Index

Ab initio calculations. *See* Potential energy surfaces
Absolute rate theory. *See* Transition state theory
Acetaldehyde. *See* CH_3CHO
Acid-catalyzed hydrolysis, 150
Activated complex. *See* Transition state
Activation, entropy of, 221, 301–302, 318
Activation energy, 14–16, 269, 472
  relation to Arrhenius parameter, 14
  relation to enthalpy of reaction, 16, 302
  Tolman definition, 294
Activity coefficients, 133–136, 392 (*see also* Debye-
  Huckel theory)
Adams-Bashforth method, 64
Adiabatic approximation, 210 (*see also* Nonadiabatic)
Adiabatic lapse rate, 470
Adsorption, 163ff., 407–415
  activated, 408
  dissociative, 166, 414–415
Adsorption isotherms 164, 407–410
Affine transformation, 233, 272
Analytical techniques for kinetic measurements,
  93–95
Angular momentum, 207, 225, 235, 242, 345, 361, 445
Anharmonic potential, 186, 366
Anharmonic bottleneck, 364, 368
Antarctic ozone hole, 485ff.
Apparent non-RRKM. *See* RRKM
Approximation methods, 37–41
Arrhenius equation, 14, 94, 221ff., 301–302, 342
  for desorption kinetics, 412
  deviations from, 17, 299–300
Arrhenius parameters, 14, 115, 343
  relation to thermodynamic parameters, 301–302
  sensitivity analysis, 117–119
  solvent effects, 393
Arrhenius plot, 14
Atmospheric chemistry, 470–498

Atmospheric models, 488–489
Autocatalysis, 151–154
Avogadro's Number, 135
Avoided crossing, 208

Beams. *See* Molecular beams
BEBO. *See* Potential function
Beer's Law, 113, 473
Belousov-Zhabotinsky reaction, 156–159
Benson's rules, 47
Bessel functions, 31, 55, 441–442
Bicycloheptene, rearrangement, 393
Bimolecular reactions, 4, 77
  collision theory of, 171ff., 217–254
  in solutions, 128–132
  on surfaces, 164–156
  (*see also* Transition state theory)
Bistability, 159
Boltzmann constant k_B, 91, 173
Boltzmann distribution. *See* Maxwell-Boltzmann
  distribution
Bond Order, 202
Born-Oppenheimer approximation, 176, 179, 191,
  207–208, 290
Boundary-value problems, 56
Br atom
  in atmospheric chemistry, 487, 496–498
  multiphoton ionization detection, 103–104
Branched-chain reaction, 453ff.
Branching ratio, 34, 423, 436
Bromine, reaction with hydrogen, 41–43
Bronsted correlation, 140
Brunauer-Emmett-Teller isotherm, 410

Cage effect, 124, 391–392
Calculus of variations, 428ff., 447
Canon of molecular dynamics, 174

Canonical distribution, 285
Canonical ensemble, 310, 430
Carbon cycle, 474
Catalysis, 147–170, 415–418, 476–488
   and equilibrium, 147–148
   (*see also* Enzymes, Heterogeneous reactions,
      Inhibition, Promotion)
Center of mass, 223, 258–260
Centrifugal barrier, 246
C_2, reactions of, 51
CF_3O
   decomposition, 57–59
   IRMPD, 373
$CH_2 = CHCN$, IRMPD, 50–51
CHO, *See* HCO
CH_2O, 187, 207, 327, 376, 459–462
   +H, potential energy surface, 199
   internal coordinates, 187
   vibrational frequencies, 207
CH_3, recombination, 105–106, 113–115, 465
CH_3CHO, 5
$CH_3COOC_2H_5$, hydrolysis, 150–151
CH_3NC, isomerization, 7, 324, 349, 352, 356–357,
   388–389
   ab initio calculations, 348
   RRKM calculation 348–352
CH_4
   atmospheric, sources of, 474
   combustion, 459–469
   sum of states, 331
$C_2H_2Cl_2$, isomerization, 23
C_2H_4, reaction with F, 361, 377
C_2H_5
   decomposition, 347
   transition state, 347
C_2H_5Cl (C_2H_4DCl)
   decomposition, 347
   IRMPD, 372, 389
   transition state, 347, 385
C_4H_8, isomerization, 355
Chain carriers, 41
Chain reaction, 41, 453–457
Challenger (space shuttle), 458
Chaotic trajectory, 366–367
Chapman-Enskog expansion, 280
Chapman mechanism, for atmospheric ozone, 153,
   477–480
Charge-exchange, 222
Chemical activation. *See* Unimolecular reactions
Chemical feedback. *See* Oscillating reactions
Chemical laser, 275–276
Chemiluminescence, 100–101, 263
Chemisorption, 164, 407, 412
   dissociative, 166, 414–415
Chlorine. *See* Cl

Chlorine nitrate. *See* $ClONO_2$
Chloroethane. *See* C_2H_5Cl
Chlorofluorocarbons
   effect on stratospheric ozone, 483, 485–488, 492
$Cl + H_2$ reaction, 43
$Cl + HX$ reaction, 101
$ClONO_2$, in atmospheric chemistry, 485, 487–488
Classical mechanics of scattering, 222–239
Classical trajectory calculations, 206, 232–239
   applied to surfaces, 418–420
   for $H + H_2$ reaction, 236–238, 268
   for model potentials, 272–274
   for nonadiabatic processes, 249
   stochastic treatment, 420
   for V-T relaxation in polyatomics, 358
Closed systems. *See* Static systems
Closest approach distance, 226
CO, oxidation on surface, 166–167, 417
Collision complexes, 245
Collision diameters, 217–218, 356
Collision duration, 236, 245
Collision efficiency, 356–357
Collision dynamics, 217–254
Collision rate
   gas-kinetic, 217–219
   Lennard-Jones, 230–231
Collisions, 171ff., 217ff.
   elastic, 222
   inelastic, 222
   reactive, 222
   with surfaces, 419–420
Collisions between atoms, transition-state theory
   for, 307
Combustion, 453–469
   of methane, 459–468
   rate constants for modeling, 508
   sensitivity analysis, 465–468
Complex reactions, 17, 22–86
   analytic solutions for, 22–37
Computer programs
   for numerical integration, 76–77
Concurrent reactions. *See* Parallel reactions
Conduction-diffusion equation, 128–130
Configuration-interaction (CI) method, 195, 208
Conical intersection, 210
Consecutive reactions, 25–32
Conservation,
   of energy, 222, 225, 259
   of mass, 23, 29, 35
   of momentum, 259
Constraints, 427, 431, 435, 447
Contact radius, 128–129
Contour diagram. *See* Potential energy surfaces
Coordinate systems
   center-of-force, 223–224, 245

center-of-mass, 223, 258ff.
 mass-weighted, 233
 transformation, 233, 260
Correlation coefficient, 112
Correlation diagram, 211
Coulomb integral, 202
Coulomb potential, 180
Coverage, 164
Cows, 474
Cramer's Rule, 50, 52
Critical configuration, 344
Critical radius. *See* Contact radius
Cross section, 217ff.
 for absorption, 476
 definition, 172
 differential, 172, 224, 241
  for hard-sphere potential, 217–129, 227–228
  for Lennard-Jones potential, 228–231, 261
 Langevin, 246–247
 reactive, 260–263
 relation to rate coefficient, 173
 total, 172
Current, in quantum mechanics, 241

D, D_2. *See* H, H_2
Data analysis, 105–116
 least-squares method, 110–113
Data bases, of chemical-kinetic rate constants, 507–508
Debye-Hückel approximation, 134
Debye-Hückel theory, 133–136, 151, 392
Debye screening length, 135
Decay time, 8, 27, 101, 449
Deflection function, 225
 for Lennard-Jones potential, 229
Density of states. *See* Sum of states
Detailed balance, 24, 44, 274–275, 283, 439, 443
 and microscopic reversibility, 174–175
Deterministic equations, 66
Dichloroethylene. *See* $C_2H_2Cl_2$
Dielectric constant, 131, 135
Differential cross section, 172, 224, 241, 261–262
Differential equations
 exact solution, 6–12, 22–37, 283
 numerical integration, 55–65
 transform methods, 47–52
Differential pumping, 256
Diffusion, 126–132
 Fick's Law of, 126
Diffusion coefficient
 in gases, 90–91
 in solutions, 126
Diffusion-limited rate constant, 130–132
Dinitrogen cleavage, 15
Dipole interactions, 180–183

Dipole moments, 182
Discharge, electric, 89
Dispersion force, 183
Displacement variables, 142, 155
Dissociation energies, 42, 180, 184–186, 454, 459
Distribution functions
 constraints on, 427, 431, 435
 for lifetimes, 362ff., 375, 449
 Maxwell-Boltzmann, 173, 235, 349, 504–505
 nascent, 263
 normalization, 425, 427
 uniform, 426, 427
Dividing surface, in transition-state theory, 290, 313

Eadie-Hofstee plot, 162–163
Eckart barrier, 298
Eigenvalue solutions, 54, 142, 156
 for master equation, 286
 for normal coordinates, 189–190
 for oscillating chemical reactions, 156
 (*see also* Matrix solutions)
Einstein coefficients for radiative processes, 370
Einstein's Law. *See* Photochemical equivalence, law of
Einstein-Brillouin-Keller (EBK) semiclassical method, 205
Einstein-Stokes relationship, 131
Elastic collisions, 222
Electron transfer reactions, 397–402
Electrostatic interactions, 180–181
Elementary reaction, 2, 6–12, 87
Eley-Rideal mechanism, 165, 417
Encounters, in solution, 124–125
Encounter complex, 132, 391
Endothermic reaction, 16, 42, 424
Energy of activation. *See* Activation energy
Energy conservation, 222, 225, 259
Energy-grained master equation, 371
Energy partitioning, 359–361
Energy transfer. *See* Intramolecular energy transfer; Relaxation processes
Ensembles, 310
 canonical, 310
 microcanonical, 310, 339, 435
Enthalpy
 of activation, 149, 301–302
Enthalpy diagram, 16, 149, 391
Entropy, 149, 426ff.
 of activation, 301–302
  relation to steric factor p, 221, 318
 of mixing, 426
 statistical, 426
Entropy deficiency, 427, 447
Entropy maximization, 427
Entropy production, 447
Enzyme-catalyzed reactions, 159–164, 169

Equilibrium
    and catalysis, 147–148
    definition, 1
    and detailed balance, 24
ER-2 aircraft, 490
Errors
    in kinetic measurements, 106–110, 115–116
        random, 107–108
        systematic, 107
    in numerical integration, 57–63, 65
Errors, propagation of, 108–110
Ethyl acetate. See $CH_3COOC_2H_5$
Euler methods, 56–59
    compared with Runge-Kutta method, 61
Exchange integral, 202
Exothermic reactions, 16, 42, 258, 424, 434
Explosion, 456–459
Explosion limits, 456–457
Exponential-gap law, 444–445

$F + H_2$ reaction, 268–272
    chemical laser, 275–276
    classical trajectory calculations, 239
    possible outcomes, 424–425
    potential energy surface, 195–198, 234, 269, 316
    product distribution, 269
    selective energy consumption (H + HF), 274–275
    specific energy release, 270
    surprisal analysis, 435–437
    transition-state theory calculation, 308–310
Fall-off region, 335, 351–352
Femtochemistry, 265, 315–317
Fick's law of diffusion, 126
First-order reactions
    analytic solutions for, 7–8, 92, 117
    half-life for, 13
    pseudo-first-order, 40, 50, 92
Fischer-Tropsch synthesis, 166, 418
Flash photolysis, 94–97
Flooding, 41
Flow kinetic systems, 91–93, 351
    error analysis, 108–110
Fluctuations, 66, 70–72
Fokker-Planck equation, 403
Force constants, 186, 188
Formaldehyde. See CH_2O
Formyl radical. See HCO
Four-center reaction, 454
    for $H_2 + I_2$, 44
Fourier Transform Doppler spectroscopy, 264
Franck-Condon principle, 307, 363
Free energy
    of activation, 301
    solvent effects on, 136–140, 393–396
    of vaporization, 394

Free-radical reaction, 41, 454ff., 480ff.
Free-volume theory of liquids, 394
Freons. See Chlorofluorocarbons
Frequency factor, 14, 293–294
Freundlich isotherm, 410
Friction constant, 403ff.

Gamma function, 441
Gas-kinetic collision rate, 218
    Lennard-Jones correction, 230–231
Gas-surface interfaces, 163–165, 407ff.
Gear routine, 65
Generalized-valence-bond (GVB) method, 195
Gibbs free energy. See Free energy
Gibbs-Helmholtz equation, 301
Gillespie algorithm 70, 77–79
Global error
    in classical trajectory calculations, 234
    in numerical integration, 59
Green's function method, 120, 466

HCO, 374–375, 459ff.
$H(D) + H_2$ reaction, 237n., 266–268, 322–323
    classical trajectory calculations, 236–238
    potential energy surface, 195, 199
    quantum scattering calculations, 244
    surprisal analysis, 437
    transition state, symmetry number for, 305–307
H_2, adsorption on Cu, 414
$H_2 + Br_2$ reaction, 5–6, 41–43
$H_2 + Cl_2$ reaction, 43
$H_2 + CO$ reaction, 165–166, 418
$H_2 + F_2$ reaction, 41
    chemical laser, 275–276
$H_2 + I_2$ reaction, 43–47
$H_2 + O_2$ reaction, 453–459
H_2O
    correlation diagram, 211
    harmonic levels, 330
    normal modes, 187–189
HO_2
    in atmospheric chemistry, 481
    in combustion, 457
    dissociation, 376
Haber process, 418
Half-life, 13
Halogens, reaction with hydrogen, 41–47
Hamilton's equations of motion, 232
Hamiltonian
    for harmonic oscillator, 365, 502
    for reactive scattering,
        classical, 232
        quantum-mechanical, 239
    for translation, 501
Hammett equation, 138–140

Hard-sphere model, 183, 185, 217–219, 226–228
  (*see also* Potential function, hard-sphere;
    Reactive hard-sphere model)
Harmonic oscillators, 184
  density of states, 329ff., 502–503
  Hamiltonian, 365, 502
  master equation for relaxation, 283–286
  maximal-entropy treatment, 430–431
  partition functions, 304
  potential function, 184
Heat of vaporization, 394
Heisenberg uncertainty relation, 230, 239, 297, 374
Heterogeneous catalysis, 163–167
Heterogeneous reaction, 390, 407, 415–418
  in atmospheric chemistry, 487–88
  definition, 1
  (*see also* surface reactions)
Higgins model, 156
Hindenburg (dirigible), 458
HO$_x$ cycle
  effect on atmospheric ozone, 480–481
Homogeneous catalysis, 148–151
Homogeneous reaction
  definition, 1
Hot-atom techniques, 267
Hydrogen. *See* H, H$_2$
Hydrogen-halogen reactions, 41–47
Hydrolysis
  of esters, 150–151
  of sucrose, 160
Hydroxyl. *See* OH

Impact parameter, 219ff., 235–236, 246
Importance sampling, 236
Induction period, 152, 462
Inelastic collisions, 222
Inflection point, 152
Information theory, 424–452 (*see also* Surprisal
  analysis)
Infrared multiple-photon dissociation, 50, 90, 327,
  355, 361, 367–374
  master equation for, 370
Inhibition, 5, 42
Initial-value problems, 56
Initiation
  of free-radical reactions, 42, 454, 459
  methods of, 89–90
Intermediate, 2, 27, 37–38, 160, 194, 316
Intermolecular energy transfer, 356–358
  classical trajectory studies, 358
  stochastic models, 358
Intermolecular potentials, 183–184
Internal conversion, 327
Internal coordinates, 187–190
Intersystem crossing, 51

Intramolecular vibrational relaxation (IVR), 339,
  363, 365–367, 370
Inverted population, 275, 437
Iodine
  energy transfer processes, 445
  reaction with hydrogen, 43–47
Iodine atoms, recombination of, 10, 14, 97
Ion cyclotron resonance, 371
Ion-molecule collisions, 245–247
Ionic strength, 135–137, 151
Ionization. *See* Multiphoton ionization,
  Photoionization
IRMPA, IRMPD. *See* Infrared multiple-photon
  dissociation
Irreversible reaction, 25, 66

Jacobian transformation, 260

K + CH$_3$I reaction, 258–263
Kinetics
  definition, 1
  experimental methods, 87–105
Kramers coefficient, 404–405
Kramers theory, 402–406

Lagrange multipliers, 428, 447
Landau-Teller model, 283
Landau-Zener model, 248–249
Langevin technique, 403, 420
Langevin theory, 245–247
Langmuir-Hinshelwood mechanism, 165, 417
Langmuir isotherm, 164, 408–410
Laplace Transform, 47–52
  tables of, 74–76
Lasers, 90, 97ff., 264ff., 370
  chemical, 275–276
Laser-induced fluorescence, 51, 102–103, 264
  in atmospheric monitoring, 490
  in molecular beams, 257, 260
Laser-selective chemistry, 275, 374
Least-squares analysis, 110–113
Lennard-Jones potential, 184, 201
  correction to gas-kinetic collision rate, 230–231
  cross section, 231
  deflection function, 228–230, 261
  parameters, 186
LEPS surface, 201–202
  classical trajectory calculations, 272
Li$^+$ + H$_2$O potential energy surface, 199–200
Lifetime, 339, 362, 375, 449
Limit cycle, 157
Lindemann mechanism, 334
Lindemann-Hinshelwood mechanism, 334–339,
  386, 405
Line-of-centers energization rate constant, 335

Linear combinations, 48
Linear free-energy relations, 136–140
Lineweaver-Burk plot, 161–162
Linewidth, 375
Liquids
    free-volume theory of, 394
    reactions in, 124–146, 390–406
London forces, 182
Long-range potentials, 180–183, 245
Lotka mechanism, 154–156

Marcus theory, 399–402
Markovian processes, 66, 124, 403
Martians, 278n.
Mass conservation, 23, 29, 35
Mass spectrometry
    in molecular beams, 257
    with photoionization, 104–105
Master equation, 68, 282–286
    eigenvalue solutions, 286
    for harmonic oscillators, 283–285
    information-theoretic solutions, 446–449
    for IRMPD, 370
Matrix solutions
    for normal vibrations, 188–190
    for opposing first-order reactions, 52–55
    for relaxation kinetics, 142
    (*see also* eigenvalue solutions)
Maxwell-Boltzmann distribution, 173, 235, 349,
    504–505
Mean, 108
Mean free path, 255
Mean free time, 218–219
Mechanism, 17–18
    experimental determination, 87–89
    indeterminacy of, 47
    for gas-surface reactions, 165
    for hydrogen-halogen reactions, 41–47
Metal-ligand complexes, 140, 145, 201
Metal-semiconductor interface, 423
Methanation reaction, 166, 418 (*see also* Fischer-
    Tropsch synthesis)
Methane. *See* CH_4
Methyl isocyanide. *See* CH_3NC
Methyl radical. *See* CH_3
Michaelis constant, 161
Michaelis-Menten equation, 161–163
Microcanonical ensemble, 310, 339, 435
Microcanonical transition-state theory, 310–312,
    343
Microscopic reversibility, 174–175
Milne-Simpson method, 65
Mixing time, 90–91
Mobility, 126
Mode selectivity, 374–377

Molecular beam scattering, 171–172, 204, 255–263
    $F + D_2$ reaction, 270–272
    $K + CH_3I$ reaction, 258–263
Molecular control, 376–377
Molecularity, 4
Momentum conservation, 259
Monte Carlo method,
    for classical trajectory calculations, 235
    for reactions in solution, 396
    for stochastic simulation, 77–79
Montreal Protocol, 492–493
Morse function, 186, 202
Multiconfiguration SCF (MCSCF) method, 195
Multiphoton ionization, 92, 103–104, 264–265

Newton diagram, 258–259, 271
Newton's law of motion, 232
NO (Nitric oxide)
    adsorption on Pt, 410–411
    catalyst for SO_2 oxidation, 149–150
    discharge-flow experiment, 92
    effect on atmospheric ozone, 481–482
    laser-induced fluorescence, 103
    multiphoton ionization, 265
    potential energy curves, 281
    sticking probability, 411
NO_x cycle
    in atmospheric chemistry, 481–482, 492
Nonadiabatic processes, 247–249
Noncrossing rule, 208
Normal vibrational modes, 187–190, 205
Normalization, 425, 427, 431, 447
Numerical integration, 55–65, 232, 489

O_2, potential energy curves, 281
O_3
    atmospheric, 476–488
    autocatalytic formation mechanism, 153–154
    Chapman mechanism for, 153, 477–480
    effect of chlorine on, 483
    effect of hydroxyl on, 480–481, 491
    effect of nitrogen oxides on, 481–482
    review of kinetics, 508
Occupation probabilities, 283
Odd-oxygen species, 477, 480, 482
OH, 211, 454ff.
    OH + CO reaction, 17
    in atmospheric chemistry, 480–481, 491
    laser-induced fluorescence detection, 490
"Old collision theory" model, 219–222
Omega Integral $\Omega^{(2,2)*}$, 230
Open systems, 91, 154 (*see also* Flow systems)
Opposing reactions
    first order
        analytic solution, 23–25

Laplace Transform solution, 49–50
matrix solution, 52–54
numerical solutions, 60–63
stochastic solution, 69–71, 77–79
mixed order, 24–25
Optical Bloch equations, 368
Orbiting collisions, 245
Order, 4
determination of, by half-life, 13
determination of, by van't Hoff plot, 12
Oscillating reactions, 154–159
Overtone excitation, 328, 355, 377
Ozone. *See* O_3

Pair distribution function, 396
Parallel reactions, 32–37, 51
Partial wave analysis, 242
Partition functions, 302–305, 308–309, 499, 504–505
electronic, 303
rotational, 304–305
translational, 303–304
vibrational 304
Permeability function, 298
Perrin hypothesis, 334, 367
Phase integral, 504
Phase shift, 242
Phase space, 294, 360ff., 500
pH, effect on reaction rate, 14*n.*
pH-jump, 143
Phenanthroline, 399
Photoactivation, 89–90 (*see also* Unimolecular reactions)
Photochemical equivalence, law of, 89
Photochemical reactions, 44, 46, 475–476
Photodetachment spectroscopy, 316
Photoelectron-photoion coincidence (PEPICO), 351–353
Photoelectron spectroscopy, 167
Photoionization, 103–105
Photostationary state, 46
Physisorption, 164, 407, 412
Poisoning, 415, 418
Poisson equation, 133–134
Poisson function
for collision frequencies, 219
Polar stratospheric clouds, 487
Polarizability, 181–182, 245
Population models, 153, 168–169
Potential energy contour, 190–192, 287–288
Potential energy surface, 120, 179–216
ab initio calculation, 191–199
analytic functions, 196–204
contour diagram, 192, 288
crossing, in nonadiabatic processes, 247–249
electronically excited, 207–211

experimental determination of properties, 204–206
for $F + H_2$ reaction, 195–198, 216, 234, 269, 316
for $H + CH_3$ reaction, 195, 200, 314–315
for $H + H_2$ reaction, 195, 199
for $Li^+ + H_2O$ reaction, 195, 199
LEPS, 201
for physisorption, 412
profile along reaction coordinate, 193
switching function, 203–204
Potential function, 179ff.
BEBO, 202, 346
centrifugal, 225
Coulomb, 180
dipole-dipole, 181–182
hard-sphere, 183–185
Lennard-Jones, 184–185
Morse, 186, 202
for O_2, NO, and N_2, 281
Potential of mean force, 395
Precursor states, 414–415
Predictor-corrector method, 63–65
Pre-exponential factor, 14, 115, 302 (*see also* Arrhenius parameters; frequency factor)
Pressure jump, 143
Primary salt effect, 136, 393
Prior rate, 432, 438–443
Probability of reaction, 218, 224, 248, 261
Product imaging, 265–266
Progress variable, 9, 11, 25, 44
Promoter, for catalyzed reactions, 415–416
Propagation, in free-radical reactions, 42, 454
Propagation of errors, 108–110, 114–115
Pseudo-first-order reaction, 40–41, 51
Pyrolysis, 351

Quantum mechanics, 205
effects in transition-state theory, 297–300
of reactive scattering, 239–245
coupled-channel theory, 244
Quasicontinuum (QC), 368–369
Quasi-equilibrium hypothesis, 290, 296
Quasi-periodic trajectory, 366–367

Radiation hypothesis. *See* Perrin hypothesis
Radiative recombination, 307
Rainbow scattering, 228–229
Raman spectroscopy, 264, 267
Random lifetime distribution, 339
Rate coefficient, 6, 77, 87
relation to cross section, 173
state-to-state, 174, 276 282ff., 438
Rate constant, 4
for desorption, 411
diffusion-limited, 130–132

effect of ionic strength, 133–137
temperature dependence, 14–17
units, 4
Rate-determining step, 39, 147
in solution reactions, 132, 391
Rate equation, 2
Rate laws, integrated, for elementary reactions, 6–12
Rate-limiting step. *See* Rate-determining step
Rate of reaction, 2
Reaction coordinate, 191, 291, 325–326
Reaction mechanism. *See* Mechanism
Reaction order. *See* Order
Reaction path, 191, 206–207, 288–289
Reaction probability. *See* Probability of reaction
Reaction rate. *See* Rate of reaction
Reaction types
consecutive, 25–32
elementary, 2, 6–12
parallel, 32–37
reversible, 22–25, 49–50
Reactions on surfaces. *See* Adsorption;
    Heterogeneous reactions
Reactive hard sphere model, 219–221
Recombination reactions, 10, 14, 96
Reduced mass μ, 173, 223, 232
Redundant coordinates, 187
Relaxation methods, 140–143
Relaxation processes
intramolecular, 365–367
master equation, 283–286
rotation-translation, 445
surprisal analysis, 437–445
vibration-translation, 356–358, 438, 444
vibration-vibration, 438
REMPI. *See* Multiphoton ionization
Residence time, 413
Resonance, quantum-mechanical, 245
Reversible reactions, 22–25, 49–50
Reynolds number, 93
Rigid rotor, 503–504
partition function, 304–305
Rise time, 27, 101
rms deviation, 112
Rotational energy transfer. *See* Relaxation
RRHO approximation, 434, 440
RRK theory, 338–343
Arrhenius A-factor, 343
RRKM theory, 343–351
ab initio calculations, 347
active modes, 344
adiabatic modes, 344
example calculation, 349
measurements of k(E), 351–356
non-RRKM behavior, apparent and intrinsic,
    362–365

product energy partitioning, 359–361
rotation energy, 345, 350
thermal activation, 348–351
transition state properties, 346
variational version, 347
Runge-Kutta method, 59–63
compared with Euler method, 61
Rydberg-Klein-Rees method, 205

"S" curve, 152, 154
S_N2 reaction, 390, 397, 406
Saddle point, 191, 206–207, 288–289
Salt effect,
primary, 136, 393
secondary, 151
Sampling, of initial conditions, 236
Scaling laws,
for energy-transfer processes, 445
Scattering, 171ff.
classical treatment, 222–239
in molecular beams, 258–263
quantum-mechanical treatment, 239–245
reactive, 231–249, 260–263
Scattering angle, 259–262
Scattering matrix, 243
Schrodinger equation, 176, 193, 239
Secondary salt effect, 151
Second-order reaction
analytic solution for, 8–9
pseudo second-order, 10
Secondary-ion mass spectrometry (SIMS), 418
Selective energy consumption, 174, 273–275, 318,
    424
Self-Consistent Field method, 195, 205
Sensitivity analysis, 116–120, 465–469
Sensitivity coefficients, 117–119
Sequences, statistical, 425
Sequential reactions, 25–32
Sex, 153
SF_6
density of states, 332
IRMPD, 368
Shock tube method, 93, 351
Single molecule kinetics, 72
Skewed coordinates, 233–234
Slater theory, 338, 365, 375
Smoluchowski diffusion equation, 404
Solutions, reactions in, 124–146, 390–406
Solvent cage. *See* Cage effect
Specific activity, of enzymes, 163
Specific energy release, 174, 270, 274, 318, 424
Spline functions, 197
Standard deviation, 108, 114
of least-squares parameters, 111
Static kinetic systems, 90–91

Statistical entropy, 426
Statistical factor, 306
Statistical mechanics, 499–506
Statistical rate theory. *See* RRKM Theory;
  Transition-State Theory
Steady-state approximation, 37–40, 55
  applied to $H_2 + Br_2$ reaction, 42–43
  applied to methane oxidation, 463
  in enzyme kinetics, 160
  in unimolecular kinetics, 335
Stern-Volmer equation, 355
Step size, in numerical integration, 57, 65
Steric factor *p*, 221, 318
Steepest descent path, 206
Sticking probability, 410–411
Stimulated emission pumping (SEP), 328, 374
Stirling approximation, 426
Stiff equations, 65
Stochastic method, 66–72
  numerical simulation, 77–79
Stoichiometric coefficients, 3
Stratopause, 471
Stratosphere, 470–471
Strong collision assumption, 335, 337, 339, 356ff.
Sudden approximation, 445
Sum of states, 329–333, 499
  classical, 329, 331, 500–505
  direct count, 329
  semiclassical, 333
  Whitten-Rabinovitch approximation, 333
Sum rule, 429, 446, 448
Supermolecule, 310
Supersonic beams, 44, 257
Surface. *See* Potential energy surface
Surface imperfections, 415–416
Surface ionization, 257
Surface reactions, 163–167 (*see also* Heterogeneous
  reactions)
Surprisal analysis, 432–445
Symmetry number, 305–306

Taylor's series, 57, 59, 184, 188
Temperature
  negative, 437
  time-dependent, 284–285
Temperature dependence of reaction rates, 14–17, 94
  least-squares analysis, 113–115
Temperature jump, 142–143
Termination, of chain reactions, 42, 454, 461
Thermolecular reactions, 5
Thermal activation, 89
Thermochemical rules, 346
Thermal explosion, 458
Thermodynamic formulation,
  of transition-state theory, 300–302, 413

Third-order reactions, 10–11
Three-body recombination, 10, 477
Time, elimination of, in solving kinetic equations,
  31
Time constant τ, 8
Time-resolved absorption spectroscopy, 97–100
  error analysis, 113–115
Time reversal invariance, 175
Total cross section, 172–173, 224
Trajectory calculations. *See* Classical trajectory
  calculations
Transform methods. *See* Laplace transform
Transition state, 274, 287ff., 344
Transition state theory, 287–323
  for adsorption kinetics, 410–414
  dynamical derivation, 294–296
  for $F + H_2$, 308–310
  for $H + H_2$, 322–323
  fundamental assumptions, 290
  microcanonical, 310–312
  quantum-mechanical effects in, 297–300
  relation between gas-phase and solution-phase
    formulations, 318, 390–402
  thermodynamic formulation, 300–302
  variational, 312–314
Translation,
  Hamiltonian for, 501
  partition function, 303–304
Transmission coefficient, 318, 404
Triangle plot, 270–271, 275
Trifluoromethoxy. *See* CF_3O
Tropopause, 470
Troposphere, 470
Tunneling, 244, 298–300, 317
Turbulent flow, 93

Uncertainty principle. *See* Heisenberg uncertainty
  relation
Unimolecular reactions, 4, 77, 324–389, 405
  chemical activation, 327, 352ff., 363–364
  photoactivation, 327, 352ff., 363
  thermal activation, 335, 348–351
  (*see also* Lindemann-Hinshelwood mechanism;
    RRKM Theory)
Units
  of activation energy, 14
  of collision rate, 218
  of cross section, 172
  of frequency factor, 14
  of rate constant, 4
  of reaction rate, 3

van der Waals
  forces, 14, 407
  molecules, 371, 374

van't Hoff
  equation, 143, 301
  plot, 12
Variance
  in stochastic simulation, 68
Variational transition state theory, 312–314
Velocity selection, 257–258
Very-low-pressure-pyrolysis (VLPP), 351
Vibrational frequency
  harmonic oscillator, 184, 190
  normal mode, 187–190
Vibrational relaxation. *See* Relaxation
Vibrational/rotational adiabatic theory, 378

Vinylcyanide. *See* $CH_2 = CHCN$
Viscosity, 93, 131
Volume of activation, 302, 394

Water. *See* H_2O
Wave mechanics. *See* Quantum mechanics
Wentzel-Kramer-Brillouin (WKB) method, 205
Whitten-Rabinovitch formula, 333
Wien radiation law, 383

Zero-order reaction
  analytic solution, 6